# Springer Praxis Books

Geophysical Sciences

For further volumes:
http://www.springer.com/series/4110

Ralph D. Lorenz • James R. Zimbelman

# Dune Worlds

How Windblown Sand Shapes Planetary Landscapes

Springer

Ralph D. Lorenz
Johns Hopkins University Applied
  Physics Laboratory
Laurel, MD
USA

James R. Zimbelman
National Air and Space Museum
  Smithsonian Institution
Washington, DC
USA

Published in association with Praxis Publishing, Chichester, UK

ISSN 1615-9748
ISBN 978-3-540-89724-8     ISBN 978-3-540-89725-5  (eBook)
DOI 10.1007/978-3-540-89725-5
Springer Heidelberg New York Dordrecht London

Library of Congress Control Number: 2014933788

© Springer-Verlag Berlin Heidelberg 2014

This work is subject to copyright. All rights are reserved by the Publisher, whether the whole or part of the material is concerned, specifically the rights of translation, reprinting, reuse of illustrations, recitation, broadcasting, reproduction on microfilms or in any other physical way, and transmission or information storage and retrieval, electronic adaptation, computer software, or by similar or dissimilar methodology now known or hereafter developed. Exempted from this legal reservation are brief excerpts in connection with reviews or scholarly analysis or material supplied specifically for the purpose of being entered and executed on a computer system, for exclusive use by the purchaser of the work. Duplication of this publication or parts thereof is permitted only under the provisions of the Copyright Law of the Publisher's location, in its current version, and permission for use must always be obtained from Springer. Permissions for use may be obtained through RightsLink at the Copyright Clearance Center. Violations are liable to prosecution under the respective Copyright Law.

The use of general descriptive names, registered names, trademarks, service marks, etc. in this publication does not imply, even in the absence of a specific statement, that such names are exempt from the relevant protective laws and regulations and therefore free for general use.

While the advice and information in this book are believed to be true and accurate at the date of publication, neither the authors nor the editors nor the publisher can accept any legal responsibility for any errors or omissions that may be made. The publisher makes no warranty, express or implied, with respect to the material contained herein.

Printed on acid-free paper

Springer is part of Springer Science+Business Media (www.springer.com)

# Preface

This book arose from the evident need to sum up the breathtaking progress made in the last 25 years in the study of dunes and ripples on Earth and the other worlds of the solar system. Not only observations by orbiting spacecraft, but the close-up views and physical interaction with landers and rovers, have transformed our views of the other planets (and we count Titan in that category, however politically incorrect that may be!). Our knowledge of the Earth, too, has dramatically improved, with global satellite data able to reveal the topography of dunes and measure their motion, and ground-penetrating radar able to profile the internal structure of dunes and diagnose their history. New instrumentation and data acquisition equipment, both in the laboratory and in the field, exposes the turbulent variations in wind and sand's response with high time resolution, freeing us from the tyranny of the average and showing ripples moving in timelapse. And all these data are being synthesized into a better understanding with ever-more sophisticated computer models. All this has left the "classic" books—Bagnold, McKee, Greeley, and Lancaster—behind.

Bagnold's *Physics of Wind-Blown Sand and Desert Dunes* (1941) has been long considered the bible of aeolian studies, although 70 years on it must be considered, perhaps, the Old Testament. That said, we pay homage throughout this book to its pioneering approaches. McKee's formidable space-age survey of Earth, *Global Sand Seas* (1979), remains a tremendous resource, but much superior satellite datasets now exist in the age of Google Earth. Greeley and Iverson's *Wind as a Geological Process* (1984) is perhaps most comparable in scope to what we have attempted, insofar as it considers other planets in some detail. However, planetary science is a field that moves quickly, and Mars has now been much better explored, Venus's few dunefields were discovered, and Titan opened up a whole new arena for planetary dunes. We have aimed for the breezy clarity of exposition exemplified by Nick Lancaster's (sadly earthbound) *Geomorphology of Desert Dunes* (1995). We draw inspiration from all these sources.

This book is not intended as a textbook; a textbook explains how to do something, whereas we see this book as more useful in suggesting what might be worth doing, and where to start—a travel guide to the research landscape, if you will. We have attempted to be reasonably comprehensive in providing references to allow the reader to follow up details close to the frontiers of present research in the various topics, but this book is not intended to be a complete review of each—that would entail far more work for us and far more pages from our publisher than either party can afford. We have tried to be fairly rigorous in covering the planetary dune literature, at least, such that this book can complement texts on (terrestrial) aeolian geomorphology, of which there are many examples such as Pye and Tsoar's *Aeolian Sand and Sand Dunes* (1990) and Cooke and Warren's *Geomorphology in Deserts* (1973). Even so, we crave the indulgence of our many colleagues, some of whose vast body of work we have inevitably failed to acknowledge.

On the subject of indulgence, one of the few pleasures available to the author is the ability to decide what to write about. We have included some more quirky topics that just seemed fun and worth discussing in a scientific manner, such as booming dunes, locomotion on dunes, and dunes portrayed in other worlds in science fiction. We hope our readers enjoy these diversions but find them instructive nonetheless.

We have pursued our research into dunes with the support of NASA, via the Cassini mission (RL) and through various Mars-related research and analysis programs (JZ). The compilation of this book, however, was very much a "spare time" affair, although RL acknowledges a Janney Publication Fellowship from the Johns Hopkins University Applied Physics Laboratory, and JZ the support of the Smithsonian Institution.

We thank Clive Horwood of Praxis Publishing who has been a patient supporter of this project from the beginning, and Mary Bourke who helped us shape the idea. We are deeply indebted to our many colleagues who unhesitatingly gave us permission to use various graphics, and indeed for finding photos for us: we have acknowledged their contributions where they occur. Prof. Jani Radebaugh of Brigham Young University deserves special mention, however, for also participating in a number of our desert adventures, and for giving us critical feedback on an early draft of the book.

Last, but by no means least, we note our appreciation of the patience and love of our respective spouses, Zibi Turtle and Cheryl Zimbelman, who have tolerated our physical absences off in desert places, and our mental absences over many weekends and evenings spent at home but hunched over our laptops working on the text you have before you.

Columbia, May 2013  Ralph D. Lorenz
Manassas  Jim R. Zimbelman

# Contents

**Part I  Introduction**

**1  Introduction** . . . . . . . . . . . . . . . . . . . . . . . . . . . . . . . . . . . . . . . . . . . . . . . . . . . 3

**Part II  Dune Basics**

**2  Sand** . . . . . . . . . . . . . . . . . . . . . . . . . . . . . . . . . . . . . . . . . . . . . . . . . . . . . . . . . 17
    2.1  Sand Size and Shape . . . . . . . . . . . . . . . . . . . . . . . . . . . . . . . . . . . . . . 19
    2.2  Compositional Considerations . . . . . . . . . . . . . . . . . . . . . . . . . . . . . 22

**3  Winds and Atmospheres** . . . . . . . . . . . . . . . . . . . . . . . . . . . . . . . . . . . . . . 27
    3.1  Atmospheres . . . . . . . . . . . . . . . . . . . . . . . . . . . . . . . . . . . . . . . . . . . . 27
    3.2  The Planetary Boundary Layer . . . . . . . . . . . . . . . . . . . . . . . . . . . . . 28
    3.3  Planetary Wind Patterns . . . . . . . . . . . . . . . . . . . . . . . . . . . . . . . . . . 29
    3.4  Wind Speed Statistics . . . . . . . . . . . . . . . . . . . . . . . . . . . . . . . . . . . . 35
    3.5  Wind Direction: The Wind Rose . . . . . . . . . . . . . . . . . . . . . . . . . . . 36
    3.6  Turbulence . . . . . . . . . . . . . . . . . . . . . . . . . . . . . . . . . . . . . . . . . . . . . 36

**4  Mechanics of Sand Transport** . . . . . . . . . . . . . . . . . . . . . . . . . . . . . . . . . 39
    4.1  Styles of Particle Motion . . . . . . . . . . . . . . . . . . . . . . . . . . . . . . . . . 39
    4.2  Fluid Forces on Particles . . . . . . . . . . . . . . . . . . . . . . . . . . . . . . . . . 40
    4.3  The Boundary Layer . . . . . . . . . . . . . . . . . . . . . . . . . . . . . . . . . . . . . 43
    4.4  Launching the Sand . . . . . . . . . . . . . . . . . . . . . . . . . . . . . . . . . . . . . 45
    4.5  Sand Grain Trajectories . . . . . . . . . . . . . . . . . . . . . . . . . . . . . . . . . . 48
    4.6  Influence of Saltation on the Boundary Layer . . . . . . . . . . . . . . . . . 49
    4.7  The Saturation Length and Controls on Dune Size . . . . . . . . . . . . . 50

**5  Deposits of Sand: Ripples Versus Dunes** . . . . . . . . . . . . . . . . . . . . . . . . 55
    5.1  Flow Separation and Recirculation . . . . . . . . . . . . . . . . . . . . . . . . . 55
    5.2  Angle of Repose: The slip face . . . . . . . . . . . . . . . . . . . . . . . . . . . . 55
    5.3  Grain Sorting . . . . . . . . . . . . . . . . . . . . . . . . . . . . . . . . . . . . . . . . . . 58
    5.4  Dunes Versus Ripples . . . . . . . . . . . . . . . . . . . . . . . . . . . . . . . . . . . 58
    5.5  Controls on Feature Scale . . . . . . . . . . . . . . . . . . . . . . . . . . . . . . . . 61
    5.6  Sedimentation and Dune Structure . . . . . . . . . . . . . . . . . . . . . . . . . 62
    5.7  Niveo-Aeolian Processes . . . . . . . . . . . . . . . . . . . . . . . . . . . . . . . . . 69

| | | | |
|---|---|---|---|
| **6** | **Basic Types of Dunes** | | 75 |
| | 6.1 | Dome | 75 |
| | 6.2 | Barchan | 77 |
| | 6.3 | Barchanoid Ridge | 80 |
| | 6.4 | Transverse Ridge | 81 |
| | 6.5 | Linear or Longitudinal (Sief) | 81 |
| | 6.6 | Reversing | 82 |
| | 6.7 | Star | 83 |
| **7** | **Other Dunes and Other Sand Deposits** | | 93 |
| | 7.1 | Sheet | 93 |
| | 7.2 | Streak | 93 |
| | 7.3 | Shadow Dune (Lee Dune) | 93 |
| | 7.4 | Climbing Dunes (and Falling Dunes) | 94 |
| | 7.5 | Echo Dune (Reflection Dune) | 94 |
| | 7.6 | Lunette | 94 |
| | 7.7 | Nebka | 94 |
| | 7.8 | Parabolic Dune | 95 |
| | 7.9 | Blowout Dune | 97 |
| | 7.10 | Compound Dunes and Complex Dunes | 99 |
| **8** | **Dune Fields, Sand Seas and Transport Pathways** | | 103 |
| | 8.1 | Sand Sources and Sinks | 103 |
| | 8.2 | Dune Stabilization | 105 |
| **9** | **Rates of Geomorphic Change** | | 107 |
| | 9.1 | The Sand Rose | 107 |
| | 9.2 | Sand Fluxes | 107 |
| | 9.3 | Observed Dune Migration Rates | 108 |
| | 9.4 | Ripple Migration Rates | 113 |
| **10** | **Booming or Singing Dunes** | | 115 |

**Part III Dune Worlds**

| | | | |
|---|---|---|---|
| **11** | **Earth Dunes** | | 121 |
| | 11.1 | Introduction | 121 |
| | 11.2 | The Terrestrial Atmosphere | 121 |
| | 11.3 | Major Deserts and Dune Fields | 122 |
| | | 11.3.1 Sahara | 123 |
| | | 11.3.2 Rub' al Khali | 124 |
| | | 11.3.3 Taklamakan | 125 |
| | | 11.3.4 Gobi/Badain Jaran | 125 |
| | | 11.3.5 Namib | 125 |
| | | 11.3.6 Kalahari | 126 |
| | | 11.3.7 Atacama | 127 |
| | | 11.3.8 Lençóis Maranhenses | 127 |
| | | 11.3.9 Great Victoria | 127 |
| | | 11.3.10 Simpson | 127 |
| | | 11.3.11 Great Sand Dunes | 130 |

|  |  | 11.3.12 | White Sands | 130 |
|---|---|---|---|---|
|  |  | 11.3.13 | Bruneau | 130 |
|  | 11.4 | Snow Dunes and Megadunes | | 131 |
|  | 11.5 | The Role of Vegetation in Dune Motion | | 131 |
|  | 11.6 | Links Between Sand Dunes and Climate | | 134 |

**12 Mars Dunes** ... 135
- 12.1 Introduction ... 135
- 12.2 The Martian Atmosphere ... 138
- 12.3 Sediments on Mars ... 140
- 12.4 Types of Dunes on Mars ... 141
- 12.5 Examples of Dune Localities on Mars ... 144
- 12.6 Dunes in Relation to Topography ... 152
- 12.7 Links Between Dunes and Climate? ... 153
- 12.8 Dune and Ripple Migration ... 153
- 12.9 Future Possibilities ... 154

**13 Titan Dunes** ... 157
- 13.1 Pre-Cassini Expectations ... 157
- 13.2 Cassini Observations ... 158
- 13.3 Sand Source, Amount and Composition ... 163
- 13.4 Implications for Meteorology ... 165
- 13.5 Future Exploration of Titan ... 166

**14 Venus Dunes** ... 169
- 14.1 Introduction ... 169
- 14.2 History of Venus Exploration ... 169
- 14.3 Venus Dunes ... 170
- 14.4 Aeolian Transport Under Venus Conditions: Experiments ... 170
- 14.5 Aeolian Transport Under Venus Conditions: Theory ... 172
- 14.6 Venus Winds ... 172
- 14.7 Venus Sand ... 175
- 14.8 Future Exploration of Venus ... 175

**15 Other Dune Worlds** ... 177
- 15.1 Triton ... 177
- 15.2 Io ... 178

**Part IV  Dune Studies**

**16 Field Studies** ... 183
- 16.1 Sand Measurements in the Field ... 183
- 16.2 Dune Shape and Structure ... 184
  - 16.2.1 GPS ... 184
  - 16.2.2 Imaging ... 185
  - 16.2.3 Thermal Imaging ... 187
  - 16.2.4 Laser Scanning ... 188
  - 16.2.5 Ground Penetrating Radar ... 189
  - 16.2.6 Gravity and Other Geophysical Methods ... 190

|  | 16.3 | Measuring Wind | 191 |
|---|---|---|---|
|  |  | 16.3.1 Cup and Propeller Anemometers | 191 |
|  |  | 16.3.2 Hot Wire and Hot Film | 192 |
|  |  | 16.3.3 Pitot Tubes | 193 |
|  |  | 16.3.4 Ultrasound | 194 |
|  |  | 16.3.5 Optical Tracer Methods | 194 |
|  |  | 16.3.6 Large-Scale Tracer Methods | 194 |
|  | 16.4 | Measuring Saltation | 194 |
|  | 16.5 | Data Acquisition | 195 |
|  | 16.6 | Aeolian Abrasion | 195 |
|  | 16.7 | In Situ Spacecraft Observations | 197 |
| **17** | **Laboratory Studies** | | **203** |
|  | 17.1 | Sand Studies | 203 |
|  | 17.2 | Process: Terrestrial Wind Tunnels | 206 |
|  | 17.3 | Process: Planetary Wind Tunnels | 208 |
|  | 17.4 | Water Tank Experiments | 213 |
|  | 17.5 | Abrasion Experiments | 213 |
|  | 17.6 | Avalanching and Granular Flow | 215 |
| **18** | **Remote Sensing** | | **217** |
|  | 18.1 | Aerial Photography | 217 |
|  | 18.2 | Orbital Imaging | 218 |
|  | 18.3 | Thermal Imaging | 223 |
|  | 18.4 | Hyperspectral Imaging | 227 |
|  | 18.5 | Topography From Imaging | 228 |
|  | 18.6 | Radar Studies of Dunes | 230 |
| **19** | **Numerical Models** | | **241** |
|  | 19.1 | Modeling the Wind | 241 |
|  | 19.2 | Modeling the Sand | 244 |
|  | 19.3 | Dynamics at the Dune and Ripple Level: Pattern Formation | 248 |
|  | 19.4 | Summary | 251 |

## Part V  Why Dunes Matter

| **20** | **Dunes as Physical Systems** | | **255** |
|---|---|---|---|
| **21** | **Dunes and Climate** | | **257** |
| **22** | **Moving on Sand** | | **259** |
|  | 22.1 | Walking on Dunes | 259 |
|  | 22.2 | Driving on Dunes | 261 |
|  | 22.3 | Vehicles for Dune Driving | 263 |
|  | 22.4 | Getting Unstuck | 266 |
|  | 22.5 | Getting Stuck and Unstuck on Mars or Titan | 268 |
| **23** | **Dunes and People** | | **273** |

**24 Fictional Dune Worlds** . . . . . . . . . . . . . . . . . . . . . . . . . . . . . . . . . . . . 283
    24.1   Arrakis . . . . . . . . . . . . . . . . . . . . . . . . . . . . . . . . . . . . . . . . . 283
    24.2   Tatooine . . . . . . . . . . . . . . . . . . . . . . . . . . . . . . . . . . . . . . . . 284

## Part VI   Conclusions

**25 Conclusions** . . . . . . . . . . . . . . . . . . . . . . . . . . . . . . . . . . . . . . . . . . 289

**References** . . . . . . . . . . . . . . . . . . . . . . . . . . . . . . . . . . . . . . . . . . . . . 291

**Index** . . . . . . . . . . . . . . . . . . . . . . . . . . . . . . . . . . . . . . . . . . . . . . . . . 303

# Figures

| | | |
|---|---|---|
| Fig. 1.1 | Moon over Southern Hemisphere dune | 4 |
| Fig. 1.2 | Mariner 9 image of Hellespontus dunes: first dune image | 4 |
| Fig. 1.3 | Titan dunes | 5 |
| Fig. 1.4 | Aeolian parameters on planets | 5 |
| Fig. 1.5 | Proctor dunes at ever-improved resolution | 6 |
| Fig. 1.6 | Hidden ripples imaged by Sojourner rover | 6 |
| Fig. 1.7 | Rover tracks across meridiani ripples | 7 |
| Fig. 1.8 | Kite camera image of lagoons at Lencois Maranhenses | 7 |
| Fig. 1.9 | Ground Penetrating Radar Study of Barchan Dune | 8 |
| Fig. 1.10 | Flow simulation over a star dune | 8 |
| Fig. 1.11 | Longitudinal dunes simulated in water tank | 9 |
| Fig. 1.12 | Sand transport across the Arounga impact crater, Chad, from Space Station | 9 |
| Fig. 1.13 | Erwin Rommel's half-track | 10 |
| Fig. 1.14 | Sandbox study of Mars Exploration Rover | 10 |
| Fig. 1.15 | South Pole station on stilts above the snow flux | 11 |
| Fig. 1.16 | Tatooine film set, Tunisia, with looming barchan dune | 12 |
| Fig. 1.17 | Linear arrangement of star dunes, seen from the Space Station | 13 |
| Fig. 1.18 | Four-pointed dune on Mars | 14 |
| Fig. 2.1 | Snow dune in the lee of a tent, Antarctica | 18 |
| Fig. 2.2 | Histograms of sand sizes from different locations | 20 |
| Fig. 2.3 | Sand size histograms portrayed on different axes | 21 |
| Fig. 2.4 | Microscopic view of sand grains from coral pink sand dunes, Utah | 21 |
| Fig. 2.5 | Microscopic imager view of basaltic sand, Mars | 22 |
| Fig. 2.6 | Sand grain shape classification | 23 |
| Fig. 2.7 | Parabolic ejecta blanket on Venus | 24 |
| Fig. 3.1 | Earth's general circulation: Hadley cells | 30 |
| Fig. 3.2 | GCM simulation of Titan's winds | 30 |
| Fig. 3.3 | Variation of Mars' obliquity with time | 31 |
| Fig. 3.4 | Dunes encountering mountain ridge, Saudi Arabia | 31 |
| Fig. 3.5 | Dunes diverting around inselbergs, Namib Desert | 31 |
| Fig. 3.6 | Dunes diverting around obstacles, Titan | 32 |
| Fig. 3.7 | Hodograph of Mars' surface winds | 32 |
| Fig. 3.8 | Wind statistics at Great Sand Dunes National Park and Preserve | 33 |
| Fig. 3.9 | Weibull curves for windspeed statistics | 34 |
| Fig. 3.10 | Wind rose at Bruneau Dunes | 34 |
| Fig. 3.11 | Intermittent wind speeds and sand transport | 35 |
| Fig. 3.12 | Wind turbulence on Mars from the Phoenix lander | 36 |

| | | |
|---|---|---|
| Fig. 3.13 | Aeolian streamers | 37 |
| Fig. 4.1 | Types of particle transportation by the wind | 40 |
| Fig. 4.2 | Dust storm in Iraq | 40 |
| Fig. 4.3 | The balance of forces | 41 |
| Fig. 4.4 | Drag coefficient of a sphere as a function of Reynolds number | 42 |
| Fig. 4.5 | Terminal velocities of particles on planets | 44 |
| Fig. 4.6 | Mars pathfinder windsock and logarithmic wind profile | 45 |
| Fig. 4.7 | Friction speed correlated with wind speed | 46 |
| Fig. 4.8 | Schematic of threshold velocity with particle size | 46 |
| Fig. 4.9 | Fluid transport threshold on planets | 47 |
| Fig. 4.10 | Fluid threshold friction speeds | 47 |
| Fig. 4.11 | Saltating particles observed with laser dheet in eind tunnel | 48 |
| Fig. 4.12 | Relative saltation paths on planets | 50 |
| Fig. 4.13 | Impact threshold vs fluid threshold: Earth and Mars | 50 |
| Fig. 4.14 | Speed profiles in clear and daltating boundary layers | 51 |
| Fig. 4.15 | Drag length on different planets | 51 |
| Fig. 4.16 | Bedform wavelengths at different heights on Tharsis Mountains, Mars | 52 |
| Fig. 4.17 | Stability of sand transport on a low hill | 52 |
| Fig. 4.18 | Bedforms as a function of speed in Venus wind tunnel | 53 |
| Fig. 4.19 | Microdunes observed in Venus wind tunnel | 53 |
| Fig. 5.1 | Flow separation over Dumont Dune, visualized with streamers | 56 |
| Fig. 5.2 | Computation fluid dynamics simulation of airflow separation at crest of dune | 56 |
| Fig. 5.3 | 'Smoking dune'—sand transport at crest of Palen dune | 56 |
| Fig. 5.4 | Sand piles made with Viking sampler arm to measure angle of repose | 57 |
| Fig. 5.5 | Sand surfing | 57 |
| Fig. 5.6 | Avalanche lobes on dune slipface | 58 |
| Fig. 5.7 | Mars dune avalanching | 59 |
| Fig. 5.8 | Megaripples | 60 |
| Fig. 5.9 | Layers exposed in Meridiani ripple by impact crater | 61 |
| Fig. 5.10 | Granule surface of Mars ripple, observed by Curiosity Mastcam | 62 |
| Fig. 5.11 | Magnetite sorting on Eureka dune | 63 |
| Fig. 5.12 | Simple ripples on coral pink dand funes, Utah | 63 |
| Fig. 5.13 | Transverse aeolian ridges in Nirgal Vallis, Mars | 64 |
| Fig. 5.14 | Shadow mechanism for ripple formation | 65 |
| Fig. 5.15 | Giant pumice granule ripple, Puna, Argentina | 65 |
| Fig. 5.16 | Close-up of pumice granule ripple | 66 |
| Fig. 5.17 | Scaled ripple topographic profiles | 66 |
| Fig. 5.18 | Grain size and bedform scales | 66 |
| Fig. 5.19 | Layers in white sands dune | 67 |
| Fig. 5.20 | Exposed layer in Salton Sea barchan dune | 67 |
| Fig. 5.21 | Slip face layers in Lencois Maranhenses | 68 |
| Fig. 5.22 | 3D simulation of cross-bedding | 68 |
| Fig. 5.23 | Cross-bedding in Utah sandstone | 69 |
| Fig. 5.24 | Cross-beds in bedrock exposure at Cape St Vincent, Mars | 70 |
| Fig. 5.25 | Zoom of cross-beds at Cape St Mary, Mars | 70 |
| Fig. 5.26 | Remnant of slip face position of gypsum dune, White Sands | 71 |
| Fig. 5.27 | Scars of barchanoid slip face location, Mars | 72 |
| Fig. 5.28 | Niveo—aeolian rransport, Great Sand Dunes National Park and Preserve | 72 |
| Fig. 5.29 | Exposed snow layer beneath sand cover | 72 |
| Fig. 6.1 | Basic dune types | 76 |
| Fig. 6.2 | Dome dune, United Arab Emirates | 76 |

| | | |
|---|---|---|
| Fig. 6.3 | Dome and linear dunes, Noachis Terra, Mars | 77 |
| Fig. 6.4 | Tiny barchan, Danikil Depression, Ethiopia | 78 |
| Fig. 6.5 | Egyptian barchans, airplane window view | 78 |
| Fig. 6.6 | Oregon barchans, G. K. Gilbert circa 1900 | 79 |
| Fig. 6.7 | Hooked barchan on Mars | 79 |
| Fig. 6.8 | Barchan morphology vs flow diversity, from flume experiments | 80 |
| Fig. 6.9 | Teardrop barchans on Mars | 80 |
| Fig. 6.10 | Tadpole barchan on Mars | 80 |
| Fig. 6.11 | Dark dunes on bright ripples, Mars | 81 |
| Fig. 6.12 | Wide Emirates barchan viewed from kite | 82 |
| Fig. 6.13 | Barchan–barchanoid ridge transition on Mars | 83 |
| Fig. 6.14 | Defrosting barchans and barchanoid dunes on Mars | 83 |
| Fig. 6.15 | Titan linear/transverse dunes | 84 |
| Fig. 6.16 | Transverse ridges on Mars | 84 |
| Fig. 6.17 | Transverse dune interaction with crater, Mars | 85 |
| Fig. 6.18 | SPOT image of Yemen linear dunes | 86 |
| Fig. 6.19 | Namib sand sea from Space Shuttle | 87 |
| Fig. 6.20 | Egyptian linear dune from kite | 87 |
| Fig. 6.21 | Titan linear dunes | 88 |
| Fig. 6.22 | Linear dunes, domes, barchans on Mars | 88 |
| Fig. 6.23 | Hooked barchans in the Rub' Al Khali | 89 |
| Fig. 6.24 | Possible hooked barchans, Titan | 90 |
| Fig. 6.25 | Reversing barchans, United Arab Emirates, from a kite | 90 |
| Fig. 6.26 | Star dunes in Algeria from the International Space Station | 91 |
| Fig. 6.27 | Akle dunes on Mars | 92 |
| Fig. 7.1 | Sand streaks, Mars and Earth | 94 |
| Fig. 7.2 | Sand streak on Venus | 94 |
| Fig. 7.3 | Superstition Mountains, lee dunes | 95 |
| Fig. 7.4 | Moenkopi clifftop linear dunes from the air | 95 |
| Fig. 7.5 | Wadi Rum climbing dune | 96 |
| Fig. 7.6 | Cat dune in the Mojave Desert—an example of a falling dune | 96 |
| Fig. 7.7 | Mars lee linear dunes and barchans | 96 |
| Fig. 7.8 | Echo ripple at Stimpy Rock, from Sojourner rover | 97 |
| Fig. 7.9 | Nebkhas on the Chott El Djerid | 97 |
| Fig. 7.10 | Parabolic and blowout dunes | 98 |
| Fig. 7.11 | ASTER image of large linears in the Rub' Al Khali | 98 |
| Fig. 7.12 | SPOT image of oman linears with superposed linear pattern | 99 |
| Fig. 7.13 | Field photo of transverse dunes on flank of a linear | 99 |
| Fig. 7.14 | SPOT image of linear dunes with transverse corridors | 100 |
| Fig. 7.15 | ASTER image of Badain Jaran dunes and lakes | 100 |
| Fig. 7.16 | Liwa megabarchans from the International Space Station | 101 |
| Fig. 7.17 | Algodones dunes from the International Space Station | 101 |
| Fig. 8.1 | Great sand dunes, lapping against mountains, from Space Station | 104 |
| Fig. 8.2 | HiRISE of Victoria Crater, Mars, showing dunes on floor and dark streaks | 104 |
| Fig. 8.3 | Airliner view of Mojave sand transport system and Kelso dunes | 105 |
| Fig. 8.4 | Colorful sand transport pathways in Terkezi Oasis, Libyan Sahara | 105 |
| Fig. 8.5 | Field photo of Arica Hills vegetated linear dunes, California | 106 |
| Fig. 8.6 | Airliner view of dunes and farmland, Australia | 106 |
| Fig. 9.1 | Sand rose diagrams after Fryberger | 108 |
| Fig. 9.2 | Drift potential/dune morphology regime diagram | 108 |
| Fig. 9.3 | Barchan migration rates on Earth, linear axes | 109 |
| Fig. 9.4 | Barchan migration at Star Wars film set, observed with Google Earth | 110 |

| | | |
|---|---|---|
| Fig. 9.5 | GPS outlines of Liwa dome and barchan dunes, compared with Google Earth. | 111 |
| Fig. 9.6 | Barchan and megabarchan migration, logarithmic axes | 111 |
| Fig. 9.7 | HiRISE image of Niili barchan, showing fiducial interdune features | 111 |
| Fig. 9.8 | Barchan migration rates on Earth and Mars | 112 |
| Fig. 9.9 | Migration rate of ripples on Earth. | 112 |
| Fig. 9.10 | HiRISE observation of ripple pattern evolution on Nili dunes | 113 |
| Fig. 10.1 | Frequency spectrum of sand mountain avalanche | 116 |
| Fig. 10.2 | Booming frequencies—The art of chartmanship | 117 |
| Fig. 11.1 | Clementine mosaic of Earth seen from the Moon | 122 |
| Fig. 11.2 | Envisat mosaic of Earth teflectivity showing fesert latitudes | 123 |
| Fig. 11.3 | Map showing principal fune sreas on Earth | 124 |
| Fig. 11.4 | MODIS image of dust plumes from Sahara into Mediterranean | 125 |
| Fig. 11.5 | Dunes, lake and palms in the Ubari Oasis, Libya | 125 |
| Fig. 11.6 | SeaWIFS image of Arabia, showing Rub Al'Khali sand pathways | 126 |
| Fig. 11.7 | Field photo of linear funes and barchans on the UAE–Oman border | 127 |
| Fig. 11.8 | Field photo of giant dunes and lakes of Badain Jaran desert, China | 128 |
| Fig. 11.9 | The Long Wall — Aerial view of the Atlantic/Namib boundary | 128 |
| Fig. 11.10 | ALOS image of the Lencois Maranhenses, Brazil | 129 |
| Fig. 11.11 | Field photo of Lencois Maranhenses | 129 |
| Fig. 11.12 | Vegetated linear dune in the Simpson desert, with kangaroo trail | 129 |
| Fig. 11.13 | Simpson linear dune diversion by topographic obstacles, aerial view | 130 |
| Fig. 11.14 | Gypsum pillar in White Sands dunefield | 130 |
| Fig. 11.15 | Kite camera view of White Sands barchans and barchanoid ridge | 131 |
| Fig. 11.16 | Sand transport on Bruneau dune | 132 |
| Fig. 11.17 | Snow dunes from the air | 132 |
| Fig. 11.18 | Shallow snow megadunes in Antarctica, aerial view | 132 |
| Fig. 11.19 | RADARSAT view of snow megadunes | 133 |
| Fig. 11.20 | MODIS optical view of snow megadunes | 133 |
| Fig. 12.1 | Mars at opposition, by the Hubble Space Telescope | 136 |
| Fig. 12.2 | Mariner 9 view of Proctor crater dunes, with MOC/HiRISE insets marked | 136 |
| Fig. 12.3 | Viking 1 panorama | 137 |
| Fig. 12.4 | Location map of Mars landers and rovers | 137 |
| Fig. 12.5 | MOC closeup of Proctor dunes showing improved resolution | 138 |
| Fig. 12.6 | HiRISE closeup of Proctor dune surface, showing gullies | 138 |
| Fig. 12.7 | HiRISE view of dark dust plumes from defrosting Olympia Undae | 139 |
| Fig. 12.8 | Map of wind streaks on Mars | 140 |
| Fig. 12.9 | Mars digital dune database dunefield map | 142 |
| Fig. 12.10 | HiRISE shaded relief of Kaiser crater barchan | 143 |
| Fig. 12.11 | HiRISE image of dark 'wing' barchans and TARs on floor of Proctor crater | 143 |
| Fig. 12.12 | HiRISE image of linear dunes and domes | 143 |
| Fig. 12.13 | THEMIS image of diverse barchan forms in Bunge crater | 144 |
| Fig. 12.14 | Mariner 9 Mosaic of Mars North Polar Cap and Olympia Undae | 144 |
| Fig. 12.15 | Viking image of Kaiser crater dunes | 146 |
| Fig. 12.16 | MOC image of Kaiser barchans and Yin-Yang crater sand deposit | 146 |
| Fig. 12.17 | Rippled dunes on the floor of Herschel crater | 147 |
| Fig. 12.18 | THEMIS thermal image of dunes in Rabe crater | 148 |
| Fig. 12.19 | Defrosting dunes in Richardson crater | 148 |
| Fig. 12.20 | CRISM Spectral Image showing Sand Composition on Mars | 148 |
| Fig. 12.21 | Moving barchan on Nili Patera | 149 |
| Fig. 12.22 | Curiosity Mastcam mosaic of Aeolus Mons and Gale crater dunes | 150 |

| Fig. 12.23 | Curiosity MAHLI view of Gale crater sand | 151 |
| Fig. 12.24 | Dunes on the floor of Victoria crater | 151 |
| Fig. 12.25 | El Dorado dunes | 152 |
| Fig. 12.26 | MOC view of TARs on the floor of Auqakuh Vallis | 152 |
| Fig. 12.27 | Eroding indurated barchan on Mars | 154 |
| Fig. 13.1 | Cassini optical image of Titan, Saturn and Rings | 158 |
| Fig. 13.2 | Near-infrared map of Titan with named locations | 159 |
| Fig. 13.3 | First Cassini radar images of Titan dunes | 160 |
| Fig. 13.4 | T8 Radar glints from giant linear dunes in Belet sand sea | 160 |
| Fig. 13.5 | Radar-dark dunes and bright interdunes in Cassini radar image | 161 |
| Fig. 13.6 | Radarclinometric profile of Titan dunes | 161 |
| Fig. 13.7 | Huygens probe descent imager mosaic showing dark dunes in the distance | 162 |
| Fig. 13.8 | Cassini VIMS near-infrared image strip of Fensal dunes | 162 |
| Fig. 13.9 | Map of Titan dune orientations | 163 |
| Fig. 13.10 | Interaction of linear dunes with crater on Titan | 163 |
| Fig. 13.11 | Latitude histogram of Titan dune coverage and radar coverage | 163 |
| Fig. 13.12 | Variation of Titan interdune area fraction with latitude | 164 |
| Fig. 13.13 | Cassini radar image of possible fossil dunes | 164 |
| Fig. 13.14 | Artist's impression of Titan airplane and dune | 166 |
| Fig. 14.1 | Venus atmosphere in Ultraviolet, from Mariner 10 | 170 |
| Fig. 14.2 | Venera 14 view of Venus surface | 171 |
| Fig. 14.3 | Magellan radar reflectivity map of Venus | 171 |
| Fig. 14.4 | Magellan image of Algaonice dunefield on Venus | 172 |
| Fig. 14.5 | Magellan image of Fortuna dunefield on Venus | 173 |
| Fig. 14.6 | Asymmetric radar reflection from unresolved microdunes | 173 |
| Fig. 14.7 | Slipfaces and surfaces of Venus wind tunnel microdunes | 174 |
| Fig. 14.8 | Venus wind tunnel bedforms as a function of speed | 174 |
| Fig. 14.9 | Venus wind tunnel saltation experiment results | 175 |
| Fig. 15.1 | Triton wind streaks | 178 |
| Fig. 15.2 | Frosted ridges on Io | 179 |
| Fig. 16.1 | Bulldozing barachanoid dune at White Sands | 184 |
| Fig. 16.2 | GPS traverse over Egyptian linear dune | 185 |
| Fig. 16.3 | Precision GPS receiver, Great Sand Dunes | 186 |
| Fig. 16.4 | 123DCatch topography teconstruction of ripple | 187 |
| Fig. 16.5 | Timelapse camera study of ripple migration | 188 |
| Fig. 16.6 | Kite camera view of linear arms, Sand Mountain, Nevada | 188 |
| Fig. 16.7 | Thermal image of sand avalanche | 189 |
| Fig. 16.8 | Field laser scanner at Great Sand Dunes | 190 |
| Fig. 16.9 | Laser scanner topography | 190 |
| Fig. 16.10 | Manual GPR operations in Egypt | 191 |
| Fig. 16.11 | GPR survey of Victoria Valley dune, Antarctica | 192 |
| Fig. 16.12 | Vehicle-dragged GPR survey of linear dune, Egypt | 193 |
| Fig. 16.13 | Field instrumentation array for saltation studies | 196 |
| Fig. 16.14 | Ventifact on Mars | 197 |
| Fig. 16.15 | Ripple migration observed by Opportunity rover | 198 |
| Fig. 16.16 | Excavation of soil by Phoenix lander rocket engines | 198 |
| Fig. 16.17 | Viking sampling arm operations | 199 |
| Fig. 16.18 | Sand grain on Mars rover solar panel | 199 |
| Fig. 16.19 | Sediments captured by MER magnet | 200 |
| Fig. 16.20 | MER rover wheel scuff on aeolian ripple | 201 |
| Fig. 16.21 | Sojourner rover on Mermaid ripple | 201 |
| Fig. 16.22 | Cohesive dirt on TEGA ovens, Phoenix lander | 202 |
| Fig. 17.1 | Curiosity Alpha-Proton X-ray Spectrum (APXS) of Mars sand | 204 |
| Fig. 17.2 | X-ray diffraction pattern of Mars sand from Curiosity | 205 |

| | | |
|---|---|---|
| Fig. 17.3 | Sullivan Marswit 1236 RS-12-05 Canon 41 small.jpg | 206 |
| Fig. 17.4 | Dune simulation wind tunnel photo.jpg | 207 |
| Fig. 17.5 | Sullivan Marswitbedform 1236 RS-12-05 Canon 11.jpg | 207 |
| Fig. 17.6 | Wind tunnel crosssection.png | 208 |
| Fig. 17.7 | boundarylayercontrol.png | 208 |
| Fig. 17.8 | Mars Rover model in wind tunnel, showing scour areas | 209 |
| Fig. 17.9 | Atlas rocket and test chamber, later used as Mars wind tunnel facility | 209 |
| Fig. 17.10 | Mars wind tunnel fan, and rail covers for pressure door | 210 |
| Fig. 17.11 | Aarhus Mars wind tunnel exterior—windows and wires | 211 |
| Fig. 17.12 | Aarhus Mars wind tunnel fans | 211 |
| Fig. 17.13 | Aarhus Mars wind tunnel laser instrumentation | 212 |
| Fig. 17.14 | Venus wind tunnel | 212 |
| Fig. 17.15 | Schematic of water flume used to simulate barchan morphology | 213 |
| Fig. 17.16 | Barchan regime diagram constructed from flume experiments | 214 |
| Fig. 17.17 | Paris water tank with oscillating platform | 215 |
| Fig. 17.18 | Barchan and leaking arms, simulated in water tank | 215 |
| Fig. 18.1 | Dunes and lakes of Badain Jaran desert, from airliner | 218 |
| Fig. 18.2 | Ubari Sand sea, early aerial photograph | 218 |
| Fig. 18.3 | Heat shield and Gale crater dunes, from descending MSL/Curiosity | 219 |
| Fig. 18.4 | Linear dunes in Arabian desert, from Gemini 4 | 219 |
| Fig. 18.5 | ASTER image of Roter Kamm impact crater and Namib dunes | 220 |
| Fig. 18.6 | Landsat image of Namib dunes | 221 |
| Fig. 18.7 | Dunes cascading into lake, Mongolia, from Space Station | 222 |
| Fig. 18.8 | Diurnal temperature history as a function of thermal inertia of surface | 223 |
| Fig. 18.9 | THEMIS thermal image of dunefield on Mars | 224 |
| Fig. 18.10 | Visible wavelength spectra of sand-forming minerals | 224 |
| Fig. 18.11 | Near-infrared spectra of sand-forming minerals | 225 |
| Fig. 18.12 | CRISM multispectral image of Olympia Undae dunes, showing gypsum | 225 |
| Fig. 18.13 | Digital elevation model of Namib sand sea from ASTER stereo images | 226 |
| Fig. 18.14 | Topographic profiles of Namib dunes from ASTER DEM | 227 |
| Fig. 18.15 | Mars Express HRSC (nadir color draped on DEM) of dunes in Mars crater) | 228 |
| Fig. 18.16 | Airborne LiDAR DEM of White Sands barchanoid ridges and parabolic dunes | 229 |
| Fig. 18.17 | Zoom of White Sands LiDAR showing dune detail | 230 |
| Fig. 18.18 | Seasat radar image of Algodones dunes | 231 |
| Fig. 18.19 | Earth radar reflectivity at 40o incidence showing major sand seas | 232 |
| Fig. 18.20 | Radar backscatter from different surfaces vs angle | 233 |
| Fig. 18.21 | Multiwavelength radar image of the Namib sand sea | 234 |
| Fig. 18.22 | Schematic of radar reflectivity from different surfaces | 235 |
| Fig. 18.23 | Shuttle radar image of Namib dunes showing slip face glints | 236 |
| Fig. 18.24 | Transverse flank dunes on Namib linears in radar images at different resolution | 237 |
| Fig. 18.25 | Backscatter of unresolved dunes vs incidence | 238 |
| Fig. 18.26 | Radar-derived digital elevation model | 239 |
| Fig. 19.1 | Mars Global Circulation Model wind vectors | 242 |
| Fig. 19.2 | Mesoscale model of circulation at Arsia Mons, Mars | 243 |
| Fig. 19.3 | Computational fluid dynamics simulation of flow around a blowout dune | 244 |
| Fig. 19.4 | Lattice gas simulation of flow over a barchan, showing separation bubble | 244 |

| | | |
|---|---|---|
| Fig. 19.5 | Sand slab manipulation rules in Werner-type cellular automaton model | 246 |
| Fig. 19.6 | Dune diversity in Werner model | 246 |
| Fig. 19.7 | Barchan morphology and sand transport in coupled flow-transport model | 247 |
| Fig. 19.8 | Cellular automaton model incorporating vegetation | 247 |
| Fig. 19.9 | Coupled flow-transport model of star dune | 248 |
| Fig. 19.10 | Comparison of numerical model with water tank experiments and Mars dunes | 249 |
| Fig. 19.11 | Cellular automaton showing barchan-barchan interactions | 250 |
| Fig. 19.12 | Schematic of transverse ripple migration | 251 |
| Fig. 20.1 | Sand-starved ripples forming Barchan-like arrangement | 255 |
| Fig. 21.1 | Sand trap at La Jornada Experimental Range | 258 |
| Fig. 22.1 | Soil failure schematic | 260 |
| Fig. 22.2 | Camel in Egypt | 260 |
| Fig. 22.3 | Wyoming geology survey by forse-drawn cart | 261 |
| Fig. 22.4 | Snowmobile in Alaska | 262 |
| Fig. 22.5 | Long Range Desert Group truck unsticking | 263 |
| Fig. 22.6 | Navy Seal dune buggies | 264 |
| Fig. 22.7 | Motorbike with paddle tires on dune | 265 |
| Fig. 22.8 | Quad bike on dune | 266 |
| Fig. 22.9 | Dismounting to unstick quad bike | 267 |
| Fig. 22.10 | Using sand channels to unstick a truck | 268 |
| Fig. 22.11 | Opportunity Navcam image of high-sinkage ripple at Meridiani | 269 |
| Fig. 22.12 | Soft ground encountered by Spirit rover | 270 |
| Fig. 22.13 | Computer simulation of Mars rover ground interaction | 270 |
| Fig. 22.14 | High-centered underside of Spirit rover | 271 |
| Fig. 22.15 | Scarecrow rover under test at Dumont dunes | 271 |
| Fig. 23.1 | The last house in Biggs, Oregon, overwhelmed by dunes | 274 |
| Fig. 23.2 | Sand fences at railway in Oregon, by G. K. Gilbert | 275 |
| Fig. 23.3 | Space Station view of Nouakchott, Mauritania, being invaded by linear dunes | 276 |
| Fig. 23.4 | Date plantations in Liwa, United Arab Emirates, from Space Station | 277 |
| Fig. 23.5 | Trees at Liwa being overrun by megabarchan slip face | 278 |
| Fig. 23.6 | Stabilizing a dune in Libya with grid of vegetation | 279 |
| Fig. 23.7 | Sand intruding onto road in Liwa, field photo | 280 |
| Fig. 23.8 | Barchan corridor running between Kharga, Egypt and its airport | 281 |
| Fig. 23.9 | Floating fence at US–Mexico border, Algodones dunes | 282 |
| Fig. 24.1 | Egyptian linears near St. Exupery's crash site from Gemini 4 | 284 |
| Fig. 24.2 | R2D2 in 'Star Wars', at Stovepipe Dunes, Death Valley National Park | 285 |
| Fig. 24.3 | Algodones, set of 'Return of the Jedi' from airliner | 285 |
| Fig. 24.4 | Quickbird satellite image of 'Star Wars' film set in barchan field, Tunisia | 285 |
| Fig. 24.5 | Star Wars buildings damaged after being overrun by dune | 286 |
| Fig. 25.1 | Earthrise from Apollo 8 | 290 |

# Tables

| | | |
|---|---|---|
| Table 2.1 | The modified Udden-Wentworth scale of particle sizes and names.... | 19 |
| Table 2.2 | Stability of some common minerals under weathering conditions at Earth's surface | 23 |
| Table 2.3 | Densities of bulk sand-forming materials. | 23 |
| Table 3.1 | Properties of planetary atmospheres | 28 |
| Table 3.2 | Surface wind measurements. | 33 |
| Table 4.1 | Reynolds numbers for different conditions. | 43 |
| Table 4.2 | Saltation parameters | 49 |
| Table 11.1 | Major desert on Earth. | 124 |
| Table 12.1 | Major dune fields on Mars | 145 |
| Table 18.1 | Thermophysical properties of selected geological materials | 222 |

# Part I
# Introduction

# Introduction

This book is about dunes and planets, the interactions of earth and air.

Sand provides a paradox. It is a solid, and yet it moves. This paradox applies to rocks in general—moving rocks are, after all, what geology is ultimately about—but granular materials on the Earth's surface can move much faster, at rates that are important to, and observable by, humans. This is due to the peculiar conditions and processes on our planet, and results in some spectacular landscapes (e.g. Fig. 1.1) in specific regions where the sand (and sometimes snow) has apparently organized itself into magically regular structures—dunes.

The ancients knew of four planets, wandering points of light in the night sky, plus Earth and its moon. The invention of the telescope led to the discovery of a few more planets and a few moons, some of which are bigger than the smallest planet. We know of hundreds of planets (many of which will have moons too) around other stars. But more importantly than the astronomical cataloguing of celestial objects has been the Space Age and, by means of robotic exploration, the transformation of some of these planetary bodies from mere dots in the sky to worlds in their own right.

This unmanned space exploration perspective has shown us that first Mars (Fig. 1.2), then Venus, and recently and remarkably, Saturn's moon Titan have dunes similar to those on Earth, despite very different conditions and materials. Thus the formation of dunes and ripples is a general phenomenon, unified by the same physical processes, even though the conditions under which these processes operate can be quite different (Fig. 1.3). This universality is perhaps nowhere highlighted better than by the vast sand seas of Titan where, despite sands made of organic muck, in frigid air four times denser than ours, on a world with gravity only one-seventh our own, the landscape is covered in dunes of exactly the same shape, height and width of the Earth's largest sand seas. Thus Titan is almost as exotic a world as one can imagine, and yet standing on its surface are landforms (Fig. 1.4) almost indistinguishable from those on Earth.

Our aim in this book is to survey dunes on these worlds with this physical perspective, highlighting the morphological similarities and differences that are exposed by dramatically improved remote sensing instrumentation at Earth and elsewhere. Over the 40 or so years of Mars exploration since dunes were discovered, the quality of orbital images (in terms of number of pixels per square kilometer) has improved ten-thousandfold (Fig. 1.5), and after the first pioneering wanderings of the Sojourner rover in 1997 (Fig. 1.6), what is now decade of continuous roving across the surface by its successors has brought a 'field geology' perspective (Fig. 1.7) with millimeter-scale imaging and advanced scientific instruments brought to bear on individual ripples, showing us the sands of Mars at the grain level.

Dunes on Earth were first explored by field geologists and geographers. While many of their traditional techniques are still applied, new methods, such as GPS and photogrammetry using digital images from the ground or the air, allow us to rapidly and quantitatively measure the shape of dunes in the field (Fig. 1.8). We can now even probe the internal structure of dunes, revealing the layers that record the history of deposition and migration, using ground-penetrating radar (Fig. 1.9).

There have also been spectacular advances in physical and computational modeling of the processes by which sand and wind interact: it is now possible to follow the formation of dunes, and indeed the interactions between dunes, in silico. A wide range of dune morphologies can be replicated via the successive application of very simple rules, taming the bewildering diversity of the landscape under a unifying algorithmic whip (Fig. 1.10). Furthermore, while observation of real dune formation and motion can take months, years, or even millennia, their virtual counterparts can be brought to animated life in seconds on the computer screen.

**Fig. 1.1** A striking picture, setting the stage for planets and dunes, with *vibrant red sand* and *blue skies* is familiar to millions as a 'desktop' background image for the Windows operating system. The image shows the moon atop a dune. But not just a dune; after reading this book, we hope you will recognize avalanche lobes and two generations of ripples in the foreground. What millions in the northern hemisphere have probably not noticed is that the moon is 'upside down'; there the moon is seen as in the inset on the *right*, with the pattern of mare (dark impact basins). The reason that the moon above the dune looks 'wrong' is that this dune picture is from the southern hemisphere (perhaps Australia, but more likely Namibia.) Montage by J. Zimbelman

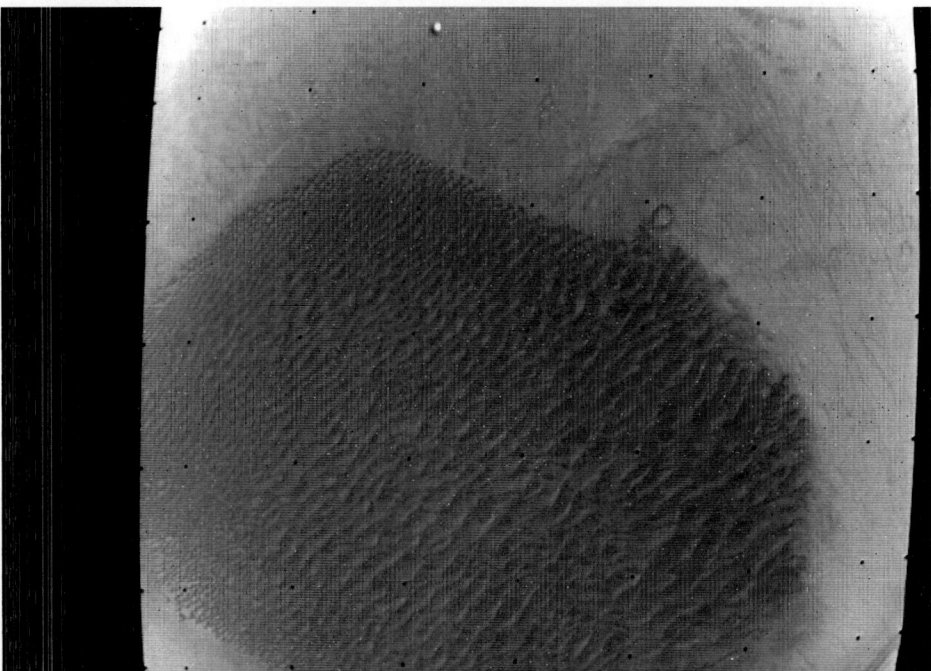

**Fig. 1.2** The first dunes recognized on another world. After the 1972 dust storm cleared, Mariner 9 snapped this image showing what at the time was called a 'suspected dune mass' in the floor of the Hellespontus crater. Note the curved edge of the image: the vidicon TV cameras of the time could cause geometric distortion of the image, which would be corrected by lining up the regular pattern of *black dots* ('reseau marks'). Note the gradient in dune size across the dunefield. Also visible towards the *upper right* is a small impact crater, and (just barely visible here, and not recognized at the time) some faint dust devil tracks. Tick marks at the left are for synchronization in data handling; the *white dots* across the dunefield are transmission errors. Mariner 9 image DAS 09807429, processed by R. Lorenz

**Fig. 1.3** A radar image of a region of Titan's surface, about 150 km across, from the Cassini spacecraft. The *dark stripes* are linear sand dunes, about 1 km wide, snaking around a bright highland (shaped a little like a snout) at the *right*. Similar deviation of dunes by topographic obstacles is seen on Earth (Fig. 12.10). Image by R. Lorenz and the Cassini RADAR Team

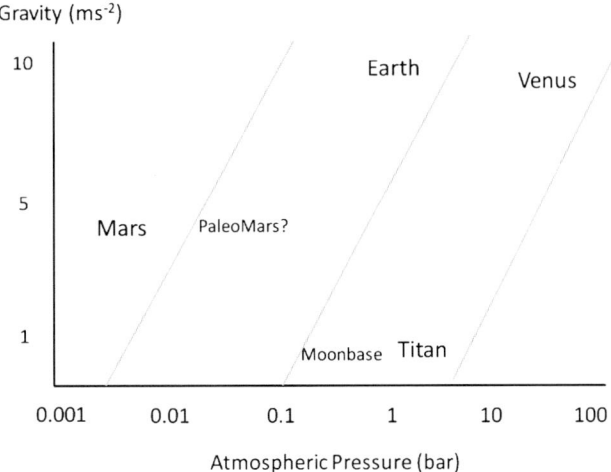

**Fig. 1.4** A schematic of the fundamental environmental features of planets with dunes. While Mars' atmosphere is thin today, it may have been substantially thicker in the past, perhaps making it easier to transport sand. A pressurized lander or base on the moon would see a gravity the same as Saturn's moon Titan, but with thinner air. The faint diagonal lines are a crude indication of equal transportability; Venus and Titan can move sediments with the gentlest winds, whereas today very violent winds are needed to launch sand into the air on Mars. These parameters are explored in more detail in Chap. 4

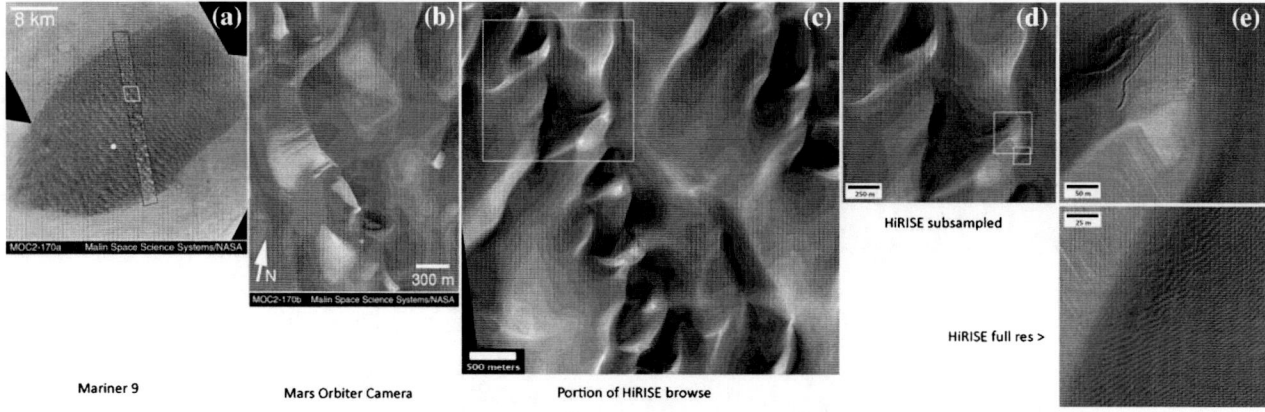

**Fig. 1.5** A mosaic of images of dunes in the Proctor crater on Mars, assembled by J. Zimbelman. **a** A Mariner 9 image from 1972 (at ~60 m/pixel); **b** a Mars Global Surveyor image at several m/pixel in 1999; **c–e** progressive zooms into a massive image (PSP_006780_1320), from the HiRISE camera on Mars Reconnaissance Orbiter in 2008, with a resolution of about 50 cm/pixel. From the dunefield perspective we can now zero in on individual dunes and even see (and observe the migration of—see Chap. 8) the small superposed ripples. (Several of these images are shown individually at a larger scale in Chap. 12)

**Fig. 1.6** First 'field trip' on Mars. The shallow bedforms in the foreground are 10 cm or so high, but were hidden from the camera on the Mars Pathfinder lander by the 'Rocknest' formation. This image was only possible because the Sojourner rover trekked away from the lander, revealing new territory. The familiar 'twin peaks' hills (also visible from the lander) can be seen on the horizon. NASA/JPL image PIA00965

These computer models also allow us to bridge the real world and laboratory experiments (which themselves have reached an impressive level of fidelity (Fig. 1.11)).

This tremendous arsenal of data and tools has brought a new level of understanding to planetary dune studies and lets us offer answers to some basic questions.

1. **How do sand dunes develop, move, and change shape?** In order to understand how sand moves, we first need to establish what sand is, which can be very different in other planetary environments. But once the different fluid and gravitational forces that act on a sand grain are identified, some simple math can tell us the basics about what winds are required to set grains in motion, although a prominent theme in modern aeolian studies is how to tackle the complex and highly fluctuating character of transport in turbulent winds.

2. **Are sand dunes the same everywhere?** Like the canonical snowflake, no two dunes are exactly alike. And yet there is clearly order in the bewitching infinity of duneforms. The first job in science is usually to classify objects into groups of more-or-less the same. While that always entails some subjectivity, it is an essential simplification. But once that is embraced, the same sets of forms can be recognized on different planets, and in the computer, and in water tank experiments, paving the way for a quantitative link between what we see on dune worlds, and what wind and time was needed to make what we see.

3. **So what do the dunes tell us?** There is a 'Goldilocks' element to the formation of dunes. The particles involved must in general not be moving, otherwise they do not really define a landscape. And yet they must move often enough to assemble into a dune. 'Often enough' is

**Fig. 1.7** Image from the Mars Exploration Rover Opportunity. The rover's solar cells and the Planetary Society's sundial are visible at bottom. The tracks show wheel slippage as it negotiates wind-blown ripples; some coarse granules can be seen on the flanks of the ripples, and cracked white bedrock is exposed in some places. JPL/NASA

**Fig. 1.8** Small digital cameras make the recording of field conditions exceptionally easy. New computational tools (Chap. 17) make it possible to build quantitative shape models of dunes and ripples with only minutes of effort. Digital cameras are small and light enough to take to the air; this example, showing barchanoid ridges with interdune lagoons in the Lencois Maranhenses in Brazil, was lofted by a parafoil kite that could scrunch up into a coat pocket. *Photo R. Lorenz*

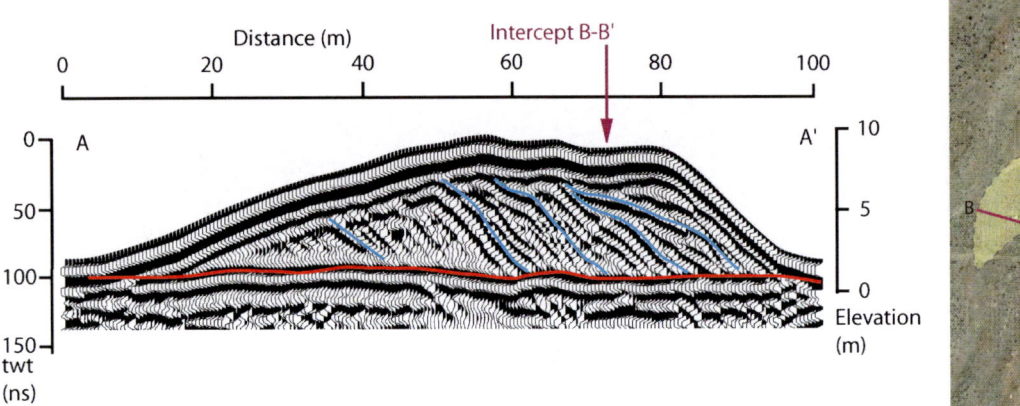
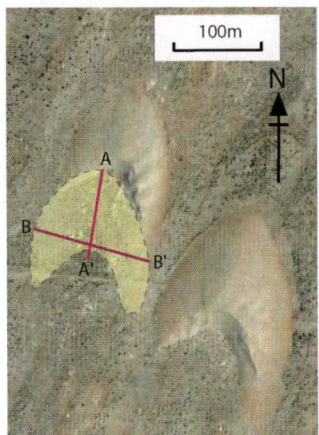

**Fig. 1.9** The internal structure of a crescentic barchan dune in Morocco is revealed in this Ground-Penetrating Radar (GPR) survey. The transect (A-A') moves left to right in the downwind direction (top to bottom in the orbital view) and shows the steeply-dipping former slip faces that marked the previous position of the dune. Enough time has elapsed between this survey (where the dune outline and transects were measured by GPS) and the orbital image at right that the dune has moved southwards. Image courtesy Charlie Bristow

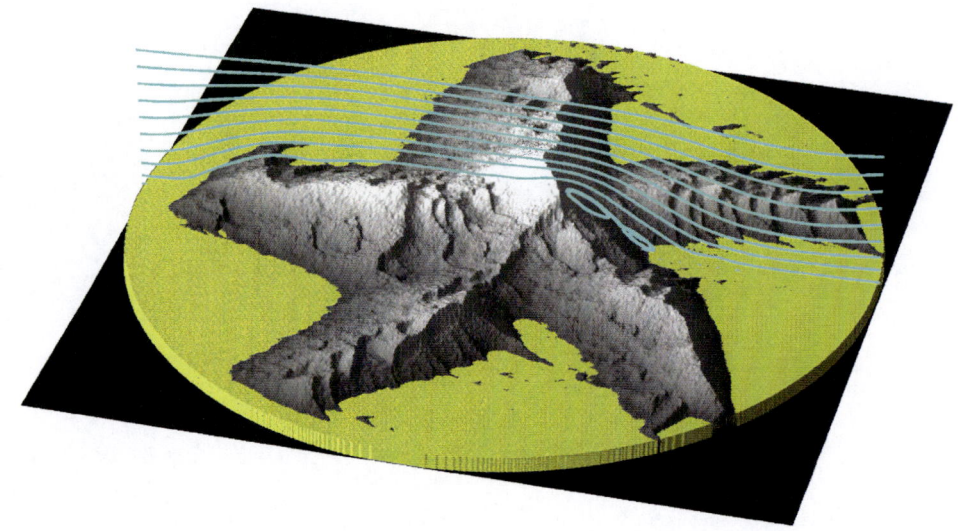

**Fig. 1.10** Modern coupled simulation of the airflow and sand transport, using a cellular automaton sand model (showing here a 5-pointed star dune). Streamlines in a lattice gas flow model are shown, influenced by the evolving topography underneath; the streamlines compress at the dune crest, and separate, leaving a vortex in the lee of the dune. Image courtesy Clement Narteau

a relative term—it depends on what else is going on. For a dune to exist, it must form or move or repair itself faster than other processes destroy it. Various geological processes prevail to different degrees on different planets: on Earth, for example, there are relatively few impact craters because fluvial or glacial erosion or other processes have removed them. There are just a couple where craters recognizable from orbit are in deserts with prominent dunes (Fig. 1.12), yet this sort of interaction is common on Titan where erosive activity, while present, is less vigorous than on Earth. On Mars, in contrast, rain has not been widespread for billions of years (if ever) and thousands of craters are visible, acting as traps for the sand.

On Earth, at least, there are dunes present where dunes do not presently form or move—they are fossils of a past climate. This brings another paradox, resurrected into the wider planetary arena from eighteenth and nineteenth century Earth science more generally: to what extent does the landscape we see today represent processes happening today, versus what processes may have happened in a catastrophically-different past? This question has come to the fore in studies of Martian dunes, a few of which have only recently been observed to move, but others of which may be very old and cemented by ice or evaporite minerals, and even on Titan where some dune-like features stand distinct in location and orientation from the majority.

**Fig. 1.11** Morphologically identical to dunes a thousand times larger, these cm-scale bedforms are generated in a water tank in which a sediment-covered platform (see Chap. 17) is moved back and forth, rotated between cycles to impose a desired 'wind regime' to quantitatively explore the dependence of morphology on the angular diversity of winds. Picture courtesy of Sylvain Courrech du Pont

**Fig. 1.12** The sand streaks and falling dunes show a clear corridor of sand transport across the Arounga impact structure in Chad in this image from the International Space Station. The crater is 12.6 km in diameter. On Earth's young surface with relatively few impact craters there are few similar examples; on Mars, craters tend to be traps for sand. On Titan, many craters appear to be partly overrun by dunes. NASA Image

**Fig. 1.13** Erwin Rommel (with binoculars) in his Sd.Kfz 250/3 half-track 'Greif' during the African campaigns. Half-tracks are a useful compromise vehicle for difficult terrain, balancing traction and manoeuvrability. Bundesarchiv—Wikimedia Commons

**Fig. 1.14** Mike Seibert and Sharon Laubach, engineers on the Mars Exploration Rover team at NASA's Jet Propulsion Laboratory, Pasadena, check the exact position of a test rover in preparation for the next test of a possible maneuver for Spirit to use on Mars. The test setup at JPL simulates the situation where Spirit is embedded in a patch of soft soil dubbed 'Troy', in Mars' Gusev Crater. The preparation shown here on July 7, 2009, preceded an assessment of straight-backward driving the next day, one of several possible maneuvers assessed in the test sandbox. *Image Credit* NASA/JPL-Caltech

**Fig. 1.15** Living in the Antarctic is perhaps a close analogy to how people might one day live on other planets. Quartz is not the only material on Earth to be blown by wind and form dunes—snow can do so as well, especially when very cold and dry. As the old dome at the South Pole became buried by blowing snow, the new Amundsen-Scott South Pole Station was built 1999–2008 on stilts to allow snow to blow under it, and the building was somewhat aerodynamically-shaped, with the wedge-shaped underside oriented into the prevailing wind specifically to inhibit snow accumulation. *Photo* National Science Foundation

Whether active today, or fossils of the past, dunes tell us certain things about the conditions of their formation. There must be sand—whatever it happens to be made of, it means particles of a size with the right mobility. In some cases, this means particles that have been ground down from massive rock, sometimes evaporite crystals formed as a lake dries up, sometimes it means sand re-liberated from sandstone rock that was once in the form of dunes. Yet in other cases (as at Titan) it may mean grains built up from smaller particles somehow, or falling from the sky as with snow on Earth. Like the dunes it can form, sand itself evolves—often initially jagged or sharp-cornered from its origin in fracture or precipitation, the innumerable collisions with its fellows as it bounces along the ground may grind it smaller or make it round. New analytical methods, and even remote sensing, can let us match the minerals from which a grain or dune is made to a locale where the grain must have come from. Other laboratory techniques can even tell us how long a sand grain has been buried inside a dune, letting us gauge the dune's age. Yet the microscopic properties of sand grains can influence such experiential phenomena as how the sand sounds when jump on a dune and make the sand slide.

In addition to telling us about past climates, sand and dunes are a major factor in how humans live and work in some desert environments. Large deserts have been barriers to commerce and military operations and many different machines and/or techniques have been developed to enable vehicular travel across sand dunes. Some of the first traverses of the Sahara were made with half-track vehicles, also a favourite[1] of World War II's 'Desert Fox', Erwin Rommel (Fig. 1.13).

---

[1] During the preparation of this book, we searched for images in the Bundesarchiv but sadly failed to find any of this vehicle on an actual dune. Most operations were (prudently) on the flat coastal plains and in the mountains.

**Fig. 1.16** The 'Star Wars' film set near Tozeur, Tunisia. A set of buildings (used to portray the city of Mos Espa on the planet Tatooine in 'Star Wars, Episode 1') makes convenient reference marks to track the migration of nearby dunes. The slip face of a barchan dune, about 20 m away from the set when this picture was taken in 2009, can be seen behind the first author who is standing next to a 'moisture vaporator' (sic). The dunes have been measured (Fig. 9.8) to move at about 15 m/year, threatening the site with erosion and burial. *Photo R. Lorenz/J. Barnes*

In fact, it was one of Rommel's most daring adversaries, Brigadier Ralph Bagnold, who first made a systematic study of sand transport and dunes as a physical process. Bagnold explored the deserts prior to WWII, and his initial curiosity about desert landforms motivated him to investigate the physics underlying wind-induced particle motion, both in a laboratory of his own design and through fieldwork in deserts around the world. Bagnold's seminal book, *The Physics of Wind Blown Sand and Desert Dunes* (1941), is the starting point for all modern investigations of the geomorphology of sand dunes and sand deposits. Bagnold instigated the famous 'Long Range Desert Group' employing special vehicles and tactics and using his desert knowledge to prevent Rommel invading British Egypt by making deep strikes behind enemy lines across desert thought to be impassable.

Similar challenges in locomotion are encountered in snow, and now in traversing the loose, dry terrain on Mars. The difficulties of negotiating noncohesive ground have even claimed their first victim on another world. The Spirit rover (Fig. 1.14) at the Gusev crater on Mars got stuck in soft ground in an orientation that did not let it generate enough solar power to survive the winter.

Sand giving way under one's wheel s or feet is one set of problems; sand moving under the action of wind presents others. Wind-blown snow presents a similar challenge, prompting architecture that is aerodynamically designed to prevent accumulation (Fig. 1.15) which can completely bury

**Fig. 1.17** An image of the Grand Erg Orientale (part of the Sahara) in Algeria, taken through the window of the International Space Station. The overall arrangement is clearly linear, but the crestlines are star-shaped. Are these linear dunes that have been morphed into star shapes, or are they star dunes that just happen to have formed in lines? The linear dunes are about 2 km wide. Image Science and Analysis Laboratory, NASA-Johnson Space Center. "The Gateway to Astronaut Photography of Earth"

buildings over some decades. Migrating dunes can block roads, railways or canals, requiring expensive mitigation; on the larger scale, the burial of fields or oases by dunes can simply force us to migrate ourselves, admitting defeat against the onslaught of sand. In fact, the battle of arid lands ecologists against dunes in Oregon inspired the science fiction author James Herbert to write his iconic *Dune* novel, and dunes feature on a number of fictional worlds (Fig. 1.16).

Our aim in this book is to survey the physical principles behind dunes, review observations and measurement techniques, emphasizing the newest developments, and to draw together what has become known about dunes on other planetary bodies with what has been learned about the Earth. Just as dunes are a feature of the interface between the ground and the air, so studies of dunes have fallen across the domains of geography, physics, geomorphology, geochemistry, planetary science and other fields. We hope the reader enjoys the adventure, as we have, of straying beyond the traditional bounds of whatever discipline they consider themselves in. This is not a textbook, merely a travel guide. Our goal is to provide breadth, and to identify pathways in data, techniques and the literature for the reader to wade in as deeply as they wish.

We have organized the book in five main sections. First is the introduction you are now reading. In the second section, we describe the principal physics behind dunes, what they are made of, and what controls their size, shape and movement. Next, we discuss the aeolian features seen on various dune worlds, namely Earth, Mars, Titan and Venus, as well as the possibility for bedforms on a few other planetary bodies. The fourth section discusses the various ways in which dunes are studied: in the laboratory or wind tunnel, in the field, and by computer simulation. The various remote sensing techniques applied to aeolian studies are also reviewed. Lastly, we survey why dunes and related landforms and processes are important—as physical systems, as environmental

**Fig. 1.18** An image from the HiRISE camera of the Martian surface (this image forms the *dark blue background* on the cover of this book) with aeolian features on three scales. The image shows a set of barchan and dome-like dunes (see Chap. 6). Dominant at the *right* of the image are sharp-crested ripples or TARs (see Chap. 5). Close examination will also reveal small ripples on the surface of the dunes. *Image* U. Arizona/JPL/NASA

indicators, and how they affect infrastructure and mobility, on Earth and other planets (including fictional ones). Along the way, we provide images to illustrate particular points, although as researchers entranced by the beauty of our subject, we have unashamedly included some images (like Figs. 1.17 and 1.18), just because they are pretty.

# Part II
# Dune Basics

In this section, we outline the basic physics behind dunes, and the different forms dunes and ripples may take, and how they move. This naturally begins with a discussion of the particulate material from which they are made. This is usually—but not always—sand, sand being a size category rather than a composition. Dune-forming material, even on Earth, can have a wide range of origins and compositions, and this range is broader still when considering other planets.

We then describe the principal features of planetary atmospheres and their winds, both on the global scale, and how winds at a given location may be statistically described.

The movement of sand by wind depends on a balance of forces: wind drag and lift pulling particles up, and gravity and cohesion holding them down. This classical balance, which depends on environmental factors such as the particle size, gravity, and the densities of the air and sand, is what once dominated discussion of sediment mobility (e.g., Bagnold 1942; Greeley and Iverson 1987). However, there has been a growing recognition of the importance of unsteady dynamics, the difference between sand starting to move and continuing to move (Kok et al. 2010; Zheng 2009).

It is this difference that leads to the distinction between two broad classes of bedforms on Earth: ripples and dunes. Whereas ripples are defined largely by the pseudoballistic trajectories of individual particles accelerated by the wind, dunes are controlled more by the airflow itself and how the airflow is modified by the dune. Such distinctions merit reconsideration in other planetary environments—for example, in the thick Venus atmosphere ripples as we consider them may simply not exist, but microdunes form on a scale that we associate only with ripples on Earth.

After exploring this distinction, we note the different morphologies of dunes and how they relate to wind regime and sand supply, a relationship that appears to be essentially universal (and is explored in numerical models in Part IV). We then introduce a discussion of the factors that control the scale of dunes, notably the saltation path length, the saturation length over which a variation in sand transport will adjust to imposed conditions, and the thickness of the atmospheric boundary layer.

Dunes are not, in general, static landforms. We summarize the rates at which dunes and ripples are observed and predicted to move on different planets. On Earth, megabarchans have been documented moving at $\sim 0.1$ m/yr, while small barchan and dome dunes can scoot along at tens of meters per year. Ripples can move at centimeters per minute. On Mars, until very recently it was puzzled why dune movement had not been observed, but observations with higher resolution and longer timespans have now detected ripples and dune evolution and migration. On Venus and Titan, migration rates are predicted to be too small to detect with current (relatively meager) observations. However, even though present-day motion is not observable, computed sand transport rates are important in considering how long the observed landscape may have taken to form.

Finally, we will reward the reader's patience with all this physics on a lyrical note, with a discussion of singing sand and booming dunes.

# Sand

When we hear the word 'sand', the visual image evoked is most likely of sand in aggregate—of a pleasant beach or a massive field of dunes. But the poet William Blake provides an apt perspective on the inspiration that can come from contemplating even a single, isolated particle:

> To see a world in a grain of sand,
> and heaven in a wild flower,
> hold infinity in the palm of your hand,
> and eternity in an hour.
> (from 'Auguries of Innocence', 1803; in The Pickering Manuscript, published 1863)

For an enchanting and accessible discussion of sand and its relationship to our planet and its inhabitants, we recommend Michael Welland's book, *Sand: The Never Ending Story* (2009). A good paragraph-length summary of the microcosm exposed in grains is this by Raymond Siever who expanded more scientifically upon Blake's poetic sentiment:

> A single sand grain, an irregularly shaped fragment of rock, is the mute record of former mountains, rivers, and deserts, and of millions of years of the Earth's upheavals and quiescence. To make a grain tell us its history, we tear it apart bit by bit to find out its crystal structure, its chemical composition, its radioactive age, its external shape, and its internal strain. Yet we cannot tell all we want to know of a sand grain's origin from its composition alone, any more than we can deduce political history from human physiology. The context of the state of the world's continents and oceans at a particular time is the background. That grain was produced by forces that made the rock it was eroded from, by the Earth's surface environment that eroded it from its parent and carried it to a resting place, and by the internal deformation of the Earth's crust that buried it. (Siever 1988, p. 1)

Here we attempt to expand upon Siever's terrestrial point of view, to consider sand and its associated landforms identified during spacecraft exploration that has been conducted in the other worlds of the solar system.

We can start by returning to the simple question: what is sand? A typical definition for sand is 'a hard, granular rock material finer than gravel and coarser than dust'. As this short statement points out, it is important to realize that 'sand' formally refers to sediment particle size (which we will quantify shortly), not to particle composition.

It is conventional too to think of sand as something—usually quartz (see later)—that is *broken down* from a larger mass of bedrock. However, from the viewpoint of sand being particles of a given size or, broader yet in the context of this book, dune-forming material in general, this perspective is somewhat parochial. Snow forms dunes, yet is crystallized in the air from water vapor; Titan's sand may start in a similar way, perhaps agglutinated somehow on the surface. Agglutination of dust is often responsible for the formation of granules that form the crest of ripples on Earth, and dustballs are suspected as particles in high-altitude martian bedforms (where the air density is so low that it would be hard to move solid particles). So material can grow into sand, not just be broken down into it.

Let's start with snow, as it defines an instructive end-member in the spectrum of dune-forming material. Because it often has a very low density (bulk snow can be 100 kg/m$^3$, about one-tenth that of water, although this is typically due to porosity between grains or flakes; the effective density of a fractal snow particle may be typically more like 300 kg/m$^3$, of course only a third as dense as a solid piece of ice), the drag at even low air speeds is comparable with its weight, and thus snow swirls in every eddy of wind, rather than leaping in little ballistic hops. In this sense, snow dynamics are sometimes more akin to sand underwater than to sand in air. A distinctive feature of snow, of course, is that it can be sticky, and so rather than clean slip faces on the lee side of dunes or drifts, it can form dramatic overhanging cornices, or accumulate into long streaks (Fig. 2.1).

Evaporite minerals can form dunes, most notably gypsum sand, famous at New Mexico's White Sands. This material, being both water-soluble and much softer than quartz, cannot migrate for long before it either becomes cemented or ground into dust (see also Synkiewicz et al. 2010). Gypsum appears to be a major component of some Martian dunes, too.

**Fig. 2.1** We deliberately challenge the reader's expectations by starting the illustrations in this sand section with a dune made from material that we do not typically call sand, but that meets our criterion of 'dune-forming'. Here, the lee of a stack of supply cases on a meteorite-hunting expedition in Antarctica has allowed blowing snow to form a somewhat linear dome dune overnight. A slight halo of saltating snow can be seen above the dune against the tent background. *Photo* Jani Radebaugh

Dunes can form on Earth in volcanic ash, a couple of celebrated examples being a dark ash barchan that has been marching away from the Ol Doinyo Lengai volcano in Tanzania at about 10 m per year, and the colorful painted dunes in Lassen Volcanic Field. Indeed, cross-bedding textures can often be seen in ash deposits; some may be caused not by conventional winds but by the strong outflowing surge when an ash column collapses. Bedforms are not limited to forming from sand-sized particles: pumice gravel in the Puna of Argentina forms the largest wind-ripples known on Earth. These hard-to-move lumps define the other end of the dynamical spectrum of particulates—stuff that needs the strongest winds to be launched into the air, and is only modestly affected by the airflow once that happens. This is true, generally, for sands on Mars, many of which have a volcanic (basaltic) composition.

Some sands are quite literally grown. Shell fragments of mollusks are a common component of many beach sands, and limestones in particular (but carbonate rocks on Earth in general—e.g. Brooke 2001) may have a substantial amount of tiny shells, which if eroded out and transported by wind can be referred to as an eolianite. When these rocks break down, these tiny fossil shells are the natural result; oolitic limestone is defined by spheroids ('ooids', from the Greek word for 'egg') between 0.25 and 1.25 mm. The tiny but resistant silica corpses of smaller living things yet, hard-shelled algae known as diatoms, form sediments (diatomite, or diatomaceous earth) which can break back down into hollow particles typically 10–200 $\mu$ across. Their small size and very low density makes them easy to transport by wind, and makes them responsible for the dustiest places on Earth, like the Bodélé Depression in Chad (which also has some of the fastest-moving dunes made of these little shells, see Chap. 8).

At the densest end of the spectrum, sorting by fluvial or aeolian processes can concentrate minerals. Often, bands of dark magnetite can be found on sand dunes, and Gay (1999) describes a small dune in Peru where aeolian sorting led to it having a concentration of some 46 % magnetite. The density of magnetite is some 4900 kg/m$^3$, around double that of quartz. Of course, human activities are better yet at sorting: there are doubtless piles of metal ores at mines and foundries worldwide that beg for aeolian experimentation. And in a cruel imprint of human history on geology, the sands of certain beaches in Normandy have a high fraction of steel particles.

With such a wide range of 'sand' and formation processes on Earth, one might expect that an even wider range needs to be considered for other worlds. Yet in fact, by and large what we know of Mars at least suggests the processes that make sand, and the resultant compositions, are the same as, but just a subset of, what happens on Earth. Titan, at least, likely has sands of an exotic composition (probably containing such organic chemical compounds as phenanthrene, coronene and other exotica), although they may be processed into saltating sand in much the same way as evaporitic sands on Earth. However they are made, it is important to adopt a wide

**Table 2.1** The modified Udden-Wentworth scale of particle sizes and names

| Name | Size range (grain diameter) | $\phi$-scale |
|---|---|---|
| Boulder | >256 mm | ($\phi < -8$) |
| Cobble | 64–256 mm | ($\phi$ −6 to −8) |
| Gravel | 2–64 mm | ($\phi$ −1 to −6) |
| Granule | 2–4 mm | ($\phi$ −1 to −2) |
| Sand | 1/16–2 mm | ($\phi$ 4 to −1) |
| Very coarse | 1–2 mm | ($\phi$ 0 to −1) |
| Coarse | 1/2–1 mm | ($\phi$ 1 to 0) |
| Medium | 1/4–1/2 mm | ($\phi$ 2 to 1) |
| Fine | 1/8–1/4 mm | ($\phi$ 3 to 2) |
| Very fine | 1/16–1/8 mm | ($\phi$ 4 to 3) |
| Silt | 1/256–1/16 mm | ($\phi$ 8 to 4) |
| Clay | <1/256 mm | ($\phi > 8$) |

perspective, as Titan's discoverer Christiaan Huygens did in his book *Cosmotheoros* (1698)—it matters less what the stuff is made of, than how it behaves.

> Since 'tis certain that Earth *and Jupiter* have their Water and Clouds, there is no reason why the other Planets should be without them. I can't say that they are exactly of the same nature with our Water; but that they should be liquid their use requires, as their beauty does that they be clear. This Water of ours, in Jupiter *or Saturn*, would be frozen up instantly by reason of the vast distance of the Sun. Every Planet therefore must have its own Waters of such a temper not liable to Frost.

## 2.1 Sand Size and Shape

A challenge with something as mundane as dirt is to formally systematize something that everyone thinks they understand. Sand is small stuff, but not really small.[1] Jon Udden, a professor of geology at the small midwestern college of Augustana, was the first person to publish a statistical analysis of sand grain sizes that occur within

---

[1] The number of grains of sand on Earth has long been a metaphor for a quantity beyond human comprehension. Scientists—at least since Archimedes' 'The Sand Reckoner'—however, have still attempted estimates of the number of sand grains after making certain assumptions. For example, one estimate for the number of sand grains on the beaches of the world is the staggering number of $\sim$5000 billion billion ($5 \times 10^{21}$) grains of sand, assuming that the average sand grain is 0.25 mm in diameter, the 'average' beach includes 50 m of sandy beach that is 1 m deep, the sand grains are perfectly packed together, and that the world has 1.5 million km of shoreline (Greenberg 2008, p. 39). The vast sandy deserts present on several continents (e.g., the Sahara), along with the sand that is now stored within sandstone deposits exposed around the world, suggest that even this estimate of the number of beach sand grains is likely more than a factor of ten too small to encompass all of the sand grains on Earth.

windblown deposits (Udden 1898); he subsequently compared aeolian sands to the sands found in rivers and on beaches (Udden 1914). He chose to follow the lead of soil scientists, who used sieves to separate soils into mass fractions based on how the materials passed through a stacked array of progressively smaller meshes. Charles Wentworth (1922) codified the progressive size scale using divisions based on powers of 2, what is now termed the 'phi scale' (where $\phi = \log_2 d$, with the grain diameter measured in millimeters). The Modified Udden-Wentworth scale, still in wide use throughout sedimentary geology, lists several particle size divisions (Table 2.1). Many additional size descriptors and parameters have been applied to particle size measurements through the years (e.g., see Folk 1966; Blott and Pye 2001), but the subdivisions within the sand size fraction have remained constant since the introduction of the Udden-Wentworth scale.

It is common among geologists and geographers to document the size distribution recovered by sieving as a histogram, and it often looks somewhat like the bell-shaped Gaussian curve that elementary statistics is so full of. Such a plot in linear form is a useful guide, but a more broad-ranging and broadminded mathematical investigation can be more instructive yet.

Note that the factor-of-two binning is a logarithmic scale, which is what is needed to grapple with a wide range (much as stellar magnitudes, or earthquake magnitudes, or intensity of sound). The utility of logarithms is often forgotten, and in fact Bagnold himself found interest in the use of logarithmic axes in plotting not just the size (the x-axis) but also the the number of sand particles in each sieve (the y-axis). When Bagnold plotted the results, he got something like a bell-shaped curve. But it wasn't quite 'right'—at very small and very large sizes, there were more particles than a Gaussian or 'normal' distribution suggested there should be.

A more conventional scientist—better indoctrinated in the tyranny of the Gaussian distribution—would have shrugged his or her shoulders at the obvious and insignificant experimental error. But Bagnold, originally educated in Cambridge as an engineer, was too practical a man to ignore his own measurements.

> The difficulty was that the measured frequencies along the tails became so small that they were unplottable. Not being a statistician, I concluded that this difficulty could be overcome simply by plotting the frequencies indirectly as their logarithms, thus giving every frequency, however small, an equal prominence. The precise pattern of the complete size distribution of natural sand at once appeared. It consisted of two converging straight lines joined by a curved summit. The curve resembled a simple hyperbola.

A Gaussian, on logarithmic axes, falls away rapidly like a parabola. As the numbers involved tend towards zero, their logarithms tend towards minus infinity, and the tails of

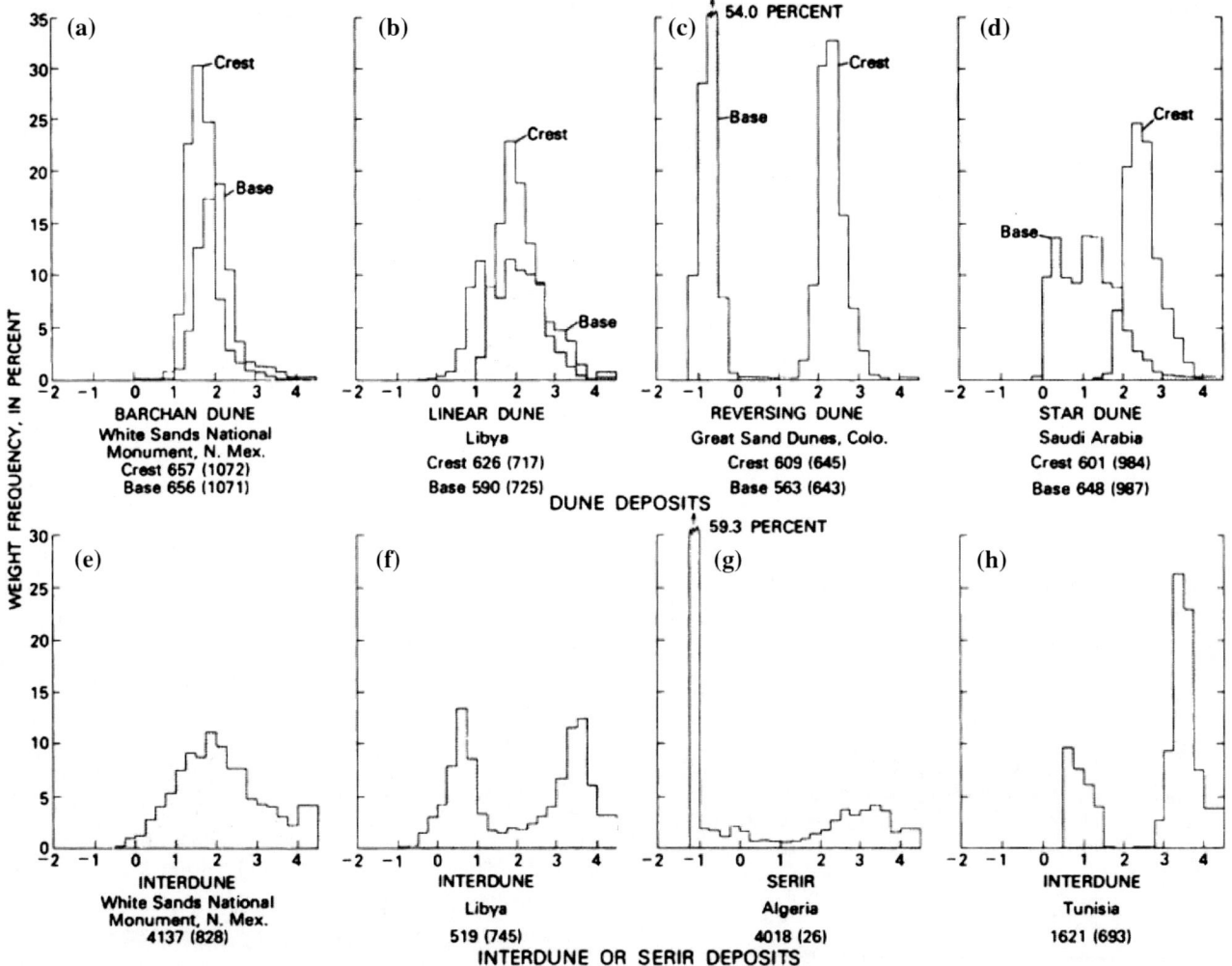

**Fig. 2.2** Histograms of sand size: amount (in weight %) against the logarithm of particle size (expressed as the f number in Table 2.1). Note the differentiation between the base and crest of the dunes; the sharp peaks indicate strong sorting of the particles. Note also the generally broader sizes in the interdune areas. These data from Ahlbrandt (1979), USGS image

the distribution droop vertically. But Bagnold's sand numbers looked at this way were startlingly different: the points fell on two inclined straight lines, indicating two power laws. Such a functional form likely says something about the probabilities of moving a particle, or the process by which grains are broken down (Gaussian processes result from random additions, but successive fracturing, for example, leads to a power law). More generally, logarithmic plots avoid suppressing attention to the extremes of the distribution—we highlight the importance of this aspect in Chap. 3: Wind. In addition to logarithmic axes, sometimes data of this sort are plotted with 'probability' axes (essentially a way of stretching the plot under the assumption it is Gaussian); such approaches have been used in dune pattern analysis, although they seem less general and no better than logarithmic ones.

Note that much of the literature on sand sizes considers the statistical moments of the distribution—not only the mean size and the standard deviation, but also the skewness (how much one tail is fat relative to the other) and kurtosis (how spread the tails are). Sometimes these statistical measures are plotted for collections of sand samples in an effort to draw conclusions about the sand's provenance. Such conclusions may or may not be terribly robust (Figs. 2.2 and 2.3).

While the Udden-Wentworth scale is essentially configured by the measurement tool (sieves) it should be noted there are other ways to measure particle size. One is to observe the terminal velocity, to infer how long material takes to fall out of the air (or out of liquid suspension). Laboratory particle-sizers can use this technique, and the longevity of impact-raised dust clouds was used to constrain surface particle size at Venus and Titan. But electronic

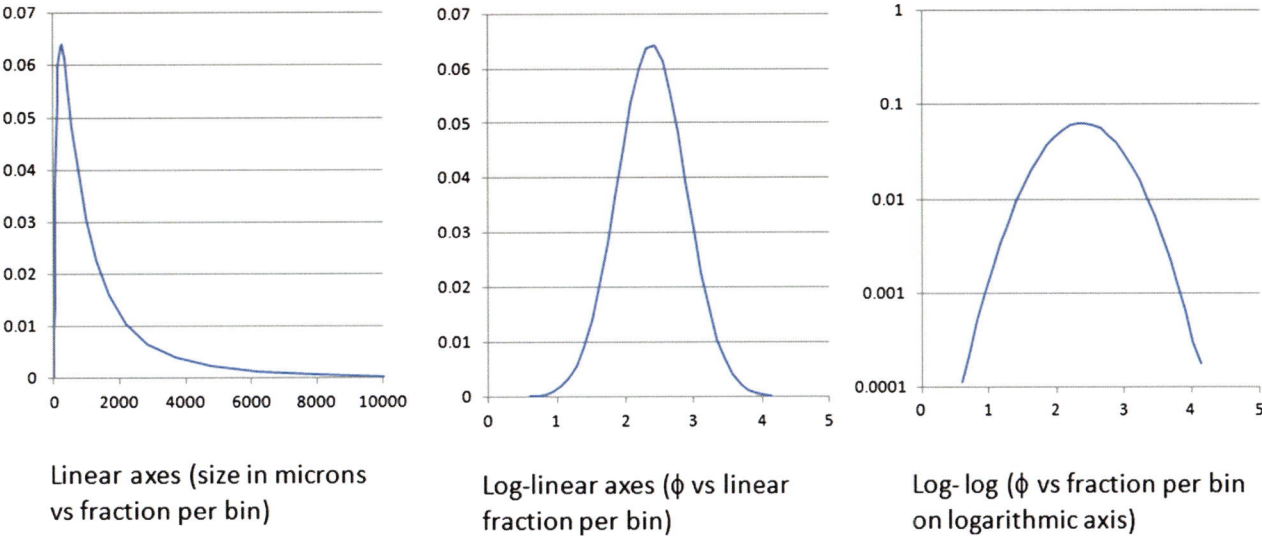

**Fig. 2.3** Lies, damn lies, and statistics. These three graphs show exactly the same (synthetic) data, notionally a log-normal size distribution of sand. The extreme skewness of the distribution is evident in the leftmost (linear) plot, which loses detail at the small-particle end of the plot. That problem is restored by using a logarithmic size axis (one could simply use a logarithmic axis label, easy to do in a modern plotting program, or transform the size into a logarithmic measure such as f). This presentation makes it difficult to see quantitatively the abundance at the tails of the distribution, to which Bagnold noted some clarity could be brought by using a logarithmic ordinate. Some other aspects of chartmanship are highlighted in Chap. 10

**Fig. 2.4** Microscopic images of sand. These are of sand from coral pink sand dunes in southern Utah, low res, high res; grid is 1 mm. *Photo* J. Zimbelman

imaging makes it now rather easy to measure the number, size and shape of sand grains optically (Fig. 2.4), and this has now become almost routine at Mars (Fig. 2.5).

There is more than one definition of particle size (e.g., the cube root of volume may be different from the largest or smallest dimension). But size is not the only aspect of a sand particle's dimensions that are of interest. For example, a rod-shaped grain can pass through a sieve opening that is smaller than its longest dimension if it happens to encounter the sieve in just the right orientation. Clearly, additional descriptive terms were needed to categorize sand grains derived from diverse source rocks and environments. Scientists struggled to determine the most utilitarian methods for categorizing sand beyond that of a sieve-derived measure of its size. The shape of the particle is an attribute that is related to the particle as a whole; it encompasses the three-dimensional aspects of the entire grain. Once sand was routinely observed through a microscope, various shape-related terms were used to describe the grains, among which descriptors such as spherical, cylindrical, tabular, blade-like, or sheet-like came into common usage. Early on, there was little consensus on how best to measure or quantify the shape of sand grains. At a finer scale, the microscope also revealed that the grains varied greatly in the smoothness or roughness of their surfaces, to which various modifiers could be added, such as well-rounded or poorly rounded (Fig. 2.6). By 1940, the basic definitions were in place for describing various attributes of grains, but a variety of methods were explored to quantify and measure these different properties (Siever 1998).

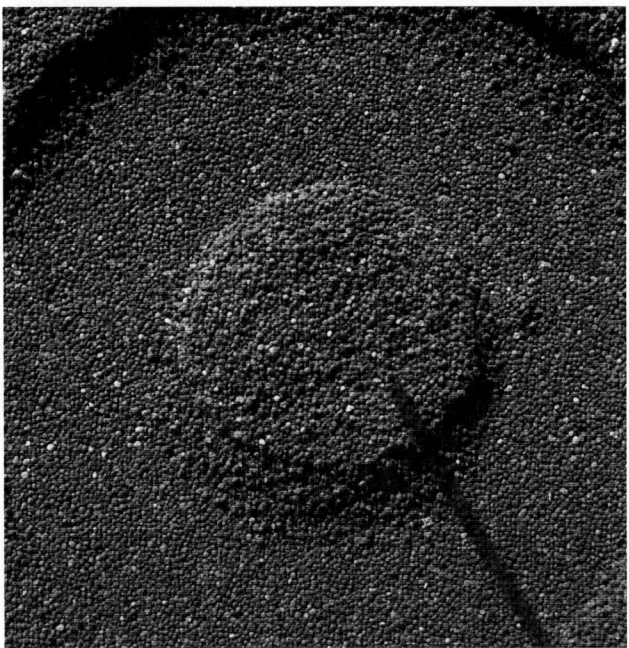

**Fig. 2.5** MER microscopic imager view of basaltic sand (Spirit at El Dorado—see Chap. 12) showing individual grains; view is ~3 cm across. The annular depression is from the contact plate of an APXS instrument (see Chap. 17)

Haakon Wadell devised a simple method whereby the two-dimensional view of grains derived from photographs taken through microscopes could be systematically measured, which also provided a good way to distinguish between shape and roundness (Siever 1998, p. 52). Wadell defined shape as the ratio of the cross-sectional area of the grain to the smallest circumscribed circle, and roundness as the ratio of the average radius of curvature circles inscribed within corners on the grain to the radius of the maximum inscribed circle within the entire grain. These ratios led to both shape and roundness ranging from 0 to 1, where both a shape and a roundness of 1 would result from a perfect sphere. Rather than always having to carry out detailed measurements for each individual grain, calibrated silhouettes of measured grains allowed researchers to estimate the numerical value of both parameters quickly through visual inspection.

These shape parameters are important first in giving a quantitative, or at least consistent, basis for comparing different samples. Is one rounder than another? And what does this mean about how far it has been transported? For example, the sands of the Sahara, which have swirled around in North Africa for ages, are notably fine and round.

The shape influences how grains interact, defining the friction coefficient and thus the angle of repose. This is something that has been measured on Mars by making little piles of sand using the sampling arm of the Viking lander, for example. Grain shape is also important in generating sound by shearing (see Chap. 10). Grain shape also affects the fluid-dynamic properties, such that the drag and thus the saltation threshold depend on the shape. Generally, however, the shape effects on threshold are small compared with the influence of particle density, and the range of angles of repose encountered in nature is actually quite small.

## 2.2 Compositional Considerations

As discussed in the introduction to this section, sand grains consist of a wide variety of compositions. It is worth discussing, however, why we so often think of sand on Earth as referring to the mineral quartz. The abundance of quartz in sand on Earth is not a direct indicator of the quantity of silica in crustal rocks, and in particular the granitic rocks that dominate the continental crust (as opposed to the basaltic ocean crust). Rather, it is the result of the strong resistance of quartz to chemical weathering, at least compared with most other common rock-forming minerals. The combination of field observations of rock weathering processes and chemical experimentation on the stability of common minerals leads to a ranking of mineral in the order of their stability under terrestrial weathering conditions (Press and Siever 1974, p. 208; Table 2.2), with quartz by far the most stable common mineral, and olivine the least stable There are nonetheless two locations on Earth—Hawaii (e.g., Tirsch et al. 2012) and the Galapagos (both sites of basaltic volcanism)—where olivine-rich grains are present in such abundance as to form perceptibly green sand beaches. These two beaches (Iceland may also have some similar beaches) amount to perhaps a millionth of the surface area covered by predominantly quartz sands on Earth. Among other exotic sand compositions, there are reportedly a few garnet sand dunes in Namibia, and wave-ground steel particles are abundant on some beaches in Normandy.

It is the chemical instability of the silicate minerals which dominate igneous rocks on Earth that leads to the formation of clays, which are a major component of all of the different types of soils present around our planet. Extreme weathering conditions, such as the torrential rainfall that is common in equatorial jungles, can even leach away the quartz, which then leads to the formation of lateritic soils that are mined for their enhanced concentrations of aluminum and iron oxides. Thus, with increasing length or intensity of exposure to weathering (particularly by water), the mineral abundances within sand on Earth steadily evolves as the least stable materials progressively disappear, which also results in the steadily increasing relative abundance of the most stable mineral, quartz.

Measurements of composition, as well as of the size and shape of the sand, have been major tools in tracing the origins of sands in Earth's deserts. This approach

**Fig. 2.6** A schematic of shape descriptors for sand grains

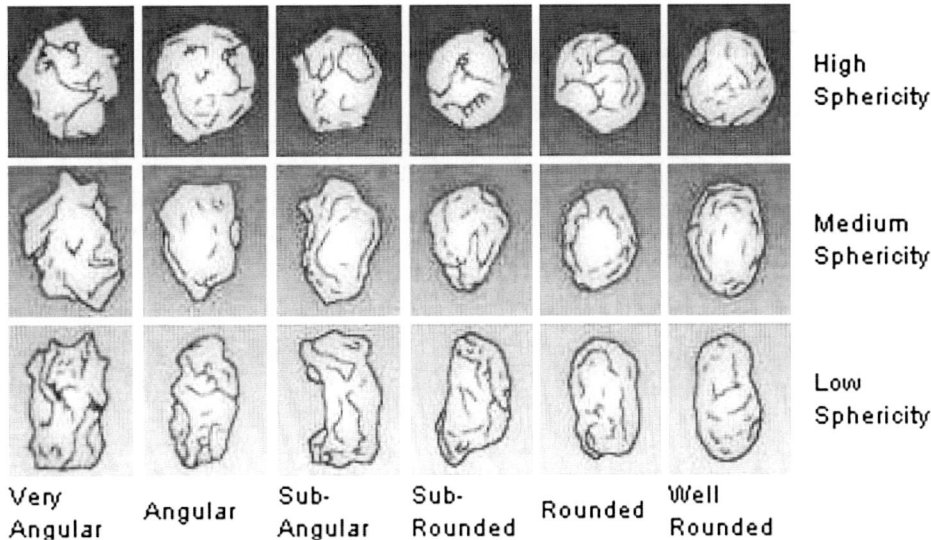

**Table 2.2** Stability of some common minerals under weathering conditions at Earth's surface

|  | Mineral | Product of weathering |
|---|---|---|
| Most stable |  | Fe-oxides |
|  |  | Al-oxides |
|  | Quartz |  |
|  |  | Clay minerals |
|  | Muscovite |  |
|  | K-feldspar (orthoclase) |  |
|  | Biotite |  |
|  | Na-feldspar (albite) |  |
|  | Amphibole |  |
|  | Pyroxene |  |
|  | Ca-feldspar (anorthite) |  |
| Least stable | Olivine |  |

**Table 2.3** Densities of bulk sand-forming materials. Porous materials (e.g., snowflakes, or ooids made of thin shells of calcite) can have much lower effective densities (see Chap. 4). The composition of Titan's sands is not exactly known, but simple alkanes and paraffins at Titan surface temperatures (94 K) have densities below 1000 kg/m$^3$. PAHs (polycyclic aromatic hydrocarbons) like Pyrene and Phenanthrene have densities rather higher than this, so Titan's sand density is somewhat uncertain at present

| Material | Density (kg/m$^3$) |
|---|---|
| Ice | 900 |
| Gypsum | 2300 |
| Quartz | 2600 |
| Calcite | 2700 |
| Basalt | 3000 |
| Olivine | 3300 |
| Garnet | 3100–4000 |
| Magnetite | 5180 |
| Hydrocarbons | ~800 |
| PAHs | ~1300 |

dominates, for example, the formidable study of the Rub' Al Khali by Helga Besler (Besler 2008), and the studies of the US and other deserts by Daniel Muhs (e.g., Muhs 2004, notes that, on the basis of the quartz vs feldspar abundance, the Algodones dunes were not derived from Whitewater river sands).

The terrestrial weathering sequence is not necessarily followed on the surface of other planets, where environmental conditions are very non-Earthlike, particularly with regard to the abundance and stability of liquid water. For example, spectroscopic data from Mars have revealed very little quartz on the Red Planet (Bandfield 2002; Bibring et al. 2005; Smith and Bandfield 2012), let alone in the abundant sand deposits present in many places on Mars, while at the same time the dark sand dunes in the Nili Patera region show an olivine signature (Mangold et al. 2007); both results are precisely the opposite of what would be expected from our terrestrial weathering experience. Rover investigations of dark sand in the Columbia Hills area show that the grains are predominantly comprised of fragments of basalt (Fig. A, Sullivan et al. 2008), but the Martian dunes in some areas have been shown to include olivine and pyroxene (Tirsch 2008) and even gypsum (Fishbaugh et al. 2007; Feldman et al. 2008; Calvin et al. 2009). Also, whereas terrestrial sand evolution may generally be

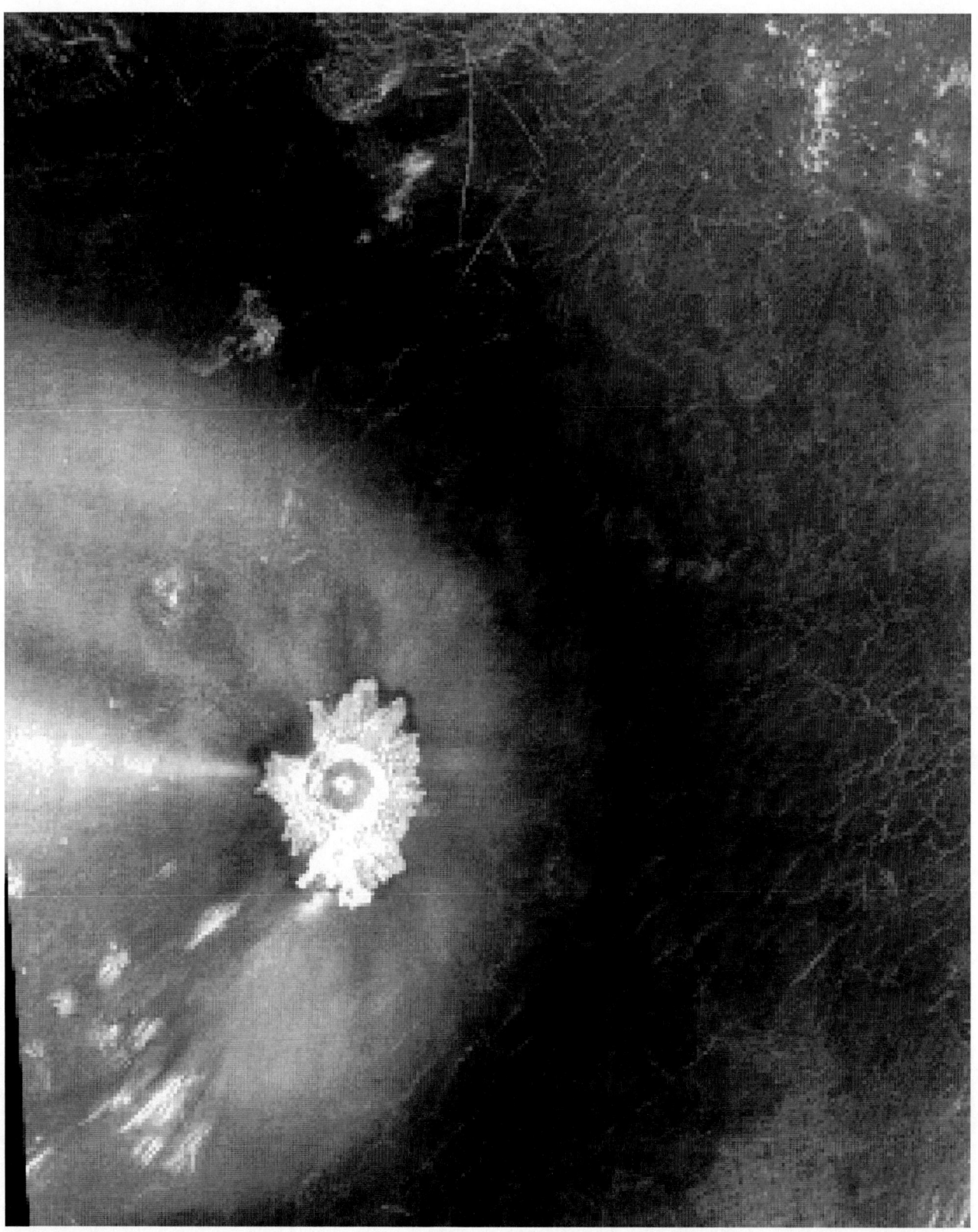

**Fig. 2.7** Parabola on Venus formed by ejecta deposited above the atmosphere in an initially circularly symmetrical pattern, then winnowed eastwards by zonal winds. The inner region is bright due to blockier ejecta: the dark parabola is poorly radar-backscattering, requiring a thickness of several centimeters of fine ejecta. Source is the crater Adivar at the center ($8.93 \pm$ N, $76.22 \pm$ E, 30 km across). *Image* Magellan Image F-MIDR-10N076, NASA

dominated by even rare exposures of water, Mars sands may be dry for millions of years and processes that are not usually noticed on Earth may become dominant. An interesting example is the experiment by Merrison et al. (2010) who found red coloration emerging in sand simply by rolling the sand around in a flask of simulated Mars atmosphere ($CO_2$ at 10 mbar pressure) for a few months, perhaps indicating a tribochemical process.

Different minerals have different densities, leading to different mobility in fluids such as water or air. Occasionally this effect is deliberately exploited to concentrate materials—panning for gold is one example—but more generally it contributes to an often striking uniformity of composition in a dune, or at least of layers within a dune (see Chap. 5). The range of densities encountered for sands known in our solar system at least is actually less (Table 2.3) than the range of densities of sand-forming material on Earth, although perhaps on some exotic world there are dense diamond sands!

Unfortunately, we have absolutely no observational constraints on the composition of sand on Venus. Although several Soviet landers provided clear evidence that basalt is the dominant rock type encountered at most of the landing sites, even then, the process by which sand-size particles are generated is not obvious. It may even be dominated by impact; large radar-dark (and thus likely fine-grained or smooth—see Chap. 18) parabolic deposits (Fig. 2.7) are seen around recent impact craters that may contain an abundance of sand-sized microtektite particles (launched above the atmosphere from the impact fireball and winnowed by the atmosphere). The compositional variations of sand throughout the solar system are the result of the unique chemical and atmospheric conditions present on each planetary body, where different groups of materials end up being either more or less stable within the corresponding surface environment. We have no idea, for example, how methane rain might progressively alter the composition of Titan's organic sands, if at all.

# Winds and Atmospheres

Essentially, all solid bodies in the solar system, and likely most of those in other planetary systems, have some fine-grained materials on their surface. But few have atmospheres to blow these grains around to make dunes. Entire libraries exist on the subject of meteorology; here, we can only highlight the principal characteristics of these atmospheres and their winds. For further reading, the text on desert meteorology by Warner (2004) is recommended for the Earth, while students of Mars should refer to Read and Lewis (2004).

## 3.1 Atmospheres

The principal parameters of interest in aeolian studies are the air density rho_a, and its viscosity. Density is by far the more important of these. On Earth at sea level, air has a density of 1.25 kg/m$^3$—your lungs hold about 5 grams of air. This density is 800 times less than that of water, and thus about 2000 times less than that of sand-forming materials. The density of Titan air is a factor of 4 higher than Earth's, but the planetary range on dune worlds is quite dramatic—from torrid Venus with air 50 times denser than Earth, to the whistling thin atmosphere of Mars 50 times (or in places, several hundred times) less dense (Table 3.1).

The atmospheric pressure is an indication of the total amount of gas in the atmosphere; it is the weight of the column of gas (and hence has dimensions of force per unit area). Indeed, we used to determine our atmospheric pressure by balancing it against the weight of a column of a different fluid—mercury. The weight of a $\sim$10 km column of air with a density of 1.25 kg/m$^3$ is the same as a 760 mm column of mercury with a density of 13,600 kg/m$^3$ (or, of interest to divers, a 10 m column of water with a density of 1000 kg/m$^3$). These are equivalent to one 'atmosphere', or 1 bar (= 1000 mb): the formal SI unit of pressure is the Pascal (Pa) and 1 bar = 101325 Pa (or 1013 hPa, a hectopascal being about the same as 1 millibar).

Density and pressure P are related. For ideal gases, this relationship is simple: $\rho a = PM/R_o T$, where $R_o$ is the universal gas constant 8314 J/K/mole, T is the temperature in Kelvin, g is gravity and M is the relative molecular mass (28 for nitrogen, 29 for air, 44 for $CO_2$). If we kept the total mass of the atmosphere (and thus the pressure) the same, but raised the temperature, then the density will get lower. Substituting values for the summit of Mauna Kea on Hawaii (a familiar location to astronomers, with an altitude of 4200 m) P = 611mb, T = 260 K, we find the density to be about 0.8 kg/m$^3$, or a third less than at sea level, which accounts for the symptoms of altitude sickness often encountered there.

Note that on Titan, the atmosphere is thick and cold enough to be somewhat close to partial condensation (the air on Titan is only 100 times denser than the methane-nitrogen-ethane liquids there which have a density of 450–700 kg/m$^3$) and the ideal gas law is several percent off and a more complex equation of state should be used for accurate calculations. For most aeolian studies, a few percent accuracy is good enough, however.

This 10 km column mentioned above is the 'scale height' of the atmosphere. Because, unlike liquids, gases tend to be compressible, the density declines gradually with height and there is no sudden 'top of the atmosphere'. The decline is usually an exponential function of height, and is usually written $\rho(z) = \rho_o \exp(-z/H)$ where $\rho(z)$ is the density at height z, $\rho o = \rho(0)$, the density at the surface, and H is the height distance over which the density falls by a factor of e = 2.718. It can be shown that for an isothermal atmosphere in hydrostatic equilibrium, $H = R_o T/gM$, with definitions as above. Thus for Earth, $T \sim 287$ K, $g = 9.81$ m/s$^2$ and M = 28, we have H $\sim$ 9000 m. The values for other worlds are similar, although for Titan with its low gravity, H $\sim$ 21 km. We see that the scale height is proportional to temperature; thus, if we warmed the atmosphere up and made the density lower as above, the scale height increases by the same factor, so the product of column height and

**Table 3.1** Properties of planetary atmospheres

| Body | Venus | Earth | Mars | Titan |
|---|---|---|---|---|
| Planetary radius (km) | 6052 | 6370 | 4470 | 2575 |
| Gravitational acceleration (m/s$^2$) | 8.9 | 9.8 | 3.7 | 1.35 |
| Rotation period (sidereal day) (days) | 223 | 1 | 1.04 | 16 |
| Diurnal period (solar day) (days) | 119 | 1 | 1.04 | 16 |
| Surface atmospheric pressure (bar) | ~90 | 1 | 0.006 | 1.46 |
| Surface atmospheric temperature (K) | 740 | 283 | 200 | 94 |
| Dominant gas species | $CO_2$ | $N_2, O_2$ | $CO_2$ | $N_2, CH_4$ |
| Atmospheric density (kg/m$^3$) | 64 | 1.25 | 0.02 | 5.4 |
| Dynamic viscosity ($10^{-6}$ Pa-s) | 35 | 17 | 13 | 6 |
| Planetary boundary layer (km) | 0.2 ? | 0.3–3 | >10 | 2–3 |

density (i.e., the total atmospheric mass) remains constant, as it should—the atmosphere just puffs up.

It is important to note that on Mars in particular the atmospheric parameters are not global constants. The topographic range on Mars is large compared with the scale height, such that the pressure and, more importantly, density of the atmosphere can vary significantly. A notable example is that on the Tharsis volcanos, moving from the base where the pressure is about 6mb to the peak around 22 km, the pressure and density drop by almost a factor of 6, which is reflected in the size of the bedforms observed there (Lorenz et al. 2014).

Atmospheric pressure on planets may change through geologic time, depending on how volatiles are delivered to and lost from the planet (although variations on Earth have probably been quite small). Variations may be dramatic if the dominant atmospheric component can condense, as is the case at Mars. Indeed, over the course of a Martian year, the atmospheric pressure varies by about 30 % as part of the thin atmosphere freezes as a seasonal polar cap of $CO_2$ frost. The Martian atmosphere may well have been a factor of several times thicker in the deep past than at present.

The other static property of a gas that is of interest is viscosity ($\mu$), which is a measure of how well momentum diffuses through a fluid. This is a function both of composition and temperature. The property is the ratio of shear stress to strain rate (or velocity gradient) and has SI units of Pa-s. Confusingly, there is also a unit named the Poise (P, with 0.1 Pa-s = 100,000 $\mu$Pa-s = 1 P = 100 cP). Further confusion arises because in some fields it is common to refer to a 'kinematic viscosity', which is just dynamic viscosity divided by density and has units of its own; generally, if unspecified then dynamic viscosity is meant, but to be safe one should state explicitly which is referred to. In air, $\mu$ = 18 uPa-s, whereas on Titan (also mostly nitrogen, but rather colder) the viscosity is almost three times lower. On the other hand, carbon dioxide has a slightly higher viscosity than nitrogen. Liquid water has a much higher viscosity, 900 µPa-s, and liquid hydrocarbons on Titan may be a factor of 2 lower or higher than that, depending on composition.

The viscosity is a property that affects how thick a layer of fluid is slowed down as it flows across or around a body. We will discuss its implications, particularly via the Reynolds' number (the ratio of viscous to inertial forces in a fluid) in Chap. 4.

A final property that occasionally is important is the mean free path, the distance an air molecule may statistically travel before it hits another air molecule. Generally, this distance is very short (in sea level air on Earth it is about 70 nm, much smaller than any object we are considering). The distance increases at lower pressures (densities), and can be several microns at Mars' surface conditions, which is comparable with the size of fine dust particles.

## 3.2 The Planetary Boundary Layer

A distinct feature in the vertical structure of a world's atmosphere is the planetary boundary layer (PBL). One must avoid confusion between this (a meteorological structure typically 300 m–10 km deep) and the friction layer (which aerodynamicists might term a boundary layer) of a few centimeters to a few meters near the surface where the windspeed falls: the friction layer is discussed in Sect. 4.6 on the mechanics of sand. The PBL is the layer which is well-mixed, and so has a uniform profile of certain thermodynamic properties, notably the potential temperature. The PBL typically grows in thickness throughout the day, as solar heating stirs the air near the surface.

The PBL is of profound importance for dune construction, in that the top of the PBL acts as a cap; it is energetically demanding to push the PBL upwards. Thus, in many ways the flow affecting dunes can be thought of not as the flow of an infinite atmosphere that progressively thins upwards into space, but rather like a layer of fixed depth as if it were a liquid held under an elastic sheet. Airflow encountering an obstacle that is small compared with the PBL thickness can go over and around. However, if the obstacle is thick (i.e., an appreciable fraction of the PBL), the air flow must accelerate strongly to get over it, and this acceleration at the top will increase the shear velocity enough to prevent sand from accumulating at the crest. Thus dunes cannot grow to be higher than some fraction of the PBL thickness and, given that dunes have a constant maximum height-to-wavelength ratio, limits the ultimate size and spacing of sand dunes to approximately the PBL depth (Andreotti et al. 2008).

Over the oceans on Earth, where the heat capacity of the sea is large, the PBL is typically a few hundred meters thick. Over the continents, and in large deserts especially, the PBL can grow to be 2–3 km deep. This difference is made visible in mature coastal dune fields such as the Namib, where coastal dunes are somewhat small but, further inland, much larger and higher dunes are seen where the PBL is thicker.

On Titan, the PBL (controlled in part by the length of the day, the solar heating, and the heat capacity of the land and atmosphere) is about 2–3 km deep. This may account for the general uniformity in dune size observed—on Titan the seas of hydrocarbon liquid are found only near the poles—and thus the boundary layer is not likely to vary dramatically in thickness across the equatorial areas where dunes are found.

On Mars, the thin atmosphere needs little heat to be warmed appreciably, and so the PBL can grow to be quite deep—as much as 10 km. One way in which the depth of the PBL is sometimes indicated on Mars and Earth is that dust devils can rise up to the PBL depth; Martian dust devils have been observed to reach much greater heights than their terrestrial counterparts.

On Venus, the PBL depth is not constrained by observations. Even more so than Titan, the thick atmosphere takes a lot of heat to stir, although the length of the solar day ($\sim$120 Earth days) on Venus gives the PBL time to grow. Thus a depth of the order of a few hundred meters to $\sim$1 km seems plausible.

## 3.3 Planetary Wind Patterns

Clearly, no discussion of planetary dunes can be complete without consideration of the winds that form these features. A variety of factors affects winds at different temporal and spatial scales; these variations are superposed, and how they interact is well beyond the scope of this book (although we review briefly the models used to simulate flow at the global and local level in Chap. 18).

With the possible exception of the role of gravitational tides in shaping Titan's near-surface winds, the ultimate power behind wind, and thus dune-building and migration, comes from the sun. In a nutshell, solar heating is stronger in some places than others, and causes air to rise. Continuity requires a horizontal movement of air to replace this rising cell causing winds, and this horizontal motion can be affected by the planetary rotation and by topographic obstacles. This process occurs at a range of different scales that are superposed. First is the natural result of planets being spheres, such that some parts geometrically receive more sunlight than others: this is a function of the tilt of their equator with respect to their orbit around their parent star, as recognized by Edmond Halley in the 1600 s. In fact, all the dune worlds have modest obliquities so that the polar regions on average receive less sunlight (which is not true for worlds with obliquities above 40°). With modest obliquity, solar heating is concentrated at low latitude and leads to rising air over the equator. Where it comes down— the mean meridional circulation, sometimes referred to as the Hadley circulation—depends on planetary rotation, and the downwelling flow is typically dry and thus is where deserts tend to form. On Earth, this occurs in two bands roughly 20–30° from the equator (Fig. 3.1). On slowly-rotating Titan, a wide equatorial band tends to dry out.

The interaction of planetary rotation with the meridional flow leads to overall surface wind patterns which, because of their utility for exploration and commerce, are named the 'trade winds'. The first global map of these was published by Halley. They are belts of diagonal winds where the upwelling and downwelling flows (which are statistical averages—the air does not move in one large coherent global way) meet the surface and go north or south, but are diverted by the so-called Coriolis effect due to the planet's rotation.

Because of Titan's slow rotation, the meridional circulation is pole-to-pole for each half of the year (reversing each equinox), without the intermediate cells seen on Earth (Fig. 3.2).

Note that even subtle changes in these astronomical drivers of the atmosphere can drive climate change. The Croll-Milankovich variations in the rotational and orbital parameters of a planet (and their arrangement with respect to each other—specifically the timing of perihelion with respect to the seasonal cycle) play a role, perhaps even a dominant one, in forcing change in the extent of ice sheets, sea level, and the extent of deserts on Earth. These changes, on timescales of tens to hundreds of thousands of years (i.e., large compared with the construction time of a typical dune) can therefore form and leave dunes in locations where present conditions do not allow them to form. Thus dunes

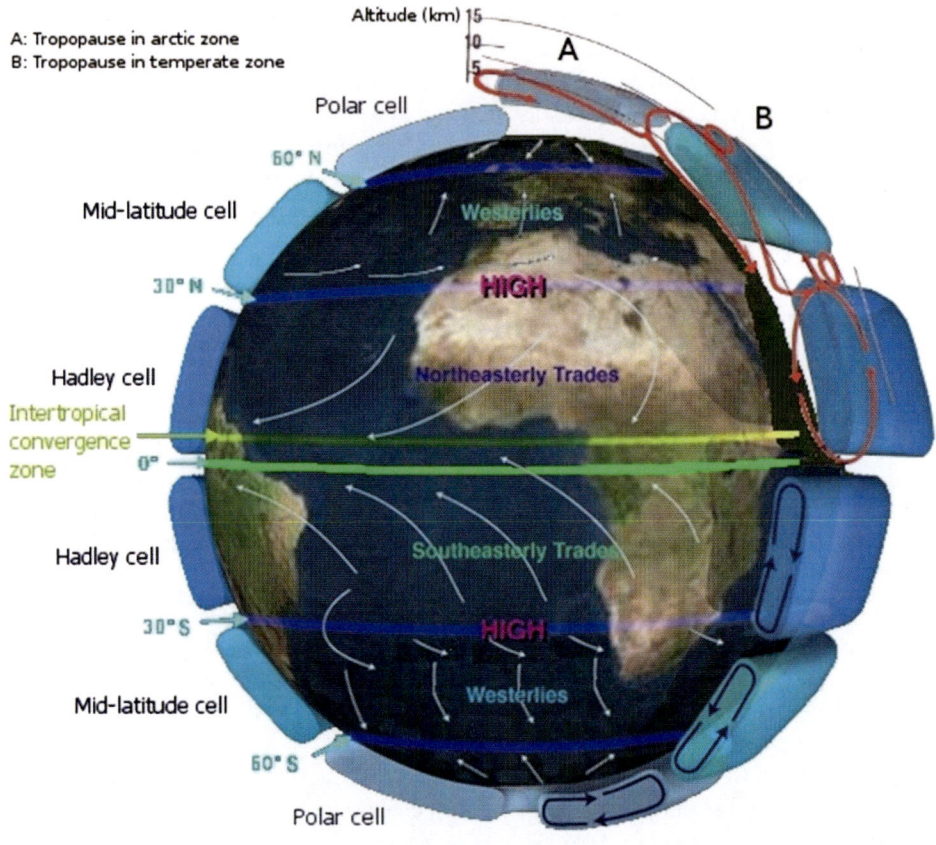

**Fig. 3.1** Schematic of the Earth's general circulation: the trade winds and the mean meridional circulation. Downwelling at around 30° latitude leads to consistent high pressures and dry air, hence our deserts (see Chap. 11). On another world of different size and/or rotation rate, the latitudinal extent of the cells would be different (as is the case on Titan). NASA image

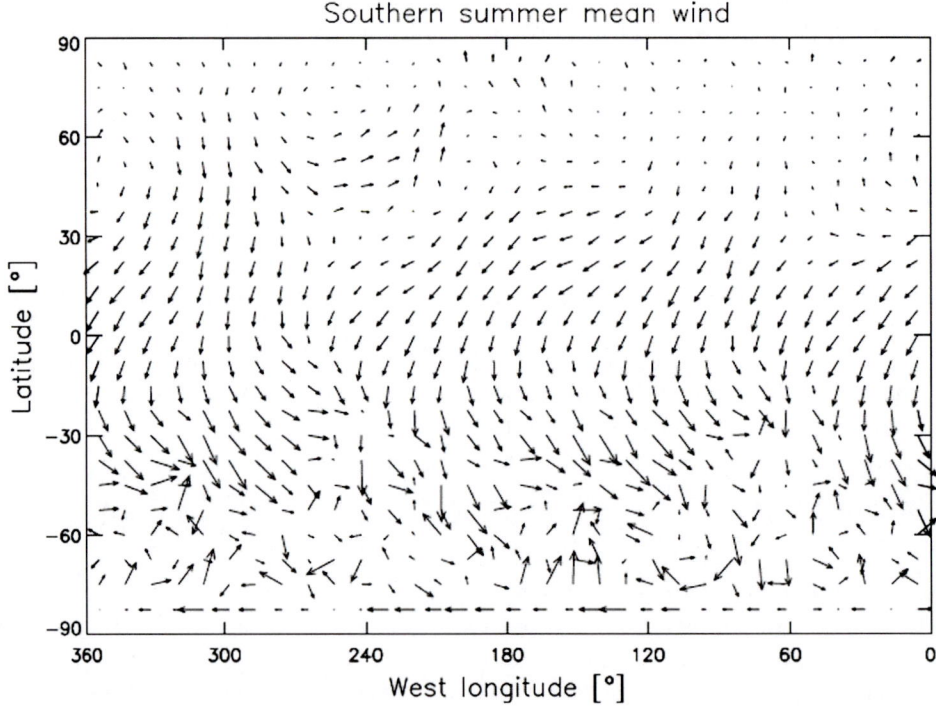

**Fig. 3.2** Mean surface winds on Titan predicted by a Global Circulation Model (an arrow length of one grid spacing equals 1 m/s) in southern summer. Rising air in the summer hemisphere draws in a surface flow from the north. GCM winds for Mars are shown in Fig. 19.1. Image courtesy Tetsuya Tokano

## 3.3 Planetary Wind Patterns

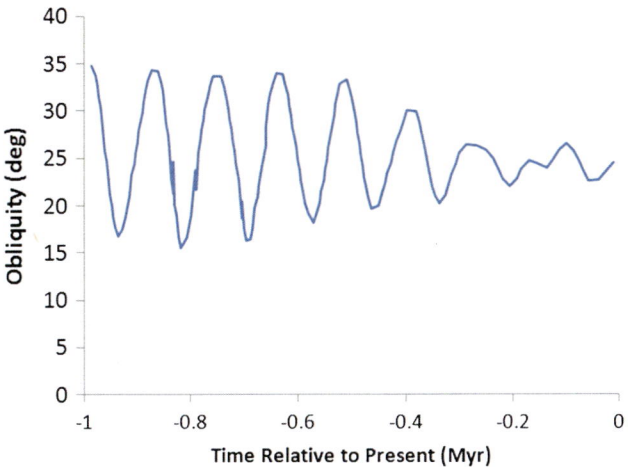

**Fig. 3.3** Astronomical variations in the rotational and orbital parameters can drive climate changes—the so-called Croll-Milankovich cycles. The variations on Mars are particularly strong, as shown here can be a window into paleoclimate. Croll-Milankovich changes on similar timescales are also suggested to be responsible for the intricate Martian polar layered terrain and, perhaps, for the concentration of liquid hydrocarbons at Titan's northern polar regions. On Mars, the effects are stronger than on Earth, in part because the Earth's large moon gives some 'gyroscopic' stability to the system. As shown in Fig. 3.3, Mars' obliquity has seen major excursions: 500,000 years ago, the equatorial tilt varied from about 17 to 32° over a ~50,000 year period. Its orbital eccentricity also underwent significant change—all these factors drive the climate significantly.

The second major influence on wind patterns is the distribution of surface materials which affect the temperature response to the sun. On the Earth, the principal factor here is the oceans which have a much higher thermal inertia, and thus much less temperature variation than the land, on both seasonal and diurnal timescales. This leads to persistent onshore flows in warm lands, and offshore in cold places (and at night). On Mars, dust-covered areas have low thermal inertia and warm up during summer days, whereas more rocky areas are slower to respond to the sunshine (see Chap. 18). Not only the thermal inertia of the surface, but also its reflectivity or albedo, can influence the thermal response. For example, Titan's dark sand seas appear to be persistently warmer than their surrounds, which doubtless influences the wind pattern (and of course there may be an interesting feedback at work, in that the sand seas influence the winds, which may influence the transport of sand into the sand sea…).

A third effect is topography. Topography can play a blocking role, as wind diverts around obstacles (Figs. 3.4, 3.5 and 3.6). The conservation of mass in a flow (i.e., the product of the density, area and speed of a flow—which for subsonic

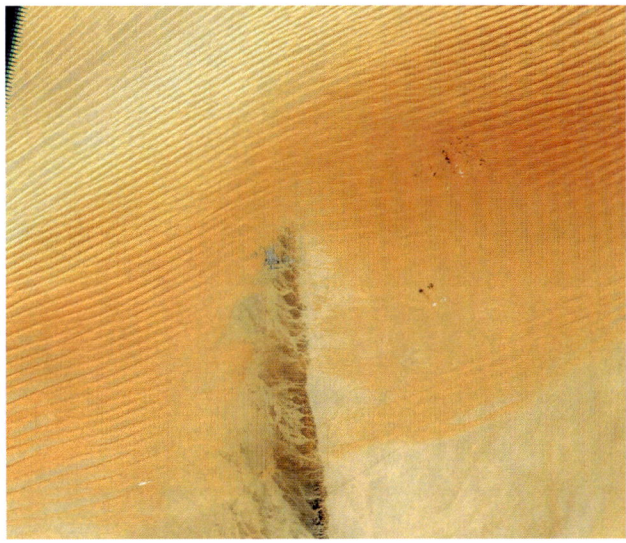

**Fig. 3.4** A mountain ridge (with the small town and airport of Sharurah near its tip) intrudes into a linear dunefield in the Rub' Al Khali. The accumulation of sand is prevented by the topography and the disrupted airflow causes a sand sheet to form in the lee (*right*) of the ridge before the flow stabilizes and linear dunes re-form further downwind. Note also that the linears (and thus presumably the wind) are diverted slightly north around the ridge: similar topographic interactions can be seen in the linear dunes of the Simpson desert (see Fig. 12.12) and the Namib and on Titan (Figs. 1.3, and Chap. 13) A couple of small clouds and their shadows can be seen. NASA Earth Observatory image using USGS Landsat 7 data

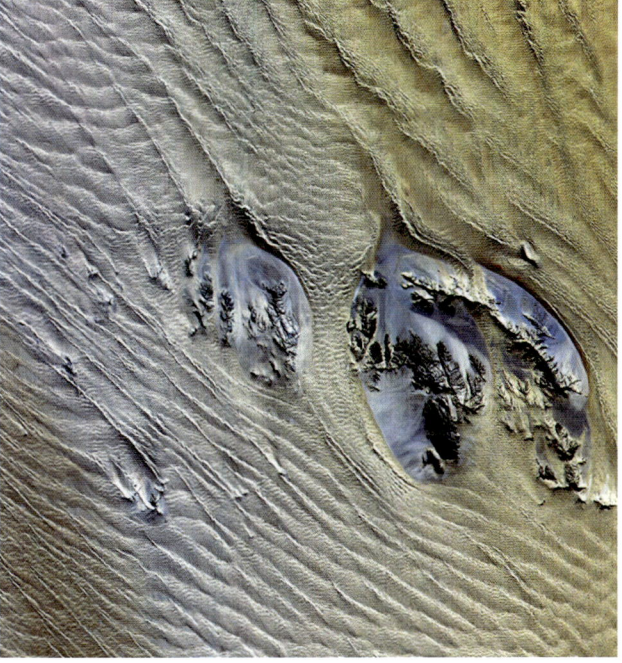

**Fig. 3.5** Linear dunes in the Namib desert sweeping around inselbergs ('island mountains'). The local modification of the wind directions causes the orientation of the dunes to resemble streamlines. ASTER image, courtesy NASA/JAXA

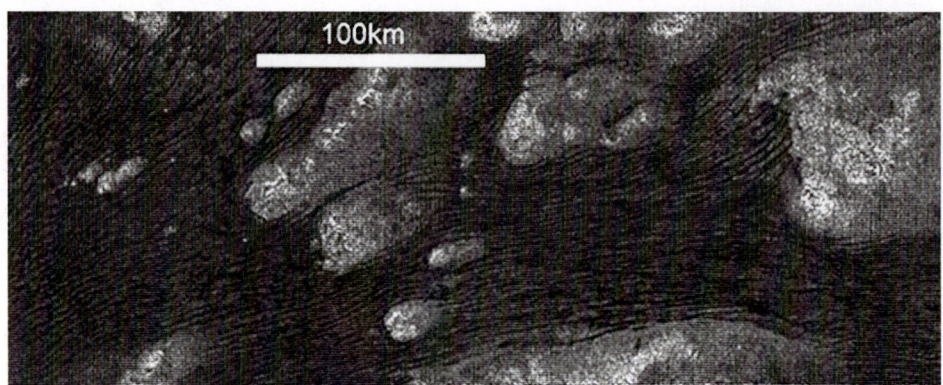

**Fig. 3.6** Titan dunes observed in a radar image from the T44 encounter. The dark dunes divert around some bright hills, and in some cases are blocked by them. NASA/JPL/Cassini Radar team

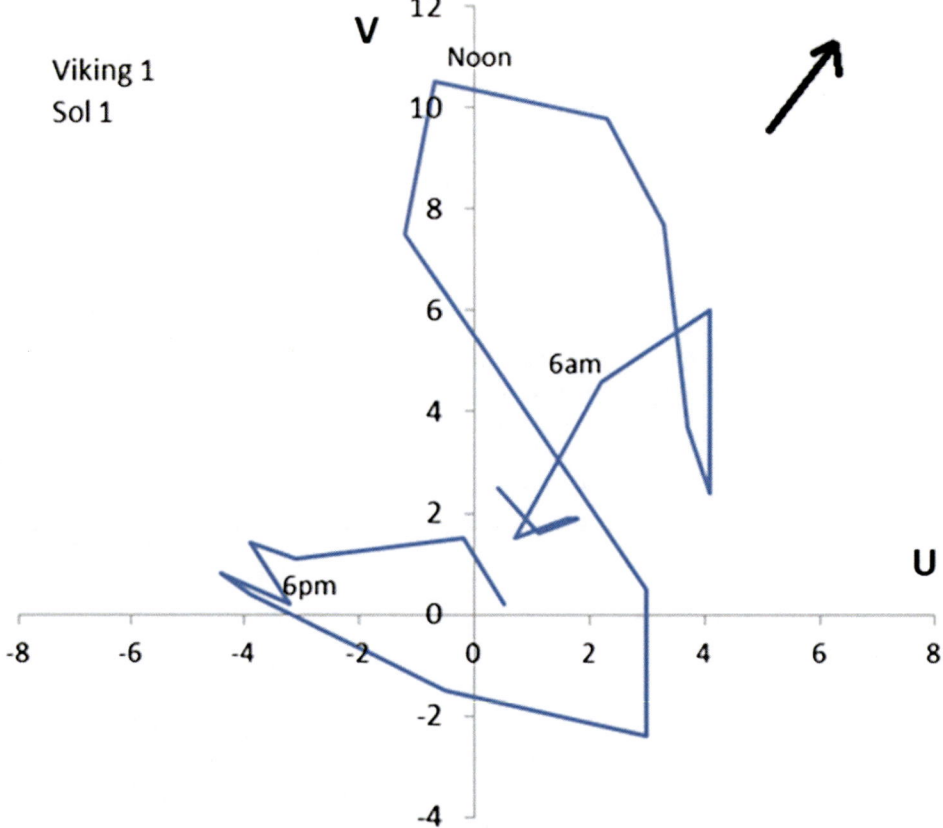

**Fig. 3.7** A hodograph of the winds measured by the Viking lander 1 on its first day on Mars. At each hourly measurement the wind velocity is described by a vector from the origin to the relevant point on the line. Note that the downslope direction (*arrow*) plays a significant role in driving morning winds

boundary layer flows effectively just means the product of width and speed) requires that flow accelerates through a constriction or over a gentle hill, but may decelerate when the flow can expand.

When a wind that may be strong enough to transport sand, e.g., through a pass, is decelerated as the flow broadens into a valley, the wind slows and the sand transport ceases. This leads to a net deposition, and some dunefields are formed this way (e.g., Coral Sands in Utah). A related effect is where wind may be collimated by topography from several different directions, leading to a net local convergence which sweeps sand into a local pile—the Stovepipe Dunes in Death Valley (see Fig. 24.2) form in this way.

The diurnal heating influence via albedo, thermal inertia, or slope effects, often leads to a rather reproducible behavior in wind direction (and to a lesser extent, speed) over the course of a day. This is often plotted in a hodograph, a vector diagram wherein the heads of vectors (representing wind speed and direction) are joined from one hour to the next (Fig. 3.7). Usually these define a loop, with the wind direction rotating through the course of the day.

Topography, even at the building or boulder level, will influence the flow, in effect biasing the distribution of u' and v' around an obstacle. Sand will accumulate around and in the lee of bushes on playas or sabkhas to form small coppice dunes or nebkhas. Streaks of sand deposition, related to

## 3.3 Planetary Wind Patterns

**Table 3.2** Surface wind measurements

| Planet, location | Wind speed | Dataset, height | Reference |
| --- | --- | --- | --- |
| Venus, Venera-9 | 0.4 ± 0.1 m/s | Cup anemometer 1.3 m<br>49 min, 0.4 Hz | Avduevskii et al. (1976) |
| Venus, Venera-10 | 0.9 ± 0.15 m/s | Cup anemometer 1.3 m<br>1.5 min, 0.4 Hz | Avduevskii et al. (1976) |
| Mars, Viking-1 | Up to 20 m/s, usually <10 m/s | Hot Wire anemometer, 1.6 m<br>45 sols, hourly | Hess et al. (1977) |
| Mars, Viking-2 | Up to 20 m/s, usually <10 m/s | Hot Wire anemometer, 1.6 m<br>1070 sols, hourly | Hess et al. (1977) |
| Mars, Pathfinder | <5–10 m/s<br>~1 m/s at night | Hot Wire anemometer[a], 1.1 m | Schofield et al. (1997) |
| Mars, Pathfinder | 7–10 m/s, usually less | Images of windsocks at 0.33, 0.62 and 0.92 m | Sullivan et al. (2000) |
| Mars, Phoenix | 2–10 m/s | Images of tell-tale at 1.6 m<br>7600 measurements over 150 sols | Holstein-Rathlou et al. (2010) |
| Titan, Huygens | 0.6 m/s<br>0.3 m/s<br><0.25 m/s | Doppler tracking at 300–1000 m<br>Parachute shadow V ~ 10 m<br>Cool-down of probe <1 m | Folkner et al. (2007)<br>Karkoschka et al. (2008)<br>Lorenz (2006) |

[a] The Mars Pathfinder anemometer suffered from inadequate superheating such that the windspeed calibration was inapplicable during strongly convective conditions when temperatures fluctuated severely

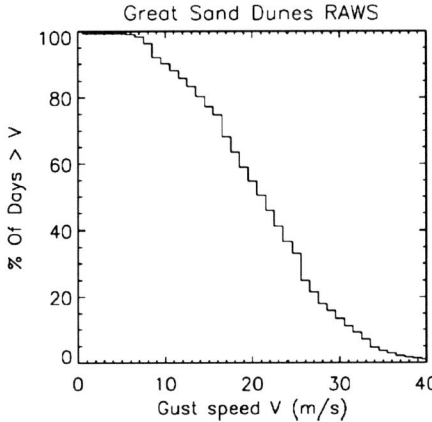
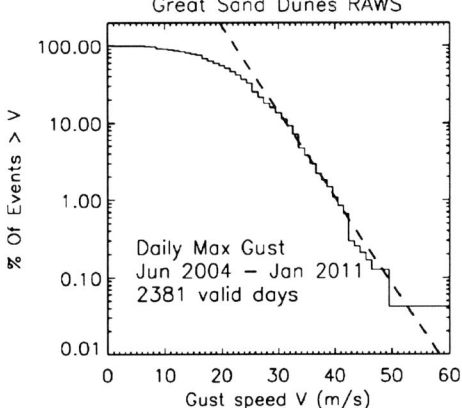

**Fig. 3.8** Plots of the number of days on which the strongest wind gust exceeds a given threshold at a weather station near Great Sand Dunes National Monument. Note that the x-axis covers a narrower range in the linear plot at *left*. Extending it to the same extent as the *right* (logarithmic) plot would not show anything useful and on a linear plot the frequency would be indistinguishable from zero. At the *right*, a logarithmic plot exposes the very rare strongest gusts

horseshoe vortices, can be similarly tied to impact craters on Mars. Artificial topography can be introduced (see Chap. 23) such as dune fences can be introduced to prevent deposition.

Topography also plays an important role in shaping winds due to heating, through so-called slope winds. Generally, solar-heated slopes will see uphill flow, and cooler slopes will see downhill flow. These patterns, influenced by the orientation of the slope relative to the sun, may be a major influence on Venus (although we have very little data on Venus winds), simply because the other factors are so small. On Earth and Mars, another prominent example is the katabatic flow that develops off the polar caps, where cold, dense air runs downhill; the Olympia Undae dunefield around the Martian north polar cap is undoubtedly affected by this pattern.

On top of these steady-state patterns, there is seasonal and diurnal forcing. Diurnal variation on Venus and Titan, with their dense, optically-thick atmospheres, is modest compared to Mars, whose thin atmosphere warms up and cools rapidly. Seasonal variation is almost nonexistent on Venus, because it has almost zero obliquity.

Finally, rotating fluids have a dynamical mind of their own. Characteristic internal dynamical modes or waves can be excited by the forcings above, and these modes are a

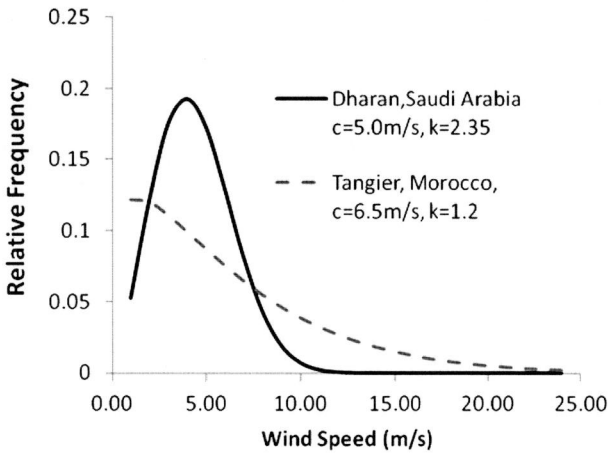
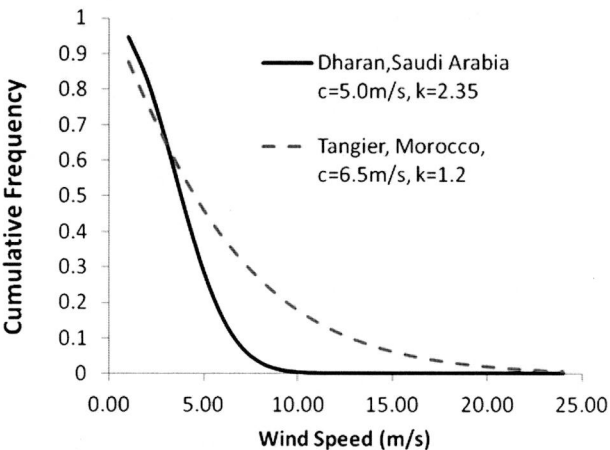

**Fig. 3.9** Weibull fits to the average windspeed distributions at two desert sites on Earth. At *left* is the differential frequency (i.e., number of occurrences in a range of wind speed, here 1 m/s wide) which shows clearest the most common speed. Note the influence of the Weibull shape parameter k—a higher value gives a more 'normal' distribution with a distinct and symmetric peak, while a lower value gives a more skewed fall-off. Often more useful for aeolian studies than the differential frequency is the cumulative frequency (the fraction of time that wind is above a given windspeed). Note that these hourly average winds are considerably lower than the peak gusts indicated in the figure above

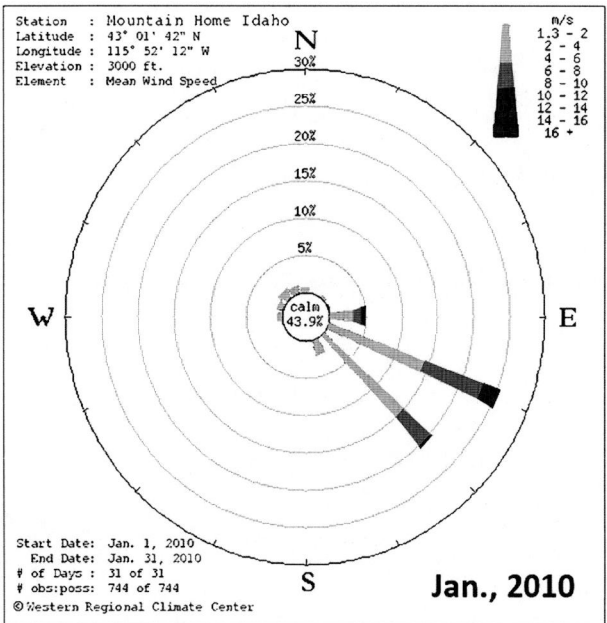
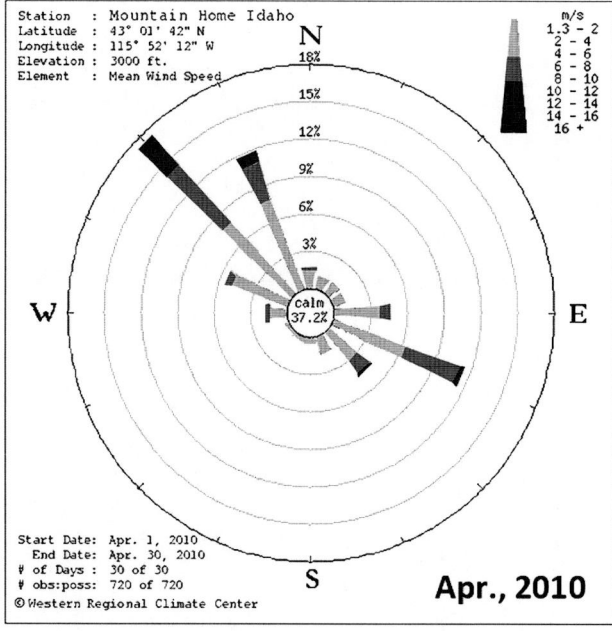

**Fig. 3.10** A wind rose diagram for the Bruneau dunes, generated by the Western Regional Climate Center website for two different seasons. It shows that strong winds (6–10 m/s) blow consistently from the SE in winter, but the spring wind regime is rather more bidirectional. See also Chap. 8. Image generated by J. Zimbelman

feature of the inertia of the atmosphere and the planetary size and rotation, and develop without specific forcing. A casual inspection of midlatitude weather records on Earth will typically show a ~7-day periodicity in pressure and wind. Other planets have different characteristic periods. Similarly, coherent weather structures can form (notably dust storms on Mars, and hurricanes and related systems on Earth).

Predicting the winds on other planets, then, is not a trivial business. Given some broad parameters like the rotation rate of a planet, the density of its atmosphere and so on, one might predict the typical amplitude of surface winds (although efforts to do so at Titan met with only limited success). But dune formation is concerned with the winds at a given location, and quite possibly just the exceptional winds at that location, and thus prediction requires sophisticated

## 3.3 Planetary Wind Patterns

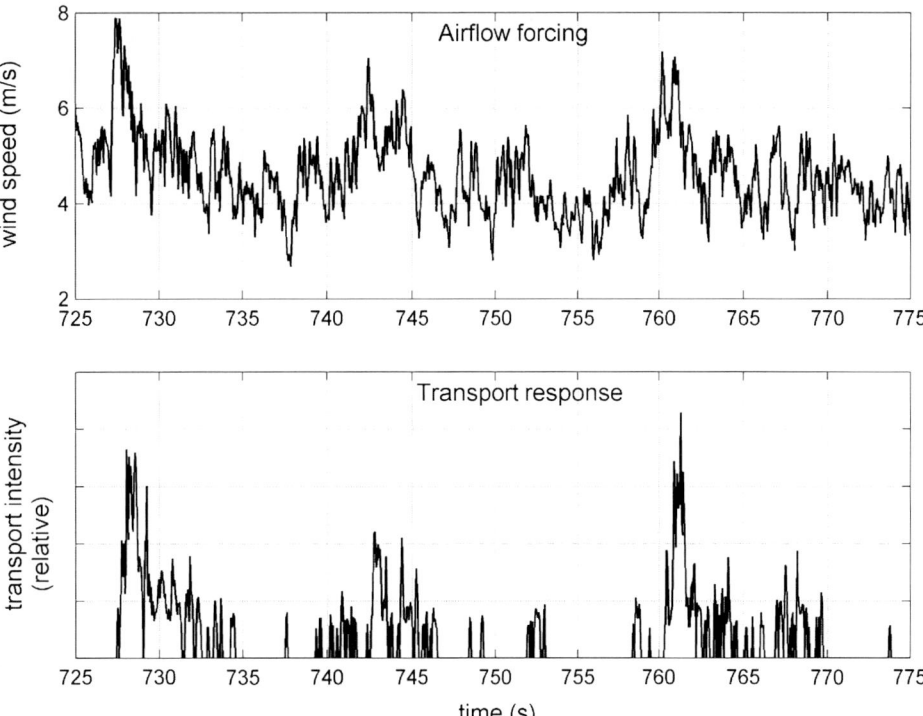

**Fig. 3.11** Plot of wind speed vs time (*upper panel*) showing fluctuating character. Such a signal is often termed '*red noise*' or '*brown noise*' by a color analogy with the spectral properties. When the wind speed exceeds a threshold of about 5 m/s, sand transport (measured by SAFIRE detectors, see Chap. 17) suddenly picks up. The transport signal is clearly very 'bursty' and episodic. Figure courtesy Andreas Baas

numerical models which are discussed in Chap. 18. Ideally, one can use a record of previous winds at a given location (or at least nearby) to predict winds in the future, but of course such records are all but nonexistent for other worlds. Low-level winds can sometimes be estimated by tracking clouds or dust devils in imaging from orbit, but generally must be accomplished by in situ measurements (see also Chap. 17). A brief summary of in situ measurements is given in Table 3.2: wind data from NASA planetary missions can be found at the PDS Planetary Atmospheres node.

## 3.4 Wind Speed Statistics

The net result of the planetary circulation is a time history of winds at a given location. Two scalar parameters are of particular concern at any place and time: namely, wind speed, and direction. These two scalars can also be represented as a vector in 2-dimensional space (for the global and regional scale at least, the third dimension and vertical velocity can be ignored—at the dune scale, of course, this is not the case).

A location's statistical wind properties over some length of time are usually expressed in a couple of ways. First is a histogram or probability distribution of wind speed. This is of interest because winds fast enough to cause saltation occur for only a small fraction of the time, so the simplest statistical description—the average windspeed—does not usefully inform how much aeolian transport occurs.

Windspeed distributions are highly asymmetric, not least because speed cannot have a negative value. Windspeed distributions are usually highly skewed; there is a long tail of fast, but infrequent winds, and some statistical functions that are often used to describe the probabilities include the Rayleigh and Weibull functions (the latter, a flexible two-parameter distribution, is widely used in the wind energy industry). Note that because winds exceeding the saltation threshold may be very infrequent (e.g., less than 1 % of the time) it is sometimes most useful to plot the relative frequencies on a logarithmic axis (Figs. 3.8 and 3.9).

Lorenz (1996) fit the in situ meteorological measurements of the Viking landers at Mars with Weibull functions. For the longest measurement sequence of 0–1040 sols recorded by Viking 2, the hourly-averaged wind was calm (below 1 m/s) for 14 % of the time, and the rest of the time could be described overall by $c = 3.85$ m/s, $k = 1.22$. Note that individual gusts could be much higher. Note also that the winds over this nearly two-Martian-year period include extended calm and windy epochs—fits to specific periods would give different values. Fits to a couple of terrestrial locations are given in figs, but note these data are at standard anemometer height (10 m) rather than the ∼1.5 m of the Viking meteorology mast. Fenton and Michaels (2010) study the Mars surface winds predicted by a Large Eddy Simulation (see Chap. 18) and explore the effectiveness of Weibull fits (as with the overall Viking fit above, Weibull fits tend to underestimate the high-wind tail of the frequency distribution).

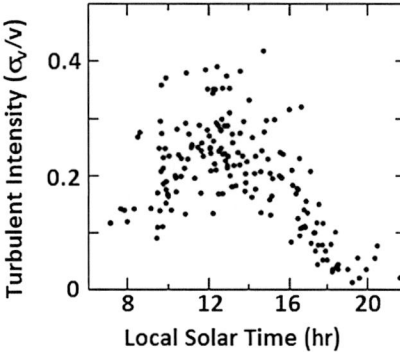

**Fig. 3.12** The turbulent intensity (standard deviation of wind speed divided by the wind speed), as a function of local solar time, as recorded by the telltale wind indicator on the Mars Phoenix lander. Few measurements were made in the early morning, but it is clear that the turbulence is highest in the early afternoon, when sunlight triggers convective activity. The turbulence level is quite high, with fluctuations in windspeed equal to about one third. (Data from Holsten-Rathlou et al. 2008.)

As for Titan, Tokano (2010) reported near-surface (~300 m) wind speed exceedance probabilities of westerlies and easterlies near the equator as calculated by a GCM. He found that while easterlies were more probable for low windspeeds (<0.6 m/s), strong winds were more likely to be westerly, consistent with the general transport direction indicated by the dune morphology. Winds of 0.8 m/s occurred only 1 % of the time, and 1.3 % of the time only 0.01 % of the time, with westerlies about twice as frequent as easterlies in each case. At a polar location for the late summer season, Lorenz et al. (2012) reported a Weibull fit for surface winds (10 m altitude) with c = 0.4 m/s and k = 2.0. Clearly, Titan winds are much gentler than their Martian and terrestrial counterparts. As yet, there are no detailed surface wind predictions for Venus, although the surface measurements of ~0.6 m/s (see Chap. 14) are rather similar to Titan.

## 3.5 Wind Direction: The Wind Rose

Wind speed alone is not enough to describe the likely morphology of dunes—direction information is usually desirable. The direction and speed frequencies over a period (usually a year) are most succinctly displayed in a wind rose. This is a polar diagram, where the wind direction is binned into 8, 12 or 16 azimuths, and the length of the radiating bar or wedge at that azimuth corresponds to the relative frequency with which that direction occurs. There are various conventions for encapsulating speed information, e.g., showing the rose only for those winds above some threshold, or (as in Fig. 3.10) dividing the bar into sections corresponding to speed ranges. As always, the caveat must be stressed that the directions shown, by meteorological convention, are those *from* which the wind blows. Although wind rose diagrams can be constructed by hand, dedicated software exists to generate them from wind data, and many meteorological organizations make such wind rose diagrams and the summary data to generate them available on the internet (Fig. 3.10).

Inspection of the wind rose for a site may allow one to predict the dune morphology that might result. First, one should only consider winds above the saltation threshold since gentle winds will not generally cause sand transport. If the fast winds are dominated by a single direction, then transverse dunes or barchans are likely to result; if bimodal, then linear dunes are to be expected, and if highly variable, star dunes will form (see Chap. 6).

While one can make a reasonable guess at the qualitative dune type and orientation by eye, to quantitatively compute the orientation and the sand fluxes (which will in turn allow estimation of the formation time or migration rate of a dune), the transport for each leg of the wind rose must be computed to determine the resultant drift potential (see Chap. 8).

## 3.6 Turbulence

The word 'turbulence' is often used in a rather dismissive sense to describe the spatiotemporal complexity of the wind field, as if invoking the term allows one not to think about the problem further. (Further thinking might not get most of us very far anyway—even the famous Richard Feynman said: "Turbulence is the most important unsolved problem of classical physics."). However, short-term fluctuations in wind speed and direction are instrumental in sand transport, and in contrast to the Bagnold-era consideration of mean properties, many modern studies seek to explicitly quantify the turbulence of the wind field and its effects, and measurements Fig. 3.11 and modeling tools can now handle this.

The usual way to treat turbulence is to consider the mean flow (usually assumed horizontal in one axis, so denoted by U(y)), with two orthogonal fluctuating components (u', v') superposed on it (we'll consider only two dimensions here). The first general quantity to consider is the 'turbulent intensity'—how big those fluctuations are. For this to be meaningful, the fluctuations have to be averaged over some space or time. The root-mean-squared values $\sigma_u$, $\sigma_v$ are often used and divided by the mean speed U to express the intensity $I = \sigma_u/U$. For a well-designed wind tunnel that aims to minimize turbulence, I might be 1 % or lower. In the open, or where flow obstacles are present, it can be much higher: Fig. 3.12 shows measurements of I for the Martian surface. At night when conditions are quiescent (but windspeed is still a few m/s) I ~ 5 %, but builds

## 3.6 Turbulence

**Fig. 3.13** Aeolian streamers—concentrations of saltating sand. These are snaking towards the camera, and are rendered visible by their contrast against the *dark road*. They arise from the turbulent structure of the air itself, coupled with the hysteresis in sand transport (see also Fig. 12.13). Similar structures are often visible in snow. *Photo* R. Lorenz, United Arab Emirates

during the late morning and early afternoon to 30–40 %. Clearly, when considering a threshold phenomenon such as saltation, turbulence is very important. Obviously, measuring turbulent fluctuations requires instrumentation (see Sect. 16.3) with a very fast response time, and so ultrasonic or hot-wire anemometers are typically used, with the former particularly suitable for measuring 2- or 3D winds.

These fluctuating components have power spectra that portray how much variation occurs at different length scales as energy cascades down from the large-scale flow through ever-smaller eddies to the molecular scale where the kinetic energy is dissipated by viscosity into heat. The spectrum of turbulence was studied significantly by the Soviet mathematician Kolmogorov, who found that a power law with an exponent of $(-5/3)$ describes the typical scale variation in free air. In free air, the variations are often similar in all directions, so-called isotropic turbulence.

At a flow boundary, such as the surface of a planet modifies the spectrum somewhat, and of course significantly influences the directionality of turbulent components of flow (which cannot cross the boundary). The turbulent flows are usually discussed in terms of in which quadrant of a plot of (u', v') they appear. A sweep, which introduces a fast flow down into the surface (v' negative, u' positive) will apply a strong shear to the surface, and thus may be effective at initiating surface particle movement. Flows away from the surface (u' negative—ejection, and u' positive 'outward interaction') may help bring particles away from the surface and into the stronger flow U at higher z.

Spatial and temporal correlations often exist due to dominant scales in the eddy spectrum, presumably related to the size of obstacles or to the size of the boundary or friction layers. These persistent flow structures can sometimes be rendered visible by moving sand. The most obvious is the formation of aeolian streamers (Fig. 3.13), somewhat regularly-spaced streaks that can often be seen as light snow blows across a dark road. Some examples of aeolian streamers in sand at a coastal dune are described with a width of 0.2 m and spacing of $\sim 1$ m (Baas and Sherman 2005).

Thus a full description (if anything like one even exists) of a turbulent flow must invoke the spatial and temporal variations in velocity, at a range of different scales, and must recognize that next to a boundary like the ground these variations will also depend on direction. We nonetheless hope we have at least made the reader a little literate in turbulence, and that at least part of the wind's turbulent character can be measured and quantified.

# Mechanics of Sand Transport

The character of an aeolian landform must depend on how the sand moves. This, of course, depends on the interaction of the particles with the fluid: where can particles start moving, how far do they move, how fast do they hit the ground again. These all depend on the forces acting on the grains, and thus in turn on the physical environment—gravity and the properties of the fluid. The study of this problem has traditionally been addressed somewhat separately of underwater particles, and for particles in air, in fluvial/marine and aeolian geomorphology respectively. However, we can see these cases as merely two points in a continuous space of environmental parameters.

Bagnold's (1941) account of the problem is lucid and still worth reading. For the masochistic reader determined to subject themselves to the full modern glory of this problem, there are exhaustive, and sometimes exhausting, treatments with countless pages of equations in several texts, especially Greeley and Iversen (1985) and Zheng (2009), and the review by Kok et al. (2012). Our aim here is only to give the basic equations and numbers, and beyond that a familiarity with the principal controlling effects, and how they depend on the planetary setting.

## 4.1 Styles of Particle Motion

Movement of particles by the wind can take several different forms (Fig. 4.1). The most common form of movement of sand by the wind is called **saltation**, which refers to the 'hopping' motion of the individual grains that results when lifted grains gradually attain an increasing fraction of the wind velocity before impacting the surface downwind at a shallow angle (generally <10° with respect to horizontal). Saltation path length can be affected by a variety of factors, but generally medium-to-fine sand on Earth tends to saltate along paths ∼10 cm in length (Bagnold 1941). If a saltating sand grain happens to impact onto a large rock rather than onto other sand grains, it may rebound to a much greater height than can be attained through aerodynamic lift forces alone. Saltating sand grains that impact onto resting sand grains cause a 'splash' of several of the impacted grains, each of which follow paths much smaller than that of an individual saltating sand grain. The lower energy of ejection for the impacted grains is the primary reason for the reduced scale of the splashed-out grain trajectories. Splashed sand grains follow a kind of 'staggered random walk' pattern across the surface of the sand, a process distinct from saltation that is now referred to by the term **reptation** (Anderson 1987). Reptation tends to occur at a scale that is about one-sixth that of a typical saltation path length; both theory and experiment have shown that reptation is a crucial element to the growth of aerodynamic ripples on sand (Anderson 1988). Saltating sand grains that impact onto small gravels and pebbles can produce a gradual rolling movement of the coarse particles in the downwind direction, in a form of particle motion termed **creep** or **traction**; granule and pebble ripples are examples of aeolian bedforms that result from this form of wind-induced particle motion where the large particles rarely lose contact with the surface. Note that in fluvial transport, the term 'bedload' is used for particles transported by creep and saltation.

If the velocity in turbulent eddies within the flow is larger than the terminal velocity of the particle, the particle is said to be in **suspension**, and such particles (usually silt or dust) can be transported very great distances (even hundreds to thousands of kilometers); large deposits of loess found worldwide on Earth are prime examples of the transportation of suspended small particles by the wind. Note that fine silt and clay-sized particles can stick together effectively and are generally too small to be lifted out of the laminar boundary layer by the airflow alone. Often they may be ejected by the impact of saltating sand grains, or sucked out of the surface by the pressure drop in a vortex such as a dust devil.

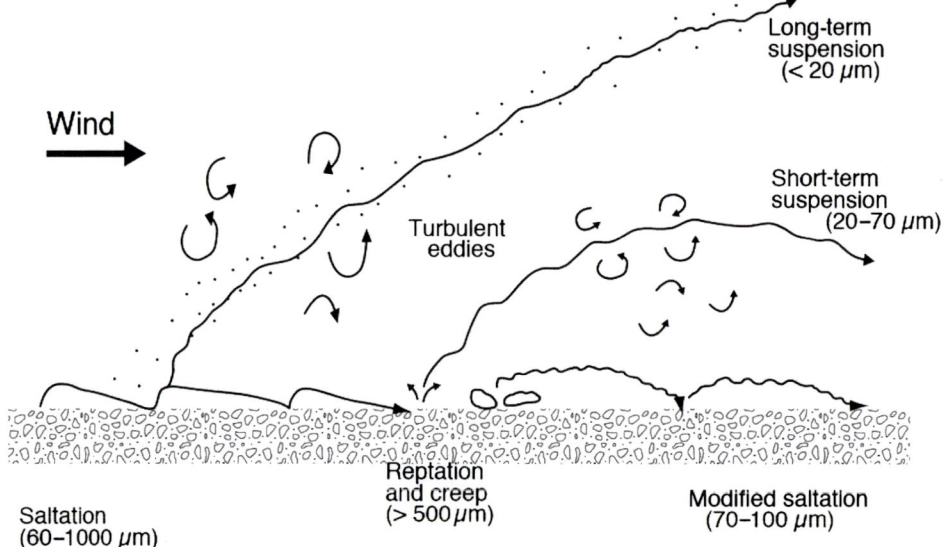

**Fig. 4.1** Cartoon of different types of particle transportation by the wind. *Image* courtesy of Nick Lancaster

**Fig. 4.2** A dramatic photo of an approaching haboob (dust storm front), taken by a soldier in Iraq. Here dust is in suspension. Wikimedia Commons

Although not the subject of this book, airborne dust and sand transport are often considered together (conferences and conference sessions on aeolian processes cover them both). Airborne dust is an important factor in air quality and the nutrient balance, Saharan dust is an important factor in providing iron to the oceans and even to the Amazon basin. Dust is a significant factor in controlling climate (on both Earth and Mars) and in the use of solar power on both worlds. Mars is, of course, famous for both dust devils and for dust storms, which also happen occasionally on Earth (Fig. 4.2).

## 4.2 Fluid Forces on Particles

The interaction of objects with moving fluids is fundamental in a number of fields, from gravel moving in a river to a fighter jet moving in air. Each one has developed a set of conventions in formulation and nomenclature, even though the underlying process is the same. Thus some works in aeolian studies refer to a 'threshold parameter', a 'Shields parameter' is used in hydraulic sedimentology, while aerospace engineers have a 'drag coefficient'. All of

**Fig. 4.3** The balance of forces

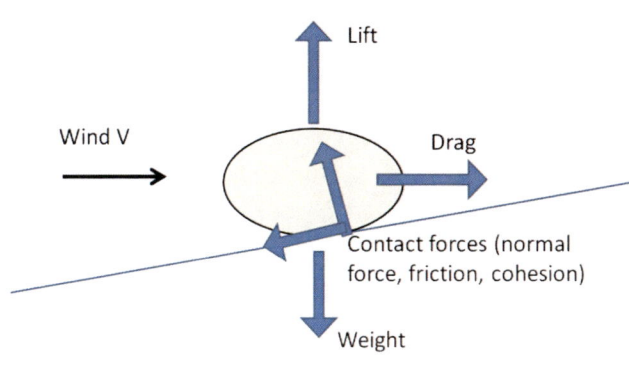

these are different permutations—in principle related and ultimately equivalent—to the problem of relating the forces seen by an object to the density and speed of the fluid relative to it. In fact, the approach we will take is based on the aerospace convention.

It is an interesting point of trivia that the relation of force to the density of a fluid, and the square of the velocity was recognized theoretically by Isaac Newton, although the subtleties of fluid mechanics are such that his idealizations are actually most accurate for the interaction of rarified atmospheres with spacecraft. The first practical calculations on fluid body forces were made in 1691 by Newton's friend and advocate, Edmond Halley. He recognized—viscerally, from his experiences wearing a diving helmet he invented himself and attempting salvage on the seabed—the overwhelming role of fluid density, that forces in water running at a given speed are 800 times higher than wind at the same speed. Halley calculated how fast a sailing ship should move, and even how large wings a human would need in order to fly (Lorenz 2012).

The balance of forces on a particle (Fig. 4.3) determine whether it will move. First is its weight W. The effective weight, by Archimedes' principle, is really the difference between the vacuum weight of the particle (its mass times the gravitational acceleration g), and buoyancy, the weight of fluid that the particle displaces. For a sphere of diameter d and density $\rho_p$ immersed in a fluid density $\rho_f$, this is then $W = (\pi/6)d^3(\rho_p - \rho_f)$.

The drag force arising from fluid motion is defined to apply in the direction of the relative wind (i.e., opposite to the direction of motion relative to the fluid). For a perfectly spherical nonrotating body in a uniform flow field, drag is the only fluid force that will act.

In general, due to irregularities in particle shape and the gradient in fluid speed near the ground, and sometimes due to particle rotation, the fluid force will have components orthogonal to the fluid motion. Usually only the component in the plane containing the fluid velocity and the gravitational acceleration vectors is of interest—the lift (A side-force component can exist, and is of interest in many sports—see e.g. Lorenz 2006.). Lift manifests itself as a reduced pressure on the upper surface of the object compared with the lower surface. The application of a force to the object (like an airplane wing) must be accompanied by a corresponding force applied to the airstream, which is deviated downwards as a result. The fluid may or may not flow more quickly on the upper side of a lifting body: it is not required to (although many textbooks make the mistake of saying so) as an inclined flat plate can generate lift without forcing the air over a longer path. Lift can also be produced by rotation of the particle, the Robins-Magnus effect which is important, for example, in lofting golf ball trajectories.

Finally, forces exist at the contact between a particle and the surface it rests on (or between the particle and other particles). In the high-school physics block-on-a-plane problem, we would think of these as the normal force and friction; between particles we might think of it as cohesion. These forces will try to balance the others up to some maximum value, which they cannot exceed.

If these forces are in balance, the particle will not accelerate. If it is at rest, it will stay at rest. The weight of the particle is the product of its volume (and thus the third power of its diameter), its density, and gravity. Lift and drag vary as the particle cross-sectional area (and thus the square of diameter). Thus, all else being equal, the balance becomes dominated by weight as diameter increases, and so big rocks tend to sit on the ground.

The lift and drag are also proportional to the fluid density rf, and the square of the fluid speed U. The constant of proportionality is a dimensionless number, termed the lift coefficient $C_l$ (or drag coefficient, $C_d$). The lift coefficient is a complex function of shape; for many particles it may be close to zero, but can be of the order of 0.1. The drag coefficient is also somewhat dependent on shape and surface roughness, but less so than the lift coefficient. For large objects (we will return shortly to what determines 'large'), $C_d$ is often in the range 0.5–1, although it can be much higher for very small particles, as we discuss shortly.

**Fig. 4.4** The drag coefficient of a sphere as a function of Reynolds number. At *left*, viscous forces dominate, and the drag coefficient varies as (24/Re) and yields Stokes' law. Over ~100<Re<100,000, $C_d$ is fairly constant (changing the shape dramatically changes $C_d$ by only a factor of 2), but $C_d$ drops sharply (the 'drag crisis') at a roughness/turbulence-dependent Re value of ~100,000

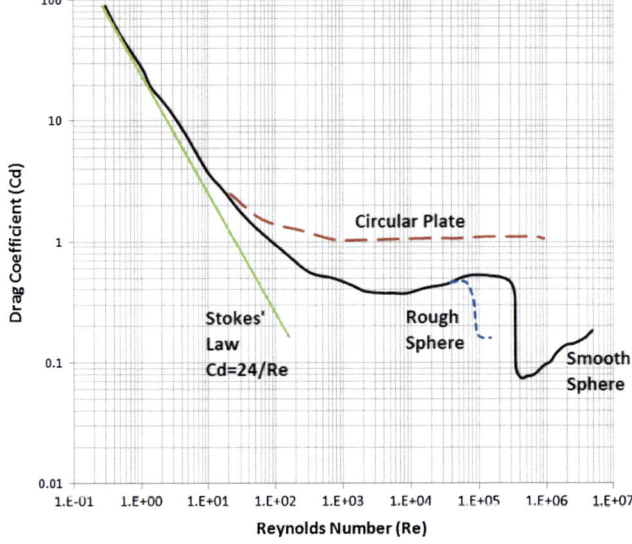

Thus $L = 1/2SC_l\rho_f U^2$ and $D = 1/2SC_d\rho_f U^2$, with S the particle area ($=\pi d^2/4$ for a sphere). The parameter group $1/2\rho_f U^2$ is called the dynamic pressure, and is equal to the kinetic energy per unit volume of the airflow.

If we consider a particle away from others and away from the planet surface, it will be exposed to weight and lift/drag only. In the absence of wind, it will be accelerated downwards by its weight until it is descending relative to the fluid at a speed Vs (terminal velocity) where the drag balances the weight (since the particle is falling vertically, the drag is upwards). Equating the weight and drag we find:

$$D = (1/2)SC_d r_f U^2 = (1/8)\pi d^2 C_d \rho_f V_s^2 = W$$
$$= (\pi/6)d^3(\rho_p - \rho_f) \text{ and thus} \quad (4.1)$$
$$V_s = ((4/3)d(\rho_p - \rho_f)/C_d\rho_f)^{0.5}$$

All else being equal, the terminal velocity varies as the square root of diameter.

The above is all pretty much 17th-century Newton stuff, and looks straightforward because all the complexity is hidden in the innocent-looking $C_l$ and $C_d$ lift and drag coefficients. These are functions of shape (which is why car and aeroplane manufacturers spend lots of time with wind tunnels and computer simulations) and also on three dimensionless parameters which define the way the fluid behaves, whether it is gas or liquid.

We can dismiss a couple of these rather quickly. First is the Knudsen number, the ratio of the mean free path of air molecules to the particle size—for everything we'll discuss this number is very small and fluids behave as a continuum. For satellites in the upper atmosphere, the mean free path is large and air molecules behave as unconnected billiard balls travelling in basically straight lines ('free molecular flow'—how Newton imagined things, where the drag coefficient $C_d$ is exactly 2.0). The borderline between these regimes is approached for small dust grains in the thin Mars atmosphere, but for everything else the fluid behaves as a continuum.

The second parameter is the Mach number, the ratio of fluid velocity to the sound speed in the medium. The 'sound barrier' is the phenomenon where the drag coefficient rises sharply (and the lift can change) as an aircraft approaches the speed of sound. This was actually recognized (Lorenz 2006) in 1740 by an astute artilleryman, Benjamin Robins, whose cannonballs did not go exactly where he expected; he also recognized the role of spin, hence the Robins-Magnus effect. For all the situations we'll consider, the Mach number is small and the flow is subsonic.

But the last parameter, the Reynolds number Re (named after Osborne Reynolds, a British physicist), is what causes complications. This is the ratio of inertial forces to viscous forces, and defines whether a flow is laminar or turbulent. In laminar flow, the fluid follows streamlines which stay constant in time. In turbulent flow, streamlines can be drawn to define the average flow path, but individual parcels of fluid wander and swirl across these average lines, causing much more vigorous mixing. Reynolds performed a series of experiments in glass-walled pipes to determine what caused flow to become either laminar or turbulent and found that the criterion related to the ratio of the following parameters:

$$Re = u\, d\rho_f/\mu \quad (4.2)$$

where u is velocity, d is particle dimension, $\rho_f$ is the fluid density, and $\mu$ is the fluid (dynamic) viscosity. From many laboratory experiments in over a century of work, the drag coefficient of various shapes as a function of this quantity is very well known (Fig. 4.4). In sedimentology, the Shields curve is essentially a redrawing of this drag coefficient relation, but with slightly different axes.

**Table 4.1** Reynolds numbers for different conditions. Aerospace vehicles and thrown objects have high Re and viscosity does not matter; for dust, Re is typically much less than 1 and viscosity dominates. Sands in atmospheres tend to be in between with Re of ~100, although sands in liquids tend to have low Re

| Setting | Density (kg/m$^3$) | Viscosity (μPa-s) | Size (mm) | Speed (m/s) | Re |
|---|---|---|---|---|---|
| Sand grain in water | 1000 | 1000 | 0.2 | 0.01 | 2.0E+00 |
| Silt particle in water | 1000 | 1000 | 0.02 | 0.001 | 2.0E−02 |
| Pebble in water | 1000 | 1000 | 20 | 0.5 | 1.0E+04 |
| Dust in air at sea level | 1.25 | 17 | 0.02 | 0.1 | 1.5E−01 |
| Sand grain in air at 5000 m | 0.8 | 16 | 0.25 | 10 | 1.3E+02 |
| Sand grain in air at sea level | 1.25 | 17 | 0.25 | 10 | 1.8E+02 |
| Baseball | 1.25 | 17 | 70 | 40 | 2.1E+05 |
| 747 airliner in cruise | 0.3 | 15 | 1E+05 | 250 | 5.0E+08[a] |
| Sand grain on Venus | 64 | 35 | 0.2 | 1 | 3.7E+02 |
| Dust grain on Mars (6 mbar) | 0.02 | 13 | 0.002 | 0.1 | 3.1E−04[b] |
| Sand grain on Mars (6 mbar) | 0.02 | 13 | 0.2 | 10 | 3.1E+00 |
| Sand grain on Mars (Tharsis) | 0.003 | 13 | 0.2 | 25 | 1.2E+00 |
| Sand grain in Titan air | 5.4 | 6 | 0.2 | 1 | 1.8E+02 |
| Sand grain in Titan sea | 660 | 2000 | 0.2 | 0.1 | 6.6E+00 |
| Huygens probe at touchdown | 5.4 | 6 | 1300 | 5 | 5.9E+06 |

[a] For jet aircraft, Mach number effects can be important
[b] For dust on Mars, Knudsen number effects can be important

In general, low Reynolds numbers mean viscous forces dominate and the flow is laminar; when inertial forces overpower the resisting viscous forces, the flow is turbulent. The borderline between the two is somewhat shape- and condition-dependent, but if Re > ~100, then fluid is turbulent. When we calculate the value of Re for various conditions on different planetary bodies (Table 4.1) we find, somewhat inconveniently, that many situations of interest are close to this boundary, so we require some care in evaluating the drag—neither the Stokes limit nor the large-Re limit are universally applicable.

Because much of the early quantitative work on particle transport by fluids was in water, where the Reynolds number turns out to be low, much of the literature on sedimentology is shaped by the low-Re limit of fluid behavior. This in turn focuses on Stokes' law (named after George Stokes) which estimates the velocity that results when frictional (drag) and buoyant forces are exactly balanced by gravitational force (weight) for a spherical particle falling vertically through a viscous fluid:

$$u_s = 2(\rho_s - \rho_f) g r^2 / 9\mu \quad (4.3)$$

where $u_s$ is the settling velocity, $\rho_s$ is the particle density, $\rho_f$ is the fluid density, g is the acceleration of gravity, r is the particle radius, and $\mu$ is the dynamic viscosity of the fluid. The Stokes law in Eq. 4.3 is valid only for laminar flow around the particle, so it should be restricted to situations where Re < 1 (the algebraically-minded will see that this equation is equivalent to Eq. 4.1 above, when the low Re limit of $C_d = 24/Re$ is substituted). One can see that when viscosity dominates, the force depends linearly on flow speed, not on the square of flow speed (put another way, because velocity is in the Reynolds number, and the drag coefficient varies as the reciprocal of Reynolds number, one of the velocities in Eq. 4.3 gets cancelled out). This complicating difference in flow regime proved a challenge in the historical development of fluid mechanics, with debate raging among the foremost mathematicians of the 17th and 18th centuries over whether drag depended on speed or the square of speed, when in fact 'it depends…'.

So, armed with Eq. 4.3 we can calculate the terminal velocity of particles (Fig. 4.5). While for many situations Stokes law works, in a few cases it is not applicable and the more general form of Cd must be used.

In general, the settling velocity in air for particles of silt and clay (or 'dust') is much smaller than the upward velocities available from turbulent eddies in wind flow (see Sect. 3.6) which is the explanation for why these particles can be transported great distances in suspension.

## 4.3 The Boundary Layer

The wind speed near the ground approaches zero (in fact aerodynamicists refer to a 'no slip' condition wherein the speed exactly adjacent to a surface must be zero, otherwise the shear stress would be infinite). The layer above the surface in which the fluid velocity is 99 % or less of its

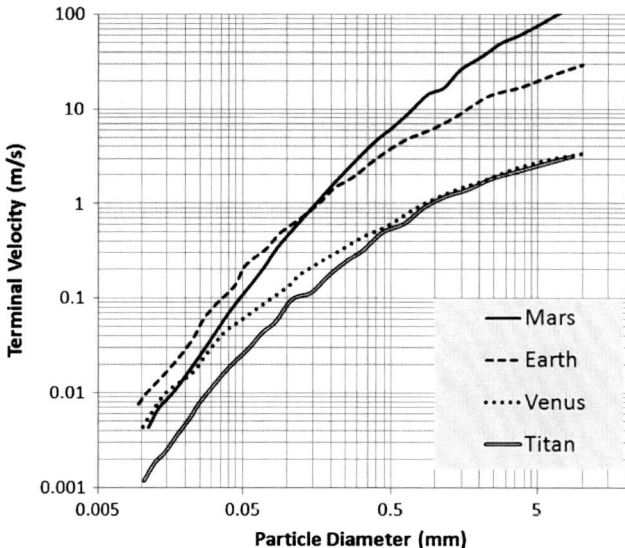

**Fig. 4.5** Terminal velocity of particles in air on different bodies. The *curves* have slightly different slopes because of the varying roles of viscosity versus density. In the high-Reynolds number limit, the largest particles fall faster on Mars since the air is very thin and density is dominant; Venus and Titan with thickest atmospheres tie for slowest fall (Titan's weak gravity making up for the fact that its air is thinner than Venus). However, at the small particle end (low Reynolds numbers) the ordering changes

freestream value is termed the (velocity) boundary layer (not to be confused with the planetary boundary layer, or the thermal boundary layer of heated surfaces). Usually the wind profile can be approximated by a function of height referred to as the Prandlt-von Karman equation or, more commonly, 'the Law of the Wall'.

$$u/u_* = (1/K)\ln(z/z_o) \quad (4.4)$$

where u is the wind velocity at height z, $z_o$ is the aerodynamic roughness length of the surface, $u_*$ is the shear velocity (or friction velocity) of the wind, and K is the von Karman constant, which is $\sim 0.4$. This logarithmic velocity profile is a consequence of friction on the wind flow caused by the rough surface. The aerodynamic roughness length $z_o$ is defined as the height at which the wind velocity reaches 0 on a semi-log plot of velocity above the surface (Fig. 4.6). In principle, this line can be defined by two points, but it is the general (and good) practice to use at least three, to judge how good a line fit really is.

This profile is why it is standard practice to record wind measurements at a standard 'anemometer' height, typically 10 m. Corrections need to be applied to measurements made closer to the ground. It is often the case in field measurements, and usually the case on planetary landers, that non-ideal measurements must be made, closer to the ground than the meteorological standard. Global Circulation Models (GCM) may report friction speed, or sometimes only the windspeed in their lowest grid cell, which could be 100 m up or higher. It is good practice to report any wind speed measurement (or simulation) with the corresponding height as a subscript, e.g., $U_{10}$ is the windspeed at 10 m.

The roughness length is related to, but is not exactly equal to, the typical scale of roughness elements. Field measurements (e.g., Greeley et al. 1991b) have shown that the aerodynamic roughness length correlates with the roughness that one might infer from radar backscatter observations—see Chap. 18. Measurement of $z_o$ both in wind tunnels and in the field is a very important determinant for how effectively the wind can cause work to be done upon a natural surface, and when sand on the natural surface can be set into motion by the wind. It is also important for correcting wind measurements at low and/or different heights to allow them to be intercompared, using (for example) the Prandtl-von Karman equation.

The somewhat rocky Viking 2 lander site on Mars has a roughness length estimated at $\sim 1$ cm, while the Mars Pathfinder landing site (an alluvial outflow area chosen precisely because there would be rocks!) had $z_o \sim 3$ cm (Sullivan et al. 2000). The rather rock-free permafrost surface visited by the Phoenix lander at 68°N had a roughness estimated at $5 \sim 6$ mm (Holstein-Rathlou et al. 2010). Smooth playa surfaces on Earth can have sub-mm roughness lengths.

This logarithmic wind profile isn't quite the full story, however. One can see that as the speed falls near the surface, the local Reynolds number will also decrease. When this approaches unity, viscosity will take over and damps out turbulent fluctuations altogether, forming a viscous sublayer

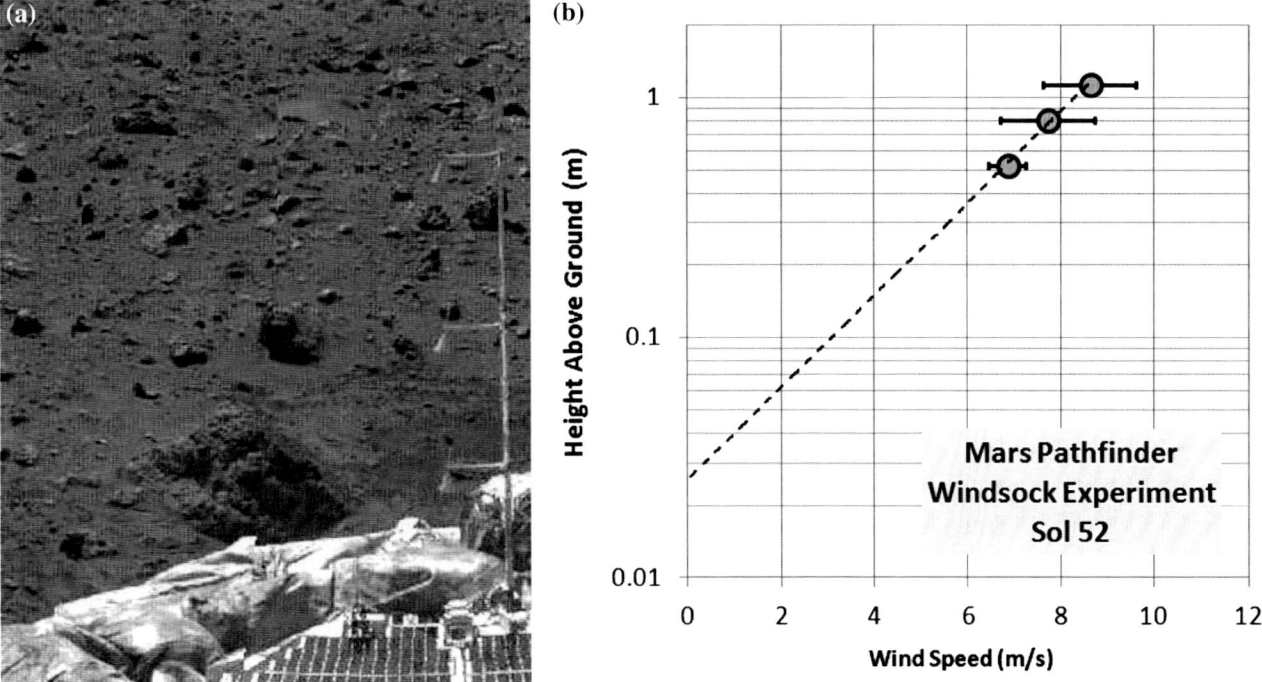

**Fig. 4.6 a** Image mosaic from the Mars Pathfinder Lander, showing the three dangling metal cones that made the Windsock Experiment. Wind tunnel calibrations allowed the orientation of these to be related to windspeed (U. Arizona/NASA). **b** The three measurements at three separate heights follow the Prandtl-von Karman equation (a *straight line* on a plot of speed against logarithm of height). The slope of this line is the shear velocity (or friction speed) and the intercept at zero velocity is the roughness length. (Data from Sullivan et al. 2000)

in which the airflow is laminar. Such a laminar, horizontal flow has little hope of lifting particles up, and so dust particles must be somehow lifted off or pushed through this layer in order to get into suspension. Due to the low Reynolds number typically encountered at Mars (compared with the Earth, where the air is 50 times denser) this viscous sublayer is rather thicker—and thus more important—at Mars. Its importance depends on the particle size on the surface, if these are large compared with the viscous sublayer thickness then the surface is considered 'aerodynamically rough' and the flow can impart stress on the particles. When particles are small (like silt on Earth), the particles can hide in the sublayer and do not feel the full force of the flow. On Earth the dividing line is usually taken to be about 80 $\mu$.

Of course, dust is raised from surfaces on Earth, despite this effect. This can happen when there are even a few roughness elements (e.g., rocks) that are large compared with the viscous sublayer, and stir the flow into the ground in their wake. There is speculation that pressure drops in dust devils can suck particles out of the ground, or even other factors like solar heating and electrostatic effects can loft the dust to the point where turbulent suspension can take over. But possibly the most significant mechanism is where sand or larger particles smash into the ground and splash dust upwards. This still leaves us with how to get the sand moving, however.

## 4.4 Launching the Sand

Sand grains poking out through the viscous sublayer will see wind stresses. If these can overcome weight, friction and adhesion, then they can get the ball rolling—literally. Rigorous treatments will consider the direction of the forces (and the turning moments) acting on a grain sitting among other grains, but we will avoid bogging ourselves down in these distinctions.

Shear velocity (also known as friction velocity) $u_*$ is proportional to the slope of the wind velocity profile when plotted on a logarithmic height scale (Fig. 4.7); it is also directly related to the shear stress $\tau$ at the bed surface and to the air density $\rho_f$:

$$u_* = (\tau/\rho_f)^{0.5} \qquad (4.5)$$

where $\rho_a$ is the air density. While $u_*$ has units of velocity, it is actually related to the rate at which the velocity is changing within the boundary conditions expressed by the Prandtl-von Karman expression in Eq. 4.1.

$u^*$ is a less intuitive quantity than the actual windspeed U. The two are connected, via the stress exerted on the surface (working through the equation above) in a somewhat convoluted way. A simpler way to at least get a feel for the numbers is to consider that the stress on a surface

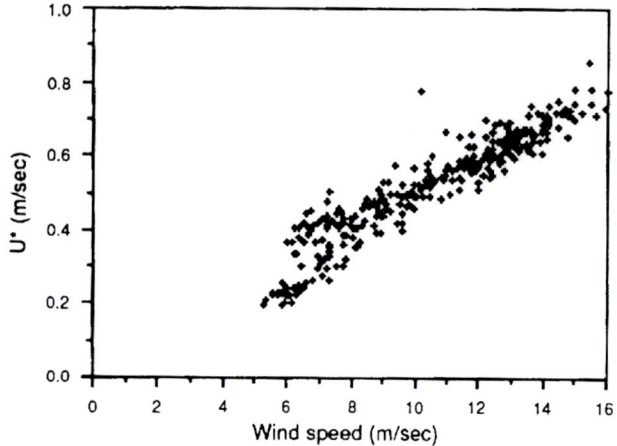

**Fig. 4.7** Measurements of the friction velocity (i.e., the slope of speed vs. logarithm of height) are linearly correlated with winds at 9.5 m height. These measurements are from Death Valley (from Greeley et al. (1991b)), and show $u* \sim 0.05 U_{9.5}$. *Image* NASA

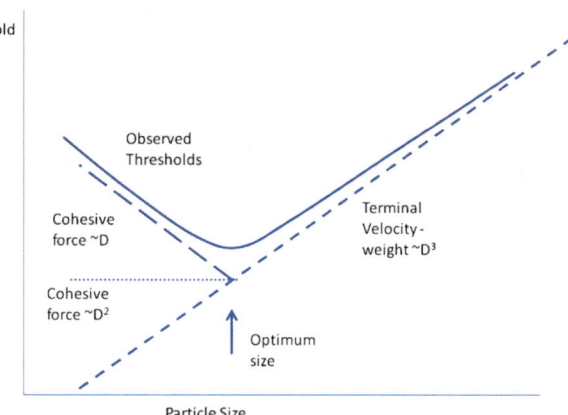

**Fig. 4.8** A schematic of the threshold speed and how it depends on diameter (shown on unlabelled, but logarithmic, axes). Weight varies as the cube of diameter, whereas lift and drag vary roughly as the square, thus the threshold windspeed that balances them is proportional to size. At very small sizes, a cohesive force proportional to diameter (i.e., a cohesive stress inversely proportional to diameter) will dominate and the threshold will be inversely proportional. Where the cohesion and weight are about the same, the overall resistance to motion is a minimum, and thus the threshold speed is the lowest: the saltation 'optimum'. For different gravity, different cohesion parameters, different fluid density, etc., the numbers will all be different, but the overall shape of the curve will be the same

(like a wing) can be written as Eq. 4.1, i.e., D/S = $0.5 C_d \rho_f U^2$ with all the parameters as before, except here the drag coefficient is not that of a spherical or similar body, but is the friction coefficient of a surface. That (as before) depends on roughness and Reynolds number, but broadly speaking is around 0.001 for 'smooth' surfaces and maybe 0.01 for rough ones. Now, we can rearrange Eq. 4.1 to calculate the wind stress on the ground as $\tau$ = D/S, and rearrange Eq. 4.3 to say $\tau = u_*^2/\rho_f$, and we find then simply that $u*/U \sim (0.5 C_d)^{0.5}$ (some definitions of $C_d$ may result in the 0.5 not appearing, but the square root means this doesn't matter terribly). In other words, $u_*/U \sim 20$, give or take a little. While this rule of thumb lacks rigor, it is a handy way to grapple with the numbers at least approximately. And, indeed, field measurements show just this sort of relationship.

For large particles, it is the weight that dominates, and so the threshold stress (i.e., friction velocity, or the corresponding windspeed) relates closely to the terminal velocities shown in Sect. 4.1. Bigger particles have larger weights and thus higher threshold speeds, and in principle arbitrarily small particles would have arbitrarily small threshold speeds.

However, at small sizes, adhesion (or cohesion) forces take over. This tendency is evident in the kitchen: icing sugar and granular sugar are the same stuff, just with different particle size. Dig some out with a spoon, and granular sugar (with a grain size usually 0.5 mm or more) will slump, making a conical pit with an angle of about 30°, set by the friction between particles. But dig a hole in icing sugar, and it can have vertical walls—the cohesion between the smaller grains is enough to prevent the walls failing. This cohesion is also what means small particles with slow terminal velocities do not easily get lifted of a surface.

A variety of mechanisms can play a role, with different behavior. If we imagine the cohesion is due to some sort of weak strength like a glue between grains, then it would have a constant force per unit area. In that case, the threshold speed as a function of size would be flat—constant—corresponding to the speed at which the aerodynamic stress equals the failure stress of the glue. On the other hand, and damp sand may behave a bit like this, imagine the cohesion force is a linear function of diameter. Surface tension of liquids can behave this way. In this case, the cohesion force grows linearly with diameter, whereas the aerodynamic stress (ignoring the Reynolds number effect) grows with the square of diameter. Thus the windspeed at which these two will balance, and thus the threshold windspeed for motion, will vary as the reciprocal of size. This is shown schematically in Fig. 4.8.

Bagnold conducted extensive wind tunnel measurements, both in the field and in the laboratory, to understand the flow conditions that led to motion. Greeley, Iversen and White in the 1970s conducted many further experiments, using particles made of different materials and using air of different density. These latter experiments tend to report their results with the drag coefficient formulation, and the cohesion formulation, buried together in a 'threshold parameter' A.

They use a rather ugly empirical formulation for this—it fits the data, although it is not clear how applicable it may be to ices or organics on Titan. Since both the drag and

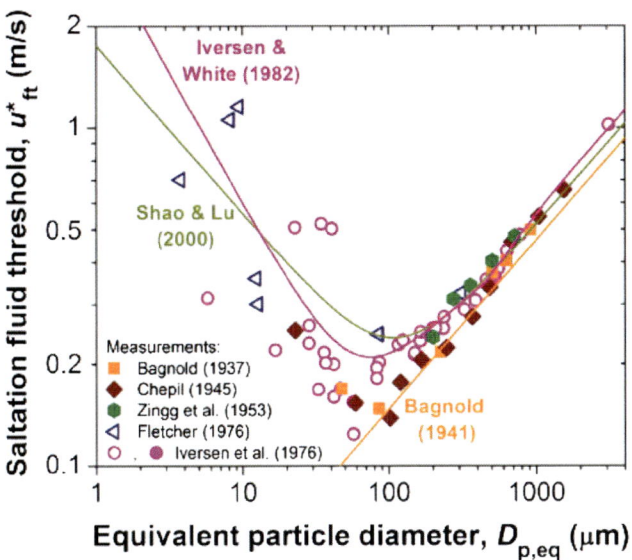

**Fig. 4.9** Wind tunnel measurements of fluid threshold as a function of size compared with various analytic expressions. The scatter in the data is comparable with much of the difference between the different models. (from Kok et al. 2012, with permission)

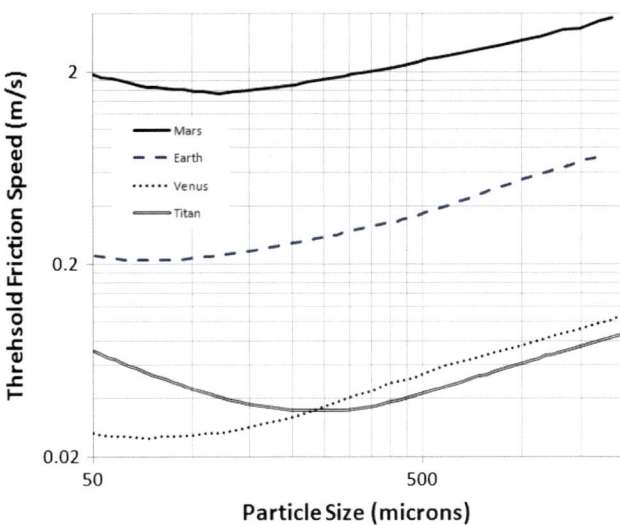

**Fig. 4.10** Fluid threshold friction speeds

cohesion are buried in the same parameter, it is difficult to judge what the effects might be. The discussion above exposes the factors a little more clearly.

We have in all this discussion neglected possible electrostatic forces on particles. These could be quite important, especially on Titan. One easy experiment is to place your hand inside a 'beanbag' chair, filled with small polystyrene foam spheres. These will stick quite well to your hand, attesting to the electrostatic attraction. The triboelectric processes that generate electrical charges (Zheng 2009; Kok et al. 2012; Merrison et al. 2012) discuss these effects at some length: they could cause saltation thresholds to go up, or go down. Electric fields generated in blowing snow or sand (or dust) can cause radio interference, and may offer a means of measuring the transport of granular materials.

Figure 4.9 shows a compilation of these wind tunnel results; it is seen that indeed the overall behavior is rather as indicated in 2.3.9, with a minimum threshold. One way of collapsing data made with different materials (and Greeley used everything from lead to sugar to walnut shells) of different densities is to scale the particle diameter. In other words, for the 'normal' quartz sand with $\rho_p = 2650$ kg/m$^3$, the scaled diameter is just the diameter. But for gypsum (say) particles of 300 μm diameter with density 1700 kg/m$^3$, the diameter to use on these plots is 300*1700/2650 ~ 195 μm. In other words, denser materials behave like bigger quartz particles. This approach is a bit of a fudge, but is reasonably effective for the so-called 'fluid threshold', the speed at which particles begin to move. As we shall see, this is not the whole story.

So, we now have the theory, backed by experiment, to support prediction of the fluid threshold on other planets, with other sands and atmospheres. This overall picture is laid out in Greeley and Iversen (1985) with a more updated discussion in Kok et al. (2012); investigations for specific planetary bodies are called out in the respective chapters of part 3 of this book. Figure 4.10 shows the results, from Lorenz et al. (1995), and quite similar results are shown by Greeley and Iversen (1985). The density of the atmosphere is unsurprisingly the dominant effect, as the labels on the respective curves follow the order of the surface pressures on the bodies shown—the threshold friction speeds in the thin barely-an-atmosphere of Triton are colossal (and likely supersonic), are still high for Mars, and get lower for Earth, Titan and Venus. Titan and Venus are not too different, the low gravity of Titan partly compensating for the thicker Venus atmosphere.

The effect of gravity can be discerned on the plot. On lower-gravity worlds like Triton and Titan, weight becomes significant with cohesion at a larger diameter than on the terrestrial planets, and thus the optimum size for saltation is at larger diameters. All this assumes, of course, that the cohesion is the same for Titan/Triton materials as for terrestrial sands. The effect of varying cohesion is readily seen by varying the moisture content of sand—measurements summarized in Greeley and Iversen (1985) suggest that even 0.3–0.6 % moisture can double the threshold friction speeds for sand; see also McKenna-Neuman and Nickling 1989.

As an amusing aside, it is pertinent to note that any future human presence on the moon may experience somewhat Titan-like mechanics. The moon's gravity is about the same as Titan (1.3 m/s$^2$), and a pressurized moonbase or landed spacecraft will have an air density of perhaps 0.9 kg/m$^3$ (if filled with air, like an airliner) or 0.5 kg/m$^3$ (if enriched with

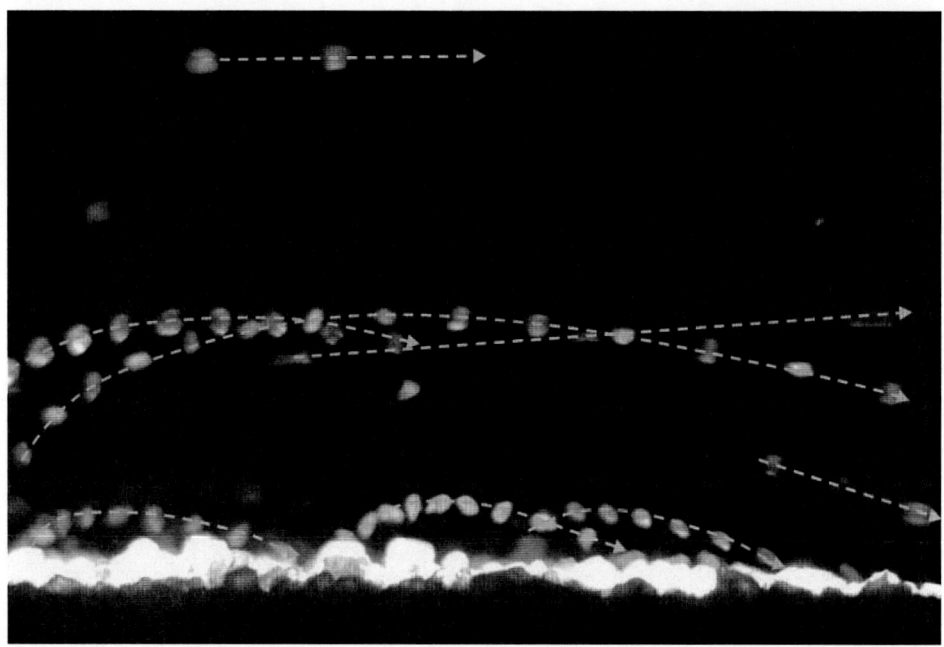

**Fig. 4.11** Beautiful high-speed photography of saltating sand grains in a wind tunnel. The grains (about 350 μm in size) are illuminated by a laser sheet, strobing at 380 flashes/s. The rotation of the grains can be clearly seen—some are spinning at least at 100 revs/s, which may affect their dynamics. The arrows show the direction of motion—the downwind flight results in a rather shallower angle of impact than their steep launch. Image courtesy of Susan McKenna-Neuman

oxygen). Thus the dynamics of sugar sprinkled onto the food of astronauts or space tourists may be rather similar to the mechanics of sand on Titan (see Fig. 1.4).

The perspective afforded by this plot is all very well; the overall mobility of material is reasonably well conveyed. This analysis also forms the principal basis for assuming that Titan's sand dunes are made of particles of 0.3 mm diameter (they have not been measured!). For Titan and Venus, the approach above is probably about right even if the cohesion may need adjusting, since the atmospheres on these worlds are dense and so fluid stresses really determine the mobility. However, a vexing problem for some time is that the threshold winds it predicts are necessary to move sand on Mars are rather high, and yet sand (and even larger granules) sometimes move.

The resolution to this paradox was somewhat anticipated by Bagnold, who pointed out that the threshold under static conditions (where no saltating grains are impacting the surface) is higher than the threshold observed when active saltation is under way; the so-called impact threshold is about 80 % of the static threshold value. The difference between starting saltation and sustaining it turns out to be critical at Mars (e.g., Kok 2010) and relies on understanding what happens to grains once they do get moving.

## 4.5 Sand Grain Trajectories

The trajectory followed by a saltating sand grain is the result of the interplay between the various aerodynamic and inertial forces acting on the sand grain. In broad terms, the initial part of the trajectory is dominated by lift forces that raise the grain from the surface, but as the grain moves higher into the wind profile, it steadily picks up horizontal velocity obtained at the expense of the wind. Bagnold (1941) considered the initial movement of the sand grains when lifted off the surface to be essentially vertical, but high-speed photography of sand motion in wind tunnels (e.g., Fig. 4.11) has demonstrated that the average ejection angle ranged between 34° and 41° (from horizontal) for sand of from 90 to 300 μm in diameter (Table 1 of Nalpanis et al. 1993). It appears that even deep within the boundary layer of the wind, sand grains lifted from the surface very quickly acquire a horizontal velocity derived from the particle's interaction with the wind.

In the absence of forces other than gravity, the height achieved by a saltating sand grain is directly related to the square of the ejection velocity divided by twice the acceleration of gravity, a straightforward ballistic relationship, and the time of flight is just twice the vertical component of speed divided by gravity. Aerodynamic lift and drag can modify these times and heights somewhat (more for Venus/Titan than Earth/Mars), see below. The sand grain is accelerated horizontally by the drag induced by the wind, and if the grain climbs appreciably through the boundary layer, it experiences ever stronger winds.

The grain may attain a high fraction of the windspeed if it remains aloft long enough. One simple metric is the so-called 'drag length' $L_d$, the horizontal distance required to achieve this, which is written simply as $L_d \sim D_p(\rho_p/\rho_a)$. Bigger particles have more inertia ($D^3$) compared to their drag area ($D^2$), and so take further to accelerate. Similarly, thinner atmospheres and denser particles take longer to accelerate. Some numbers are indicated in Table 4.2; as we

**Table 4.2** Saltation parameters, as computed by Kok et al. (2012)

| Body | Threshold friction speed (m/s) | Ratio of impact to fluid threshold speed | Typical saltation height (cm) | Typical saltation length (cm) |
|---|---|---|---|---|
| Venus | ~0.02 | >1 | ~0.2 | ~1 |
| Earth | ~0.2 | ~0.8 | ~3 | ~30 |
| Mars | ~1.5 | ~0.1 | ~10 | ~100 |
| Titan | ~0.04 | >1 | ~0.8 | ~8 |

discuss in a following section, this parameter may be the principal determinant of the minimum size of sand dunes on any given planet.

In any case, in general a saltating grain will impact the surface both faster, and more shallowly than it was launched—the grain has extracted energy from the airstream. If this grain can then launch other grains, they too will accelerate and a cascading process will build up a saltating cloud of grains. This 'splash' process (see, e.g., Gordon and McKenna-Neuman 2011; Kok et al. 2012) depends somewhat on the particle properties: it may be that softer materials are less efficient at splash than hard rocks. It seems possible that the influence of impacting saltating grains may extend beyond that of the initiation of reptation motion through low-energy splash of surface grains from near the impact site, perhaps also inducing a wider range of ejection conditions than has been previously considered.

Many laboratory and theoretical studies have expanded upon Bagnold's initial studies, including consideration of saltation under conditions representative of present-day Mars (Greeley et al. 1974a; White 1979; Almeida et al. 2008) and Venus (White 1981; Greeley et al. 1984a). If the moving grain attains a high rotation rate, which is certainly possible from the many interactions taking place during saltation, then the grain trajectory may be significantly influenced by lifting forces induced through the Robins-Magnus effect (White and Schulz 1977) or electrostatic effects (e.g., Schmidt et al. (1998) show that typical charges of 60 µC/kg on saltating sand can modify the length of the saltation path by some 60 %).

Calculated saltation trajectories are highly dependent upon the many assumptions that go into any simulation scheme but, in general, the horizontal length attained by a saltating sand grain is much longer on Mars (meters) and shorter on Venus (centimeters) than is typical on Earth (~20 cm) (Fig. 4.12). The heights attained by the saltating grains scale roughly similar to the differences in saltation length noted above (Greeley and Iversen 1985, pp. 96–98).

Almeida et al. (2008) give the following expression for saltation length at friction speeds close to the threshold as

$$L_{salt} = 1100 \left(\mu^2/\rho_f^2 g\right)^{0.33} (u_* - u_{*t}) \Big/ \left(gD_p\right)^{0.5} \quad (4.6)$$

They determine that the height has the same form, but with a smaller prefactor (81 instead of 1100). This analysis and others (e.g., White 1979; Kok et al. 2012) have concluded that saltating sand on Mars quite likely achieves horizontal lengths between lift-off and return to the surface of from around one to some tens of meters. Note that Eq. 4.6 shows that the saltation path length varies with windspeed as well as other parameters—notably the path is shorter in fluid of higher density.

These simulations can also track how effectively the saltation cloud will build up (or not) as a function of windspeed. If the growth of the cloud is positive (a grain once launched into saltation will extract enough energy from the airstream, and pass that on via splash to other grains efficiently enough) then saltation can be self-sustaining. The windspeed needed to do this is termed the 'impact threshold' as opposed to the classical 'fluid threshold' in Sect. 4.4.

Curiously, this impact threshold (Fig. 4.13) is higher than the fluid threshold on Venus and Titan. On these worlds, somewhat like sand under water on Earth, grains are rapidly slowed by drag so splash does not help much. Thus the classic fluid threshold is what matters.

The situation for Earth seems to be somewhat borderline, with the impact threshold speed about 80 % of the fluid value, as suggested by Bagnold. But on Mars, the situation is dramatically different. Kok's simulations (2010) show that the wind speed required to maintain on-going saltation is lower by as much as an order of magnitude from the wind speed needed to initiate saltation. Thus, recent documentation of dune and ripple movement on Mars (Bridges et al. 2012a, b) may be more easily understood if the maintenance of saltation requires less wind than is needed to start the movement of the sand. Of course, like the now-proverbial flap of a butterfly's wings over Brazil, saltation needs to get kicked off somewhere by perhaps a meteorite impact or an exceptional gust, but this new understanding of the hysteresis in saltation and the distinction between impact and fluid thresholds may have resolved the puzzle about Mars.

## 4.6 Influence of Saltation on the Boundary Layer

Since energy is now being extracted from the airstream, this saltating cloud itself modifies the boundary layer wind profile. Once again, Bagnold (1941, pp. 57–61) pioneered the investigation of how a cloud of saltating sand grains affected the wind velocity profile above a mobile surface of loose sand. Measurements made by Bagnold in his wind tunnel showed him that the wind profile during fully developed saltation differed significantly from the wind profile above a sand surface before the initiation of sand

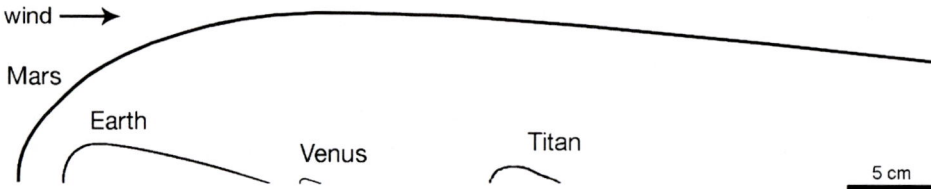

**Fig. 4.12** Schematic of typical saltation trajectories and scales on Venus, Earth, and Mars

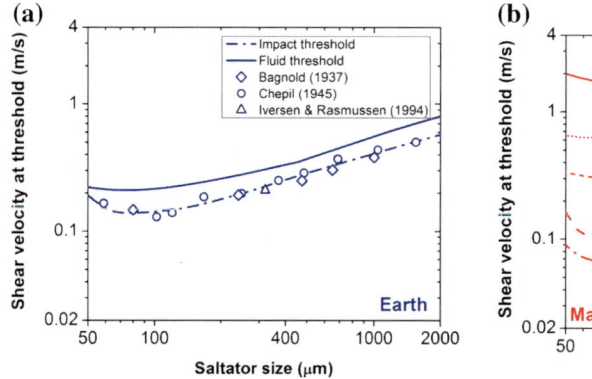

**Fig. 4.13** Fluid saltation thresholds (*solid lines*) on Earth and Mars. The *dash-dot lines* show the so-called impact threshold, the friction speed at which saltation can be sustained if it starts. For Earth (*left*), experimental data (symbols) support the idea that the effective (impact) threshold is about 20 % lower than the classic fluid threshold. For Mars (*right*), however, the impact threshold is a factor of several lower than the fluid threshold, perhaps explaining why bedforms are widespread on Mars. Note that for Venus and Titan (not shown) the impact threshold is higher than the fluid threshold, and is therefore not of interest. From Kok et al. (2012), with permission

motion (Fig. 4.14). Before the sand begins to move, the wind reaches zero velocity at height $z_o$, which is the roughness height (corresponding, in his opinion, to 1/30 of the mean height of natural surface irregularities, rather than the size of an individual sand grain). However, after saltation is well established, the wind velocity profiles has a new focus point that corresponds to a height much higher than k, and also a non-zero wind velocity. Bagnold called the new focus point z' (which he inferred to be somehow related to the height of ripples moving along the surface) and the velocity at that point $V_t$, the threshold velocity as measured at height z'. The wind velocity u at any height z above the ground during sand-driving can then be expressed in a modified version of Eq. 4.1:

$$u/u_* = 1/K \ \log(z/z') + V_t \quad (4.7)$$

The drag induced on the wind by the saltating particles thus alters the lower part of the boundary layer, but once above the new dynamic roughness height k' the wind once again follows its normal logarithmic behavior. An interesting attribute of this situation is that no matter how hard the wind blows, the boundary layer is modified so that the wind always decreases to a velocity of $V_t$ at height k'. Saltation thus establishes a new minimum height above which the wind can exert its influence on the surface.

Bagnold's measurements, both in the wind tunnel and in the field, showed roughness heights of $z_o = 0.002$ cm and z' = 0.3 cm for sand 0.25 mm in diameter.

Clearly, since the boundary layer affects saltating sand, and the saltating sand affects the boundary layer, simple algebraic models are challenged to make predictions in planetary environments. However, simulation tools (see Chap. 18) such as multiphase Computational Fluid Dynamics (CFD) are now able to model the processes accurately (e.g., Almeida et al. 2009) so considerable progress can be expected.

## 4.7 The Saturation Length and Controls on Dune Size

One consequence of the drag on the wind induced by sand-driving associated with saltation is to provide a way to estimate the minimum horizontal scale of a sand dune on Earth. Bagnold (1941) determined that the growth of the drag effect on the wind induced by saltation required from 6 to 7 m, measured from where saltation starts, for the drag effect to become fully developed, or if one fits an exponential curve, that the saltation responds over a length scale of $L_{sat} = 2.3$ m. He made measurements of where sand was either added or removed from various segments within his

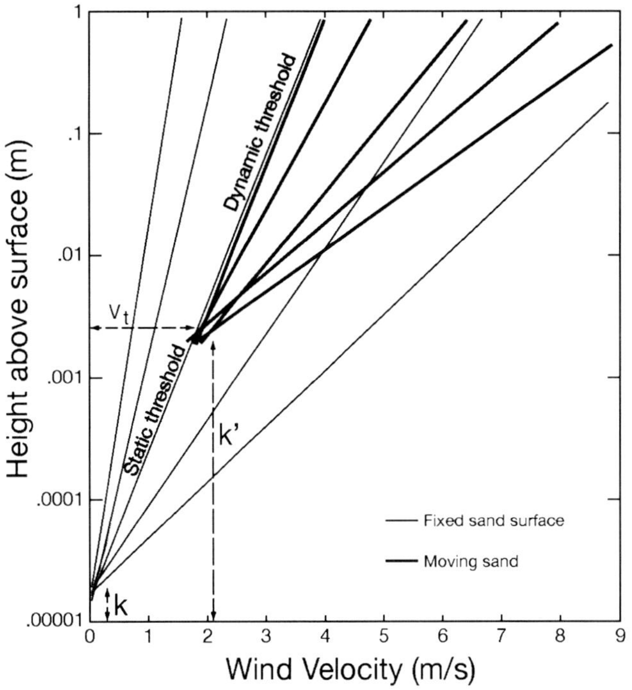

## 4.7 The Saturation Length

**Fig. 4.14** Wind velocity distributions for loose sand, both prior to the start of motion, and after sand-driving has become fully developed (modified from Fig. 18 of Bagnold 1941)

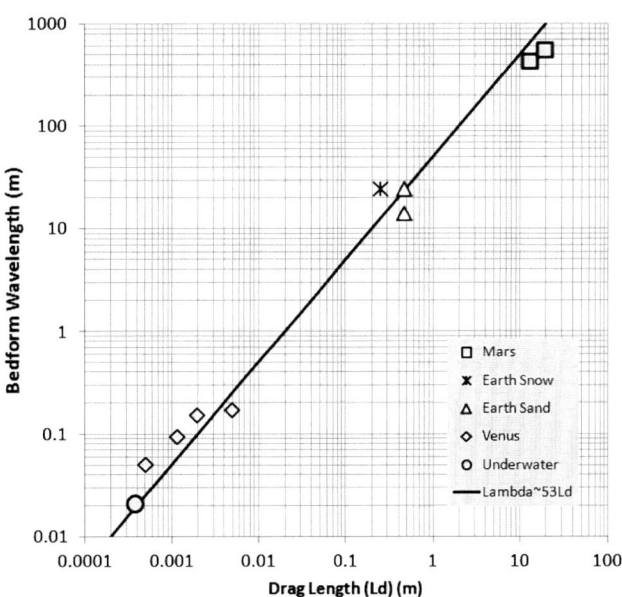

wind tunnel, and he was able to document that sand was removed from those areas where the drag effect was not fully developed while sand was deposited downwind of the place where the drag effect was fully developed. This led him to conclude that sand patches less than ~6 m in width will tend to lose sand, whereas sand patches larger than this size will tend to gain sand, gradually growing to the point of becoming a regular sand dune.

Andreotti et al. (2010) have measured the saturation length Lsat in careful wind tunnel experiments, and find a value of 0.55 m, essentially independent of wind speed. They suggest that Bagnold erred in that he estimated the removal rate (by weighing) averaging over distances comparable with the length they measured Lsat to be. Field measurements by Elbelrhiti et al. (2005) suggest $L_{sat}$ to be 1–1.7 m. These and other measurements seem to suggest Lsat being at least a little shorter than Bagnold's pioneering measurement. Andreotti et al. (2010) evaluate Lsat for sediments of different size and windspeed, and find that within 50 %, $L_{sat} \sim 2L_d$.

Claudin and Andreotti (2006) have argued that the drag length similarly (perhaps via the saturation length) defines the scale at which a flat sand bed will destabilize—essentially, the wavelength of perturbation that will grow the fastest and thus the size of 'typical' dunes that form first. Plotting (Fig. 4.15) the wavelength $\lambda$ of the smallest dunes in snow and sand on Earth, as well as dunes on Mars and seen in the Venus wind tunnel (see Chap. 14), they suggest a global scaling of

$$\lambda \sim 53L_{drag} \sim 53D_p \rho_a / \rho_p \quad (4.8)$$

**Fig. 4.15** The length scale of 'elemental' dunes in a wide range of fluid mechanical settings seems to relate to the drag length. Data from Claudin and Andreotti (2006)

This scaling suggests that, unlike on Mars, where the smallest 'proper' dunes (see next section) are many meters across, the smallest dunes on Titan will be tiny. However, once dunes form, they can grow, probably until they reach a wavelength roughly equal to the height of the planetary boundary layer (Sect. 3.2).

Equation 4.8 implies a scaling (for a given sediment size and density) with atmospheric density; this effect is nicely seen not only in the difference between typical ripples on

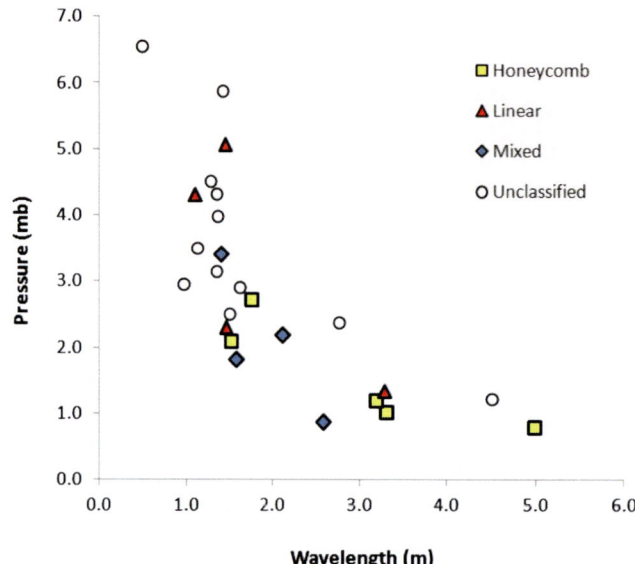

**Fig. 4.16** Bedform wavelength on the flanks of the Tharsis Montes on Mars, from Lorenz et al. (2014). These mountains span a sufficient range of elevation that the atmospheric pressure varies by a factor of almost 10. Although these bedforms are more likely to be ripples (controlled by the shallowness of the saltation path, see Chap. 5) than dunes (controlled by a saturation length), it is evident that there is an approximately reciprocal relationship between wavelength and atmospheric density (or pressure)—a trend that holds locally on Mars as here, but also between planets

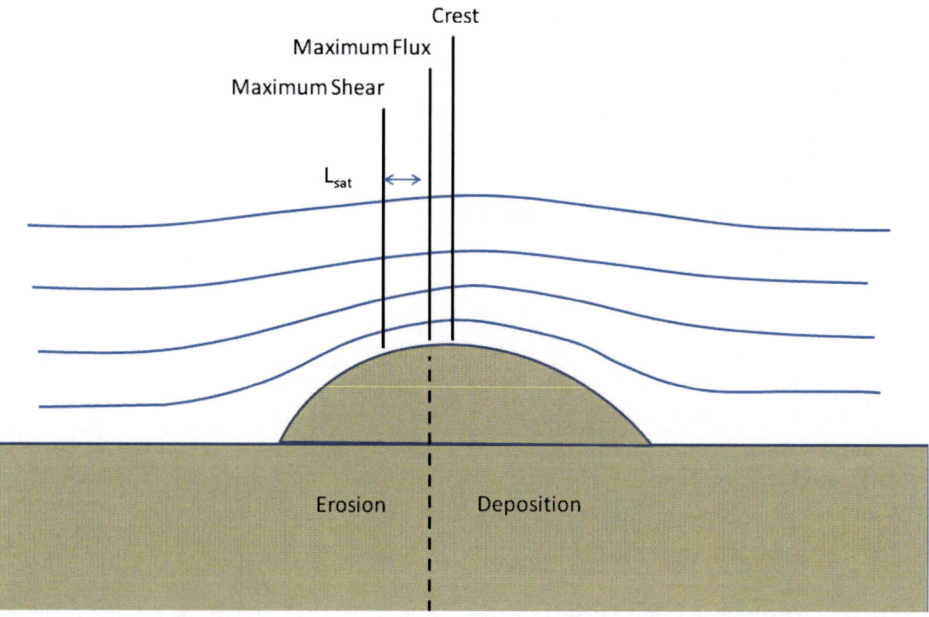

**Fig. 4.17** Flux instability criterion for bump growth. The streamline compression by a low hill like a dune causes the maximum stress $\tau_{max}$ to be upwind of the crest. If the peak flux ($q_{max}$), downwind of $\tau_{max}$ by a distance that scales with $L_{sat}$, is upwind of the crest and thus the crest is within the deposition zone, the dune will grow (as here), but if downwind, the dune will shrink. Thus a dune must be a critical size to grow

Mars and Earth, but also over a range of elevations (and thus atmospheric density) on Martian volcanos (Fig. 4.16). That said, the mechanism of formation of these bedforms is defined by the shallowness of the saltation path (see Sect. 5.4) rather than the drag length directly, although these quantities share many factors.

Although the detailed physics of the controlling influence of the saturation length (and drag length) on dunes is still being fully explored, it seems to be important. The influence can be visualized as follows: a dune acts as a gentle obstacle to the wind, which accelerates (usually by 10–100 %, depending on height and steepness) and thus

## 4.7 The Saturation Length

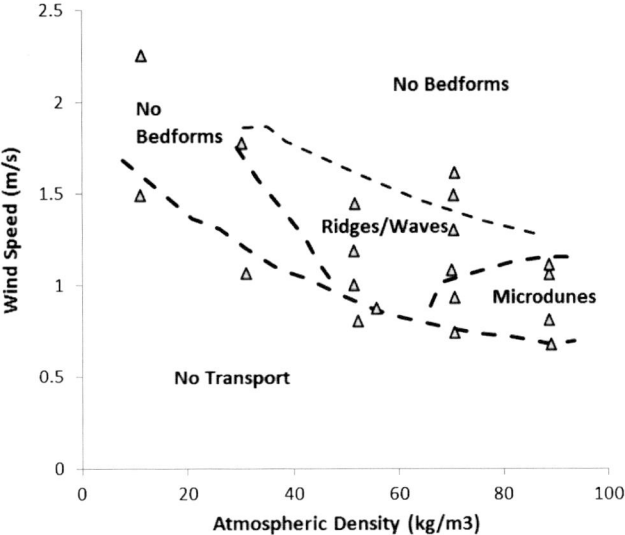

**Fig. 4.18** Parameter space of density and speed in the Venus wind tunnel as measured by Marshall et al. (1992). The triangles denote individual tunnel runs. At low speed/low density conditions, the sand (175–250 $\mu$) does not move, and at high speed/high density conditions, the bed stays plane with no bedforms. Only in a 'sweet spot' do ridge/wave bedforms with no slip faces, and microdunes (Fig. 4.19) form

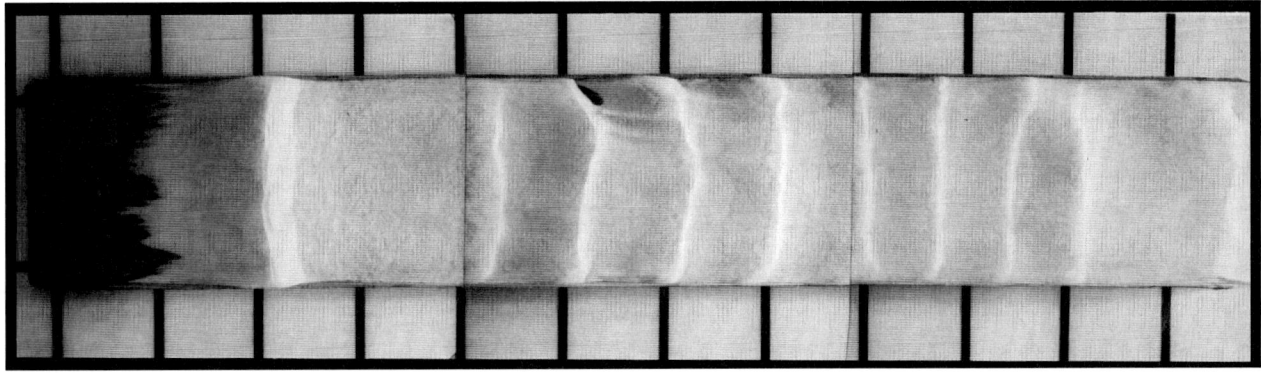

**Fig. 4.19** Periodic microdunes formed in quartz sand (50–200 $\mu$) in the working section of the NASA Venus wind tunnel at a speed of 0.79 m/s. Image courtesy of Ron Greeley

increases the wind stress moving up the stoss (upwind) slope. Because the stress depends on height and slope, the stress in fact peaks ($\tau_{max}$) before the highest point. Now, if the sand flux responded instantly to the wind stress, the maximum flux ($q_{max}$) would be upwind of the crest, and thus dunes would always grow, regardless of size. However, the lag ($L_{sat}$) in response means the peak sand flux is downwind of the peak—if it is downwind of the crest, then the crest will be eroded (see Fig. 4.17).

Bagnold (1941) suggested it would control the minimum size of dunes, and Kok et al. (2012) suggest $L_{sat}$ is 'the only relevant length scale in the physics of dunes', although that may be going a little far (not least since the atmospheric boundary layer thickness seems to be a fundamental limit to the growth of dunes). The modern physics perspective, able to simulate computationally a set of arbitrary flow conditions, and the trajectory of particles in such a flow (and their effect on it), might be profitably applied to experiments made with planetary wind tunnels. One example parameter sweep (Marshall and Greeley 1992) is in density/speed space (Fig. 4.18) in the Venus wind tunnel at NASA Ames (see Chap. 17) where bedforms appear (Fig. 4.19) in only a limited part of the space—too low speed and no transport happens, but too large a speed and bedforms are swept away.

The existence of a finite saturation length also implies an 'inertia' to turbulent changes in wind stress, which has important implications for transport overall and for field measurements in particular. Since it takes a finite time for the saltation cloud to respond to a change in windspeed, and the threshold of saltation is lower to maintain it than to initiate it, the time history of sand flux measured at the local scale—which can be highly intermittent (e.g., Stout and Zobeck 1997) has a somewhat complex relationship to the wind history. High-time-resolution measurements (e.g., Jackson and McCloskey 1997) show a delay of a few seconds between drops in wind speed and the resultant drop in sand flux (Fig. 3.11).

Having discussed how sand can move, and what factors control how far it goes, we now proceed to discuss the shapes of the resultant landforms and how they relate to the wind regime, and how wind interacts with the dune structure.

# 5. Deposits of Sand: Ripples Versus Dunes

We have seen that sediment can form mounds on the surface and that these mounds can grow through the instability associated with the saturation length. However, several effects lead to more complex structures—the recirculation bubble of airflow, the avalanching slip face of the dune, and the possibility that multiple particle sizes or densities may be present. These lead to some distinct types of structures—ripples rather than dunes. The way sediment is deposited can also leave traces in the dune (or in a sandstone if the dune becomes lithified), or on the surface on which the dune sits.

## 5.1 Flow Separation and Recirculation

First, airflow going around a surface will try to follow it. However, like tea poured from the spout of a well-designed teapot, if the surface peels away too sharply, the airflow cannot follow it and detaches from it. This flow separation means the lee side of the dune can be rather sheltered, experiencing much less wind stress. Often there is a recirculating flow behind the dune, in the so-called separation bubble, before the airflow re-attaches downwind. This recirculating vortex leads to a reverse flow on the lee side, which can sweep sand back towards the dune. This recirculating vortex and its role in dune dynamics was recognized over a century ago (Cornish 1897).

This recirculation can be studied with field instruments, or visualized with smoke or streamers (e.g., Fig. 5.1), or sometimes shown by the sand itself. It has also been studied in wind tunnels and with computer simulations (Fig. 5.2; Schatz and Hermann 2005; Hermann et al. 2005; see also Chap. 19). Occasionally, the sand itself acts (imperfectly) to visualize the flow, as saltating sand flies off the crest of a saltating dune (Fig. 5.3). Typically, the reattachment point is downwind of the dune toe by 4–6 times the dune height, although a wider range may be encountered depending on the surface roughness upwind, and the sharpness of the separation point.

The recirculating flow can sweep sediment back towards the slip face of the dune if the floor is smooth. The separation bubble can strongly influence how dunes mutually interact, notably in the case of barchans—once the downwind barchan enters the lee of the upwind one, the airflow on it will be considerably disturbed.

## 5.2 Angle of Repose: The slip face

The second effect involves the sand, whether or not an atmosphere is present. Sand generally has small cohesion, and the sideways forces at sand–sand contact surfaces is due to friction alone. This means that a pile of sand will tend to have slopes that do not exceed the angle of repose—in fact, when granular materials, whether sand, gravel, cobbles, or boulders, come to rest on or adjacent to comparable particles, they tend to form piles with uniform slopes. If sand is added to the top of a pile or slope, it may steepen initially, but soon an avalanche will occur, where the weight of sand has overcome friction and allows the sand to move until the slope has shallowed to the angle of repose once more. The progressive deposition of sand on the lee side of a dune means that the lee side sees continual avalanching and is generally at the angle of repose.

For well-rounded, dry sand grains, the angle of repose is around 33°. Accumulations of irregular coarse particles (pebbles or cobbles) can attain an angle of repose of 41°, but such a slope is not typical for sand accumulations that do not also include a considerable quantity of silt or clay. The strength of the acceleration of gravity is not a prominent factor in determining the angle of repose, which is more dependent upon systematic trends in particle shape or particle surface roughness. Since the angle of repose is weakly diagnostic of grain shape, constraints on the latter

**Fig. 5.1** Streamers on a linear dune (no slip face) at the Dumont dunes, California. Wind is blowing from *left* to *right*, as shown by streamers on sticks at stations 1–4 (stakes are about 70 cm high). The strong *left–right* wind resumes at stations 9 and 10 where the flow has re-attached. At stations 5–8, the streamers droop, indicating low wind, and at stations 6 and 7 indicate an uphill flow due to the recirculating vortex in the lee of the dune. *Photo* R Lorenz

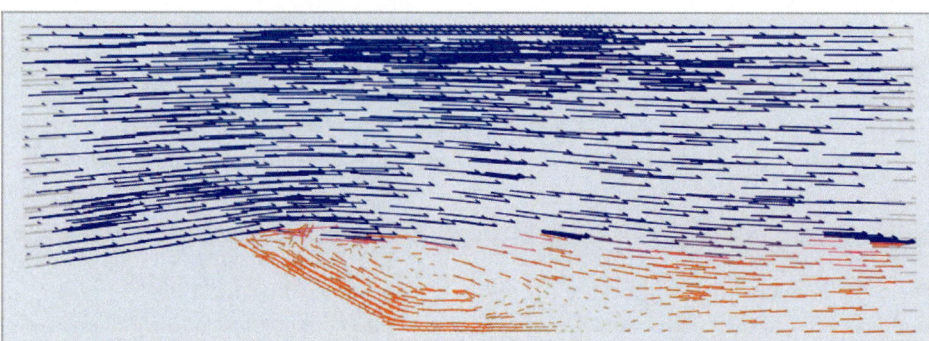

**Fig. 5.2** Airflow vectors over a transverse dune, simulated using computational fluid dynamics (CFD—see Chap. 19) showing the flow separation at the crest and the recirculating vortex in the lee. (From Schatz and Hermann 2004, image courtesy of Hans Jurgen Herrmann.)

can be derived by measuring the angle, which is readily done by drizzling sand to form a conical pile. This experiment was performed (e.g., Moore 1987) by the sampling arm of the Viking lander on Mars in 1976 (see Fig. 5.4). Experimental studies have demonstrated that changing the strength of the acceleration of gravity has a minimal to non-detectable effect upon a dynamic angle of repose, as is the case within a rotating drum.

The grains on the slip face are usually loosely packed and thus have little cohesive strength. The combination of this weak packing and the steep slopes makes it difficult to climb the slip face on foot or in vehicles (see Sect. 1.4). On the other hand, it is easy to slide down—dune skiing and surfing is possible (Fig. 5.5).

Because the entire slip face is essentially at the avalanching threshold, it is a 'critical' system wherein avalanches can take place over the whole size range from a single grain

**Fig. 5.3** 'Smoking dune': wind from *left* to *right* is launching sand which initially follows the separating streamline at the crest, but drizzles down in clouds onto the lee side. Lee deposits are thus often a layered combination of these airfall deposits and avalanche flows. See also Fig. 12.13. Dune near Palen: *Photo* J. Zimbelman

**Fig. 5.4** Small conical piles of dirt (*arrowed*) created by the Viking Lander 1 sampling arm. The pile was made to estimate the angle of repose of the sediment, and to observe any changes. The image on the *left* was taken on Sol 921; the composite image on the *right* (where the sandpiles have been blown away) was assembled from images taken on Sols 2068 and 2209. This excellent comparison was assembled by Phil Stooke

**Fig. 5.5** The slip face of a dune can be steep and loose enough to permit rapid descent, as in this case (Jani Radebaugh dune-surfing on a megabarchan near Liwa in the United Arab Emirates). However, the friction is still rather higher than for snow, and the experience is not always totally exhilarating. *Photo* R. Lorenz

dropping one grain width, to the entire slip face sliding. It is found, both in reality and with computer simulations (see Chap. 19) that the relative frequency of avalanches varies as an inverse power law, with avalanches that are 10 times bigger occurring 10 times less often. The same sort of power law statistics are seen with earthquakes, forest fires, mountain avalanches and so on, and are studied under the physics umbrella of 'Self-Organized Criticality', e.g., Bak (1999).

**Fig. 5.6** Lobe-shaped avalanches on the slipface of a barchan dune in the United Arab Emirates. The slip face is about 2 m high. The alcoves, with vertical gouges, from which the avalanching sand started are clearly visible—evidently some sorting results in the surface sand having a slightly different color. *Photo* Jani Radebaugh

Individual avalanches typically occur over fairly narrow spans of a dune, forming a narrow lobe (see Fig. 5.6). Avalanching is sometimes accompanied by a hissing sound, or sometimes much louder sounds and seismic vibrations (see Chap. 10: Booming dunes). Such avalanche lobes have been observed on Mars (e.g., Horgan et al. 2012; see Fig. 5.7).

## 5.3 Grain Sorting

A final issue is that sand can often be a mixture of materials or particle sizes. Since these are transported with different effectiveness by the processes of saltation and reputation for any given windspeed, they can be segregated, leading to different concentrations of them on different parts of an aeolian bedform. It is, for example, easy to observe that the crests of dunes often have coarser sand. The sorting is most evident on ripples, which often form (see Sect. 5.4) when fine sediment can be armoured by coarser grains. The coarse grains tend to accumulate on the upwind face, and especially the crest, of the ripples (e.g., Fig. 5.8). Exactly the same phenomenon is observed on Mars (e.g., Figs. 5.9 and 5.10)—see also Jerolmack et al. (2006).

The preferential accumulation is due to the reduced mobility of the larger particles. Similar differential mobility can result from grain size or the density of material from which the grain is made. This is often noticeable in terrestrial dunes, where the iron mineral magnetite is both appreciably denser than the quartz that makes up the bulk of the dune, but is also much darker and so accumulations are readily visible (Fig. 5.11). In the Namib sand sea, dark patches seen on dunes are concentrations of not only magnetite, but also ilmenite and garnet (e.g., Schneider 2008).

## 5.4 Dunes Versus Ripples

What is the difference between a ripple and a dune? Typically, but not always, there is a scale difference—ripples are small (e.g., Fig. 5.12). But the largest ripples on Earth can be as large as small dunes, and the distinction may blur even further on other worlds. Ripples do not have slip faces (and although no bigger than terrestrial ripples, it was the presence of slip faces on dunes formed under Venus-like conditions in a wind tunnel that led to the bedforms being called 'microdunes'—see Chap 14). However, since many dunes also lack slip faces, this is not a discriminator either. The challenge here is that the different terms are really genetic rather than purely morphologically descriptive. Both features are the result of sand grains being transported by the wind, but the processes involved in the formation of these two features are quite different. To avoid falling into the trap of misidentifying features on Mars without knowing

## 5.4 Dunes Versus Ripples

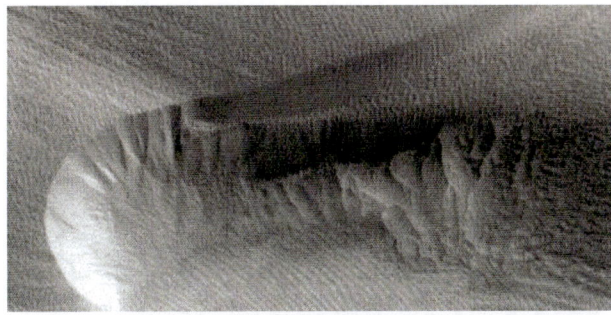

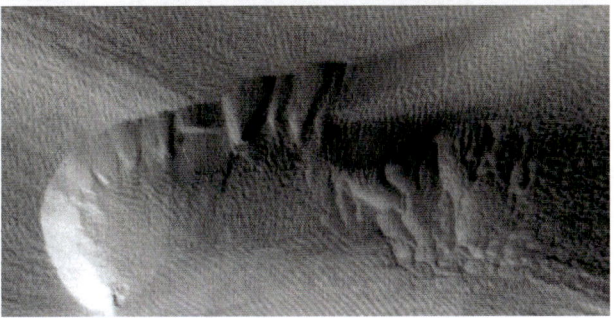

**Fig. 5.7** Mars dune avalanching. Three images of the same location (taken by the High Resolution Imaging Science Experiment (HiRISE) camera on NASA's Mars Reconnaissance Orbiter) taken at different times on Mars show seasonal activity causing sand avalanches and ripple changes on a Martian dune. Time sequence of the images progresses from *top* to *bottom*. Each image covers an area 285 × 140 m. The crest of a dune curves across the *upper* and *left* portions of the image. The site is at 84° north latitude, 233° east longitude, in a vast region of dunes at the edge of Mars' north polar ice cap. The area is covered by carbon-dioxide ice in winter but is ice-free in summer. The *top* and *bottom* images show part of one dune about one Mars year apart, at a time of year when all the seasonal ice has disappeared: in late spring of one year (*top*) and early summer of the following year (*bottom*). The middle image is from the second year's mid-spring, when the region was still covered by seasonal carbon-dioxide ice

for sure how they formed, a new (if ungainly) morphological term was introduced: Transverse Aeolian Ridges (TARs—Fig. 5.13). To help better understand why a ripple is not just a very small dune, we must go back to what several researchers had to say about both ripples and dunes.

Bagnold (1941, p. 62) described a 'characteristic path' for well-sorted, wind-blown sand grains, which he argued was the result of momentum removed from the wind at the height reached by most of the saltating grains (see discussion above regarding wind profiles during saltation). He further argued that the wavelength of wind ripples (typically ∼15 cm for simple ripples in fine sand, Fig. 5.12) is therefore the physical manifestation of the characteristic path of the wind-driven sand.

Sharp (1963) inferred that because aerodynamic (formed by the wind, as opposed to ripples formed by flowing water) ripples are observed to start off as small-amplitude, short-wavelength irregularities in a sand surface, which subsequently increase in wavelength as they evolve toward a steady state, Bagnold's 'characteristic path' concept for a wind ripple wavelength was suspect. Through geometric arguments, Sharp (1963) proposed the alternative concept that ripple wavelength depends on ripple amplitude and the angle at which the saltating grains approach the sand bed, so that ripple wavelength is controlled by the length of the zone shadowed (Fig. 5.14) downwind of the ripple from significant sand grain impacts, and also by the wind velocity that is the cause of the saltation action.

Anderson (1987) developed an analytical model for the initiation of sand ripples that result from the growth of perturbations on an initially flat sand surface. Using the 'splash function' concept derived from wind tunnel experiments by Unger and Haff (1987), Anderson (1987) modeled the wind tunnel measurements of relatively low-velocity ejection of grains caused by the impact of faster-moving saltating grains; he termed this splash process as 'reptation', from Latin for 'to crawl', to distinguish this mode of sand transport from saltation, suspension, and impact creep. Subsequent numerical modeling including both saltation and reptation (Anderson and Haff 1988) showed that small, fast-moving ripples overtake and merge with larger, slower ripples, resulting in the growth of the mean wavelength for the entire ripple field, along with a decline in the dispersion of wavelengths (that is, the observed range of wavelengths decreases as the ripple field evolves toward a steady state). Both reptation and saltation have been incorporated into subsequent continuum analytical models developed for the study of aeolian features on Earth and Mars (e.g., Momiji et al. 2000; Yizhaq et al. 2004; Yizhaq 2005), and numerical models (e.g., Landry and Werner 1994).

Bagnold (1941) described some important distinctions between ripple and dune formation processes: dunes usually display a slip face dominated by avalanche processes; ripples have the coarsest material collected at the crests with the finest material in the troughs, whereas the opposite is true for dunes. When aeolian sand is poorly sorted (that is, a wide range of particle sizes is present), the coarse grains become concentrated at the crests of large features that Bagnold (1941, pp. 153–156) termed 'ridges' rather than ripples, arguing that the coarse grains are moved by the impact of saltating sand. Other terms have been used to

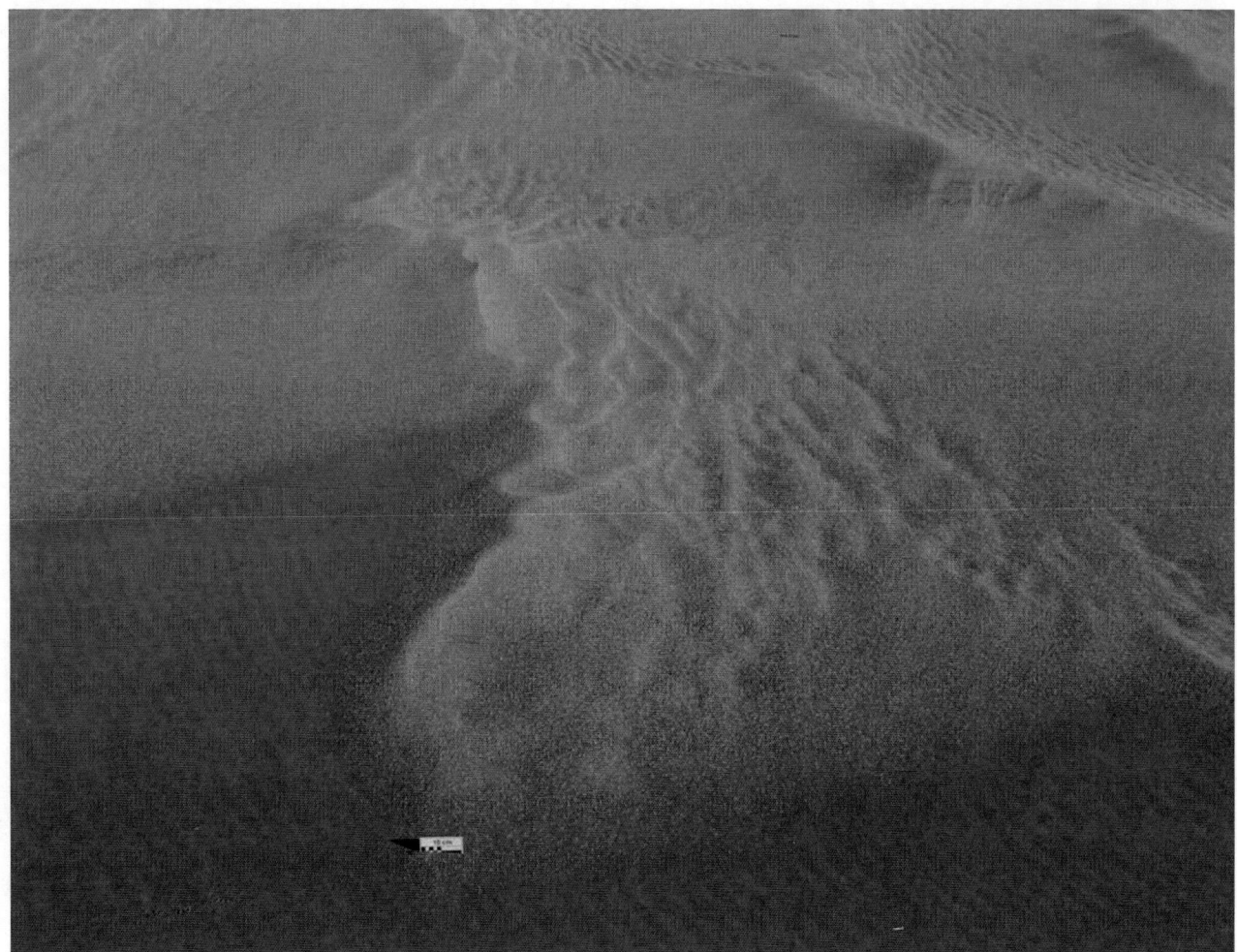

**Fig. 5.8** Granule megaripples, with smaller accumulations of coarse granules on lee of each megaripple, Great Sand Dunes National Park and Preserve. Megaripples are coated by mm-sized granule particles. Normal aerodynamic sand ripples are encroaching onto the megaripples, from *lower left* in this view. Scale is 10 cm long. *Photo* J. Zimbelman

describe large wind ripples coated by coarse particles, but the term 'megaripple' is perhaps the one carrying the least interpretive 'baggage' with its use.

Sharp (1963) described 'granule ripples' as consisting of a surface concentration of grains >1 mm (Fig. 5.7), roughly equivalent to Bagnold's 'ridges'. Sharp concurred with Bagnold that impact creep, accompanied by significant deflation of the sand surface, was the dominant process involved in the formation of granule ripples. Ripples coated by coarse-grained particles can grow to great size; wavelengths of 20 m and heights of 60 cm are reported by Bagnold (1941, p. 155), and recently wavelengths of 43 m and heights of >2 m are reported for gravel-covered ripples in Argentina (Milana 2009; de Silva et al. 2011; see Fig. 5.15). More typically, granule ripples have wavelengths in the range of ~3 to 10 m and heights of ~25 to 40 cm (Sharp 1963; Zimbelman and Williams 2006; Zimbelman et al. 2012; see Fig. 5.7).

The coating of coarse particles, whether granules or gravel, tends to be concentrated at the surface of the bedform, usually as a monolayer of the larger particles, with the interior of the bedform comprised primarily of sand and only occasional coarse particles (Fig. 5.16). For granule ripples, the granules are most thickly accumulated at the crest of the ripple (Sharp 1963).

Profiles measured perpendicular to the crest of the features turn out to be valuable for distinguishing between aerodynamic sand ripples, megaripples, and sand dunes. Ripple profiles can be measured in the field from photographs (see Chap. 16)—a neat field trick is to use a straight edge to cast a shadow on the ripple surface. The profile shapes are distinct for each of the three classes of

**Fig. 5.9** Navcam image from the Opportunity rover of a small impact crater ('Rayleigh') which cuts into a ripple, neatly showing the cross-section of the ripple and exposing the angle of the layers in it. The intersection of the layers with the ripple surface can be seen behind the crater, and on the ripple to the *right*. *Image* NASA

features; when both the horizontal and vertical dimension are scaled by the width of the feature, profile shapes can be compared directly even for features that differ in scale by more than three orders of magnitude (Zimbelman et al. 2012; see Fig. 5.17). Profile shape can be assessed from remote sensing imaging data (see Chap. 18), and if combined with an assessment of the dominant particle size associated with a particular aeolian bedform, the probable formational mechanism involved in the development of both large ripples and small dunes should be identifiable even when detailed field examination may not be possible. Balme et al. (2008) show that at least some TARs observed on Mars have identical shapes to some wind ripples on Earth.

## 5.5 Controls on Feature Scale

We have discussed various factors controlling the scale of aeolian features: it is worth a brief recapitulation at this point. Ripples are defined by the shadow zone of saltating particles—it is somewhat similar to but not quite the same as the saltation path length. The ripple wavelength therefore relates to the windspeed (as demonstrated in wind tunnel experiments, e.g., Seppala and Linde 1978) as well as the particle size and other factors. The largest ripples form when the grains are coarse and thus can only be moved by the fastest winds, giving the longest saltation path length. Similarly, the long, shallow trajectories of grains on Mars yield large ripples or TARs. On the other hand, the saltation paths are short enough on Venus and Titan (Kok et al. 2012) that ripples may be insignificantly small.

The smallest ('elemental') dunes, on the other hand, are defined by the saturation length, which scales with the drag length. This in turn depends on the particle size, and the particle and atmospheric density, but does not depend on the windspeed. Dunes can grow in size, both by direct grain accumulation, and by coalescence with other dunes. However, dune growth ceases when the dune is large enough to 'feel' the top of the planetary boundary layer, which happens roughly when the dune spacing (typically six times the dune height) approaches the boundary layer thickness. Again, there is a consistent progression in size from Venus, through Titan then Earth, to Mars with the largest elemental

**Fig. 5.10** Portion of a Curiosity MastCam image of a ripple surface, after a sample was scooped out. Note abundant fine brown sand exposed beneath mm-sized granule coating. Scoop is 4.5 cm wide. *Image* NASA/JPL

dunes. The same progression, more or less, arises for the largest dunes. The root cause is the same (atmospheric density), although the pathway is a little different: denser atmospheres take more heat to warm up by sunlight, so the thinnest atmosphere (Mars) has the thickest boundary layer and thus the largest dunes. The boundary layer depends on the thermal inertia of the surface, and on the length of the day, etc., so one could imagine a selection of exoplanets where the order of elemental dune size actually differs from the largest dune size.

Normal aerodynamic sand ripples are the smallest members in a hierarchy of aeolian landforms. Wilson (1972a, b) documented three distinct scales for aeolian landforms: ripples (wavelength 0.01–10 m), dunes (wavelength $\sim$10–500 m), and draas (wavelength $\sim$0.7–5.5 km) (Fig. 5.18). Note that Wilson's work includes both (small) sand ripples and (large) granule-coated ripples within the features he called 'ripples'; we will further explore this distinction shortly. A region of overlap exists between the measured dimensions of small sand dunes and large ripples, but the average sediment size of a 10 m dune is typically <0.2 mm while surface sediments on a ripple of 10 m wavelength are >1 mm, and importantly there are no observed transitional features between these two groups (Wilson 1972a). Therefore, particle size is an important characteristic of which class a 10 m-scale aeolian landform belongs to. Interestingly, the 10 m size of both large ripples and small dunes is comparable to the smallest aeolian features originally identified as TARs on Mars (Wilson and Zimbelman 2004). Features at all three length scales can be present at one location at the same time, but they are the result of differing wind intensities and durations (Wilson 1972b); this interpretation is consistent with the concepts presented by Sharp (1963), who stated that each aeolian bedform reaches its own quasi-equilibrium state; i.e., ripples do not grow to become dunes, nor do dunes grow to become draas (The Arabic word 'draa' means 'arm' and is often used to refer to megadunes, particularly linear ones, but also to compound dunes. Because a widely-accepted formal definition is lacking, we have tended not to use it).

## 5.6 Sedimentation and Dune Structure

The different modes of particle motion (suspension, saltation, reptation, creep) all deposit particles onto the surface in different ways. Suspended dust particles generally form a nearly uniform deposit without obvious layering. Large

## 5.6 Sedimentation and Dune Structure

**Fig. 5.11** Compositional sorting. Eureka dunes, north of Death Valley National Park, showing concentrations of dense dark grains of magnetite. Thin lamellae of magnetite are exposed as rather pretty curves and loops in the slip face of the dune at *left*, with some patches of magnetite in the background. At *right*, differential concentration of magnetite grains on the sides of aeolian ripples on the dunes shows them much more starkly. *Upper right* inset: while jabbing eyeglasses into sand dunes is not generally a good idea, these ones have small magnets, which show clearly the magnetic nature of the dark grains. Photo by R. Lorenz

**Fig. 5.12** Vertical view of typical aerodynamic ripples, coral pink sand dunes, State Park, Utah. Ripple wavelength is ∼15 cm. Notice that while the pattern is quite regular, there are some defects forming Y-junctions (see Chap. 19). *Photo* J. Zimbelman

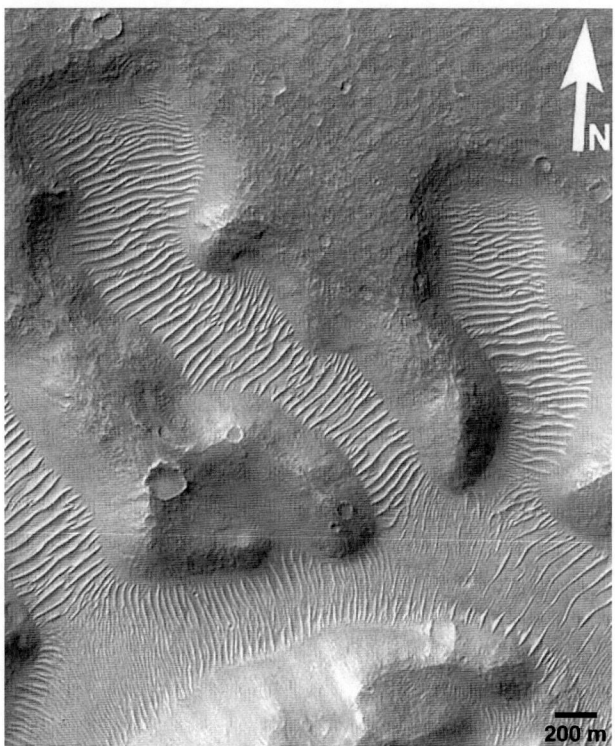

**Fig. 5.13** Ubiquitous TARs seen on Mars. From the image alone, it is possible to infer that these are ridges and they are transverse (strictly the 'aeolian' in TAR is an assumption, though it seems safe to expect that this valley system has not seen water flowing for some considerable time). In all probability these are ripples in genetic terms, but TAR is a safer term. They are appreciably larger than typical ripples on Earth. MOC image (E02-02561) *Credit* NASA/JPL/MSSS

particles migrating via creep can accumulate within a sand deposit at locations that represent a former bedform, although stringers of coarse particles preserved within aeolian sandstone (in contrast to sandstone derived from sediment deposited by a river or other fluvial process) are surprisingly rare in the terrestrial geologic record (Sullivan et al. 2012). However, saltation and avalanching in dunes and ripples can lead to magnificent subsurface structures.

On the stoss side of a dune, grains are impacted onto the upward-sloping surface, and the grains tend to be packed (making it easier to walk on the stoss side). The wind on the sheltered lee side of the dune is weak, and the sand grains tend to be dropped by the airflow. The lee side of a dune under sand-driving winds experiences almost continuous slope failure in the form of numerous avalanches as discussed earlier.

Different episodes of deposition (e.g., seasonally-dependent winds) can result in distinct layers—usually distinct by means of different compaction and/or particle size (see pictures earlier in this section). Makse (2000) shows with a numerical model how cross-bedded layers can form due to the sorting by differential hop-length and rolling of different particle sizes or compositions. Dune slip faces can result in distinctively layered sand deposits (Fig. 5.19), some layering of which may represent the advancing slip face of an active dune. Traditionally, the layered structure of dunes was explored by sectioning or trenching the dune, sometimes with a bulldozer, although ground-penetrating radar can sometimes reveal such layering with less effort (see Chap. 18). In some circumstances the positions of former slip faces are exposed in the top surface of the dune, revealed either by cementation (Fig. 5.20) or by sorting (Fig. 5.21).

David Rubin of the U.S. Geological Survey has comprehensively mapped out via simulation how a growing layer of sand might record the style of deposition. In particular, when a repeating sequence of winds generates alternating episodes of slip face avalanching and ripple movement, a 3-dimensional record of layers is left in the resultant sandstone (Fig. 5.22). Usually only one, but sometimes more than one, section of this 3-dimensional pattern is exposed, in which case the interpretation can be non-unique. Rubin and Carter's (2006) animations of the growing pattern are a mesmerizing introduction to the richness of the patterns that can be found. The striking cross-bedded sandstones in Southern Utah (Fig. 5.23) are perhaps among the best-known examples of this aeolian deposition and lithification process. Some magnificent cross-bedding can also be seen in the sandstone into which the remarkable ancient city of Petra in modern Jordan was hewn: planetary geologists have been known to get more excited by its sedimentology than its archaeology!

The advent of very high resolution imaging from Mars rovers has now enabled similar cross-bedding textures to be found on that planet. The Opportunity rover studied Victoria Crater, a 750 m-diameter crater in Meridiani Planum, from 2004 to 2008 and observed stratified sedimentary rocks in the crater walls. As explored by Hayes et al. (2011) these layers chronicle the paleo-environments of Terra Meridiani and provide glimpses into the broader history of early Mars. The stratigraphy at Victoria Crater (Fig. 5.25) includes the best examples of meter-scale cross-bedding observed on Mars to date (Fig. 5.24). The Cape St. Mary promontory is characterized by meter-scale trough-style cross bedding, suggesting sinuous-crested dunes with scour pits migrating perpendicular to the outcrop face.

Finally, another deposit type can form in certain cases, where the sand is easily cemented. This is particularly the case for gypsum. Rain can cement gypsum sands to form a resistant layer: this will prevent movement for some time until the layer is broken up. However, where the layer intersects the substrate surface it may be preferentially preserved, leaving a surface scar that marks the perimeter of the dune. These can be seen in some instances at White

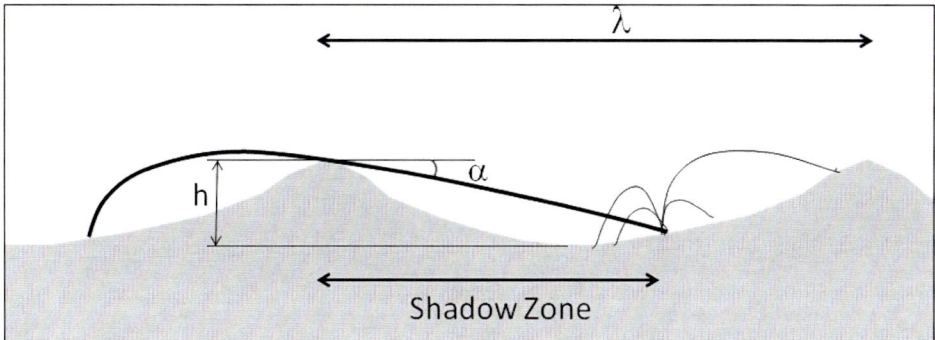

**Fig. 5.14** Shadow definition of ripple wavelength $\lambda$. For a given windspeed, the impact angle $\alpha$ will be roughly constant, and so will the aspect ratio $h/\lambda$: over time, ripples may grow in both height and wavelength. In faster winds (e.g., on Mars), the angle $\alpha$ will be shallower, and longer wavelength bedforms result. The *solid line* shows a saltating particle (note that $\lambda$ is defined in part by the length of the shadow zone and is related to, but not equal to, the saltation pathlength). The *thin curves* show typical 'splashed' particle trajectories

**Fig. 5.15** Some of the largest ripples on Earth: the pumice gravel-coated megaripples, Perulla, Argentina. See Milana (2009) and de Silva et al. (2012) for further discussion of these megaripples. Shan da Silva for scale. *Photo* J. Zimbelman

**Fig. 5.16** Concentration of cm-sized pumice clasts on the surface of a megaripple, Purulla, Argentina. The interior of the megaripple consists almost exclusively of medium sand derived from several ignimbrites exposed in the area; here particles from on the crest consistently fell onto the exposed trough dug across the ripple crest. Scale is in cm. *Photo* J. Zimbelman

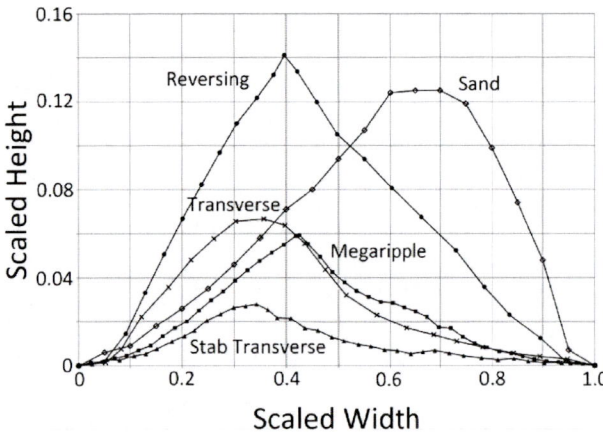

**Fig. 5.17** Scaled profiles of different types of ripple (after Fig. 6 of Zimbelman et al. 2012)

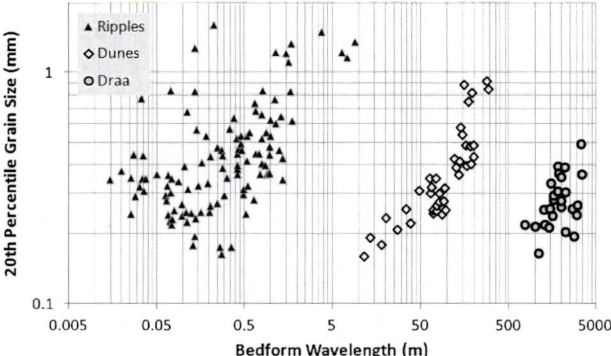

**Fig. 5.18** Plot demonstrating three distinct scales of aeolian features showing data from Wilson (1972a). Points represent measurements of features from around the world—the three sets of points are quite distinct. *Vertical axis* is the grain size of the 20th percentile of particles from the feature

## 5.6 Sedimentation and Dune Structure

**Fig. 5.19** Crossbedded layers in a dune exposed by digging: note the near-horizontal layers at *top*, and dipping slipface layers below. Ruler is 15 cm long (see also the boots at *upper right* for scale). *White* Sands National Monument, at the first Planetary Dunes workshop in 2008. Photo courtesy of Jani Radebaugh

**Fig. 5.20** Exposed cemented layers of a former slipface on a barchan near the Salton Sea (visible at *top left*) in Southern California. View is looking along one arm of the barchan, which has crossed and blocked a small road. *Photo* R. Lorenz

**Fig. 5.21** Exposed layers in the damp sand of Lencois Maranhenses (a lagoon is just visible at *top left*) showing the arcuate former positions of the slipface of the barchanoid dunes. Jani Radebaugh for scale. Some wavy elemental dunes are forming in the background as the bright sand starts to dry and organize. *Photo* R. Lorenz

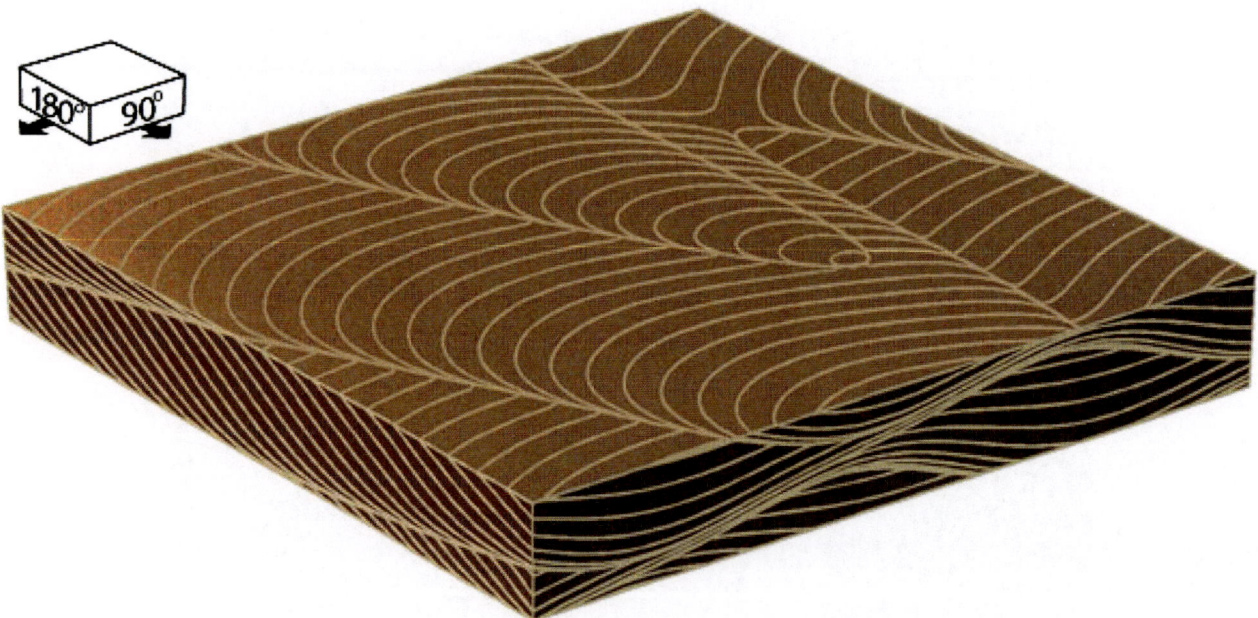

**Fig. 5.22** Simulations of bed formation by Dave Rubin of the USGS show that the bedding pattern contains rich information that maps to the sequence of deposition styles. In this example, dunes form layers with additional complexity from along-crest migrating superimposed bedforms. The bedforms have a height ratio of 0.45 and a speed ratio of 0.3. The resulting ratio of along- to across-crest transport is 0.135; the transport direction is oriented 82° from the crestline of the main bedforms. *Image* USGS

## 5.6 Sedimentation and Dune Structure

**Fig. 5.23** Cross-bedded Jurrasic sandstone in Zion National Park, Utah. Fine-scale layers represent sand deposited upon a sand dune, while the larger sets of layers, ranging from 2 to 12 m in height, record the passage of multiple sets of sand dunes. *Photo* J. Zimbelman

Sands National Monument (Fig. 5.26). Recently, high-albedo features (Fig. 5.27) seen at an equatorial dune field on Mars have been interpreted (Gardin et al. 2011) to be analogous to these White Sands 'barchan scars'. This is noteworthy since it would imply the existence of liquid water (perhaps from the subsurface) in amounts and for durations enough to cement the sand at the foot of the slipface.

## 5.7 Niveo-Aeolian Processes

An interesting complication can occur with sand dunes in cold regions on Earth in that both snow and sand can be present. Once snow falls under conditions that it is cold enough not to be sticky, it can be saltated, perhaps mixed with sand, forming intricate multiple layers (Fig. 5.28). Subsequently, further saltation may bury the snow under more sand.

The burial of snow by sand can both insulate the snow from heat, and retard the diffusion of moisture, allowing the snow to persist inside the dune. Depending on the temperature history, the snow may melt, moistening the sand and reducing its mobility (Fig. 5.29). This moisture may then refreeze, forming a very hard, cemented layer. On the other hand, if the dune accumulates rapidly, the snow may be preserved for long periods with no melting at all, and very slow sublimation if any. One example, in the Killpecker dune field in Wyoming, showed 25 cm-thick lenses of snow 1.5 m beneath the surface of a dune, some 4 months after

**Fig. 5.24** Crossbedding on Mars. Opportunity mosaic of the Cape St. Vincent promontory on the edge of Victoria crater in Meridiani Planum, showing impressive layers in aeolian sands. Image courtesy of Alexander Hayes

**Fig. 5.25** A close-up of more layers in the walls of Victoria crater, this time at Cape St. Mary. Image courtesy of Alexander Hayes

## 5.7 Niveo-Aeolian Processes

**Fig. 5.26** Arcs 0.5–1 m apart are fossil positions of barchanoid slip faces at White Sands National Monument, New Mexico, USA. The slip face position has been marked by cementation by moisture before the dune moved on. These scars are best seen with the sun low in the sky. *Photograph* R. Lorenz

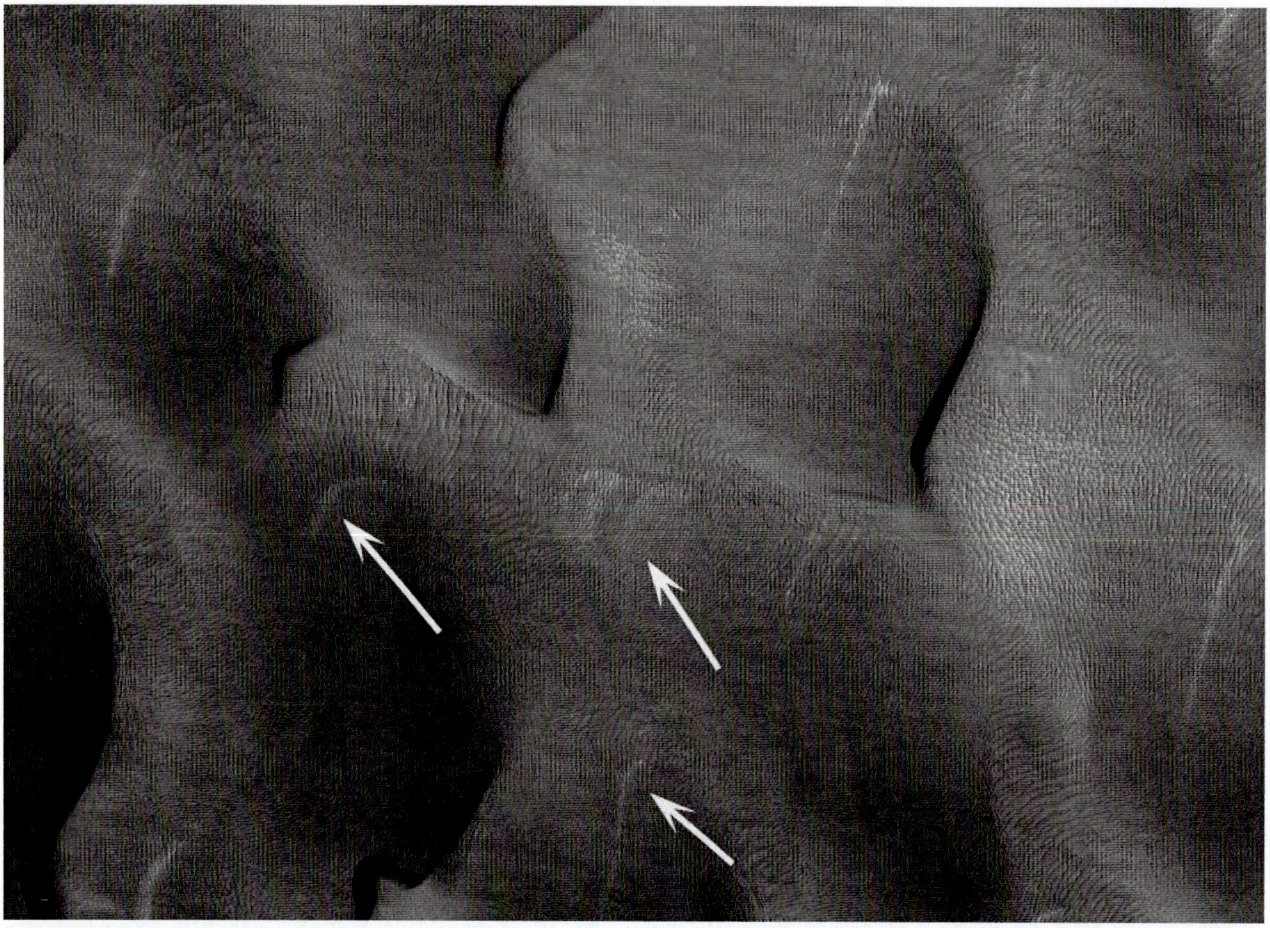

**Fig. 5.27** Barchan scars on Mars, seen in HiRISE image PSP_001882_1920 and noted by Gardin et al. (2012). Barchan dunes are readily visible, as are ripples superposed on the dunes and interdune floor. Bright arcuate streaks (*arrowed*) are interpreted as cemented marks of former slip faces—the set of three in the center is particularly indicative of progressive movement of a dune. Those arcs are about 10 m across. Related features are seen on Earth at *White Sands*, and Lencois Maranhenses. *Image* NASA/JPL/U.Arizona

**Fig. 5.28** A winter image from a timelapse camera (see Chap. 16) monitoring ripple migration and evolution at Great Sand Dunes National Park and Preserve. After a snowfall, high winds caused ~2 m-wide ripple-like structures to form in the snow—mixed sand and snow transport leads to curling layers visible at *left. Image* R. Lorenz

**Fig. 5.29** Buried snow layers, literally stumbled upon by the author, on a parabolic dune at Great Sand Dunes National Park and Preserve (Fig. 5.25 was taken on the other side of this dune, about two weeks before this picture was taken). Much of the dune here is rock hard, with moist sand frozen in place. The thickest snow layer here is about 4 cm thick. Note that it is the area where snow is present that has small ripples on the surface, where perhaps the sand mobility has been affected by moisture from the snow. *Photo* R. Lorenz

the last snowfall (Steidtmann 1973). Snow is preserved inside dunes in Alaska and Antarctica (e.g. Bourke et al. 2009).

The phenomenon may not be restricted to the Earth: Mars too may have niveo–aeolian features. The possibility of cementation by moisture was advocated for some time as a possible explanation for why Mars dunes were not observed to move. Although some examples of migration have now been seen (see Chap. 8), the possibility of incorporation of ice in Martian dunes remains. Evidence from orbital neutron spectrometry (Feldman et al. 2008) suggests that the Olympia Undae dunes on Mars contain enough hydrogen (i.e., water) that they may have a substantial ice component.

# Basic Types of Dunes

Nature has produced an amazing variety of shapes in which sand grains have accumulated to make landforms that are called 'dunes', a variety that accounts for the enduring aesthetic appeal and scientific interest in aeolian landforms, and the wide range of names in different languages (notably Arabic) for the different forms. Yet there is order in this variety, which somehow also promotes their beauty, as well as the scientific tractability of their study. In fact, there are only a handful of basic forms, which through field study, then remote sensing, and now laboratory and computation studies (see Chaps. 16, 17, 18 and 19) can be directly related to just two controlling factors: the sand supply, and the variability in the wind direction. This latter factor is arguably the most significant.

The advent of satellite monitoring of Earth's surface provided a common medium with which to observe and document sand dunes (and other features) across the planet, and it also led to a classification of dune types based on characteristics that can be derived from remote sensing data. McKee (1979) presented a general dune classification system illustrated in part using Landsat images of various sand deposits, supplemented by ongoing field studies of various sand dunes. The McKee system (Fig. 6.1) has proved to be a useful way to distinguish dune types across the Earth, as well as now on other planetary surfaces. The classification scheme is based on the overall shape and outline (planform) of the sand deposit along with the number of slip faces present on the dune, all attributes that can be determined from field investigations, satellite images and aerial photographs. A more quantitative classification of the major types (barchan, transverse, longitudinal and star) was made as a function of equivalent sand thickness and directional variability (see Chap. 8) by Wasson and Hyde (1983).

Most planetary dunes can be considered examples of these basic types which are a direct mapping between the wind and sand supply regime and the resultant duneform. This simple relationship relies on the controlling factors being constant in time. We also show examples in this section where varying conditions have led to horizontal variations in dune form.

We defer to a subsequent section discussion of where different dune forms are superposed on each other (compound or complex dunes), as well as some additional types of sand deposits (e.g., sheets) and dunes associated with topographic obstacles and vegetation. There is extensive and useful discussion about dune morphology in several texts: Bagnold, as ever, is worth reading, as is (while also somewhat dated), Cooke and Warren (1973). An excellent more modern review is Lancaster (1995). On vegetated and minor dune types, Pye and Tsoar is useful.

An important property in relating morphology to wind regime is the quantitative variability in wind direction (the ratio of resultant drift potential to total drift potential, which we discuss in Chap. 8). Another important concept is that of gross bedform normal transport (due to Rubin and Hunter 1987), roughly that the bedform will arrange itself to maximize the transport across it.

## 6.1 Dome

We begin with the simple dome (Fig. 6.2). This is a dune form that results from the simple accumulation of sparse sand on a flat bed, and can be thought of as a growing disturbance in the surface that results from the saturation length instability described in Chap. 4. Also, sand is more likely to accumulate on sand than on the interdune pan where saltating sand grains may bounce more efficiently. Dome dunes do not grow to be large, as once they reach a height of some 10 s of cm on Earth, they develop slip faces (see Chap. 5) and become barchans—in essence, a dome can be thought of as a slipfaceless barchan. Because dome dunes are small, they move quickly (see Chap. 8) and can be rather ephemeral, appearing and disappearing in response to

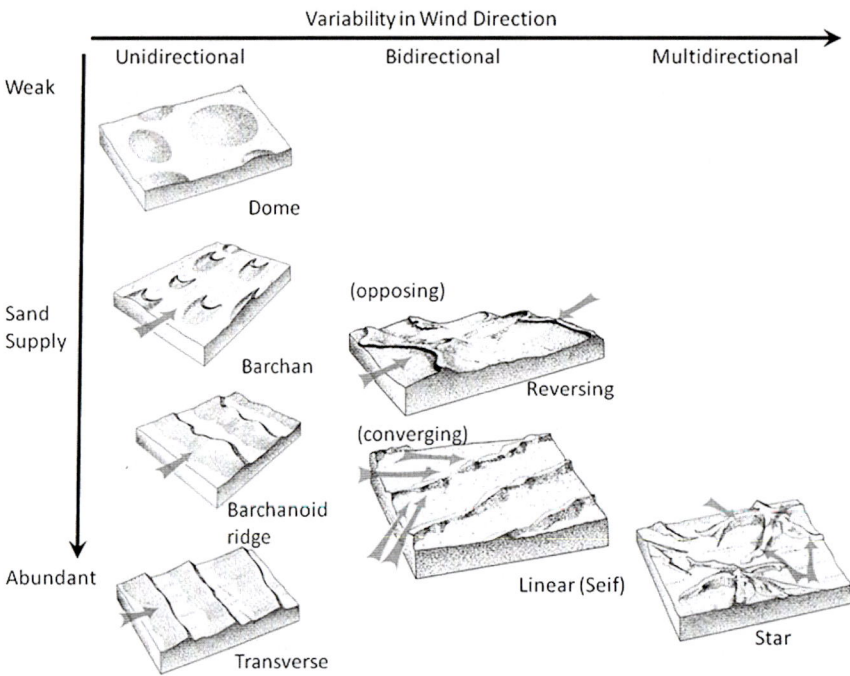

**Fig. 6.1** Major dune types, as defined by McKee (1979) mapped out on a diagram of sand supply and wind variability

**Fig. 6.2** Slipfaceless dome dune in the Rub' Al Khali, United Arab Emirates, first author for scale. *Photo* J. Radebaugh

## 6.1 Dome

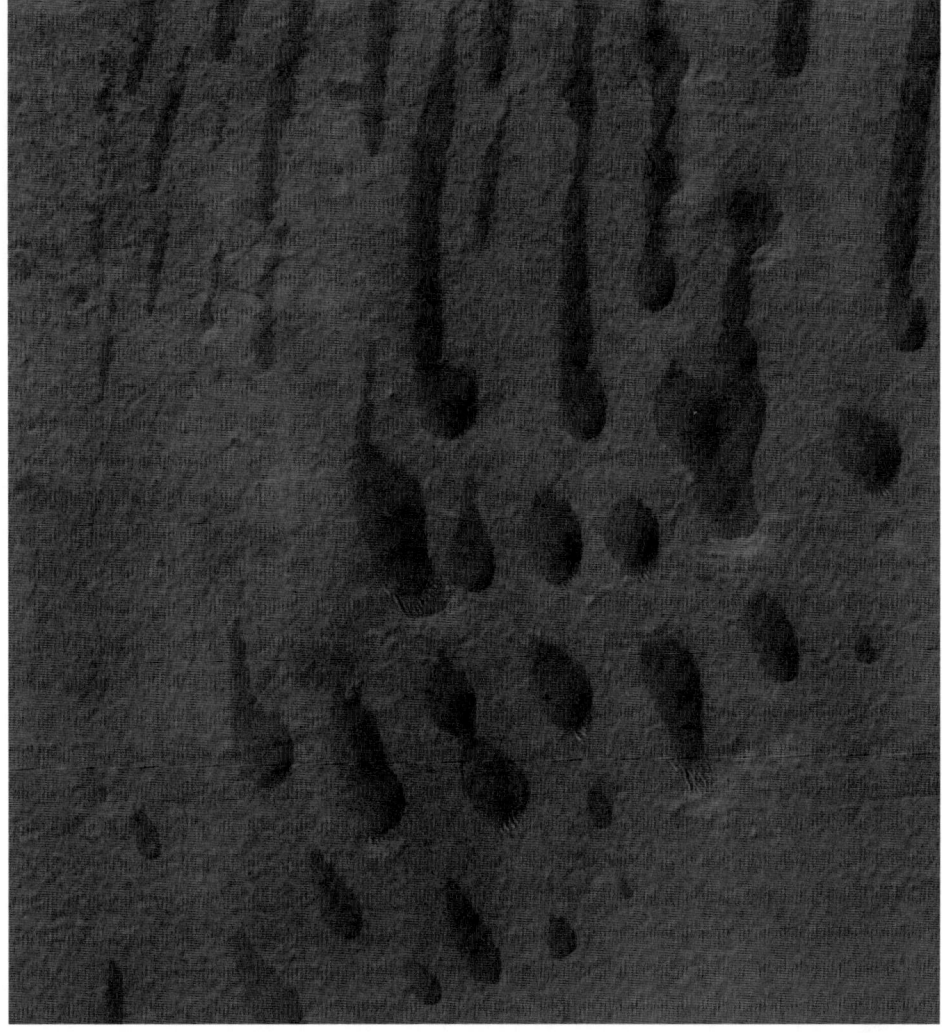

**Fig. 6.3** HiRISE image of dome dunes in Noachis Terra, Mars. These appear to be grading into linear forms, perhaps suggesting the domes are either forming into linears, or perhaps are the last remants of some linears that have now all but disappeared. Some small ridges appear at the bottom of some of the domes (which lack slip faces) and a few dust devil tracks cross the dunes. *Image* NASA/JPL/U.Arizona

changes in sand or wind conditions. Dome dunes are also widely encountered on Mars (Fig. 6.3). They have not been identified on Venus or Titan, but since they are the smallest type of dune, this may be more a function of the resolution of observations rather than an indication of their absence.

## 6.2 Barchan

The barchan is a common crescent-shaped dune form (Figs. 6.4, 6.5, 6.6 and 6.7) found in regions with both a limited sand supply and with a unimodal wind regime. The name is originally from Turkistan. A single slip face is present along the crest of the dune, and the horns of the dune point downwind. The horns of barchans can sometimes be the local source for subsequent barchans, where sand leaving the horn becomes the source of another barchan dune, and barchans often form extended corridors. Barchan movement downwind has been documented at sites around the world; the rate of movement of barchans within a given area is inversely proportional to the height of the dune—see Chap. 8.

Because they are common in Egypt, barchans feature prominently in Bagnold's book—they may have thus received more attention than they deserve on an area- or volume-weighted basis worldwide. Nonetheless, as the most dynamic of dunes, and with an appealing 2-dimensional planform, they are important and interesting. More recently, barchans have also been the subject of extensive study in water tanks/flumes (Taniguchi et al. 2012; see Fig. 6.8) as well as in numerical simulations (see the discussion in Chap. 19, and especially Andreotti et al. 2002). That work is largely stimulated by observations of barchans on Mars which have a variety of forms (Figs. 6.9, 6.10 and 6.11), from teardrop domes with two slipfaces and a tail, to wing-shaped reversing barchans (Fig. 6.12), to hooked barchans which may be intermediate between barchan and linear (longitudinal) forms. Note that barchans may develop superposed small ('elemental') duneforms (e.g., Elbelrhiti et al. 2008); while morphologically similar to the much

**Fig. 6.4** Rather tiny barchans (essentially dome dunes with a slip face bitten out) in the Danikil Depression in Ethiopia. First author is holding a kite string. *Photo* R. Lorenz/S. Diniega

**Fig. 6.5** Classic barchans in Egypt (quite possibly among those studied by Bagnold himself) viewed in the gloomy dusk from an airliner. Note the spectrum of sizes, the en echelon arrangement of close barchans. Some are almost dome-like with a small slip face, others have long arms trailing into linear-like forms. A couple of barchan sets towards the top have merged, while several of the larger barchans have small bedforms ('elemental dunes') forming on their surface. In the center of the image a few yardangs are present. *Photo* J. Radebaugh

**Fig. 6.6** Barchan dunes near Biggs, Oregon. G. K. Gilbert photo 1899 from the USGS Photographic Library

**Fig. 6.7** A beautiful HiRISE image of barchans on Mars: some of these are 'hooked' barchans, a transitional form between barchans and linears

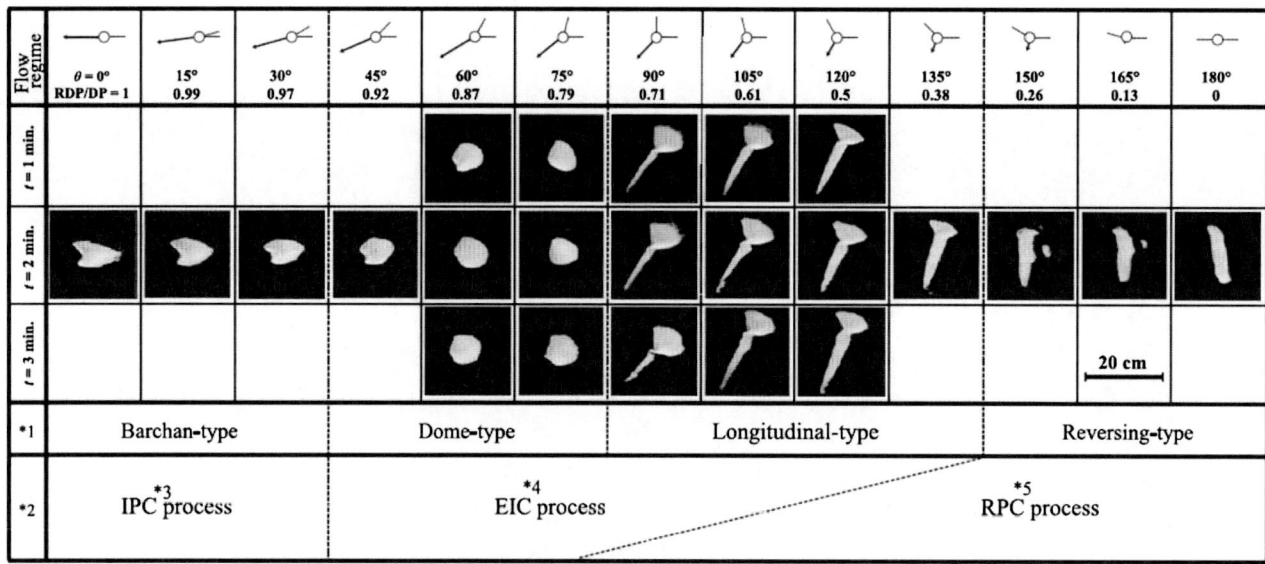

**Fig. 6.8** Elegant water-flume experiments have mapped out the different wind regimes that lead to different barchan forms. This compilation image courtesy of Keisuke Taneguchi—see also Fig. 17.16. Similar results have been obtained in numerical experiments

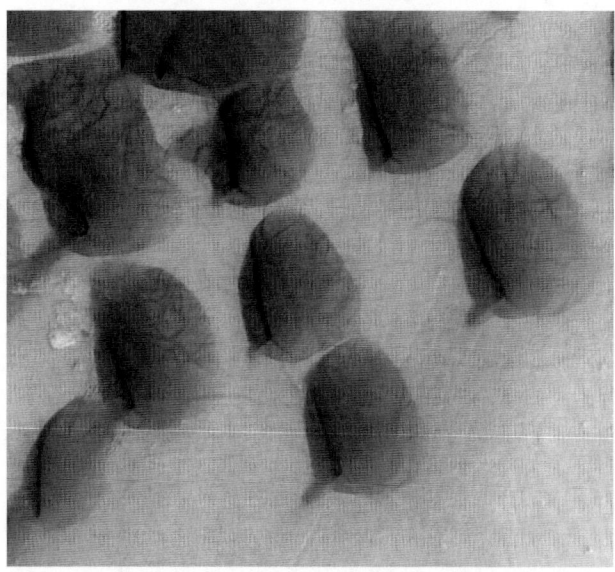

**Fig. 6.9** HiRISE image of almost dome-like 'teardrop' barchans. The presence of two slip faces attests to the multidirectionality of the wind regime (separated by 60–75°), according to the water tunnel experiments in Fig. 6.8). Note the dust devil tracks on these dunes. *Image* NASA/JPL/U.Arizona

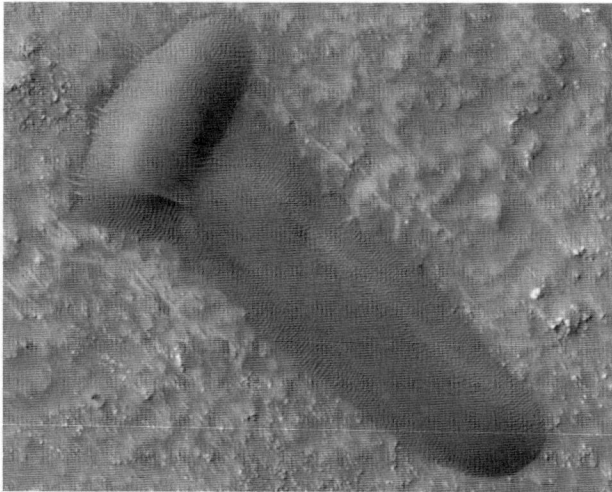

**Fig. 6.10** HiRISE image of a 'Tadpole', a hybrid between a dome and a linear dune. PSP_007663_1350, 50 cm/pixel. This type of dune appears to be much more common on Mars than Earth. *Image* NASA/JPL/U.Arizona

larger megabarchan (see next section) it may be a rather different process.

Some well-studied barchans are the corridor marching through the Kharga depression in Egypt, barchans in coastal Peru and Morocco, and near the Salton Sea in California.

## 6.3 Barchanoid Ridge

A series of connected crescentic ridges result in a broadly linear feature with a single sinuous crest that is formed by coalescing slip faces (Fig. 6.12). The crest may have multiple slip faces but they are all on the same side of the crest, indicative of a unimodal wind regime. In general, barchanoid ridges form where sand supply is greater than would be

## 6.3 Barchanoid Ridge

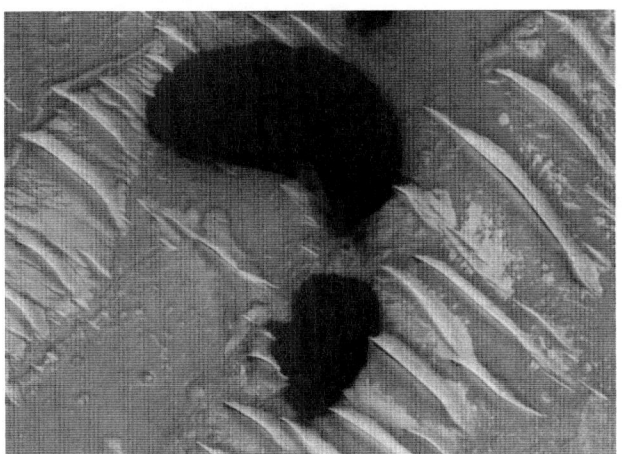

**Fig. 6.11** Dark dunes (an irregular dome and a pudgy barchan) over bright mega-ripples. HiRISE image PSP_008679_1905, 2 m/p, *credit* NASA/JPL/U.Arizona

the case for barchan dunes formed under the same wind regime (see Fig. 6 of McKee 1979).

Dunes in barchanoid fields are sometimes referred to as barchanoid (crescentic, with concave slip face downwind) and linguoid ('tongue-shaped', with convex slip face downwind) .

Barchanoid forms are very common on Mars (Figs. 6.13 and 6.14) and Earth. Among terrestrial examples are White Sands and the northern part of the Rub' Al Khali in the Arabian desert.

## 6.4 Transverse Ridge

A series of essentially linear crests that are oriented perpendicular to the unimodal wind direction. Transverse ridges tend to be asymmetric in cross-sectional profile, with the shallower face on the upwind side, in contrast to linear dunes that are more symmetric in overall shape. A single slip face is present, indicating a single dominant wind direction. In general, transverse ridges form where sand supply is greater than would be the case for barchanoid ridges formed under the same wind regime (see Fig. 6 of McKee 1979).

Because (unlike barchans) transverse dunes can be considered as purely two-dimensional structures, sectioned vertically downwind, they are popular for modeling (e.g., Melo et al. 2012) and field measurement studies (e.g. Wang et al. 2002) since results are relatively free of the complexity of 3-dimensional effects, although some secondary airflows may give 3-D effects (e.g., Walker and Nickling 2002).

Secondary airflow effects may be responsible for a grid pattern that may sometimes develop, referred to in the Sahara and French literature as 'aklé' (Cooke and Warren 1973). Morphologically, there may be challenges in discriminating this from superposed generations of linear or transverse dunes.

Transverse ridges are found in several sand seas on Earth. One set of dunes on Tital (Fig. 6.15, where a normally-bimodal wind regime may be 'straightened' by a topograhic obstacle) was identified (Lorenz et al. 2006) as possibly transverse. Transverse ridges are common on Mars, e.g., Figs. 6.16 and 6.17, as might be expected given the abundance of the rather similar barchanoid ridge form.

## 6.5 Linear or Longitudinal (Sief)

A generally symmetrical ridge (Figs. 6.18, 6.19, 6.20 and 6.21) is sometimes topped by a crest that undulates in height (Fig. 6.20).

Tsoar (1989) made a review of linear dunes, although much new work has been stimulated on them by the discovery of almost invariably linear dunes over vast areas of Titan. Notably, numerical simulations and water tank experiments (e.g., Reffet et al. 2010) have explored the variation in wind direction and its effect on morphology, an issue first systematically explored by Rubin and Hunter (1987) and Rubin and Ikeda (1990).

A certain type of linear dune can form in the lee of obstacles (see next section) but these appear to only exist on rather small scales, 10−100 m rather than the 10−100 km scale typical of the giant linears that define major sand seas on Earth and Titan. A wavy appearance often develops along the crest of linear dunes, and leads to their being referred to by the Arabic term 'seif' (sword). In north Africa especially, large linear dunes are sometimes referred to as 'draa' (arm), although this term—applied most particularly to the largest scale bedform in compound linear dunes—is not formally defined. In the Arabian desert, the word 'uruq' is more often used for large linears (Arabic is, after all, a family of regional dialects).

The McKee (1979) diagram shows winds converging at about 90° forming linear dunes in the vector-average wind direction. The experiments of Rubin and Hunter and Reffet et al. show that when the convergence is less than 90°, a transverse dune (or a row of barchans) develops, only above 90° does the longitudinal form appear. Geomorphologists often insist on caution in calling these 'longitudinal' dunes, in that it may not be certain from morphology alone that the dune is indeed aligned along the direction of net sand transport and thus the term 'linear' (without genetic connotations) is to be preferred. It should also be noted that, while in many cases the dune is indeed longitudinal in terms of the sand transport, the wind never

**Fig. 6.12** Rather wide barchans in the United Arab Emirates, viewed from a kite camera. This barchan field is at the edge of an area where the sand is more abundant, and thus grades into the distance into more continuous barchanoid ridges. *Photo* R. Lorenz

actually blows along the length of the dune (i.e., because of the vector addition, the instantaneous sand transport, in the instantaneous wind direction, is never the same as the resultant sand transport).

Because the winds are generally somewhat orthogonal to the crest, there may be some net sideways migration of at least the crest, if not the whole dune. The large size of linears means this migration, however, is generally very small (although may be indicated in the structure of the beds within the dune).

Much of the older literature discusses helical vortex winds as a means of forming linear dunes. It is certainly true that roll vortices can form (with their diameter determined more or less by the thickness of the atmospheric boundary layer), and the presence of these vortices can often be revealed by lines of cloud that form at the upwelling sides. The presence of linear dunes may help to stimulate and/or anchor these dunes. However, the extension to these vortices as having any importance at all in sand transport, and thus in defining and forming linear dunes, has never been demonstrated. That said, the boundary layer thickness may be what defines the ultimate height, and therefore spacing, of linear dunes (e.g., Andreotti et al. 2009; Lorenz et al. 2010) as well as the size of roll vortices, so the vortices and linear dunes may have a common controlling parameter. But that is not the same as saying one controls the other.

The Namib and the Arabian deserts have perhaps the best examples of giant linear dunes, the latter with perhaps the most perfectly uniform examples (see Figs. 6.18 and 6.19). Much smaller—and usually partly vegetated—linear dunes can be found in the USA, the Kalahari desert (Fig. 6.20) and the Simpson desert in Australia. Titan's massive sand seas are almost exclusively of linear dunes (Fig. 6.21), yet linear dunes are almost entirely absent on Mars.

When the bimodal wind regime that forms linear dunes changes, the dune system re-orients itself (see Chap. 7). This ultimately may lead to a disappearance of the dunes (perhaps seen in Fig. 6.22), more typically recognized via a conversion into a megabarchan form (e.g., Fig. 6.23) perhaps seen on Titan (Fig. 6.24).

## 6.6 Reversing

Symmetric to asymmetric sand ridges (depending on the relative intensities of the winds) having two slip faces, one each on both sides of the ridge crest. In a sense, these can be seen as an end member of linear dunes. Reversing dunes form primarily through vertical accretion of sand in response to a bimodal wind regime, although one wind direction can at times be stronger than the other, resulting in an asymmetric profile. Correspondingly, barchans that are

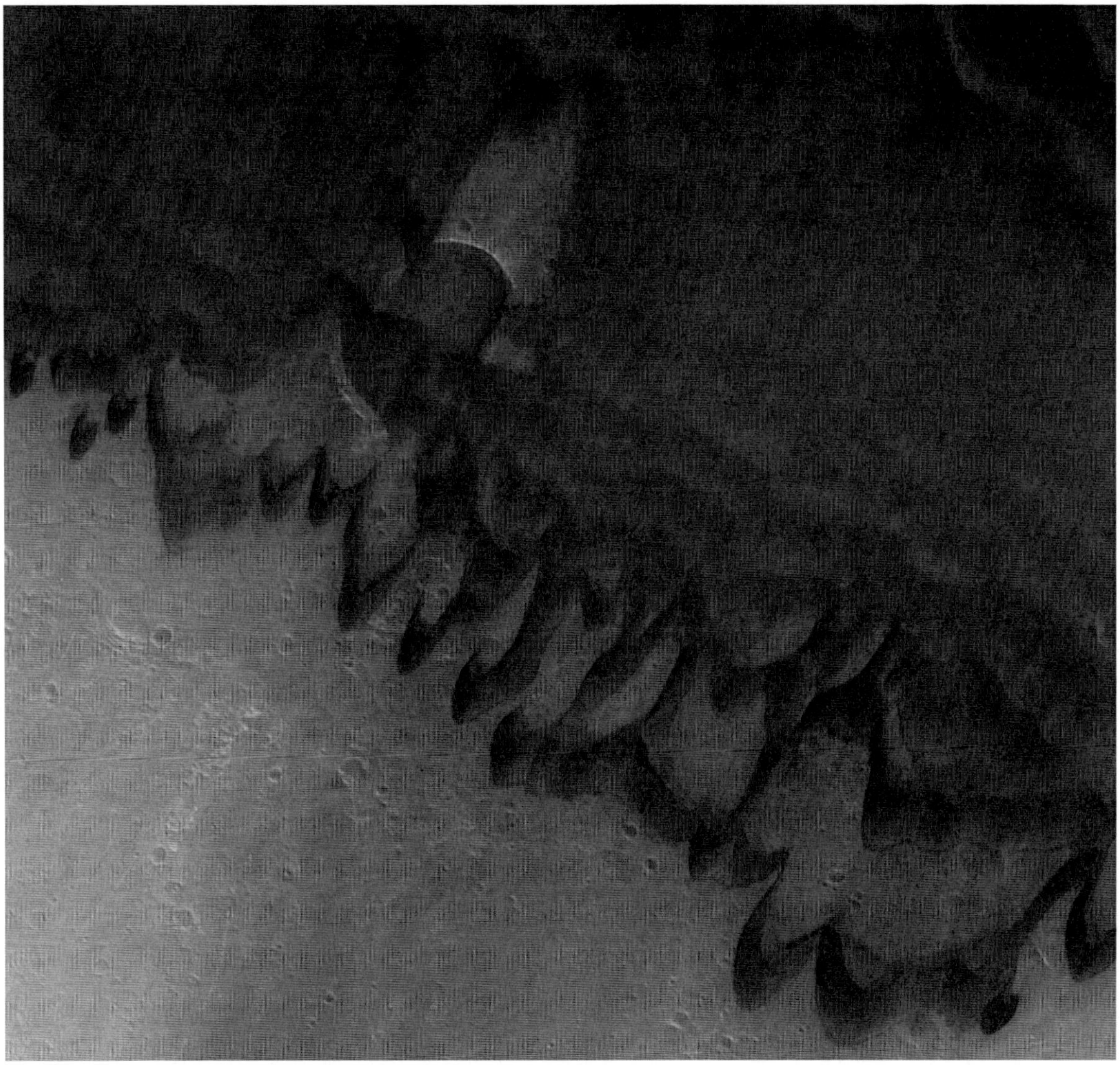

**Fig. 6.13** Barchans on Mars merging to form gentle barchanoid ridges. HiRISE image ESP_014404_1765. The slip faces indicate that these barchans are moving towards upper right, and hence colliding with the larger sand mass and barchanoid ridges. *Image* NASA/JPL/U.Arizona

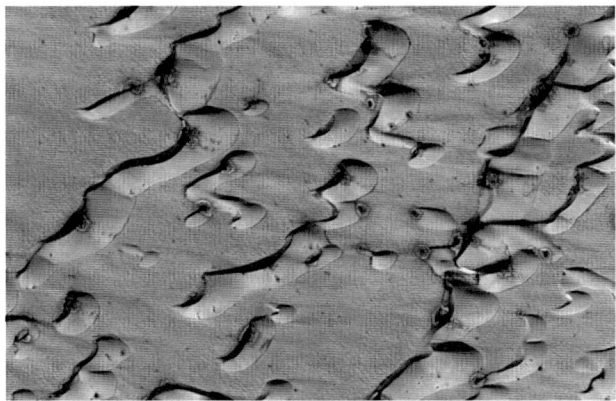

**Fig. 6.14** More barchans and barchanoid ridges, North Polar Erg. PSP_007676_2650, 2 m/p. The dunes and the interdunes here are all covered in $CO_2$ frost, although a few dark spots indicate where sand or dust has broken through the frost. *Image credit* NASA/JPL/U.Arizona

exposed to occasionally-reversed winds tend to have an open 'wing' shape and may have crests with slip faces in both directions (Fig. 6.25). Recent comparisons of the profiles of reversing dunes on Earth to TARs on Mars suggest that a reversing dune origin seems likely for large TARs (Zimbelman 2010).

## 6.7 Star

Dunes with three or more slip faces are the result of sand-driving winds from multiple directions. Star dunes (reviewed by Lancaster 1989) usually have a central peak that is higher than any other part of the dune, from which three (or more) distinct, often sinuous arms extend radially away from the central peak. Each arm can display a separate slip face

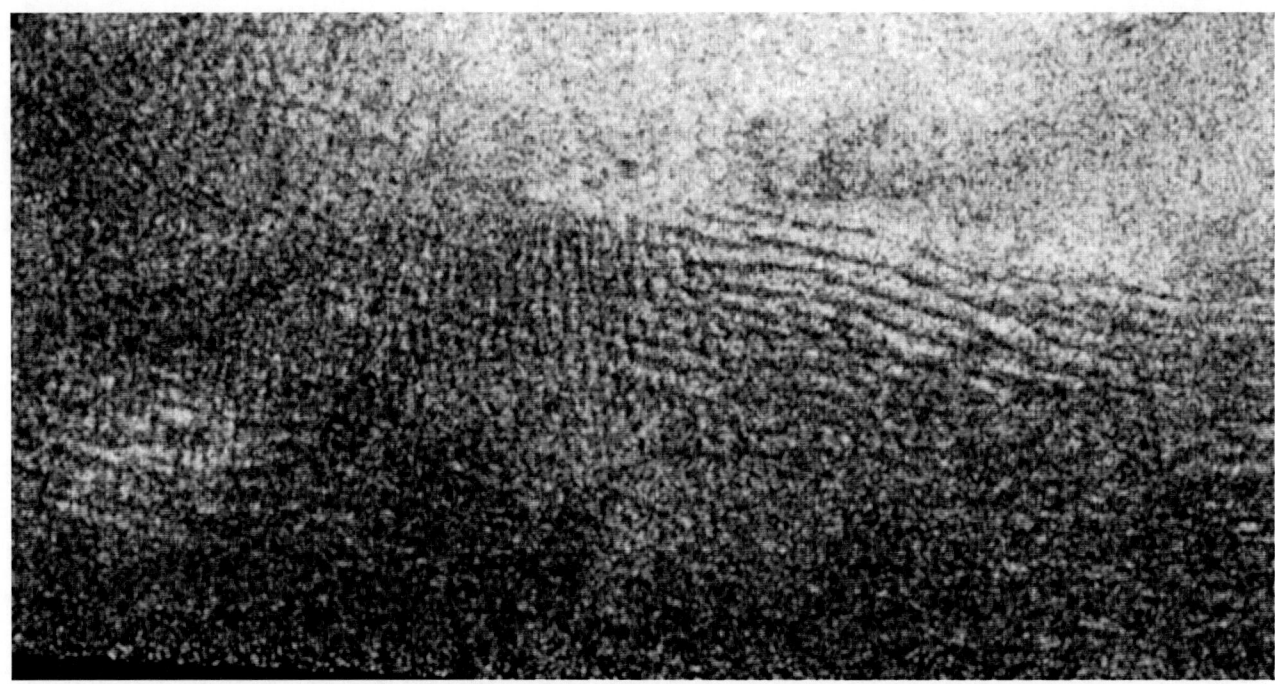

**Fig. 6.15** Part of the Belet sand sea on Titan. At the right, dark dunes interpreted to be linears are seen against the bright background of a highland terrain extending to the top. In the middle of the image, a set of near-vertical features can be seen, interpreted to be transverse dunes. *Image* R. Lorenz/Cassini Radar Team/NASA/JPL

**Fig. 6.16** Transverse ridges on Mars. HiRISE image ESP_016036 _1370. In the present epoch, it seems appropriate to call these transverse ridges, in that they have a uniform slip face orientation which is consistent with that shown by the wide 'wing' barchans at the bottom (which are departing from the main dunefield). It is a separate question whether the sand accumulated into these very parallel and regular forms is in the same wind regime as at present—one possibility is that these were linear dunes and subsequently had a unimodal wind regime imposed on them. *Image* NASA/JPL/U, Arizona

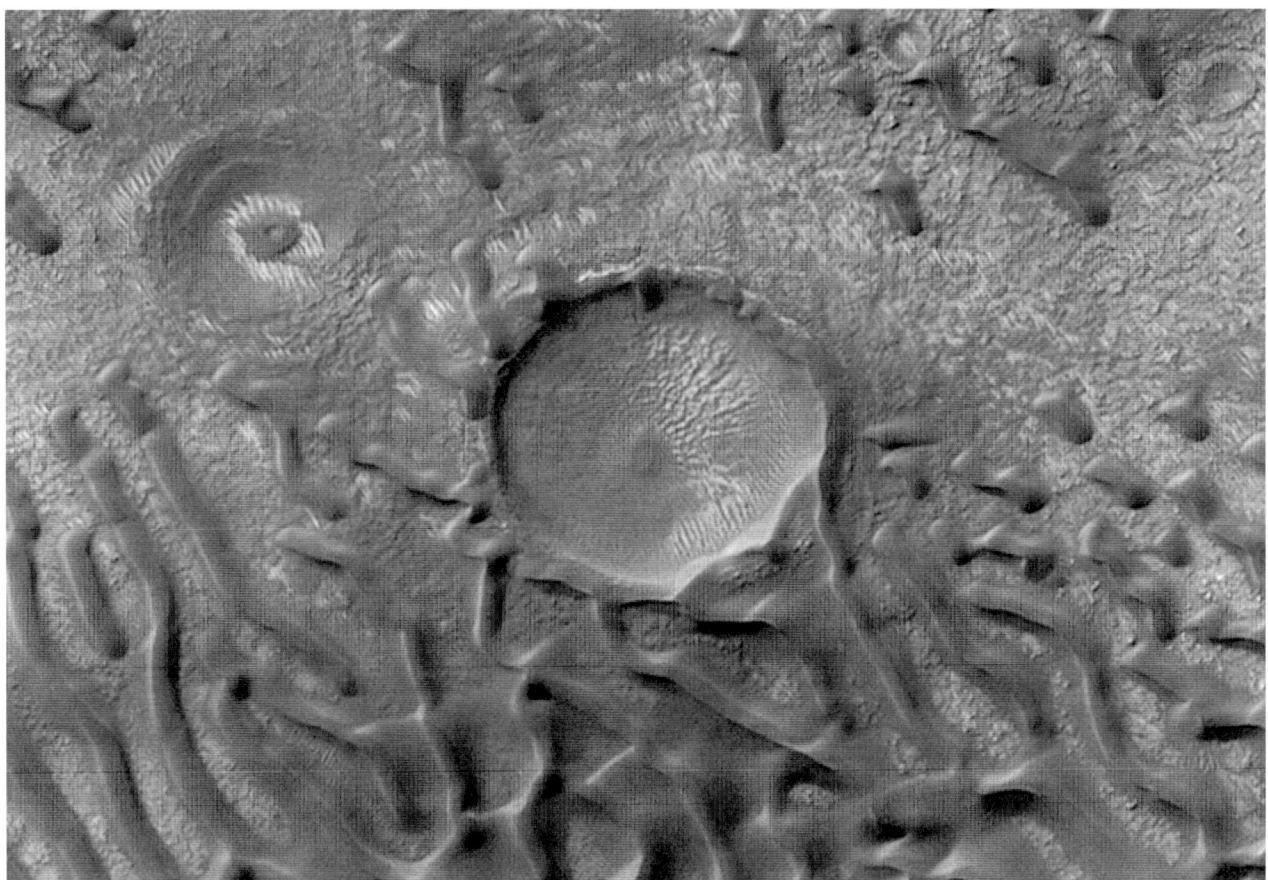

**Fig. 6.17** Transverse dune interaction with a crater rim—dunes migrating from upper right. Compare with the ripple-crater interaction in Fig. 5.8. Note bright TARS on the crater floors. HiRISE image PSP_007663_1350, 4 m/p. *Image* NASA/JPL/U, Arizona

**Fig. 6.18** A SPOT image of linear dunes in the Arabian desert. The astonishing regularity of linear dunes is often their most remarkable feature (see also Fig 18.4). *Image* courtesy of SPOT

## 6.7 Star

**Fig. 6.19** Linear dunes of the Namib desert, viewed from the Space Shuttle, looking southwest with the Atlantic in the background. It was in fact this image that inspired the interpretation of Titan's dunes as linears. *Image* NASA JSC Earth Observation Laboratory

**Fig. 6.20** View from a kite-borne camera of a linear dune at Quattaniya, Egypt (see also Sect 3.6). The meandering crest is typical of seif dunes—slip faces may develop seasonally as winds reverse. Notice the shallow plinth of sand, wider on the right side than the left, on which the dune proper sits. *Photo* R. Lorenz

**Fig. 6.21** A particularly regular stretch of linear dunes in the Belet sand sea on Titan. *Image* NASA/JPL/R. Lorenz

**Fig. 6.22** HiRISE image of tadpole dunes on Mars. These dome dunes have long tails. A plausible scenario is that these are the last vestiges of previous linear dunes, or perhaps these are an extreme example of barchans with long tails—see Figs. 6.9 and 6.10. *Image* NASA/JPL/U.Arizona

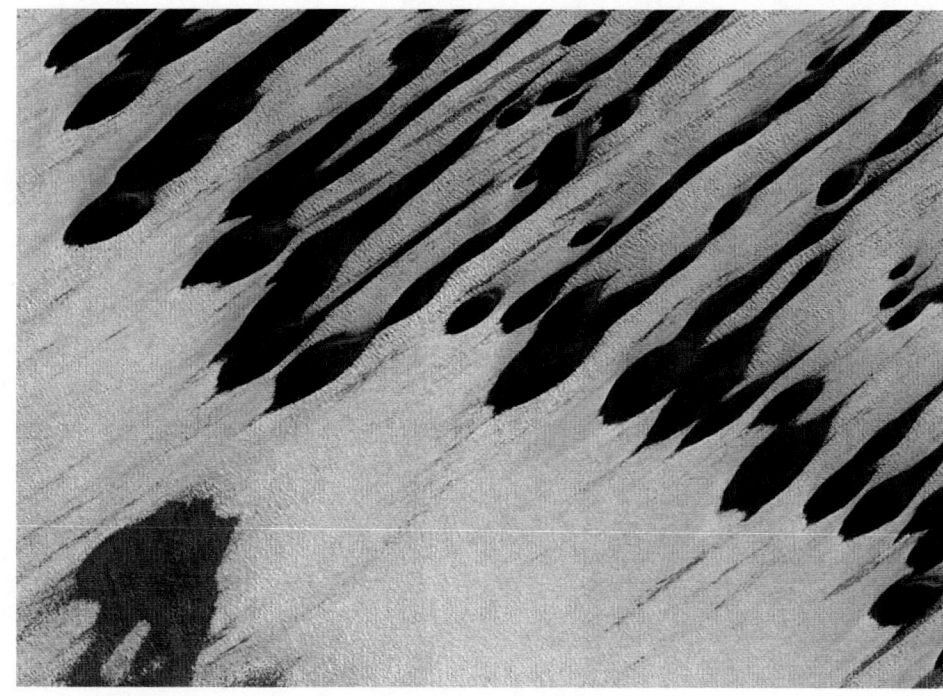

## 6.7 Star

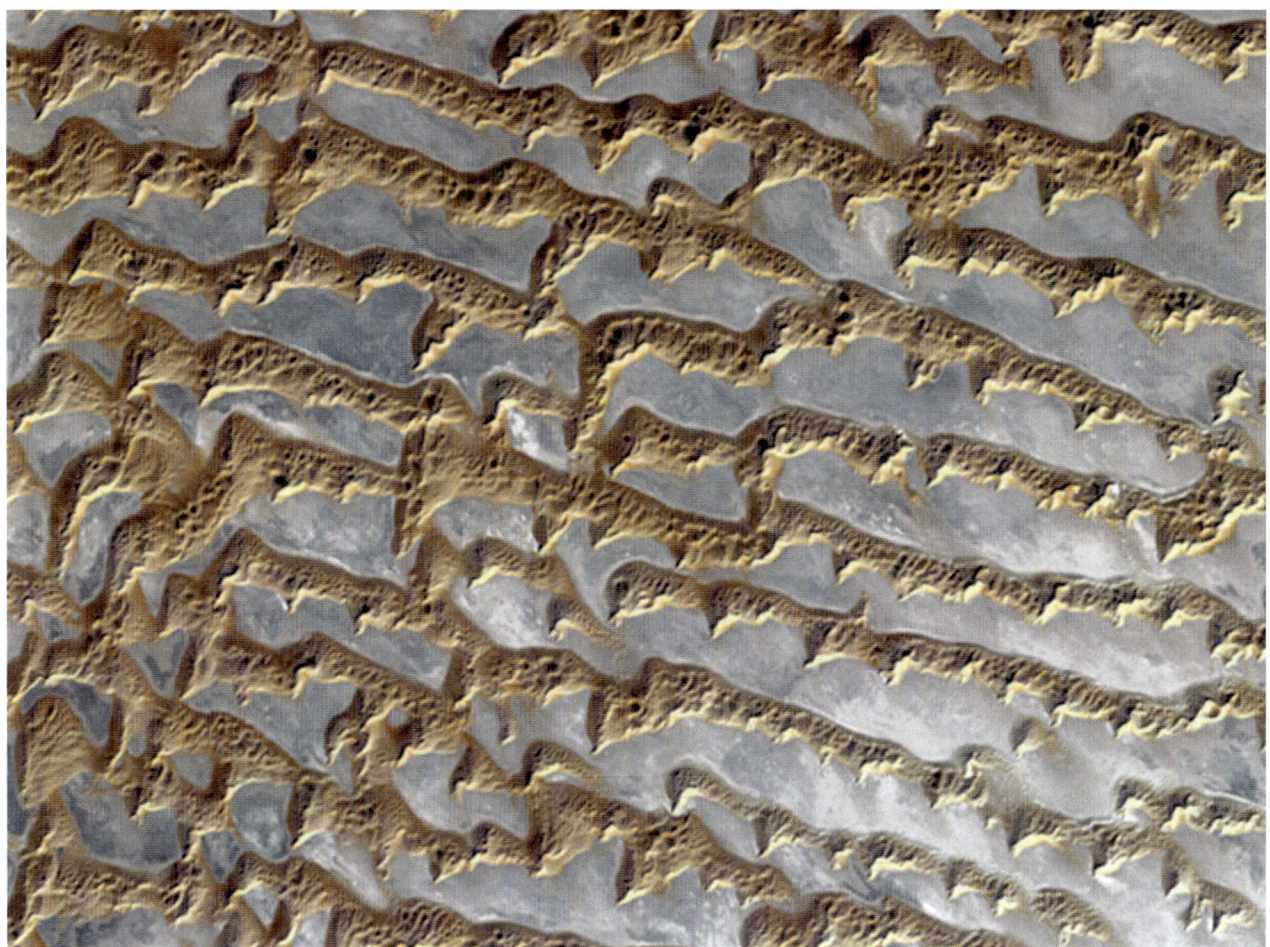

**Fig. 6.23** Linear dunes in the eastern Rub' Al Khali desert, developing into hooked megabarchans; the reddish sand dunes are stark against the gravel and clay interdune pans (see Fig. 12.7, of a nearby area on the UAE/Oman border). *Image* NASA/GSFC/METI/Japan Space Systems, and U.S./Japan ASTER Science Team

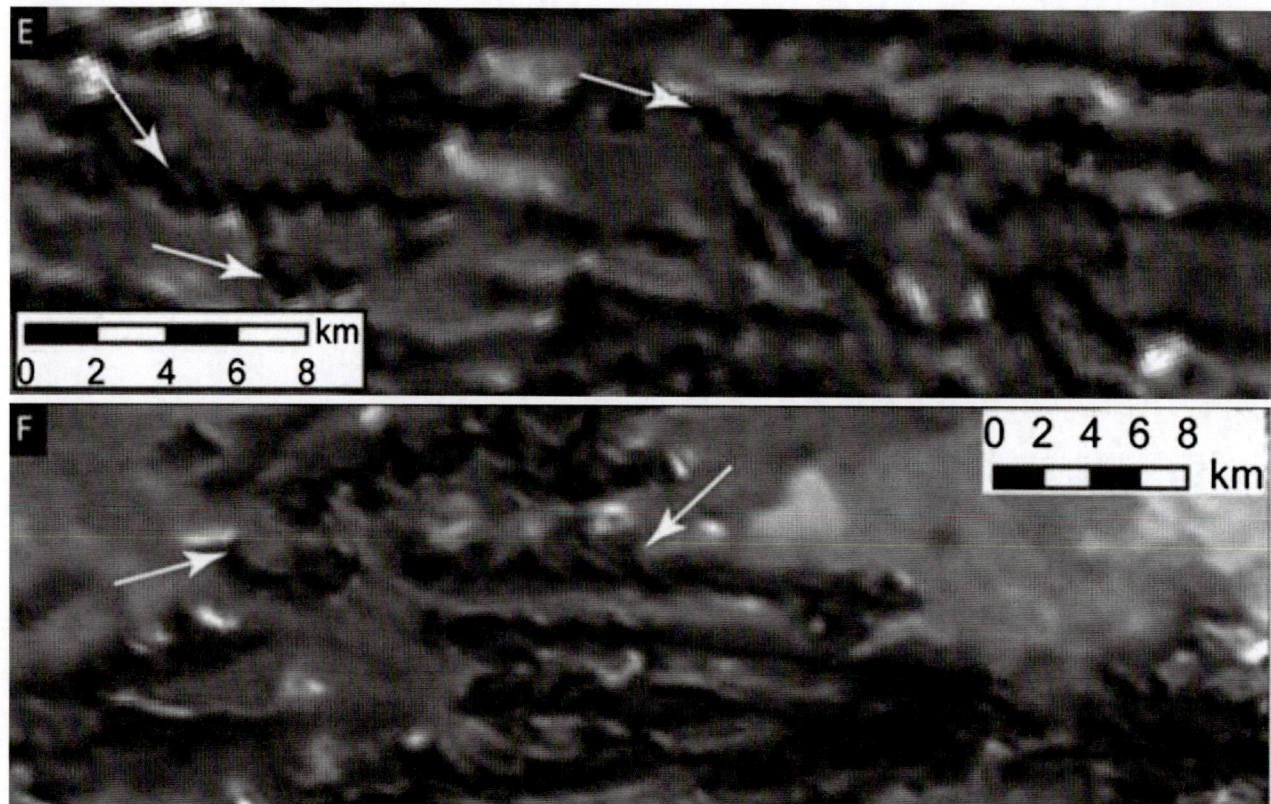

**Fig. 6.24** Reoriented crestlines of linear dunes and possible hooked barchans on Titan (Ewing et al. 2012). These features are near the limit of the features that can be resolved with the Cassini radar mapper, but appear to show crestline morphologies and orientations that are inconsistent with the overall sand accumulation in linear ridges. Compare with Fig. 6.23. *Image* courtesy of Alexander Hayes

**Fig. 6.25** A corridor of barchan-like forms seen from a kite in the United Arab Emirates. These dunes have slip faces on both sides, indicating that the wind regime here is a somewhat reversing one. *Photo* R. Lorenz

**Fig. 6.26** Magnificent star dunes in Algeria, seen from the International Space Station. Notice the long meandering arms. *Image* NASA JSC EOL

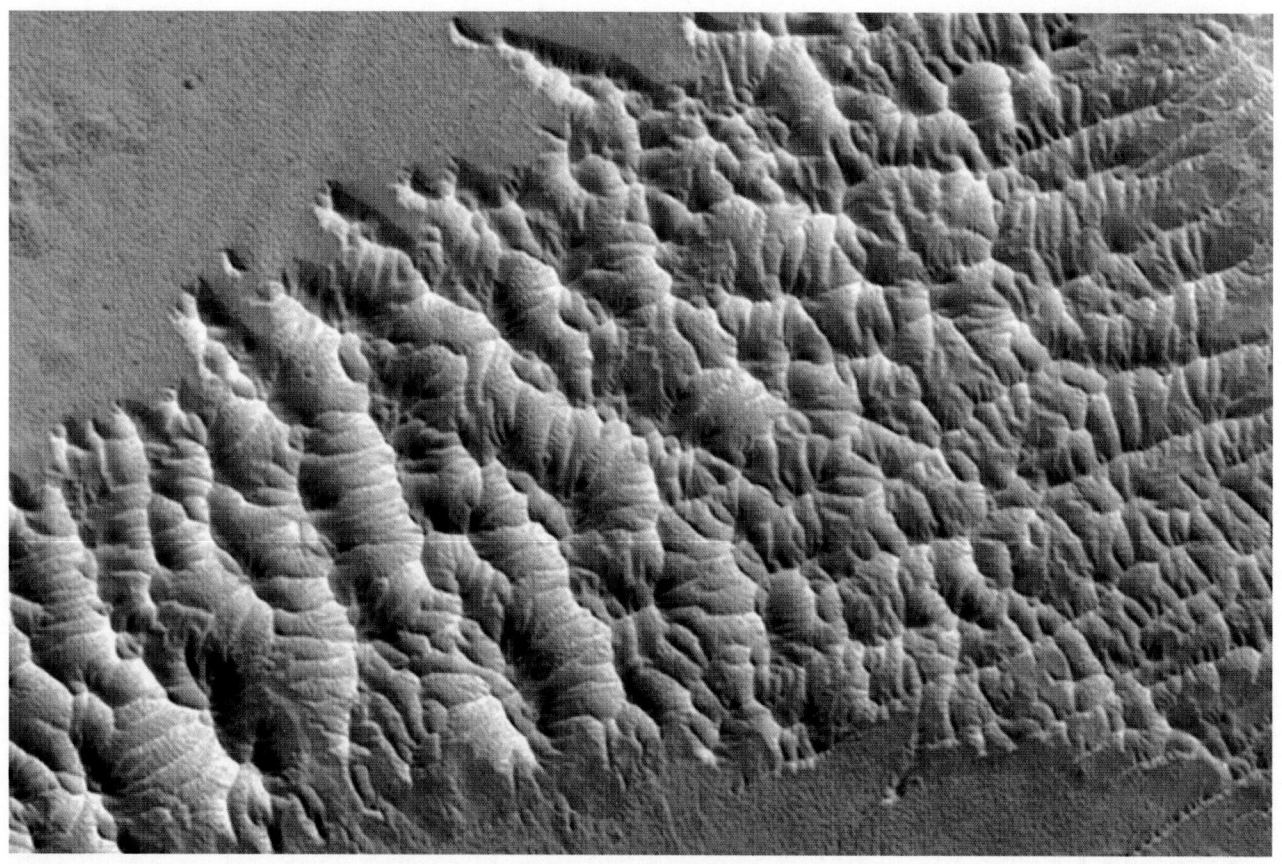

**Fig. 6.27** Star or akle dunes on Mars. HiRISE image PSP_008323_1735, NASA/JPL/U.Arizona

orientation, depending on the most recent orientation of a sand-driving wind. Star dunes tend to grow vertically through sand accretion from the sides rather than to show much movement laterally. The tallest dunes on Earth tend to be star dunes, and the best examples are to be found in the Algerian Sahara, where they are called ghourds or rhourds (Fig. 6.26).

Recent numerical simulations of star dunes (Zhang et al. 2012) suggests that in fact star dunes only grow extensive arms when the wind comes from an odd number of equal-spaced directions (typically three or five), in which case the arms grow upwind in the respective directions. When four or six wind directions are involved, a square or hexagonal pyramid—without long arms—results.

The star-like duneforms identified on Mars appear to be in general linked in a trellis-like pattern (i.e., akle, Fig. 6.27), perhaps as a result of forming as linear or transverse ridges that have then been subjected to a shift in wind direction.

## 6.7 Star

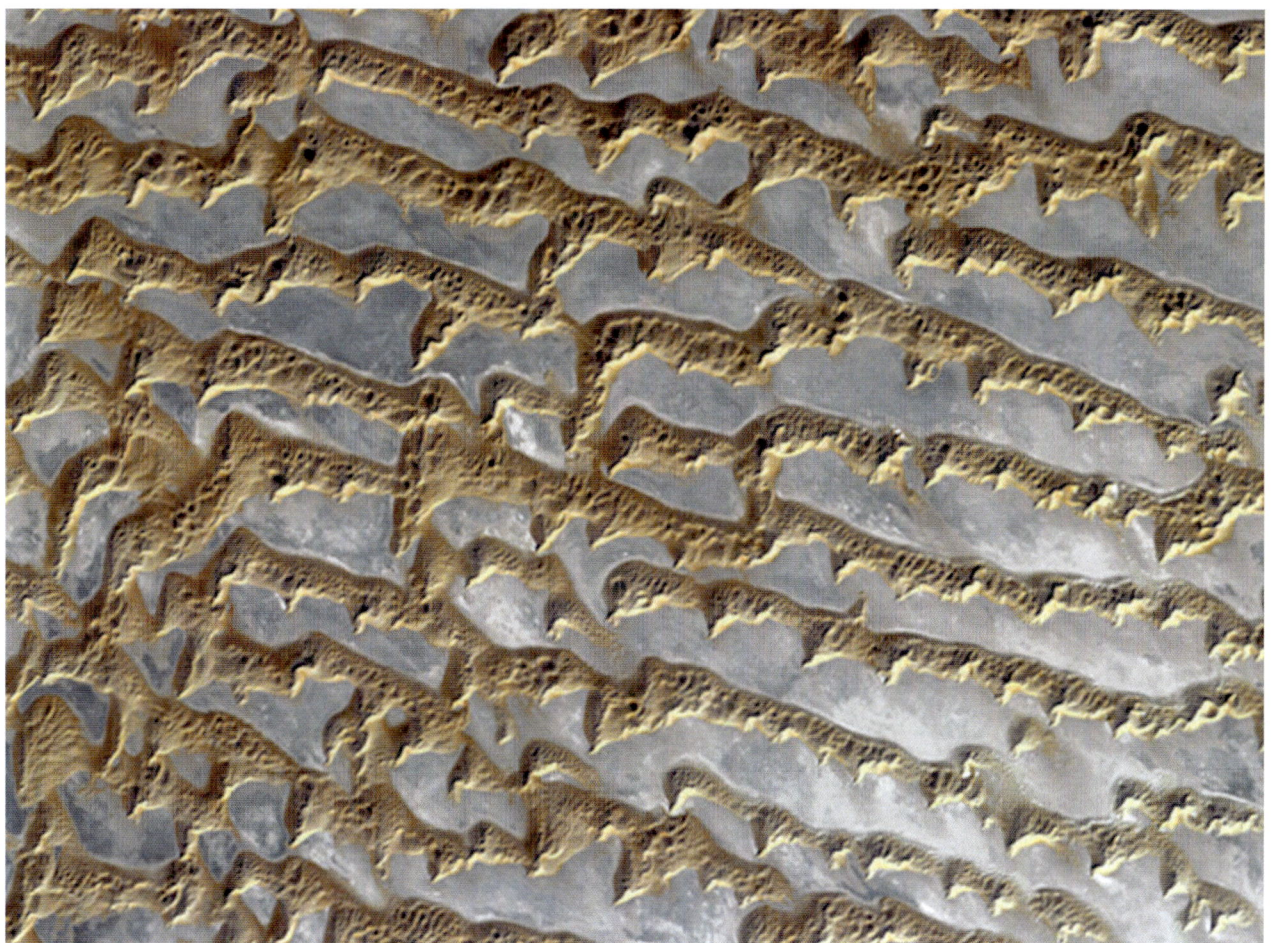

**Fig. 6.23** Linear dunes in the eastern Rub' Al Khali desert, developing into hooked megabarchans; the reddish sand dunes are stark against the gravel and clay interdune pans (see Fig. 12.7, of a nearby area on the UAE/Oman border). *Image* NASA/GSFC/METI/Japan Space Systems, and U.S./Japan ASTER Science Team

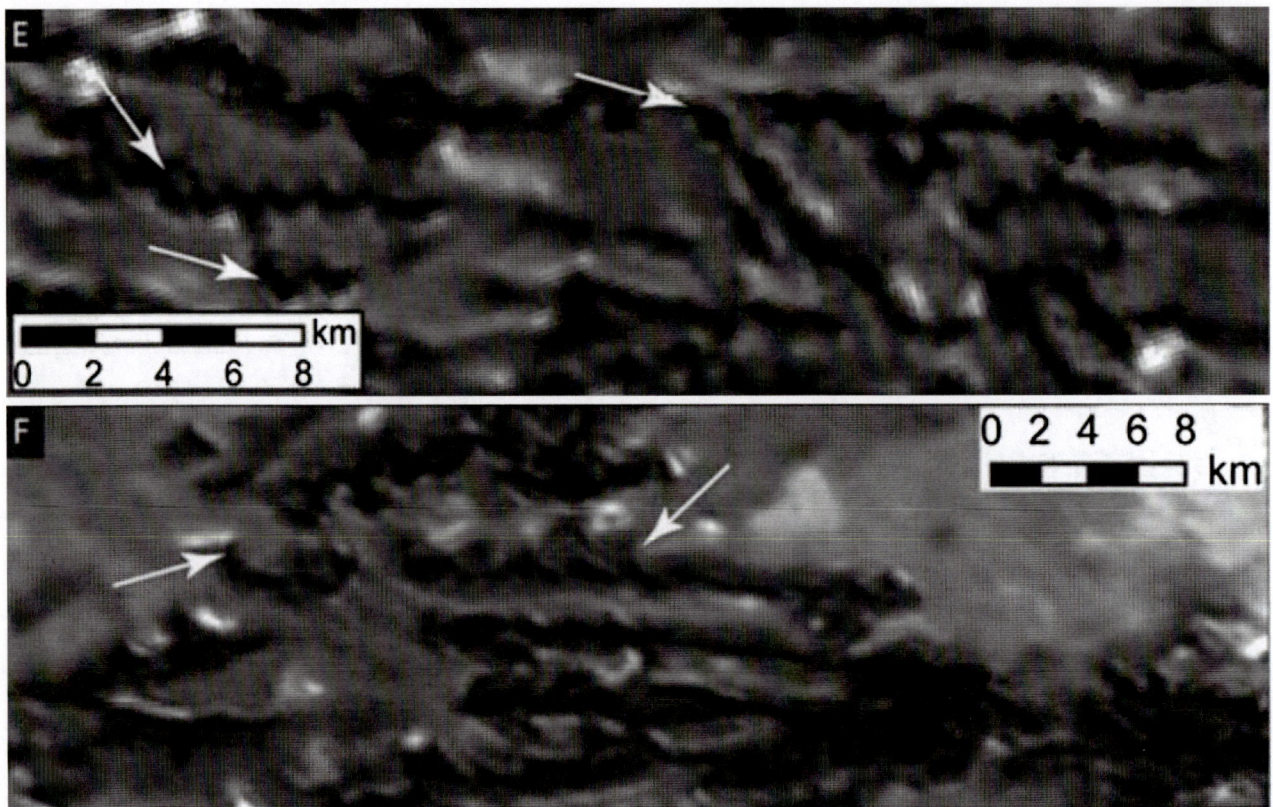

**Fig. 6.24** Reoriented crestlines of linear dunes and possible hooked barchans on Titan (Ewing et al. 2012). These features are near the limit of the features that can be resolved with the Cassini radar mapper, but appear to show crestline morphologies and orientations that are inconsistent with the overall sand accumulation in linear ridges. Compare with Fig. 6.23. *Image* courtesy of Alexander Hayes

**Fig. 6.25** A corridor of barchan-like forms seen from a kite in the United Arab Emirates. These dunes have slip faces on both sides, indicating that the wind regime here is a somewhat reversing one. *Photo* R. Lorenz

**Fig. 6.26** Magnificent star dunes in Algeria, seen from the International Space Station. Notice the long meandering arms. *Image* NASA JSC EOL

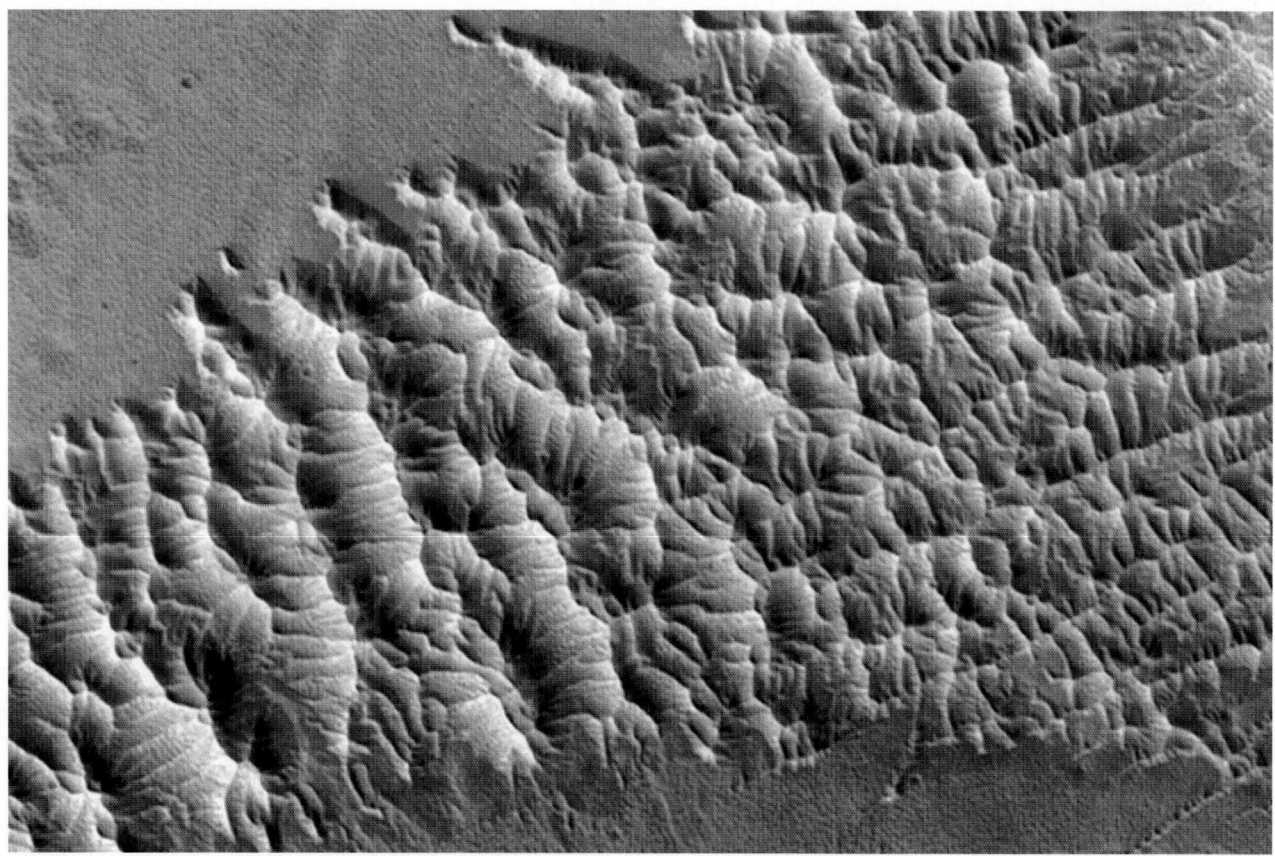

**Fig. 6.27** Star or akle dunes on Mars. HiRISE image PSP_008323_1735, NASA/JPL/U.Arizona

orientation, depending on the most recent orientation of a sand-driving wind. Star dunes tend to grow vertically through sand accretion from the sides rather than to show much movement laterally. The tallest dunes on Earth tend to be star dunes, and the best examples are to be found in the Algerian Sahara, where they are called ghourds or rhourds (Fig. 6.26).

Recent numerical simulations of star dunes (Zhang et al. 2012) suggests that in fact star dunes only grow extensive arms when the wind comes from an odd number of equal-spaced directions (typically three or five), in which case the arms grow upwind in the respective directions. When four or six wind directions are involved, a square or hexagonal pyramid—without long arms—results.

The star-like duneforms identified on Mars appear to be in general linked in a trellis-like pattern (i.e., akle, Fig. 6.27), perhaps as a result of forming as linear or transverse ridges that have then been subjected to a shift in wind direction.

# 7. Other Dunes and Other Sand Deposits

In this section we discuss some non-dune sand deposits (sheets and streaks), as well as those dune forms that are defined by topographic interaction or by vegetation. We furthermore discuss the superposition of multiple generations of dunes to create compound dunes.

## 7.1 Sheet

A sand sheet is basically a planar (flat) sand deposit lacking large individual accumulations of sand. Sand sheets are generally broad in extent, consisting of a flat surface except where sand accumulations are associated with vegetation or large blocks that are present on the sheet. No slip faces are present because topographically large sand deposits are lacking. Sand sheets are among the most aerially extensive sand deposits on Earth, and this may also be true for sand deposits on other planets. However, they of course are not prominent in remote sensing data.

## 7.2 Streak

Sand deposition and/or erosion can produce thin, elongate strips of sand, often (but not always) anchored to a topographic obstacle on the upwind end of the streak (Fig. 7.1). No slip faces are present along the elongate strips. Wind streaks are quite common on Earth, but their almost ubiquitous presence across the surface of Mars has stimulated considerable interest in studying their mechanisms of formation here on Earth. Wind streaks can result from either enhanced deposition or enhanced erosion, depending on the size of the particles involved in the creation of the brightness contrast that distinguishes the wind streak from its surroundings. Greeley et al. (1974a, b) studied such processes in the wind tunnel, noting a characteristic horseshoe-shaped vortex flow in the lee of circular obstacles. A few wind streaks have been observed in radar images of the Venus surface (Fig. 7.2) which are indicative of near-surface winds (the more obvious parabolic ejecta features (Fig. 2.7) seen around some recent craters are formed by upper atmospheric winds and are not of particular interest in this book except perhaps as a source of sand). Sand or dust streaks from the Victoria crater on Mars are shown in Fig. 8.2.

## 7.3 Shadow Dune (Lee Dune)

Linear ridges are generated by sand blowing over the top of an obstacle, where the ridges represent the paths followed by the majority of the sand as it moves into the wind shadow downwind of the obstacle. Usually no slip faces are present, and the ridges often break into barchan chains with increasing distance from the primary obstacle (Fig. 7.3).

This type of formation had been suggested by Rubin and Hesp (2009) as a possible origin of Titan's linear dunes, but the scale of the lee dunes in China they point to is orders of magnitude smaller than for Titan, and most Titan dunes lack an obvious obstacle to 'nucleate' lee forms.

An interesting variant of this dune type is in the Navajo areas of northern Arizona, where winds blow from the southwest across the Painted Desert and are driven into the steep scarp of the Moenkopi Plateau (Fig. 7.4). Linear dunes appear at the edge of the cliff, fed by sands from headward erosion of the cliff, or blown up gullies cut into it. In some places, climbing dunes form at the base of the cliff. Interestingly, the linear dunes form behind the headlands rather than behind the gullies, presumably because wind coming up the gullies sets up a possibly helical flow pattern that collimates the sand there. The dunes are typically a few meters thick, of order 10 m across, and in some cases about 1 km long. They generally lack slip faces. The dunes have some sparse vegetation but are evidently still active, having ripples orthogonal to the long axis. The linear dunes (as well as some parabolic and barchanoid forms on the plateau itself) and the vegetation associated with them are documented by Hack (1941). He notes that the longitudinal dunes are not found in areas with more than 25 cm of rainfall per year.

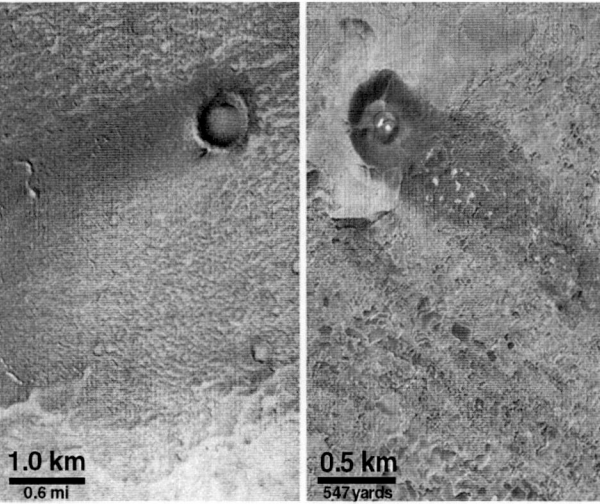

**Fig. 7.1** Dark wind streaks downwind of conical obstructions, Mars and Earth. *Left* is from the Daedalia Planum region of Mars (MOC); *right* is near Amboy in southern California (USGS); both views are from a portion of MOC2-206, 2/7/00, NASA/JPL/MSSS

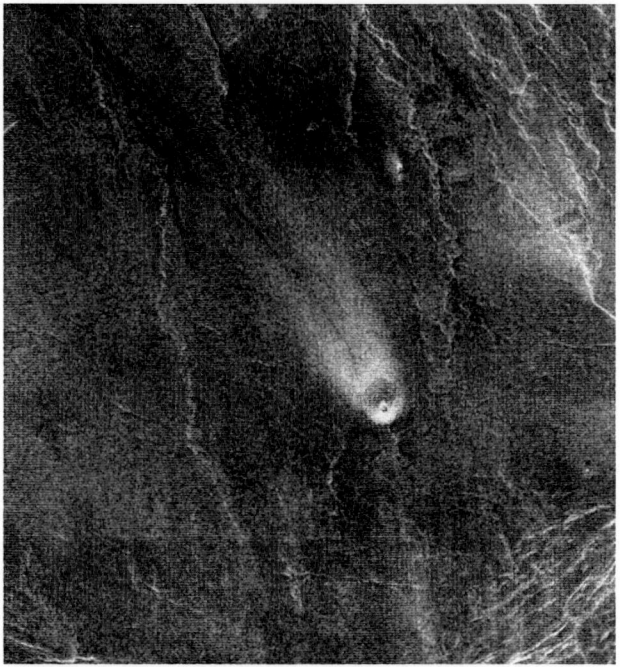

**Fig. 7.2** This Magellan radar image shows a bright wind streak associated with a small volcano. The streak is bright (rough) implying it is rough, and the dual 'horns' are characteristic of a 'horseshoe vortex' downing of an obstacle, suggesting that rather than the deposition of material from the volcano, the feature is actually due to the removal of fine material (roughening what is *left*) by scouring by winds. The volcano is about 5 km in diameter, and the wind streak is 35 km long and 10 km wide, located at 9.4S, 247.5E at the western end of Parga Chasma, Venus. Magellan press release P-38810

## 7.4 Climbing Dunes (and Falling Dunes)

This type of dune is essentially a sand deposit on the upwind side of a hill which interrupts a sand transport pathway. The sand accumulates forming a ramp (e.g., Fig. 7.5). The formation of these dunes has been simulated in wind tunnel experiments (e.g., Xianwan et al. 1999). Sand climbing a hill may accumulate on the downwind side, forming a 'falling' dune (Fig. 7.6). Obviously, a falling dune could grow to the point where it might become more properly classified as a lee dune (see Sect. 7.3) and a continuum of examples can be seen where sand transport pathways cross irregular topography (e.g., Figs. 1.12 and 7.7).

## 7.5 Echo Dune (Reflection Dune)

A ridge of sand that has accumulated near the base of a cliff or wall, where wind diverts upwards over the cliff and sand tends to pile up at a standoff distance from the base of the cliff. This standoff distance is larger the steeper the cliff (e.g., Qiam et al. 2011). This sort of deposit may be familiar when blowing snow encounters buildings and is often also seen at the small scale around boulders (e.g., Fig. 7.8).

Clearly, there is a transition between climbing dunes and echo dunes: this transition depends on the steepness of the obstacle (perhaps also on its scale). When the obstacle is sufficiently shallow, the streamlines merely divert over it, and the sand may form a shallow ramp. When the obstacle is steep, the streamlines of the flow must depart from horizontal well before the obstacle, and a separation bubble forms (with a reverse airflow—essentially the same effect as the lee vortex discussed in Chap. 20). This separation may occur between about half and double the obstacle height (e.g., Tsoar 1983) and leads to sand accumulation in this range.

## 7.6 Lunette

A crescent-shaped dune downwind of a dry lake-bed (often referred to by the Spanish word 'playa'; coastal salt flats are also sometimes called by the Arabic term 'sabkha'). As the name suggests, the overall shape is similar to that of a narrow crescent moon. Often lunettes are comprised of sand-sized aggregates of silt and clay-sized particles derived from the lakebed. The close proximity of a lunette to a playa is really all that distinguishes this crescent-shaped feature from the more common barchan dune.

## 7.7 Nebka

Nebka is sand (typically displaying some degree of soil development) around large plants, where erosion has lowered the surroundings more than that of the materials

**Fig. 7.3** A view downwind from the small Superstition Mountains in Southern California, where linear dunes several hundred meters long have formed in the lee of the mountains. When these were documented by Greeley et al. (1978), they trailed into chains of barchans (as at Mars in Fig. 7.7), but the sand mobility may have been reduced since the 1970s when that study was made. The site is presently used as an area for off-road driving (see Chap. 22). The proliferation of vehicle tracks on these dunes may influence the morphology by rounding slip faces and influencing the sand mobility. *Photo* R. Lorenz

**Fig. 7.4** Moenkopi plateau in Northern Arizona, from a commercial airliner looking East. Note that the dunes tail from headlands rather than the gullies up which the sand presumably moves. *Photo* R. Lorenz

surrounding the base of the plants. In some instances (e.g., Fig. 7.9) the attenuation of the airflow by the plant allows a lee dune to form. Nebka (another Arabic term; they can also be called 'coppice dunes') can be misinterpreted to be dome dunes with plants on top of them. On other planets, large blocks or other obstacles might take the place of the stabilizing vegetation, sheltering sand and indurated fines from wind erosion.

## 7.8 Parabolic Dune

A 'U'-shaped feature with two horns that are typically much longer than the thick arcuate deposit that connects the horns on the downwind side of the feature (Fig. 7.10). The horns of the 'U' are stabilized by vegetation, although possibly rocks or large blocks might serve as the stabilizing role on other planetary surfaces, and only a single slip face is

**Fig. 7.5** Without wind information it is hard to judge if this is a climbing dune or a falling dune, but most probably the former. In a human sense, however, it is literally a climbing dune! Sand has piled up against a rock massif in the popular Wadi Rum area in Jordan. Tourists (just visible silhouetted at the top of the dune) are invited to climb up the steep sand slope. *Photo* R. Lorenz

**Fig. 7.6** Cat dune in the Mojave Desert—an example of a falling dune. *Photo* J. Zimbelman

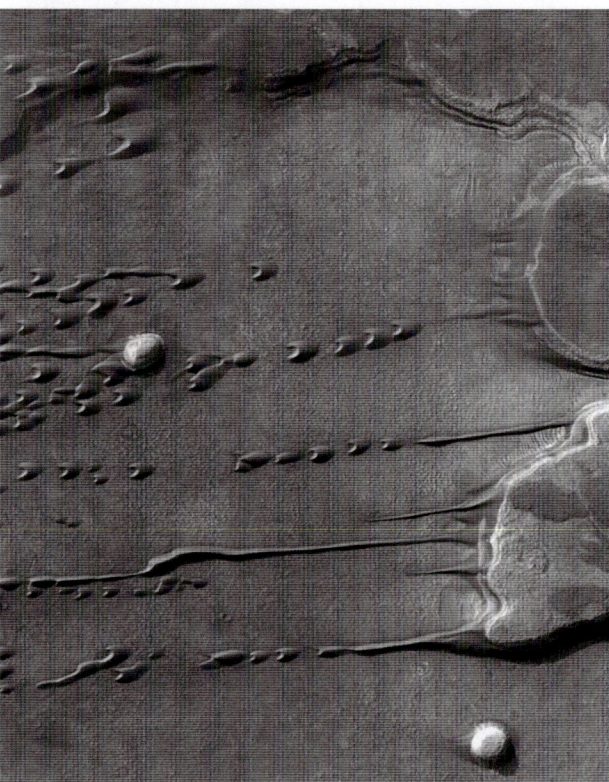

**Fig. 7.7** Dunes on the lee side of plateau (*right*) on Mars. Towards the top the dunes are short, and might be classified as falling dunes, but grade to much more definitely lee dunes towards the bottom. As the linear lee dunes peter out, a chain of barchans takes their place. *Image* NASA

present on the downwind side of the arcuate central section, indicating formation under a unimodal wind regime. The slip face on parabolic dunes occurs on the side opposite to that of the horns, so that the horns point in the upwind direction, the opposite to the wind orientation for the horns of a barchan dune.

**Fig. 7.8** A ripples showing the 'echo' effect on Mars. The innermost ripple follows the shape of, but is standing off a short distance away from, the aerodynamic obstacle, a rock named 'Stimpy'. Image from the Sojourner rover on Sol 70 of its mission. The blank line across the image is where data were lost during transmission between the rover and the lander. NASA/JPL image

**Fig. 7.9** Nebkhas on the sebkha. Small dunes forming in the lee of salt-tolerant plants on the Chott El-Djerid, in Tunisia. See also snow dune in the lee of a tent in Fig. 2.1, and sand drifts in the lee of boulders on Mars. *Photo* R. Lorenz

## 7.9 Blowout Dune

Disruption of a dune surface by locally enhanced wind scour and the alteration of any adjacent dune pattern (Fig. 7.10b). There may be one or more slip faces present on the original dune form on which the blowout developed, but the blowout itself generally lacks a slip face. Blowouts most often occur on dunes previously stabilized by vegetation, and the blowout forms where that stabilizing vegetation has been disrupted or destroyed, as by a localized fire.

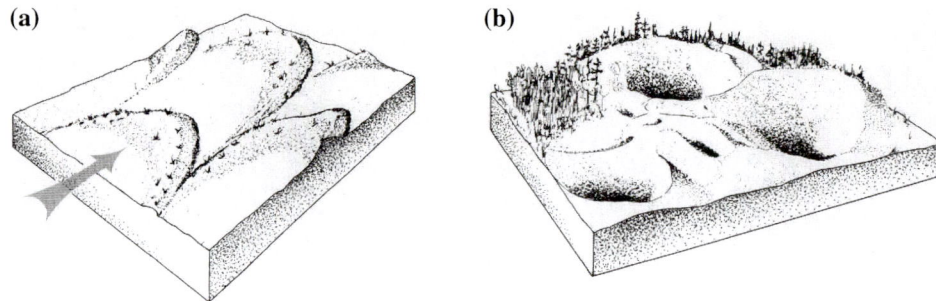

**Fig. 7.10** a, b Parabolic dunes and blowout dunes (from Fig. 9 of McKee 1979). Arrow indicates the unimodal wind regime. Horns point into the upwind direction, opposite from what the orientation would be for barchan dunes. *Image* USGS

**Fig. 7.11** ASTER image of large linears ('uruq') in the Rub' Al Khali. Superposed on the somewhat rounded giant linear dunes are a smaller set of dunes, inclined at a slight angle. Image courtesy of NASA/GSFC/METI/ERSDAC/JAROS, and U.S./Japan ASTER Science Team

## 7.10 Compound Dunes and Complex Dunes

Compound forms consist of two or more of the basic dune types formed by some combination of sand supply and wind regime changes, resulting in the close association of two distinct scales of dune forms. Examples of common compound dunes include coalescing barchanoid ridges, coalescing star dunes, small barchans on the flanks of large barchans, small parabolic dunes between the arms of a large parabolic dune, and large sand ridges superposed by smaller linear dunes. Successive generations of bedform are seen particularly on linear dunes (which may in general be older than other types), for example, Figs. 7.11 and 7.12.

Ewing and Kocurek (2010) (see Chap. 19) have developed 'pattern analysis' approaches to disentangle successive generations of superposed dunes, identifying, e.g., three generations in the Namib. The computational approaches described in Chap. 19 offer promise that it may be possible to quantitatively relate such superposed duneforms to a wind regime history.

It is common to see small duneforms develop somewhat transversal to the orientation of large linear dunes. These may result from flow along the linear forced by the presence

**Fig. 7.12** A similar superposition is seen in Oman in this SPOT image. Here the superposed smaller dunes are much more extensive and almost hide the larger underlying uruq pattern beneath at the bottom of the image

**Fig. 7.13** Linear dune in the Rub' Al Khali in eastern United Arab Emirates, with regular transverse ridges developing along its flank. *Photo* J. Radebaugh

**Fig. 7.14** Linear dunes in the Rub' Al Khali breaking up into chains of stars (or perhaps crescentic peaks; see also Fig. 6.23), with near-orthogonal trails of sand forming a trellis pattern. Field inspection shows some of these orthogonal trails of sand to be streams of reversing barchans (see Fig. 6.25). The overall arrangement of these dunes likely reflects a transitional morphology as the dune pattern slowly responds to a change in wind regime. *Credit* Spot Image

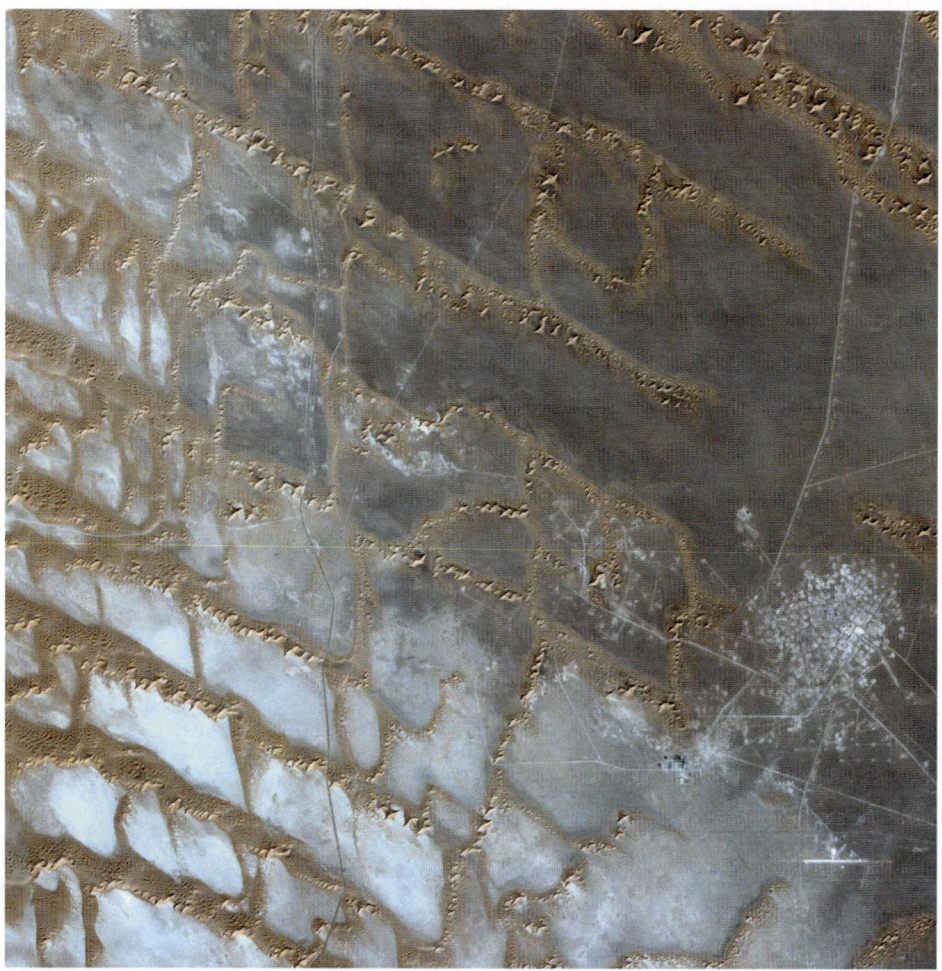

**Fig. 7.15** The monster dunes in the Badain Jaran desert in China are usually described as star dunes, although they are somewhat crescentic in form, suggesting perhaps the imposition of a more unimodal recent wind regime. They are notable in that the water table is sufficiently close to the surface that many dunes have lakes at their base. Compare with a field photo (Fig. 11.7) and with an airplane window view (Fig. 18.1). ASTER image courtesy of NASA/GSFC/METI/ERSDAC/JAROS, and U.S./Japan ASTER Science Team

## 7.10 Compound Dunes and Complex Dunes

**Fig. 7.16** Megabarchans in the Rub' Al Khali desert, seen from the International Space Station (south is *up*). These dunes have a spacing of about 2 km, with ∼100 m superposed barchanoid forms. *Credit* NASA JSC EOL

**Fig. 7.17** Complex dunes of the Algodones in southern California, seen from the International Space Station. For a view of the same dunefield from an airliner, see Fig. 24.6; for a radar view of these dunes, see Fig. 18.18. West is up. Irrigated fields in Mexico are at left, just beyond the 'sand dragon' border fence (see Fig. 23.9). The Pan American canal, and Interstate-10 cut across the dunefield, which grades from some narrow linears and small barchanoids at the top, to massive (∼1 km-spaced) compound ridges, and a sand sheet at the bottom. *Credit* NASA/JSC/EOL

of the dune itself. Some examples on Namib linears can be seen in Fig. 18.24. A field example with small incipient transverse ridges is shown in Fig. 7.13.

Dong et al. (2010) recognize this form with subsidiary ridges perpendicular to the main linear dunes in the Kumtagh desert in China as 'raked' linear dunes. It may be that these are the result of a change in wind regime—dunes built up as linears in a bidirectional regime and winds from one of those directions have since diminished, resulting in the accumulated sand now blowing in a more unidirectional fashion (e.g., Fig. 7.14). This leads to barchan-like slip faces periodically along the former crest; a similar effect is seen in water tank experiments by Reffet et al. (2010). An alternative idea is that ridges may grow as an 'orthogonal mode', creeping upwind in the manner that the arms of star dunes grow. More work is needed—likely exploiting numerical models predominantly (Chap. 19) to elucidate these evolutionary mechanisms.

It may be that an analogous evolution leads to the somewhat consistent slip face orientation in the giant dunes in the Badain Jaran desert (Fig. 7.15). A classic type of compound dune is the megabarchan. This is a draa-scale crescentic form with a large slipface (see Fig. 7.16) upon which many smaller barchanoid forms are superposed. Although we have tried to emphasize the potential simplicity in decoding superposed dune patterns, even relatively small dunefields (e.g., the Algodones, Fig. 7.17) can exhibit striking complexity.

# Dune Fields, Sand Seas and Transport Pathways

One of the most striking things about dunes is their tendency to congregate in awe-inspiring expanses, sand seas. These are reviewed in a paper (Wilson 1973) with the deliciously laconic title 'ergs', invoking the Arabic word often used to refer to a sand sea. Why should sand do this? From the process perspective, just as for the formation of an individual dune, the accumulation just means that sand has entered a region at a higher rate than it has left.

For an active system (rather than a fossil one for which see Sect. 8.2), this can be due to several factors (a good discussion of the formation of sand seas can be found in Pye and Tsoar (1990). First, topographic barriers can cause sand to accumulate because the sand cannot move as easily uphill; most Mars crater dune fields, and the Great Sand Dunes in Colorado (Fig. 8.1) are of this type. Second, a convergent wind regime can lead to reduced net sand motion. Sand is brought into a region where it moves as far forward as back each year, and thus accumulates: this is generally the process by which linear and star dunes—the largest types—form. The major sand seas are in this category, although also smaller dunefields such as the Stovepipe Dunes in Death Valley. Finally, winds passing through a topographic constriction may fall in speed as they expand away from it, causing deposition (the Coral Pink Sand Dunes in southern Utah are one example): a similar effect enhances deposition of sand in craters.

The geomorphology of dunes in the absence of topographic barriers may attest to the age of the deposit. Linear and star dunes are forms that accumulate more-or-less statically and thus can be very old, whether they are active at the present day or not. On the other hand, transverse— and especially barchan dunes—indicate a dominant transport direction, and thus if the dunes are not fossilized, they must be in motion (see Chap. 8). If they are in motion, then there must be a sand supply and active transport. Barchans are often found in corridors, defining the transport pathway from their source: an isolated patch of barchans implies that the sand supply or the conditions to move it have been short in duration.

## 8.1 Sand Sources and Sinks

The source of sands on Earth is usually river action, and many sand seas have a direct association with river sources; for example, the Namib sands are derived from rivers draining into the Atlantic. On geologically short timescales, sand may accumulate in some regions like depressions, or may swirl around in closed paths, notably in the Sahara and Australian deserts. In the long-term at the planetary scale, one might expect most sand on Earth to eventually blow into the sea. This might be thought of as a one-way trip, but over geological time, tectonic movements and other changes bring that sediment back into play either as re-exposed sands and evaporitic deposits, as sandstone or limestone which can weather out, or (via subduction) as volcanic rocks.

This rock cycle does not apply (as far as we know) to the other planets. Mars may have had seas in the past, but at present the sand is free to swirl around forever. In fact, most dunefields appear to be in local depressions (notably impact craters—e.g., Fig. 8.2; see also Fig. 18.16) with the exception of the north polar erg.

Titan was expected to have seas, which would act as sand sinks and led to some pessimistic early expectations about dunes on that world (Lorenz et al. 1995). However, at the present epoch at least, the seas cover only a few percent of the polar regions, and the low latitudes are dry and substantially covered in dunes.

A variety of forensic approaches can be used on terrestrial dunes to ascertain the provenance of the sand. In addition to remote sensing or mapping (from which the pathways can be obvious—see Figs. 8.3 and 8.4, and 12.6), the composition of

**Fig. 8.1** An image from the International Space Station, looking roughly north, showing Great Sand Dunes National Park and Preserve. The sand is sourced from rivers to the west, and is topographically trapped by the dark Rocky Mountains. Compare with sand trapped on Mars by crater walls (Fig. 18.16). Frame ISS016E006986, taken with 800 mm lens (NASA/JSC)

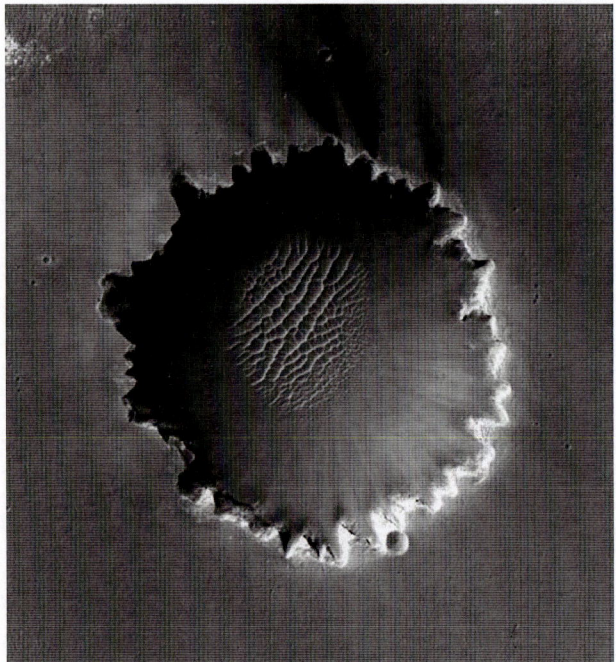

**Fig. 8.2** Victoria Crater on Mars, imaged from orbit. This crater is 800 m in diameter, and sports a network of duneforms on its floor. This crater is particularly well-studied, as the Opportunity rover drove around part of it. Note the dark streaks emanating from the crater, suggesting that material weathering out of the crater walls is being transported away. HiRISE image, *credit* NASA/JPL/U.Arizona

the sand (Chap. 2) can be evaluated to match candidate source regions, such as the study of southwestern USA dune sands and the role of the Colorado river in providing the source (Muhs et al. 2004), or the Rub' Al Khali study by Besler (2008). The sand texture is also diagnostic, in that rounded and fine sands are likely to have been transported further.

For Mars, we are forced to rely on what composition can be inferred spectroscopically from orbit, and from geomorphology. For example, the orientation of dunes can be used to infer the sand transport direction. Some global studies have been made since the Viking era (e.g., Breed et al. 1979). The only very prominent sand accumulation obvious in that data was the north polar erg, Olympia Undae (e.g., Tsoar et al. 1979). The dunes are dark, suggesting a basaltic composition, although they also contain gypsum, which may indicate evaporite deposits as a source for some material. No specific source region has been identified for Olympia, although it may be that sand comes from everywhere. A numerical experiment (Anderson et al. 1999) of putting sand all over Mars and observing where it migrated due to winds simulated in a general circulation model (Chap. 19) noted that sand accumulated in the northern polar erg.

The growing body of data from Mars missions has now permitted some regional and local studies. Fenton (2005) considered the dunefields (mostly on the floors of impact craters) in the Noachis Terra region in the southern highlands, suggesting that most of the sand must be sourced locally, since large-scale transport patterns were not visible. Silvestro et al. (2010b) have studied six dark dune fields in Aonia Terra in the eastern Thamasia region of Mars and have attempted to interpret the sand transport pathways and sources from imaging and topography data.

For Venus, there is not much to say given the present meager dune inventory: the sand source is probably the limiting factor in dune formation and is probably local to

## 8.1 Sand Sources and Sinks

**Fig. 8.3** An example of a regional sand transport system, viewed from an airliner. The Kelso dunes in southern California are seen at the bottom of the picture, looking north: the sand accumulates in part due to topographic blocking and in part due to convergent winds. A dominant pathway for the sand (sourced primarily from the Mojave river) is evident towards the top of the picture, where lee dunes and sand streaks in the Devil's Playground show sand migrating eastwards into the Kelso basin where it then turns south to pile up in the main Kelso dune system. From the tip of the arrow to the rightmost end of the dunes is ∼30 km. *Photo* R. Lorenz

**Fig. 8.4** Sand of different compositions (false color highlighting the differences) 'flowing' in a larger-scale sand transport system near the Terkezi Oasis in Libya. Image is about 120 km across. Even larger-scale sand transport systems are visible in remote sensing of Earth (see, e.g., Fig. 12.6). US Geological Survey image from Landsat 7

the dunes. With Titan there is essentially the opposite problem. The dunes form a long and nearly contiguous belt around the equator, with neither an obvious beginning nor an end. The sand source has not been identified, although one speculation is that the sand may form in the polar seas: if this is the case, then somehow the sand must not only migrate ∼4000 km from pole to equator, but must do so uphill. Clearly, much work remains to be done on this topic.

## 8.2 Dune Stabilization

Large deposits of sand can be found on Earth which are not active, or are only weakly so, and so the dunes are not especially large or prominent; examples include the Nebraska sand hills, the Kalahari, and some of the Australian deserts. This is usually because vegetation cover (see Figs. 8.5 and 8.6) becomes sufficient to halt significant movement of sand by the wind; both sand supply and sand loss essentially go to zero. The ability of vegetation to grow is largely a function of rainfall amount, which accounts for the great attention paid to that quantity in terrestrial arid land studies. As noted in Sect. 1.2, rainfall patterns can change in both directions, and it is expected that the Kalahari desert

**Fig. 8.5** Small linear dunes, stabilized by vegetation in the Arica Hills, California. *Photo* J. Zimbelman

**Fig. 8.6** Farmland in Australia seen from an airliner, productive in today's moister conditions in spite of the evident presence of sand dunes (linear dunes, in south part of the Simpson Desert). *Photo* Jani Radebaugh

may substantially remobilize. The change in sand mobility can also lead to a change in morphology; e.g., Wolfe and Hugenholtz (2009) show that what are now parabolic dunes on the Canadian prairies were barchan dunes in a drier period about 200 years ago.

Vegetation can become the dominant 'roughness elements' of the surface, causing the height of zero wind speed in the boundary layer to rise high enough so that sand grains can no longer be induced to move (and thus sand accumulates in the lee of individual plants, forming nebkhas; Fig. 7.18). Many field experiments have documented the roughness height represented by vegetation; the effect on the wind is a complex interplay between both the height of individual plants and their spacing on the surface (see also the discussion in Sect. 1.3 on deliberate introduction of vegetation to control dunes). Once vegetation becomes sufficiently abundant, the surface becomes effectively insulated from the wind, except perhaps where turbulence may be generated immediately around large obstacles, but such localized scour is not capable of carrying out regional transportation of the sand at the surface. Vegetation also has the effect of introducing organic matter into the sand, making it more cohesive.

It is possible to immobilize dunes without invoking vegetation. Enough moisture alone can reduce transport substantially. Other mechanisms on Earth include induration by ice (see Chap. 5) and cementation by soluble minerals such as gypsum.

It was speculated for some time that many dunes on Mars are immobilized via similar mechanisms—the fact that a few examples of dune and ripple migration (Chap. 9) have now been documented does not detract from the fact that dune migration has not been generally observed, and indeed observation of blocky eroding barchan forms (Fig. 12.25) shows that some dunes have been indurated.

On Titan it is assumed that the large equatorial sand seas have active dunes, but some outlier features at higher latitudes have been observed (with different orientations from the main sand seas, and somewhat different radar appearance) and it has been speculated that these may be fossil dunes from a previous climate epoch.

# 9  Rates of Geomorphic Change

In this section, we discuss the net transport and what it means in terms of bedform construction and migration. Had we written this section only 2 years ago, it could have confined itself to the Earth, but dune and ripple movement on Mars has now been directly observed and quantified.

## 9.1 The Sand Rose

The systematic formalization of the variability and strength of wind at a site was introduced by Fryberger (1978), and his convention is widely used today. For wind blowing in a given direction at speed U, the sand flux Q will be $Q = U^2(U - u_t)f$, (an expression derived from Lettau and Lettau (1969)) where $u_t$ is the threshold windspeed, and f is the fraction of time (usually expressed as a percentage) that this wind persists: if $U < u_t$, then $Q = 0$. To calculate the net sand transport in a given direction, Q is calculated for all the windspeeds encountered. Because of the near-cubic dependence on windspeed, it may be that relatively infrequent winds may dominate the transport: if $u_t = 5$, and $U = 20$ occurs only 1 % of the time while $U = 10$ occurs 10 % of the time, the $U = 20$ contribution will be larger than that of $U = 10$. The sum of these contributions is a vector in the wind direction, and is referred to as a Drift Potential (DP). It can be plotted as a line, and the set of DPs ('arms')—usually for the 16 points of the compass—plotted together (following meteorological convention, with a line defining the direction from which the wind came) is called a sand rose.

The magnitudes of all the DPs added together is also referred to as the Drift Potential, and is essentially a non-linear measure of the windiness of a place. However, of most interest for overall transport is the resultant when the DPs are added vectorially, such that an East drift cancels out a West drift, etc. This vector sum has a magnitude (the Resultant Drift Potential, RDP) and a direction (Resultant Drift Direction). This vector is displayed on the sand rose as an arrow with a head to discriminate it from the contributing DP lines. The vector is shown in the direction of the resultant sand transport: in the case of a near-unidirectional wind regime, this has the convenient effect of avoiding the close superposition of the resultant vector on top of the larger wind lines. Figure 9.1 shows some examples.

The RDP divided by the DP is a measure of the wind unidirectionality. If the quantity approaches unity, the wind is unidirectional, whereas a smaller ratio indicates winds that are variable in direction. This gives a quantitative metric against which to compare resultant morphologies. Figure 9.2 shows this mapping. It should be noted, though it is rarely stated, that the magnitudes of the individual DPs, and thus both the magnitude and direction of the resultant, are explicitly dependent on the choice of $u_t$. The methodology for computing the sand flux in Fryberger (1979) is arranged somewhat archaically towards hand computation via tables of binned windspeeds and display with a plotter, now it is straightforward to simply add the east and west components of sand transport vectors in a script or spreadsheet and compute the results from a wind measurement time series directly. Some care must be taken, however (Bullard 1987) in using units: Fryberger's methodology somewhat implicitly assumes the use of knots throughout.

## 9.2 Sand Fluxes

Fryberger suggests three broad categories of transport environments, based on the magnitude (in 'vector units' or VU) of the RDP: low energy (<200 VU), intermediate (200–399 VU) and high (>400 VU).

The sand transport rate relates directly to the RDP (Fryberger shows a plot, using the transport relation of Lettau and Lettau (1969)), with a constant of proportionality of $\sim 0.07$ m$^3$/m/year per VU (the constant depends slightly on wind threshold, and on sediment density). In other words, a low/intermediate environment with DRP = 200 would see a sand flux of about 14 m$^3$/m/year. Note that this transport relation is only one of several—Kok et al. (2012)

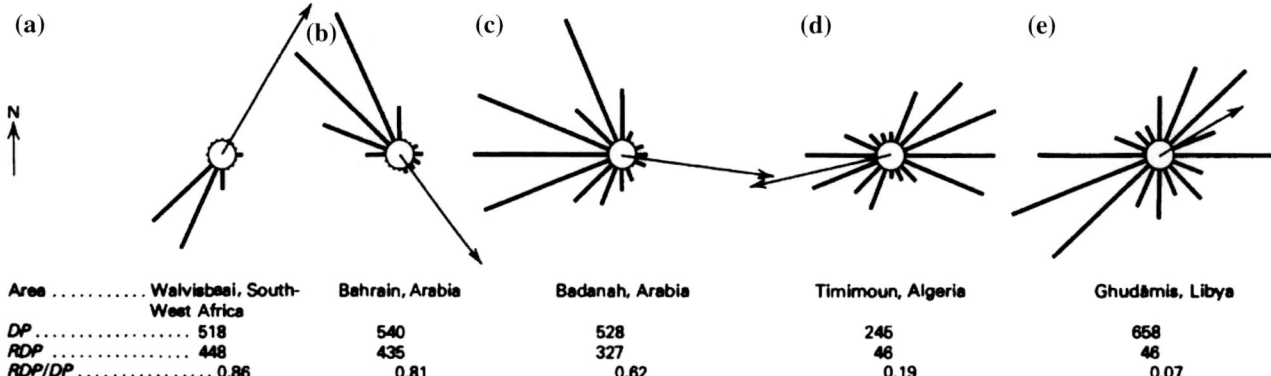

**Fig. 9.1** The sand rose diagram popularized by Fryberger (1979) in McKee's (1979a) *A Study of Global Sand Seas*. The thick bars denote the integrated drift potentials from each direction, and the arrow indicates the direction to which the resultant drift potential (RDP) will move the sand—this switch in direction usually makes the diagrams much clearer in near-unidirectional cases. The shape of the plot indicates the type of dune that should result, and the length of the arrow suggests how much sand should move over the course of the year (or whatever period the diagram refers to). The length of all the thick bars added together is the total drift potential (DP). *Image USGS*

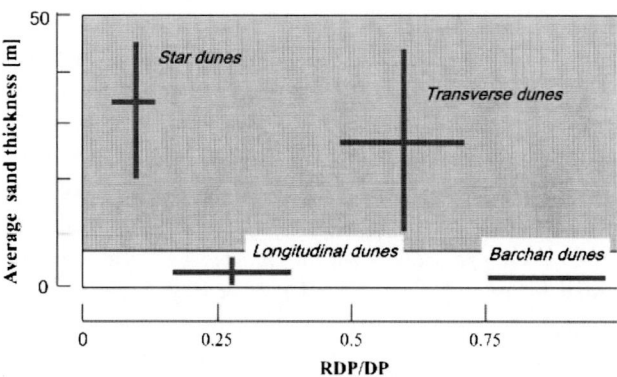

**Fig. 9.2** Quantitative regime diagram, originally due to Wasson and Hyde (1983). Compare with Fig. 6.1

list and compare no less than seven mass flux equations and compare with laboratory data. As noted in Chap. 3, wind is strongly intermittent, and sand transport even more so, so that measurements (see Chap. 16) are challenging. Over typical ranges of shear velocity (from one to a few times the threshold) experimental data have a scatter of about a factor of 4, and the various proposed expressions have similar variation. One could stick, for convention's sake, with the Lettau/Fryberger expression, or for accuracy one could choose a $Q \sim u_{*t}(u_* - u_{*t})^2$ form advocated by Kok et al. (2012), or classicists may have affection for Bagnold's original $Q \sim u_*^3$ expression. The results will not be dramatically different (prefactors have been omitted for clarity in this discussion, and the prefactors will be different for different planets).

While we will devote the rest of this section to discussing the sand flux as manifested in dune and ripple movement, we should of course note that the sand flux can be measured directly with sand traps (see Chap. 16).

The flux is the volume of sand that crosses a span of unit length in unit time: its units are thus m³ per m per year, sometimes written in the dimensionally-equivalent form of m²/year. It is a simple matter of geometry to understand that, for a given migration speed, a bigger dune implies a larger sand flux. Thus for a given sand flux, a larger dune will move more slowly than a small one.

Let us consider a flat plain with a small barchan, 20 m across with a height of 1 m. We can approximate its volume as that of a cone with these dimensions, and thus ~100 m³. The average cross-section of the dune (i.e., the volume divided by its span) is 5 m². Hence, if the sand flux is 5 m³/m/year, then the 20 m span sees a flux of 100 m³ in 1 year, and thus the dune moves by 100/5 = 20 m.

The calculation is simpler for a transverse dune (or the slipface of a megabarchan) where the cross-section does not vary over the timescale considered. For the same sand transport rate (which is that in the Rub' Al Khali desert) of 5 m³/m/year, then a slipface 50 m tall will advance 5/50 = 0.1 m. A date palm that began growing in the clear a few meters in front of a dune may become buried in a few decades, as is seen in the Liwa Oasis (see Fig. 23.5; Lorenz and Radebaugh, submitted).

## 9.3 Observed Dune Migration Rates

Virtually all the published dune migration rates (e.g., Figs. 9.3 and 9.8) are for barchan dunes. This is because they are easy to measure—they are the fastest-moving dunes, and because they form in areas of poor sand supply, the dune is usually easy to demarcate against a flat and distinct plain. Although transverse dunes may have appreciable movement, the shape of the dune is difficult to

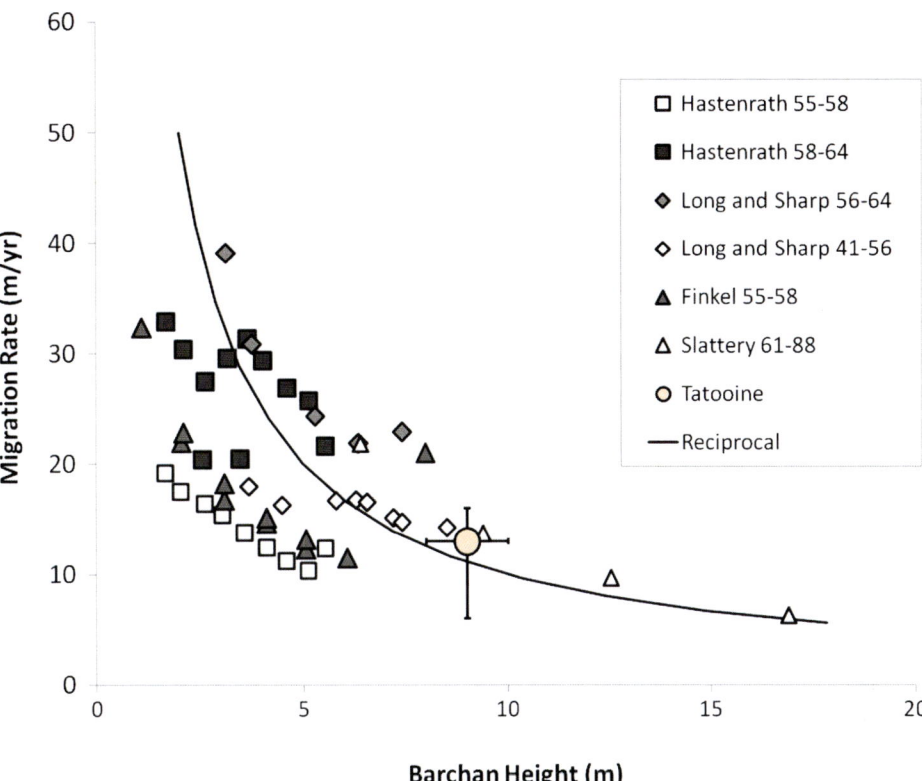

**Fig. 9.3** A compilation of barchan migration rates: numbers after the dataset authors indicate the years over which migration was observed. The Tatooine data point is a barchan at the Star Wars film set in Tunisia (see Chap. 24, and Lorenz et al. (2013)). A candidate reciprocal function as described in the text, with $Q = 300$ m$^2$/year and $H_o = 3$ m is shown. The scatter of points at a given location about such a curve is small—the scatter of the points in the plot overall is largely due to the difference in transport rate (Q) between different sites (Hastenrath and Finkel measurements are in Peru; Long and Sharp at the Salton Sea, USA, Slattery in Namibia)

discriminate against the fluid sand background. Similarly, stakes or other reference marks may be easily buried. Star dunes are, by and large, static. Linear dunes similarly accumulate in part because of a lack of a net transport direction, although may undergo a slight lateral migration. Over the course of a year, the change in direction of wind (that leads to the linear/longitudinal arrangement) may lead to a back-and-forth shift of the position and orientation of the crest, but the dune overall may not move much. The first quantitative dune migration rates (cited by Bagnold) are those of barchans at the Kharga Oasis in Egypt (see Fig. 6.5) in 1910. Bagnold notes the reciprocal relationship with height and that, at that migration rate, those barchans may have taken about 7000 years to reach their current location given an assumed source point at the beginning of the barchan stream.

Dune migration is now often measured on Earth from aerial photographs and, increasingly, satellite images can be used to measure movement on a larger scale. It is essential to have fiducial markers—in a great sea of sand where all the dunes move a comparable amount, there may be no fixed reference points to detect the motion. Modern methods use automatic correlation techniques to determine fixed references, an example being the analysis by Vermeesch and Drake (2008) showing migration rates of ∼25 m/year for quite large dunes (tens of meters high) and an implied sand flux of 600 m$^3$/m/year. It should be noted that this very high value may be due to the sand: a fine diatomite that likely has a lower density (and thus lower threshold speed) than typical quartz sands.

In some cases, bedrock exposures may be present, but often the most convenient fiducials are roads, railways or pipelines. Buildings are also effective (see, e.g., Fig. 9.4). Remote sensing is discussed in Chap. 18. One other remote technique, that merits further exploitation for dune and ripple migration studies, is radar interferometry, which can detect cm-scale changes in surfaces.

The traditional field approach to measuring dune movement is to embed stakes in the ground as reference points and measure the dune movement over the course of some years with a tape measure or theodolite, or perhaps with field photographs (see Chap. 16). Four studies of this sort are particularly notable. First, after using the astronomical identification of its location and the identification of some empty food and fuel tins to identify Camp 18 of Bagnold's 1930 expedition in the Sudan, it was possible to construct a 57-year record of a single barchan (Haynes 1989). This 16 m-high dune was observed to have a constant movement of ∼7 m/year. Another observation, by Lettau and Lettau (1969) of a barchan in Peru is notable in being time-resolved. They observed the barchan for two days, making contemporaneous wind measurements, and observed advance of 2–3 cm over a few hours in that period when the wind speed and direction was favorable. There is ample scope for repeating this type of observation using timelapse cameras today (see Sects. 9.4 and 16.2.2).

**Fig. 9.4** A sequence of commercial satellite images (GeoEye) of the Star Wars film set (see Chap. 24) in Tunisia, visualized with the Google Earth historical imaging too (see Lorenz et al. 2013). The movement of the dune, by some 14 m/year, is evident because the buildings serve as fiducial reference markers. The images were acquired on **a** 11 July 2004, **b** 21 January 2008 and **c** 25 September 2009. Image **c** was acquired the day before our field visit—see Fig. 1.16, where the dark band adjacent to the circular building is seen to be a temporarily wet area. Image **b** was acquired with a lower sun angle, and has more prominent shadows. In this image vehicle tracks are very evident on the barchan. Images courtesy of Google Earth

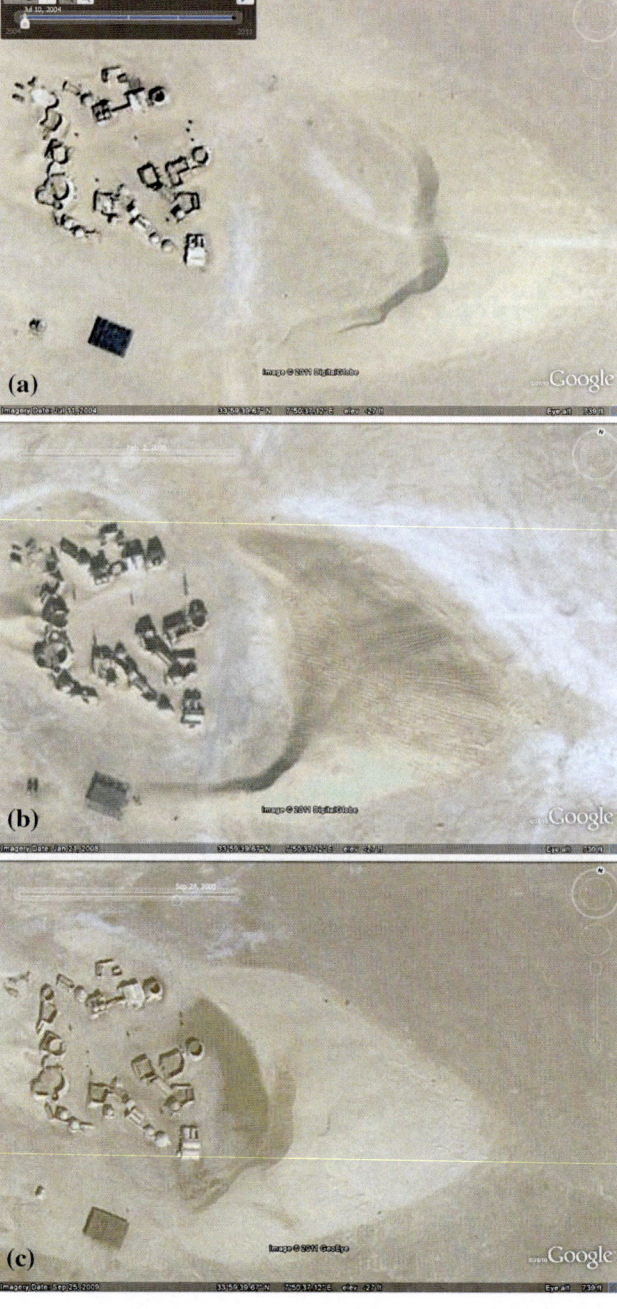

A third oft-cited barchan study, that of Long and Sharp (1964) at the Salton Sea in California, USA, is notable in that the studied dunes had somewhat diminished by the mid-1990s, perhaps due to land-use change and consequent effects on sand flux. Lastly, a relatively recent study by Bristow and Lancaster (2004) documented the movement of a small slipfaceless dome dune in the Namib between their observation in 1999, and when Ed McKee had marked its outline with stakes in 1976, 23 years earlier. The sand flux associated with the 45 m-wide, 1 m-high dune moving 90 m in this period is only ~2 % of the resultant potential sand flow expected, indicating that such shallow dunes are inefficient at trapping sand.

At the larger scale, walking the perimeter or the crest of a dune with a GPS receiver (e.g., Fig. 9.5; see also Sect. 16.2.1) is now an effective way of documenting the outline of a dune. Conventional handheld receivers are accurate to a meter or two, so displacements of the order of 10 m can be easily documented; geophysical-grade receivers with differential corrections are accurate to a few centimeters. If satellite images can be adequately referenced to GPS coordinates (e.g., by noting the position when

## 9.3 Observed Dune Migration Rates

**Fig. 9.5** A GPS track obtained simply by carrying an ordinary handheld receiver while walking the perimeter and slip faces of four dome and two barchan dunes south of Liwa in the United Arab Emirates. The track was made in December 2012, while the satellite image was obtained in November 2010. The smallest dunes have moved some 10–14 m, while the barchans have moved by 6–8 m. Image generated with Google Earth, with data from Lorenz and Radebaugh

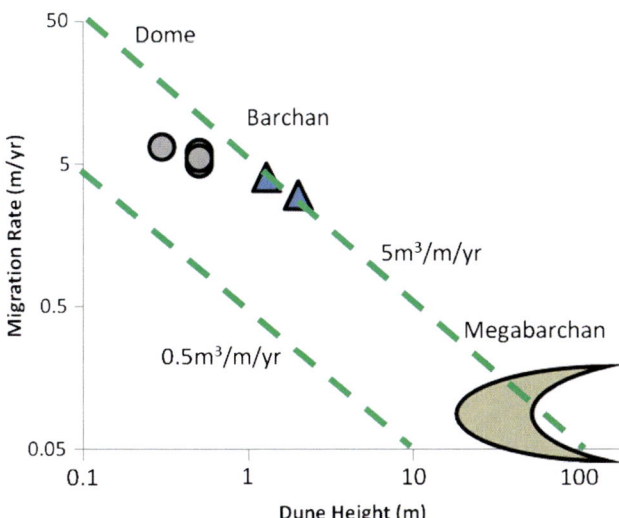

**Fig. 9.6** Migration rate versus size for very different sizes of dune at a single location (Liwa, United Arab Emirates) plotted on logarithmic axes. On such a plot a constant sand flux Q is a straight line (compare with Fig. 9.3), and it can be seen (as discussed in the text) that the migration of dunes varying two orders of magnitude in size are consistent with the same sand flux. Note that the dome sand fluxes appear slightly smaller than those for the barchans, perhaps because some sand saltates over the dome without contributing much to its movement. This plot highlights the fact that a dune migration rate is more or less meaningless without also knowing the height of the dune

standing on a road or other easily-noted image feature), then a dune migration can be compared by measuring the displacement of outline coordinates from the dune position in a satellite image.

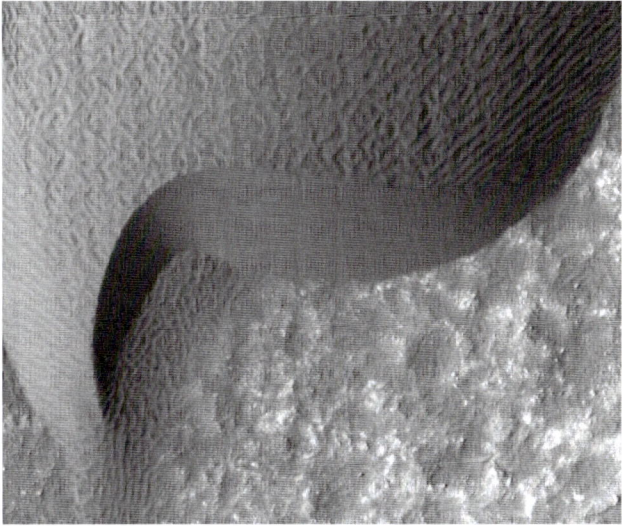

**Fig. 9.7** A HiRISE image of dunes in Niili Patera (see also Fig. 12.19) from which ripple and slipface migration rates have been measured. The key features are the intrinsically high resolution of the image, and a pattern of marks on the ground next to the dune to act as reference markers against which to refer the slipface movement. http://photojournal.jpl.nasa.gov/catalog/PIA14878. *Credit* NASA/JPL/U.Arizona

Historical migration rates can also be inferred from dune cores. For example, a core of a megabarchan in the Rub' Al Khali desert has sand exposure ages from optically-stimulated luminescence dating (see Chap. 17) that increase by 1 kyr per 20 m depth. If we imagine this accumulation rate is due to an advancing slip face, then it implies an advance

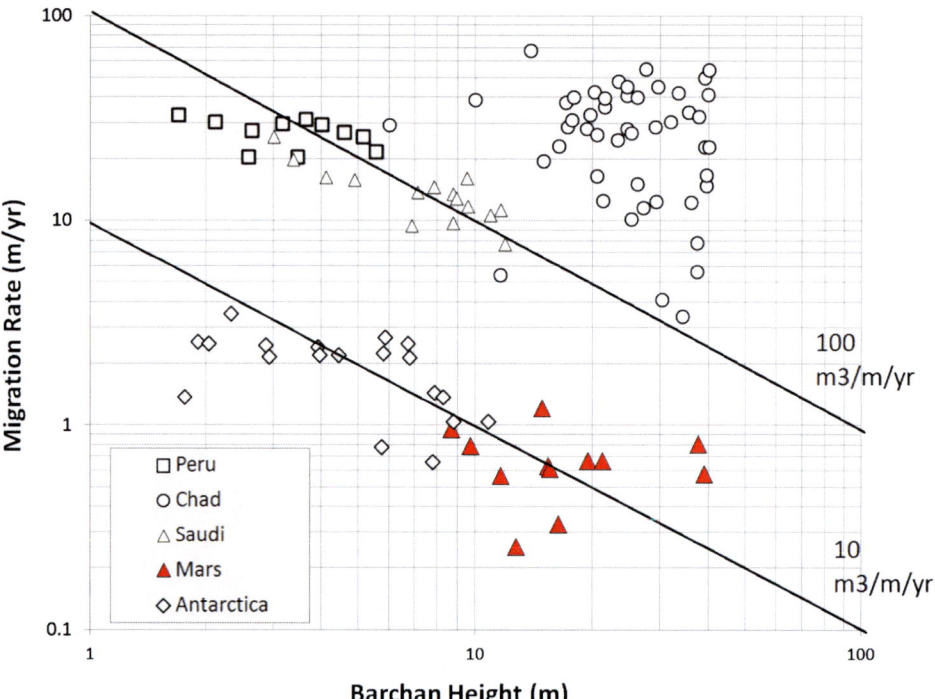

**Fig. 9.8** Dune migration rates from a variety of locations (compare with Figs. 9.6 and 9.3). The most extreme movements are seen in the low-density sands of the Bodele depression in Chad, while rather slow movement is documented for dunes in the Victoria Valley of Antarctica. These slow movements have now also been documented at Mars (Bridges et al. 2012b)

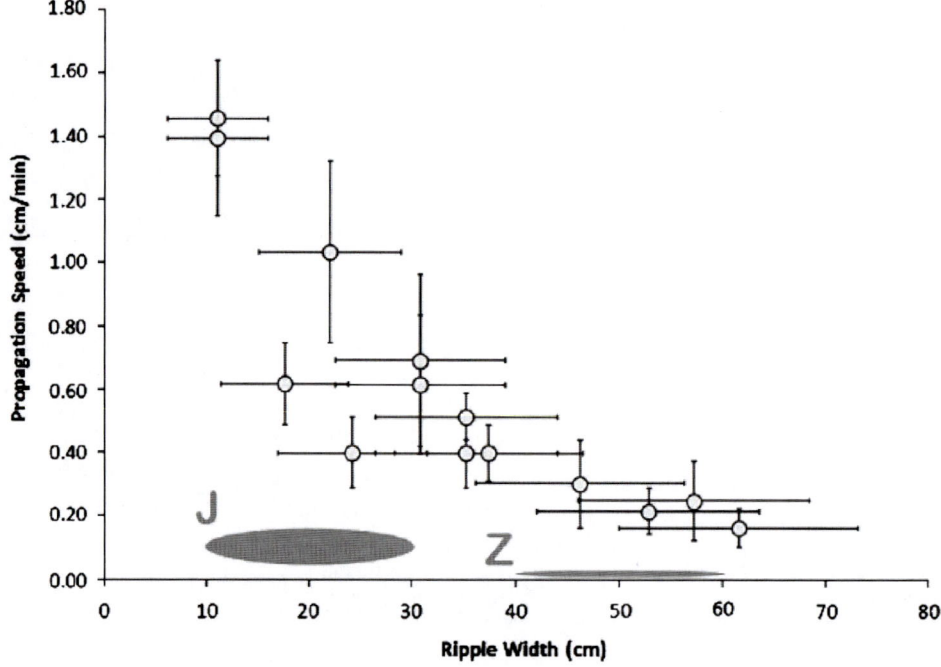

**Fig. 9.9** Circles with error bars denote ripple migration rates measured in the field from imaging obtained with a timelapse camera at Great Sand Dunes National Park and Preserve (Lorenz and Valdez 2011)

rate of 40 m/kyr, or 0.04 m/year (not too different from the ~0.1 m/year inferred from palm tree burial).

It is important to note that the sand flux will not be perfectly uniform across a dune field, or even across a single dune. Nor, for that matter, does the sand flux implied by a barchan slip face advance capture the entire sand flux, since some sand 'leaks out' of the barchan horns and some may even be launched from the brink in high winds fast enough to escape the slipface. These losses are balanced by the flux of sand intercepted by the stoss margin of the barchan. Similarly (see next section) ripple migration may capture even less of the sand flux. Nonetheless, the migration flux will usually be dominant, and is useful enough for cross-scale comparisons (see Fig. 9.6).

## 9.3 Observed Dune Migration Rates

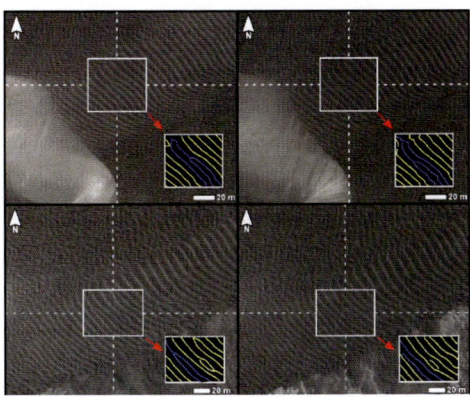

**Fig. 9.10** Two pairs of before and after images from the High Resolution Imaging Science Experiment (HiRISE) camera on NASA's Mars Reconnaissance Orbiter illustrate movement of ripples on dark sand dunes in the Nili Patera region of Mars. The three images on the left are excerpts from a June 30, 2007, observation (late autumn at the site). The three on the right are of the same ground observed 15 weeks later, on October 13, 2007 (winter at the site). Ripple crests discernable in the central portion of each image are diagrammed in the lower right portion of each image, with blue lines highlighting the largest changes. White scale bars in the bottom right of each of the images are 20 m (66 ft.) long. North is toward the top. http://www.nasa.gov/mission_pages/MRO/multimedia/pia12860.html. *Image credit* NASA/JPL/U.Arizona

Whereas for many years no quantitative measurements of aeolian change existed on Mars, higher resolution imaging from the HiRISE camera on the Mars Reconnaissance Orbiter has now found a number of sites where aeolian change can be observed. Chojnacki et al. (2011) report several dome dunes at the Endeavour crater (visited by the Opportunity rover) which had deflated away and/or migrated by 10–20 m. HiRISE imaging has also revealed displacements of the lee and stoss margins of several north polar dunes (Hansen et al. 2011), finding position differences (relative to the conveniently patterned ground beneath them) of between 2.2 and 4.7 m, corresponding to about 3 m/year. Most recently, Bridges et al. (2012b) documented both ripple and slip face movement in the barchans of Niili patera (Fig. 9.7), finding migration rates of the dune slip faces of the order of 1 m/year (Fig. 9.8). Although this is small (and comparable with some of the slowest migrations documented on Earth, in the Victoria Valley of Antarctica (Bourke et al. 2009)), given the size of the dunes, this corresponds to a respectable sand flux.

## 9.4 Ripple Migration Rates

Ripples are small enough that their migration can be noted in real time in the field. It is easy to mark the crest of a ripple with a stake like a pencil, and see that even just some minutes later the pencil is no longer at the crest. Measurements of ripple migration of the order of $\sim 1$ mm/min have been documented by examining a marker after some hours. Timelapse imaging (e.g., Lorenz 2011; Lorenz and Valdez 2011) has also been used (which can easily allow a range of ripples of different sizes at a locality to be measured) and has detected even faster motions—of the order of a few cm/min for the smallest ripples.

As with barchan observations, smaller ripples move the fastest, with an approximately reciprocal relationship between ripple height (or width) and migration rate (Fig. 9.9).

Note that the migration rate of granule ripples in particular may not completely capture the magnitude of saltation flux, since the granules that define the ripple crest only move slowly when nudged by saltating sand grains. Much sand may fly through the system without causing observable motion of the bedform.

The migration rates of mm−cm/min of course only pertain when the wind is fast enough to cause saltation, which may be only a small fraction of the time. For example, the site where the data in Fig. 9.9 were acquired saw no motion at all on 59 out of the 70-day observation period. Thus the average migration rate that might be inferred by comparing only two images widely separated in time would be considerably lower.

On Mars, ripples have been observed to move in such temporally-separated orbital imaging. For example, Silvestro et al. (2010) report ripple migration over the stoss side of dark barchan dunes in Nili Patera (Fig. 9.10). The measured average migration of $\sim 1.7$ m in four (terrestrial) months clearly indicates that saltation is active. Bridges et al. (2012) noted that the ripple migration rate is faster towards the top of the dunes, presumably due to the speed-up of wind towards the crest.

# Booming or Singing Dunes

Dunes are not silent. When the sand is blowing, you can hear a hiss. However, there is a much more dramatic and evocative acoustic emission from dunes—so-called booming or singing. This was noted in early Chinese records (there is a dune named Mingsha San, or Singing Sand Mountain), although it was first documented in the Western literature by Marco Polo in the Taklamakan desert in the 13th century. Charles Darwin noted the phenomenon in Chile at around the same time that two American scientists, Bolton and Julien, surveyed 'musical sands'. Physicists, including Bagnold, have since struggled to explain the phenomenon, which has recently become a rather active, and bitter, arena (!) of scientific debate (e.g., Chalmers 2006).

First, it is important to define the phenomenon. Sheared sand makes noise: walking on certain sands from places as far afield as the Scottish island of Eigg to Barking Sands beach on the Hawaiian island of Maui produces a squeaking noise. Such short-lived emission can also be generated in the laboratory with a pestle, and it usually has a frequency of a few hundred Hertz. However, the 'booming' of dunes is a quite distinct process. This has a much lower frequency—typically 80–120 Hz—and lasts for ten seconds or longer, in association with (and perhaps occasionally longer in duration than) sand avalanches on slip faces. These avalanches can occur naturally as the slip face oversteepens with sand saltating over the brink, or (more usually when scientists cannot wait for that) by walking on or kicking down on the slipface. The sound is often compared to the drone of a propeller plane.

Some early attempts to explain the phenomenon were made by Poynting and Thompson (1909) and Bagnold (1966), who suggested the frequency should vary as the inverse square root of the sand diameter. The first good measurements (Bolton and Julien compared the sand note to a violin to estimate frequency) were made by Lindsay and three Criswells in 1976. However, only in the last decade has substantial progress been made in characterizing the process, its environment and mechanism through better field measurements, theory and laboratory experimentation.

The Paris group of Stephane Douady and Bruno Andreotti and student Pascal Hersen observed booming from avalanching slip faces of barchans in Morocco in 2001. This led them to explore the mechanism and to realize that not only must sliding sand produce sound somehow, but this sound must be coherent in some way. As discussed in a fascinating overview by Chalmers (2006), Douady and Andreotti diverged, eventually bitterly so, on their interpretations and became unable to work together on the topic, such that Andreotti moved to a different lab. They have since pursued somewhat different aspects of the problem, e.g., Andreotti (2004) on the synchronization of the sound emission, using novel field measurements of both the acoustic emission with a microphone and the movement of the surface of the avalanche with an accelerometer, while Douady et al. (2006) explored the sound generation from sheared sand in the laboratory.

Several prominent dune structures in the Mojave are known to boom: Kelso, Dumont and Sand Mountain. These are all relatively accessible from Los Angeles, and in 2007 another set of researchers entered the fray—PhD student Nathalie Vriend and her advisor, Melany Hunt. They documented booming on a number of occasions at these different sites, but also measured seismic emissions with an array of dozens of geophones and used those to measure the structure of the dune, finding distinct layers. They argued that these layers were key to the process, and dismissed the grain size factor, showing a plot of emission frequency against grain size that essentially shows no correlation. Regardless of the interpretation, the record of the emission is the best data of its kind so far, and highlights that the emission is not monochromatic, but a superposition of harmonics. This may explain why early (pre-electronic) estimates of the frequency appear to be rather higher (200 Hz+) than modern measurements (80–140 Hz) (Fig. 10.1).

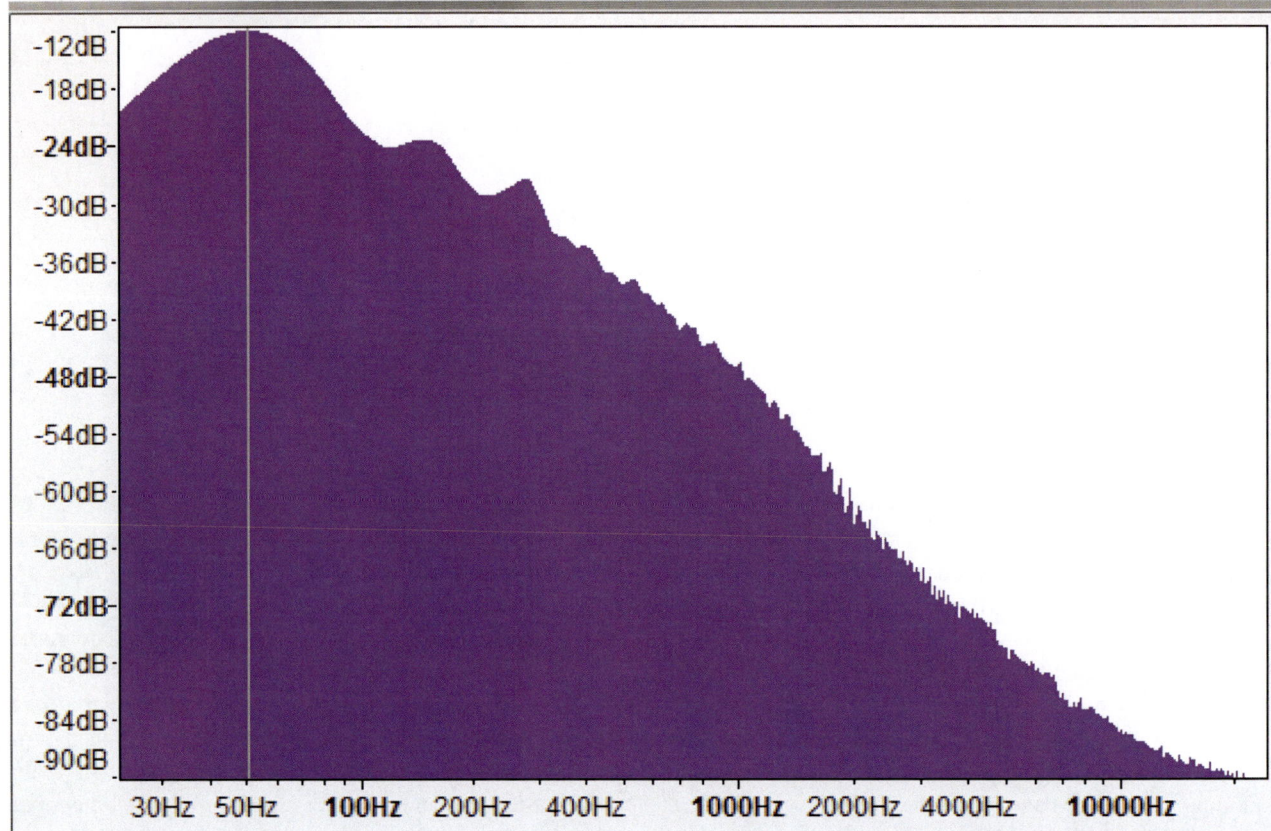

**Fig. 10.1** Power spectrum recorded by Lorenz of a short 'burp' made by jumping into sand on Sand Mountain, Nevada in October 2012—on that occasion sustained avalanching/booming was not obtained. The peak here is at about 50 Hz, rather lower than the 70 and 85 Hz peaks noted by Dagois-Bohy et al. (2012)

The emphasis on existing layers in the dune prompted a rebuttal from Andreotti (2008), who argued that the seismic interpretation of discrete layers was flawed, that gradients of porosity in the dune would refract, rather than reflect, seismic energy. To the dismay of those of us attempting to follow the literature, Andreotti shows a plot, purportedly of the same data, with a very strong dependence of emission frequency versus grain size. What to make of all this? A counter-rebuttal by Vriend shows more data, with more error bars (Fig. 10.2).

In the meantime, the Douady group has followed up with some useful laboratory work exploring the change in emission frequency during the progress of a laboratory avalanche (Dagois et al. 2010; interestingly, in an Ultrasonics journal: perhaps an effort to circumvent the grid-locked trenches of the geomorphology and physics journals). This chapter also points out some work on what sands sing and which don't. For example, glass beads do not, but after tumbling them with materials to enhance their friction, they will. Similarly, the beach sands from Morocco near their dune site do not sing, but the sand in the barchans downwind of a sebka (where the beach sands have been wetted with salt water and dust and then dried out) do.

Clearly the sand texture is important. Additional laboratory experiments and analysis has been pursued by Patsistas in Canada (e.g., Patitsas 2008) and most recently by Dagois-Bohy et al. (2012).

Two recent review chapters summarize the situation well and are the suggested starting points for serious scholars; despite the acrimony in the subject, some impressive progress has been made in the last decade. Hunt and Vriend (2010) focus on geophysical investigations of the dunes they have studied (including GPR data showing layers at Dumont—see Radar chapter) and present a nice list of some 40 or so booming dunes worldwide. Andreotti (2012) presents a thoughtful overview with a stronger physics slant, and outlines the challenges any theory must address.

It seems to us that there are elements of truth in all these works, but a key point has apparently been missed. It seems to us that the avalanche is the key thing, and the layer of sliding sand is both the energy source and the resonator. But avalanching sand does not have the same properties as stationary sand: essentially, grains behave as a gas of individually massive grains bouncing against each other and thereby inhibiting friction. These agitated properties, on the large scale, are important in allowing large landslides

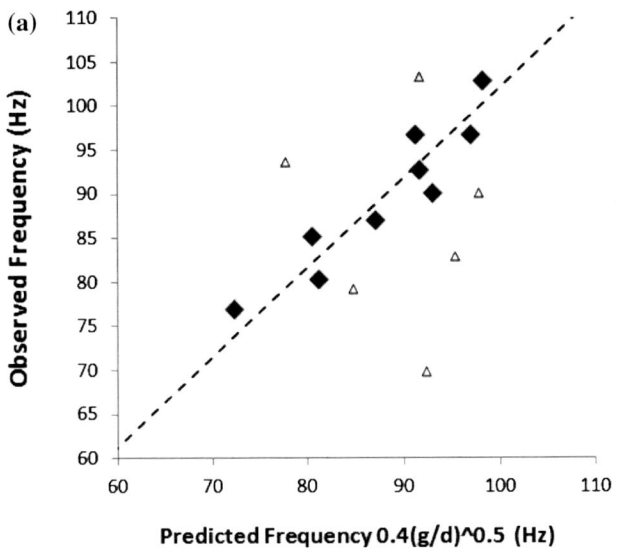
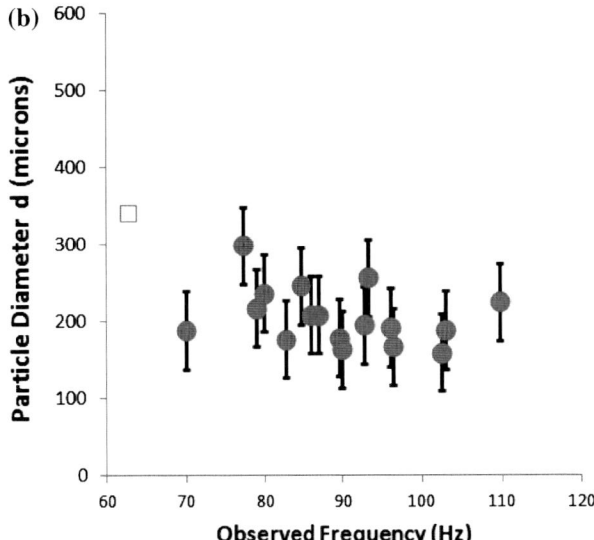
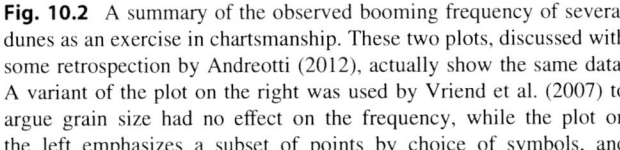

Fig. 10.2 A summary of the observed booming frequency of several dunes as an exercise in chartsmanship. These two plots, discussed with some retrospection by Andreotti (2012), actually show the same data. A variant of the plot on the right was used by Vriend et al. (2007) to argue grain size had no effect on the frequency, while the plot on the left emphasizes a subset of points by choice of symbols, and transforms the grainsize into a predicted frequency, as shown in Andreotti's (2007) comment. When all data points are considered, and especially when error bars in both dimensions are shown, it is difficult to conclude anything, except perhaps about the subjectivity of the scientific process and how chartmanship can be used to advantage

('Sturtzstroms', Collins and Melosh 2003) to run out far further horizontally than you would expect, 20–30 times the vertical drop. Similarly, this acoustic fluidization process is important in allowing a seismically-shocked planetary crust to behave locally as a liquid long enough in a large impact event to allow the crater to collapse and form a central peak or (in more energetic events where the rattling takes longer to decay) a peak ring (Melosh and Gaffney 1983).

Such a granular 'gas' has a very low sound speed (perhaps just a few m/s—Andreotti's chapter expresses surprise that their measurements indicate a propagation speed an order of magnitude lower than they'd expect), making a strongly-reflecting boundary between the avalanche and air (340 m/s) and between avalanche and the packed sand of the dune (400 m/s). Where there is a difference in sound speed and density, sound is reflected, so seismoacoustic energy will bounce back within the avalanching area. Since the 'slab' of sliding sand probably has a somewhat uniform thickness, the time for a sound wave to cross the slab and back (a 10 cm thick layer and sound speed of 10 m/s would give 50 Hz) could allow a fairly narrow band resonance.

Thus while Vriend et al. are right to point out the importance of layering in a dune, we suggest that its seismic properties are largely irrelevant in the sense of controlling the sound directly. The layering of the dune may be critical, however, in controlling how deep the avalanching layer is, and that in turn may influence the frequency of emission. It is not enough to build a dune from sand that sings in the laboratory to make a booming dune. Similarly, the grain microscale properties are likely crucial in influencing how sliding generates the acoustic excitation: even a dune with the fancy layering beloved by the Caltech group will not boom if its sand is silent.

It is probably bad form for writers to advance their half-formed hypotheses in a book without the cut and thrust of peer review, but science is a journey, not a destination. We suggest some fruitful avenues to pursue may be field experiments wherein the depth of the avalanche is controlled, or perhaps experiments with very high- or low-density grains. Similarly, if methods can be devised to measure in the field the depth of and sound speed in a flowing avalanche, such information will be of value for any theory.

Lest it be thought that the study of booming dunes is a frivolous pastime, it should be noted that the flow of granular materials is of vital import to industry, whether plastic pellets, construction material, cornflour or any of a vast array of materials. These are usually stored in tall vertical cylinders or silos, such that they can be poured from the bottom. When this flow occurs—essentially a contained avalanche—oscillations analogous to booming dunes can be excited that may influence the flow or even damage the structure; the audible manifestation of these go by the delightful name of 'silo honking' (e.g., Buick et al. 2005). Vibrations in sand are of interest in civil engineering (e.g., piledriving, or the liquefaction of sand during an earthquake), and are a matter of life and death to the desert scorpion, which uses them to find prey (Brownell 1977).

It is left as an exercise for the reader to consider the possible properties of booming dunes on Mars, Venus or Titan. Also, despite its lack of atmosphere, we would be remiss not to point out in this planetary book that a connection has been drawn (Criswell and Lindsay 1974) between booming dunes and a distinct class of lunar microquakes that occur around local sunrise, perhaps stimulated by thermal expansion and avalanching of the regolith.

# Part III
# Dune Worlds

In this section of the book, we consider various planetary bodies in our Solar System in turn, and discuss how they have been explored, and what has been learned about dunes on them. The winds and sand transport parameters have already been discussed in Part II, and thus the character of transport and the scale of dunes and ripples has been explained. But the extent of dunes depends on the planetary history, how sand may have been formed (and/or destroyed) and how it has been moved around. The resultant coverages (estimated in an excellent review article by Fenton et al. 2013) are summarized in Table 3.0.1.

It is striking, given the early discovery of dunes and blowing sand on Mars, that dunes cover only about 1 % of the surface of that world—while ubiquitous, dunes are sparse. In contrast, the Earth has a similar coverage, but concentrated in a small number of large sand seas. On Titan, the concentration is even more defined, with one almost continuous belt of sand seas around the equator, covering a whopping 12 % or so of the surface (Titan mapping is still ongoing, so this number may hop up and down by a point or two). On Venus, dunes—at least at the scale with which we have observed the surface to date—seem to be an aberrant exception, associated with particular spots where sand happens to be available.

In our Solar System, then, Titan wins as the dominant 'Dune World', although in a later Part V, we discuss some fictional worlds that challenge it (they may challenge reality too, however—among the thousand-odd extrasolar planets now discovered, it may be that none approach fictional Arrakis' dune coverage).

Although dunes are not known on Triton, Io or Pluto, we review briefly the prospects for aeolian bedforms on these worlds with tenuous atmospheres.

**Table 3.0.1** Planetary dune parameters

| Planetary body | Venus | Earth | Mars | Titan |
|---|---|---|---|---|
| Radius (km) | 6052 | 6370 | 3396 | 2575 |
| Surface gravity (m/s$^2$) | 8.9 | 9.8 | 3.7 | 1.35 |
| Atmospheric pressure (mbar) | 90,000 | 1000 | 4–10 | 1497 |
| Estimated dune field coverage (million km$^2$) | 0.0183 | 5 | 0.9 | 10 |
| Dune field coverage (%) | ~0.004 | ~1 | ~0.6 | ~12 |
| Predominant dune types | (unknown) | Linear, transverse | Barchan, transverse | Linear |

# Earth Dunes

## 11.1 Introduction

Humans have swarmed across the planet for tens of thousands of years, and oral traditions regarding dunes, how they move and how one should move across them are common to many cultures within arid or semi-arid regions. Descriptions of various sand dunes exist almost since the development of writing (see Part 5), but this early information is necessarily anecdotal in nature.

It was the European empires in the late 19th century that combined global reach with a scientific literature to record findings and debate formation processes. Early work on dunes and deserts is summarized in, e.g., Goudie (1999): the major players were the French (principally in the Western Sahara) and the British (principally in Egypt and, to a lesser extent, in Arabia and elsewhere). In the English language at least, a prominent early scholar was the geographer Vaughan Cornish who in the early 1890s began to systematically study dunes in various regions. Cornish was followed by others, culminating in Bagnold's desert explorations which took place principally in the 1930s.

Aerial photography of dunes was accomplished before 1920, but it was only after World War II that wide aerial surveys were made. However, as we review in Chap. 18, it was the view from space that enabled the systematic cataloguing of dune morphologies, scales and orientations worldwide, in the book edited by McKee (1979).

This chapter starts with a summary of some important aspects of Earth's atmosphere that enable sand to be moved by the wind, followed by a review of the main deserts and dune fields on Earth, observed migration rates for sand dunes, the influence of plant cover on dune mobility, the sources and sinks of sand. It ends with a short discussion of how the study of sand dunes can provide links to broader climate issues.

## 11.2 The Terrestrial Atmosphere

Earth's atmosphere comprises mostly (78 %) nitrogen, with 20 % oxygen, just under 1 % of argon, and traces of other gases (notably a growing amount of carbon dioxide). Water vapor is present in the lower atmosphere, in the most humid regions accounting for about 1 % of the gas volume.

The atmosphere at sea level has a column mass of 10,000 kg/m$^2$, equivalent to a column of water 10 m tall (or a column of mercury 76 cm tall), defined as 1 bar (or 1000 mbar). There is enough gas that an appreciable amount of light at short visible wavelengths is scattered—strongly scattered blue light is what our eyes perceive as the color of the sky (this is not necessarily the case on other planets). Earth's distance from the sun is such that (with a modest amount of greenhouse warming) the surface temperature is within the range in which water is a liquid. At this temperature, the 1 bar surface pressure corresponds to an atmospheric density of about 1.2 kg/m$^3$, about 800 times less than water.

The drop in temperature with altitude ($\sim$10 K/km in dry conditions, $\sim$5 K/km in moist ones) means that rising air can be cooled to the point where it becomes saturated with water vapor, which condenses into abundant clouds. Clouds above Earth appear white because the particles that make them up are large enough to effectively scatter all wavelengths of light in the same way, and the substantial ($\sim$30 %) cloud cover on Earth (Fig. 11.1) accounts for the overall high reflectivity of our planet.

Not infrequently, clouds lead to rain, which feeds rivers and is (along with glacial action) responsible for much of the generation of sand on Earth. We show in this chapter a large number of dunes in proximity to water—an interesting paradox. Warmth, directly or indirectly from solar heating, restores water vapor back into the atmosphere, driving our hydrological cycle (which has some striking parallels with that on Titan). However, the rainfall and evaporation are not uniformly distributed.

**Fig. 11.1** Earth, observed from orbit around the moon by the Clementine spacecraft. The dark Atlantic ocean dominates the scene—as indeed our planet is 60 % covered by water. Reflective clouds make the earth a fairly bright object, astronomically-speaking, and are seen in their near-permanence over the Amazon rain forest at *lower left*. Towards the *right*, the skies over the Sahara are clear, as they often are, exposing the relatively bright expanse of sand. Over Europe and the north Atlantic, clouds swirl in cyclonic systems. USGS Mosaic of images acquired on April 11, 1994, with the moon above about 20°N, 20°W

The rotation of the Earth strongly influences the direction of the solar-driven advective motions that take place in the atmosphere; over most of the northern hemisphere, the wind tends to be diverted to the right of its projected path by the so-called Coriolis force, an 'apparent' force associated with Earth's rotation (see also Chap. 3). The combination of temperature-driven and Coriolis contributions to atmospheric motion leads to the general features of our wind patterns. At the surface, this pattern corresponds to the trade winds, with broad spans of latitude in which the winds are generally in one or two directions with sufficient reliability to plan sea voyages (although most of us who travel by air are perhaps more directly affected today by the stratospheric winds, the fast-moving jet streams). These trade-wind spans of latitude are defined by the meridional circulation, with air rising at the equator and descending around 30 degrees of latitude north and south, defining what are called the Hadley cells. As the air has been dried during its ascent, these descending branches of the cell have little water vapor, and so these latitudes receive less rain than the planetary average, and surface evaporation is higher. Thus, deserts form in two main belts (see Fig. 11.2), the principal factor controlling where dunes are found on Earth.

While some parts of our planet see nearly uniform winds, the seasonal variations in solar heating, together with the uneven distribution of thermal inertia (the land warming and cooling much more than the ocean) mean that some areas of Earth see predominant wind directions that change over the course of the year, so much of the planet has a rather bimodal wind regime. And areas where the high winds are dominated by moisture-driven storm systems see highly variable wind directions. This diversity of wind regimes accounts for the variety of dune-forms on Earth.

The thermal inertia of the surface, and the air density at the surface, account for the thickness of the atmospheric boundary layer (which grows over the course of the day to a thickness of a few hundred meters to a couple of kilometers). This defines the ultimate size of dunes.

Of course, the Earth today is not the Earth that has always been (see Chap. 21). Beyond the more usually-discussed changes in moisture and wind associated with astronomically-forced glacial cycles, when one goes back billions of years in Earth's history, it is worth noting that the insolation was lower (due to the faint early sun) and the rotation period was shorter, the receding moon not yet having robbed Earth of as much angular momentum as it has today.

## 11.3 Major Deserts and Dune Fields

Deserts are defined as regions that receive less than 25 cm (10 in.) of rain per year, or where the evaporation rate is at least twice as great as the rate of precipitation. Most of the deserts on Earth are not vast regions consisting primarily of shifting sand, even though movies have tended to make this the common perception of deserts for many people. Four types of deserts are recognized by geographers: subtropical deserts, cool coastal deserts, cool winter deserts, and polar deserts. The subtropical deserts are the hottest of the four types, dominated by dry terrain that facilitates rapid evaporation. Cool coastal deserts are in the same general latitude range as the subtropical deserts, but the average temperature is generally cooler than in the inland subtropical deserts, influenced by cold off-shore ocean currents near these deserts. Cold winter deserts experience drastic temperature extremes and tend to reach much lower temperatures than those experienced in subtropical deserts. Polar areas are considered to be deserts because practically all of the water that makes its way into these regions freezes and becomes unavailable to support flora. Indeed, the largest deserts on Earth are the Antarctic and Arctic regions; these polar areas cover >28 million $km^2$, which represents about 60 % of the cumulative desert area on Earth.

## 11.3 Major Deserts and Dune Fields

**Fig. 11.2** A composite image of the Earth's surface, assembled from many weeks of observation by the European ENVISAT satellite (to observe areas at the same time of day, and when they happened to be clear of cloud). Antarctica was in darkness throughout this northern summertime observation period and is not shown. Apart from the Greenland ice cap, the most prominent bright features are the Earth's deserts at about 30° latitude. In the north, *left* to *right* we see the US and Mexican deserts, then the Saharan and Arabian deserts, Iran and Pakistan, and the Gobi. In the south, the Atacama, then the Namib and Kalahari, then Australia. The same features can be seen (but inverted—dune fields are radar-dark) in Fig. 18.19

The distribution of deserts across the Earth (already suggested by the satellite mosaic in Fig. 11.2) is indicated in a simplified map of the equatorial and mid-latitude regions of Earth (Fig. 11.3). The map is annotated with labels that correspond to entries listed in Table 11.1, where some of the named deserts included within a particular labeled region are indicated. Excluding the vast polar regions (to which we will return, as they have dunes of their own), desert regions on Earth are primarily concentrated around subtropical latitudes, where global wind circulation patterns are dominated by moisture-deprived air brought to these locations. The global Hadley cell circulation over the equatorial tropics, combined with the topographic effects of large mountain ranges, is the primary explanation for the distribution of deserts on Earth.

How much sand is included within these global deserts? This question is more difficult to determine precisely than is the global distribution of deserts. For example, it is estimated that sand dunes represent about 30 % of the area covered by all of the Sahara. The Rub' Al-Khali ('Empty Quarter'), which represents one-quarter of the Arabian desert, is considered to be largest expanse of unbroken sand cover present on Earth. Other relevant estimates include that 90 % of the Kara-Kum desert is sand-covered, whereas only about 1 % of the Great Basin winter desert in the western United States consists of sand. Within the sand-dominated portions of deserts, dunes of many types and sizes can form, depending upon the local wind conditions (see Chap. 6). The great variability of sand accumulations within deserts is strongly influenced by the abundance of even infrequent precipitation within some deserts, which in most cases is sufficient to support a wide variety of desert flora. A few deserts are dominated by unvegetated mountain ranges with intervening gravel covered basin floors, such as the Sonoroan desert. Consequently, the availability of abundant sand-sized material is a crucial factor in determining where major sand dune accumulations can occur.

Rather than attempting to discuss all of the large deserts and dune fields on Earth, next we examine some of the deserts and dune fields that have some unique or specific attribute that readily distinguishes them from other deserts and dune fields. This will not be an in-depth treatment, but rather a brief introduction to some of the more memorable dry places on Earth. This information will also provide a broad base from which to explore the dune fields and deserts on other planetary surfaces.

### 11.3.1 Sahara

The name 'Sahara' means 'The Great Desert' in Arabic, and that description is most definitely very appropriate. The Sahara is the largest non-polar desert on Earth (Table 11.1), extending from the Mediterranean Sea on its northern edge south to the Sahel, a vast tropical savanna that is home to a diverse range of both flora and fauna that have adapted to

**Fig. 11.3** Global distribution of major desert regions on Earth. See Table 11.1 for attributes of the labeled entries. Colors of labels represent subtropical (*gold*), cool coastal (*red*), and cool winter (*blue*) deserts. Polar deserts are excluded from this map. 'Le' denotes Lencois Maranhenses, which is a dunefield but not a desert

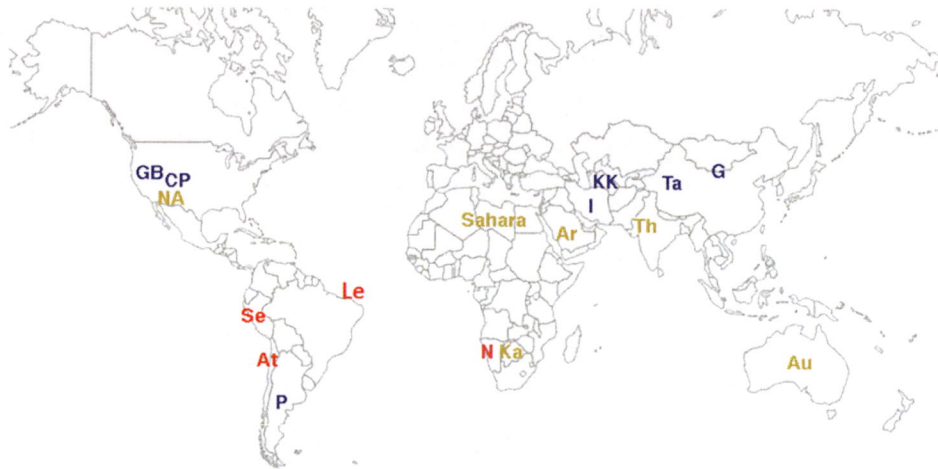

**Table 11.1** Major desert regions on Earth

| Symbol* | Location (which includes indicated named deserts) | Area ($10^6$ km$^2$) |
|---|---|---|
| Subtropical deserts | | |
| Sahara | Sahara (Libya, Nubian) | 9.1 |
| Ar | Arabia (Rub Al-Khali, Ad Dahna, An Nafud, Syrian) | 2.6 |
| Ka | Kalahari | 0.57 |
| Au | Australia (Gibson, Great Sandy, Great Victoria, Simpson) | 1.5 |
| NA | North America (Mojave, Sonoran, Chihuahuan) | 0.91 |
| Th | Thar | 0.46 |
| Cool coastal deserts | | |
| N | Namib | 0.034 |
| At | Atacama | 0.14 |
| Se | Sechura | 0.19 |
| Cold winter deserts | | |
| CP | Colorado Plateau | 0.49 |
| GB | Great Basin | 0.034 |
| G | Gobi | 0.5 |
| KK | Kara-Kum, Kyzyl-Kum | 0.56 |
| I | Iran (Dasht-e Kavir, Dasht-e Lut) | 0.26 |
| P | Patagonia | 0.68 |
| Ta | Taklamakan | 0.27 |
| Polar deserts | | |
| (Not shown) | Antarctic and Arctic | 28.3 |
| Total | | 46.6 |

* Symbols shown on map on Fig. 11.3

the semi-arid climate found south of the desert. About one-half of the Sahara receives less than 20 mm of rain annually, while the other half receives up to 100 mm of rain per year. The hyper-arid central portion of the desert has practically no vegetation, causing it to be the source of massive dust clouds swept off the African continent (Fig. 11.4); the sand dunes there can attain heights of up to 180 m. Throughout this vast sand sea, isolated patches of vegetation collect around the occasional oasis where water becomes accessible at the surface, usually because the water table intersects the surface at the lowest areas (Fig. 11.5). In contrast to the hyper-arid areas, the northern and southern parts of the Sahara consist of sparse grassland and shrubs adapted to semi-arid conditions, where both flora and fauna become more evident, but they are still not abundant. The area covered by all of the Sahara is roughly comparable to the total area of either China or of the United States. The Sahara (especially Morocco and Egypt) features many classic barchans (Figs. 1.9, 6.5 and 24.4) as well as barchanoid ridges, linear dunes (Figs. 6.21, 23.3 and 24.1).

### 11.3.2 Rub' al Khali

This name is Arabic for 'Empty Quarter', which indicates the almost total lack of natural flora and fauna within this desert in Saudi Arabia (Fig. 11.6). The Rub' al Khali is the largest sand desert in the world (Fig. 11.7), as opposed to other areas that meet the rainfall definition of a desert but are not comprised primarily of sand. Like the central portion of the Sahara, the Rub' al Khali is hyper-arid, typically receiving <30 mm of rain per year. The average daily maximum temperature is 47 °C, but individual locations have reached as high as 56 °C, thus exceeding the typical high temperature in the Gobi desert. Individual dunes within the Rub' al Khali are up to 250 m in height. The duneforms in this vast desert include some of the most 'perfect' linear dunes (Figs. 7.11 and 18.4) to hooked megabarchan forms (Fig. 6.23), megabarchans (Figs. 23.4 and 23.5), barchanoid ridges, barchans (Fig. 6.25) and domes (Figs. 6.2 and 9.5).

## 11.3 Major Deserts and Dune Fields

**Fig. 11.5** Ubari oasis in the Erg Awbari, Libya, taken by Luca Galuzzi, April 7, 2007 (Wikimedia Commons)

**Fig. 11.4** The north coast of Libya, showing plumes of wind-blown dust from the northern Sahara trailing into the Mediterranean. This highlights not only the wind transport of solid material on Earth, but also that the oceans act as at least temporary sinks for much sand and dust which would otherwise accumulate. NASA MODIS image

### 11.3.3 Taklamakan

Known as either the Taklimakan or Teklimakan, the Taklamakan desert is located within the Tarim Basin in southwestern China, which is a tectonically depressed region $\sim 1000$ km $\times$ 400 km (620 $\times$ 250 mi) in size. Although it is a local depression, it is at high elevations overall and can thus be cold in winter—in 2008 a weather station in this desert measured a winter temperature of $-26.1$ °C ($-15.0$ °F). Transverse dunes in the Taklamakan have been studied by Wang (2002).

### 11.3.4 Gobi/Badain Jaran

The name 'Gobi' is Mongolian for 'semidesert', which once again is a very apt description for the abundant vegetation that comprises what can be called Desert Steppe terrain. Average annual rainfall throughout this desert region is 194 mm, which is more than twice the rainfall in even the wettest portions of the Sahara. Located in western China, the Gobi is Asia's largest desert, placed at an elevation that ranges from 900 to 1500 m, making it also the largest terrestrial example of what could be called a 'high desert'. The high altitude of elevated deserts contributes significantly to rather large temperature extremes throughout the year, ranging from $-40$ °C in winter to $+50$ °C in the summer. Recently, the steady rise in global temperatures are causing the southern edge of the Gobi to expand into the surrounding grasslands by about 3600 km$^2$/ year. The region referred to as the Gobi includes the Badain Jaran but also several other dunefield areas.

The name 'Badain Jaran' is Mongolian for 'mysterious lakes', illustrating how early inhabitants of this area in northwestern China were surprised to find large lakes within the vast sand mountains of this desert. More than 100 spring-fed lakes give the region its name (Fig. 11.8—see also Figs. 7.15, 18.1 and 18.7), and the lakes also serve to stabilize the horizontal movement of the adjacent sand mountains (draa). This desert is best known for containing the tallest stationary dunes on Earth, with individual sand dunes attaining heights of up to 500 m.

### 11.3.5 Namib

'Namib' means 'vast place' in the Nama language. This coastal desert is located primarily in Namibia, on the southwestern edge of Africa, but arid conditions extend into adjacent countries as well. Coastal deserts like the Namib (Fig. 11.9) are the result of climatic conditions that cause decreased rainfall along some stretches of different continental margins. The annual rainfall within the Namib is actually very variable depending upon where you are located within the desert, ranging from a low of only 2 mm (0.08 in.) in the most arid portions of the desert to 200 mm (8 in.) along the Great Escarpment, which forms the eastern margin of the desert. Individual dunes in the Namib are up

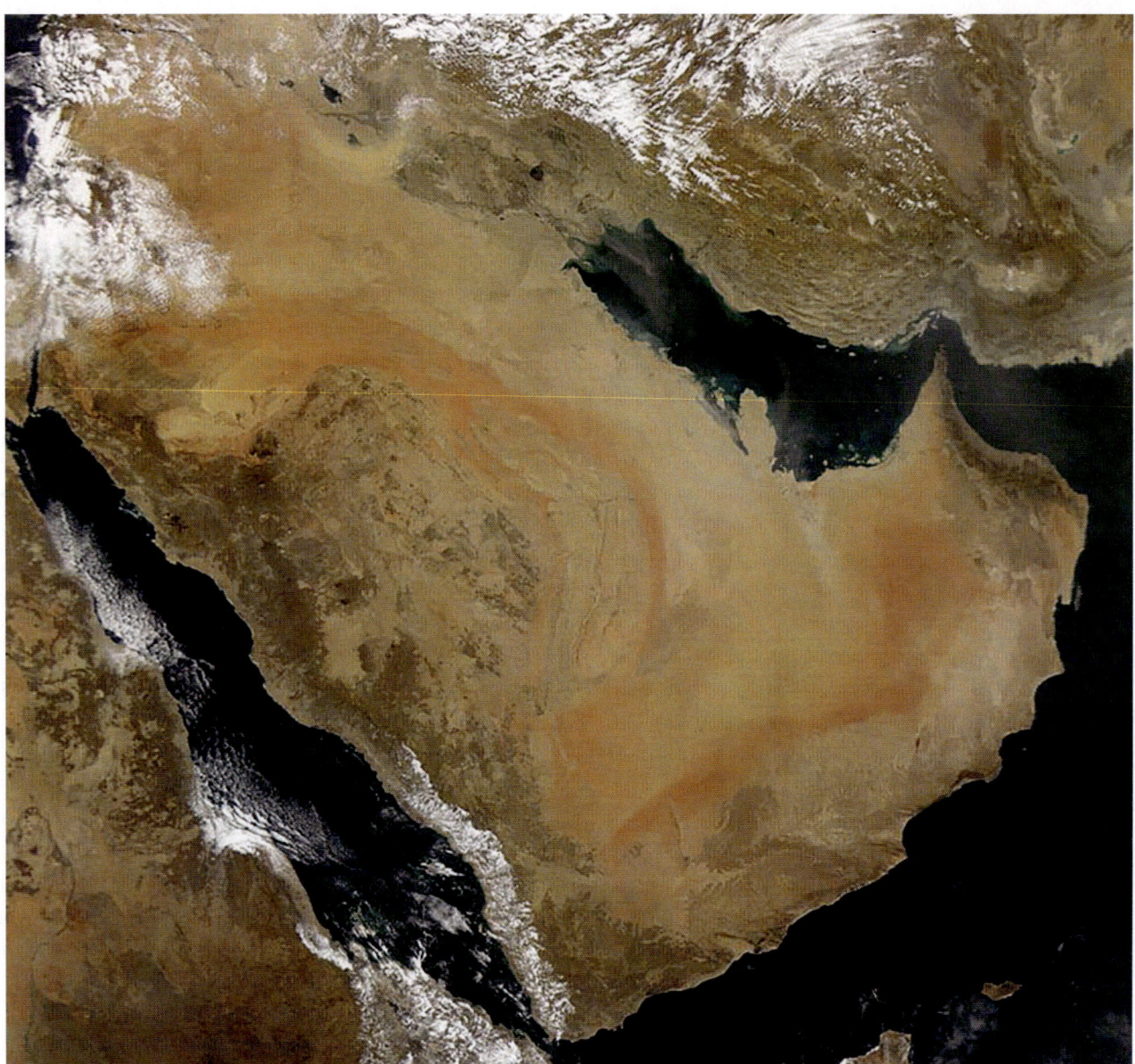

**Fig. 11.6** SeaWIFS image of the Arabian peninsula, showing the sweeping sand pathways of the Rub' Al Khali. *Image* NASA

to 300 m (980 ft) in height, and up to 32 km (20 mi) long, second in size only to the draa dunes in the Badain Jaran desert of China.

### 11.3.6 Kalahari

There are at least two possible origins for the name of this African desert. In the Tswana language, the word 'Kgala' means 'the great thirst', while the word 'Khalagari' means 'a waterless place'. Either way, you get the picture! Actually, the Kalahari is a semi-arid savanna, with considerable diversity in vegetation, plus wildlife that exist within the available vegetation cover. Perhaps some of the best known animals that live in the Kalihari are the meerkats, rodents that stand on their hind legs to scan the area surrounding their dens. The desert is located within the Kalahari Basin, which encompasses 2.5 million $km^2$ in Botswana and portions of neighboring Namibia and South Africa. Average annual rainfall ranges from 80 to 190 mm throughout the desert. Summer temperatures range from 20 to 45 °C, with local maximums of up to 50 °C.

**Fig. 11.7** The extreme eastern end of the Rub' Al Khali sees dunes straddling the UAE—Omani border (dark fence, with observation towers every couple of kilometers) in this image looking north from a linear dune. Two other linear dunes are visible, as well as some barchans on the flat desert floor, made bright by gravels containing calcrete. To the east of here the gravels (from alluvial fans coming off the Oman mountains) win and the dunes peter out: to the west and southwest across the UAE and into Saudi Arabia, the dunes extend for a thousand kilometers. *Photo* R. Lorenz

### 11.3.7 Atacama

Located along the Pacific coast of South America, the Atacama desert is a coastal desert formed on a plateau about 1000 km long, found mostly in northern Chile, but also extending into adjacent Andean countries. The Atacama has the distinction of being the driest place on Earth; average annual rainfall measured near the Chilean town of Antofagasta is only 1 mm. Some weather stations have never received any measurable rain, and written records indicate one portion of the desert perhaps received no rain between 1570 and 1971! An obvious consequence of this limited rainfall is that very little vegetation occurs within the Atacama desert, limited mostly to a few very hearty cactus. There are also very few sand dunes here; instead, the Atacama is more of a rocky desert (although there are some notable barchans on the Peruvian coast). Soils found in this hyper-arid region have been proposed as analogs to soils on Mars; in particular, perchlorates identified in the surface materials analyzed by the Phoenix lander are quite common within some Atacama soils.

### 11.3.8 Lençóis Maranhenses

This remarkable dune site is a national park on the northeast coast of Brazil. Compared with the other deserts listed here it is insignificant (1500 km$^2$ in area) but it is a useful reminder that planets are complicated places. While one would not expect a desert near the equator (and indeed LM is not a desert, having a rainy season), this locale, caught between the Atlantic and the rain forest, is remarkable for having a wide expanse of dunes up to 30 m high, with spectacular lagoons between them (Figs. 11.10, 11.11 and 1.8).

This, of course, begs the question of how the sand interacts with the lagoons (which the first author can attest are pleasant to swim in). Sand can move either by saltating across the interdunes when they are dry, or simply by marching the slip face forward. The latter process would, if the lagoons never dried, result in the dune shrinking as it marched forward. So the existence of this dunefield relies on the seasonality of rainfall, which arrives December to May. In the summer and fall, the lagoons often dry out, allowing the dunes to be built back up, and sand transport to resume (e.g., de M. Luna et al. 2012). The winds here are strongly unidirectional (hence the barchan and barchanoid forms) and cause a remarkable migration rate of 12 m or so per year into the forest.

An interesting feature of these dunes are scars (small curved ridges) that they leave behind upwind of the stoss edge of the dunes. These result from cementation and/or vegetation at the edge of the lagoon, and appear to form roughly annually (e.g., Levin et al. 2009) trace about 30 scars behind some example dunes. Although reliant on vegetation for their formation, they are an interesting analog to the scars left by gypsum cementation at White Sands and on Mars (see Chap. 5).

### 11.3.9 Great Victoria

The largest desert in Australia, the Great Victoria desert is located in the southwestern part of the continent. This desert covers 424,000 km$^2$ (164,000 mi$^2$) and it consists of individual sand hills stabilized by grasses, with grassland plains covering the interdune regions. Like portions of the Atacama desert, some areas within the Great Victoria desert are dominated by rocks rather than by sand. The Aussie word 'gibber' has become common in the geologic literature for such rocky plains, which consist of a closely spaced surface layer of pebbles and cobbles, all glazed with a distinctive iron oxide coating.

### 11.3.10 Simpson

The Simpson desert is the fourth largest desert in Australia, forming the 'red center' of the continent. The Simpson desert covers 176,000 km$^2$ (68,000 mi$^2$), less than half the area of the Great Victoria desert, but it contains some of the

**Fig. 11.8** The tallest dunes on Earth, in the Badain Jaran desert of China, are remarkable for having lakes between them, because the watertable is high enough. Field photograph courtesy of Clement Narteau, Paris. The Badain Jaran dunes and lakes are seen from space in Figs. 17.3 and 18.1

**Fig. 11.9** Lange Wand, the Long Wall; oblique aerial photo looking east onto the coast of Namibia. Sands from the Orange river are swept north by longshore drift and are then blown inland to accumulate in the Namib sand sea. The coast has a range of dune types including transverse and barchans; here the sand forms a steep cliff, 24 55' S, 14 49' E. *Photo* R. Lorenz

## 11.3 Major Deserts and Dune Fields

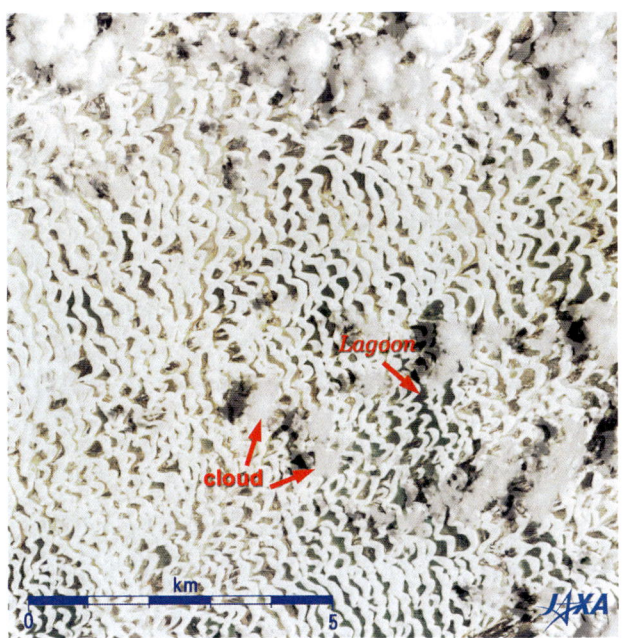

**Fig. 11.10** Image of the Lençóis Maranhenses National Park, in October 2007 from the Japanese Advanced Land Observation Satellite (ALOS, aka Daichi). The white barchans and barchanoid ridges are visible against the flat interdunes, but are very prominent against dark lagoons. Earlier in the year, especially in rainy years, the entire area can be flooded, yet evidently the region is dry enough for enough of the time for sand to move and maintain the dune shapes. At these equatorial latitudes, clouds are common. Image courtesy of JAXA

**Fig. 11.11** A number of striking features can be seen in this field photo at Lençóis Maranhenses. A cloudy sky is reflected in the placid interdune lagoon. The dunes in the distance have wavy stripes, corresponding to dry sand waves growing at the 'elemental' scale of a few meters, as the sand dries and becomes mobile. Small bright transverse ripples can be seen in the foreground, just a few cm apart (see also Fig. 20.1). Towards the *middle* of the image, the bright sand is collimated in *straight lines*, or *curved arcs* at the *right*. This seems to be where the moving bright sand has been trapped in the microtopography of bedding layers exposed in the damp dune surface. *Photo* R. Lorenz

**Fig. 11.12** A field photo of part of the dune system shown in Fig. 11.13. These dunes are still somewhat active (note the ripples) but are partly vegetated. The large alternating footprints with a central groove, indicating a large biped with a dragging tail, reveal the Australian location—they are the tracks of a kangaroo. Jani Radebaugh for scale. *Photo* R. Lorenz

longest linear dunes found on Earth. Individual linear dunes are in the Simpson range from 3 to 40 m in height (rather small compared with the giants in the Namib and Arabian desert) and are often partly vegetation-covered (Fig. 11.12). Nonetheless, some are traceable continuously for up to 200 km, and describe interesting patterns around topographic obstacles (Fig. 11.13) that may be instructive in interpreting Titan's linear dune morphology. Big Red (Nappanerica) is one of the largest and most visited linear dunes in the Simpson. Sand grains in the 'red center' are heavily stained by iron oxide, forming some of the reddest sands on the planet. Sand comprising the linear dunes generally ranges from 0.05 to 1.2 mm in size, with 0.5 mm as the average diameter on the dune crests and 0.3 mm as the average on the lower flanks of the dunes (Twidale 1980).

Because the two authors reside in the United States, and many of our readers may also, we offer some discussion of a few more locations than their extent or dune size might by itself demand. Twenty-two aeolian study sites throughout the western United States are discussed in Zimbelman and Williams (2007), along with their potential as analogs for better understanding the sand dunes on Mars. The North American deserts do not rival most of the world deserts discussed above, but the semi-arid Mojave Desert in southern California has been the site of aeolian studies for many decades, and the Nebraska Sand Hills are an impressive accumulation area for sand derived from the recent glacial epochs. Below we highlight three of the more distinctive dune locations within the western portion of the continental U.S.

**Fig. 11.13** The shallow but extensive linear dunes of the Simpson desert are seen here breaching through, and being diverted around, a ~100 m high ridge, southeast of Alice Springs. The fact that the dune pattern appears to 'sense' the ridge more than 1 km away indicates how topography can influence winds. *Photo* R. Lorenz from an airliner

**Fig. 11.14** Plant pedestal (*left* of *center*) preserved within the dunes at White Sands National Monument, with ancient gypsum-bearing sediment layers visible as white bands in the distant San Andres Mountains (NPS photo)

## 11.3.11 Great Sand Dunes

The highest dunes in North America are the Great Sand Dunes in central Colorado (Figs. 8.1 and 16.3). Located at 2400 m (8000 ft) elevation above sea level, these dunes have formed at an atmospheric pressure that is noticeably less than that at sea level (something that becomes readily apparent when a 'flat-lander' starts hiking up the dunes!). The sand is derived from the San Juan Mountains, located >100 km (>60 mi) to the west of the dunes, and they have been concentrated by westerly winds against the base of the Sangre de Cristo Mountains (see Fig. 8.1). The Sangre de Cristo Mountains are the source for granules and pebbles carried by Medano Creek along the southern edge of the dunes, where strong winds transport the granules onto the lower portions of the dunes, creating wonderful granule-coated mega-ripples (Fig. 5.8), features that are analogs to similar mega-ripples on Mars (Zimbelman et al. 2009) and whose movement has been documented (Lorenz and Valdez 2011) with timelapse imaging (see Sect. 16.2.2 and Fig. 9.9). Niveo-aeolian activity on a parabolic dune near the main dune mass is shown in Figs. 5.25 and 5.26.

## 11.3.12 White Sands

The world's largest gypsum dunefield is located in the Tularosa Basin of southern New Mexico, between the San Andres Mountains to the west and the Sacramento Mountains to the east. Gypsum is leached from thick white sedimentary beds high in the San Andres Mountains (Fig. 11.14), which precipitates in playas on the basin floor, and westerly winds transport sand-sized gypsum grains into snow-white barchans and barchanoid ridges (Figs. 11.15 and 18.16, 18.17). The gypsum dunes (e.g., Kocurek et al. 2007) cover 710 km$^2$ (275 mi$^2$) of the basin floor. Evapotranspiration is over 3 m annually, well in excess of the annual rainfall, so the gypsum remains as individual sand grains except where groundwater interacts with the base of some dunes. The roots of some plants bind together enough sand to produce pedestals within the active dune field (Fig. 11.14). These gypsum dunes are important as potential analogs to gypsum dunes identified in the north polar region of Mars.

White Sands has been an important field study site; bedding structures, exposed by brute force methods such as those in Fig. 16.1 are seen in Fig. 5.18, and gypsum scars on interdunes are shown in Fig. 5.23.

The White Sands dune field has been nicely mapped by airborne LiDAR (see Chap. 18) and has been recently considered in a model of boundary layer growth controlled by the dune roughness (Jerolmack et al. 2012).

## 11.3.13 Bruneau

The tallest 'free-standing' (that is, not buttressed against adjacent dunes or mountains) sand dunes in North America are found near Bruneau, Idaho, located within an abandoned cut-off meander of the nearby Snake River (Murphy 1973). Two semi-parallel ridges of reversing sand dunes (Fig. 11.16)

## 11.3 Major Deserts and Dune Fields

**Fig. 11.15** A kite camera image of White Sands. The brilliant white sand loses much contrast at mid-day, sunset is by far the best time to observe the morphology. The kite here is at an altitude of about 150 m, a vehicle is visible at *bottom left*. Note the serried barchanoid ridges and flat interdunes—a curved berm is seen at the *bottom*, where the sand is bulldozed to keep a loop road and parking area clear for tourists. Note also the regular wave surface on the dunes at *bottom right*, presumably due to this being the destabilization wavelength for this sediment. Compare the morphology with airborne lidar topography in Figs. 18.16 and 18.17. *Photo* R. Lorenz

comprise the largest features in the dune field, which is now a state-protected natural park. A bimodal wind regime occurs here (see Fig. 3.11), so that the dunes tend to grow vertically rather than extend horizontally. The Bruneau dunes are being investigated as analogs to Transverse Aeolian Ridges (TARs) on Mars, some of which have been shown to have topographic attributes very similar to reversing dunes on Earth (Zimbelman 2010).

## 11.4 Snow Dunes and Megadunes

Although a variety of different mineral sands on Earth can form dunes, snow is a quite distinctive material in that the bulk density of the solid is lower than other minerals (ice is 900 kg/m$^3$, compared with ~2600 kg/m$^3$ for quartz) and snow particles can be appreciably porous, lowering the density even further. This makes the dynamics of snow somewhat different—particles trajectories are less ballistic and more sensitive to turbulent fluctuations in airflow—in the scaling theory of Claudin and Andreotti (2006, see Fig. 4.15), snow plots as a separate point from sand. Nonetheless, snow can form local drifts that are essentially identical to sand dunes, and similar mitigation measures (vegetation and fences) are sometimes applied to inhibit transport onto infrastructure like roads and buildings. One subtlety in snow dynamics is that the porosity allows compaction to occur in large structures, and the enhanced stickiness of particles near the melting point means that cornices can develop at the brink, rather than slipfaces.

In Antarctica, there exist sets of quite distinctive aeolian structures, called snow megadunes. These are only 1–8 m high (measured in the field with GPS), but some 2–6 km from crest to crest and many tens to hundreds of kilometers long. Their topographically subdued character meant that they went largely unnoticed by explorers on land, although pilots did see them (e.g., Figs. 11.17 and 11.18). It was only satellite imaging (Figs. 11.19 and 11.20) that revealed the vast extent of these features, which cover some 900,000 km$^2$ of the East Antarctic Plateau.

The structures move very slowly, if at all (the ice on which they sit is itself moving at a few meters per year) and form in the near-continuous strong katabatic winds that descend across the plateau. The subtle appearance of the dunes is due to a textural difference between dune and interdune, which is also manifested in the microwave scattering properties (Lambert and Long 2006), making the dunes very distinct in radar imaging (Fig. 11.18). This textural difference is likely attributable to different recrystallization rates. The spacing of the features (they don't really meet the definition of dunes or ripples) has been speculated to be due to a standing wave in the atmospheric boundary layer, that somehow influences the recrystallization process (Fahnestock et al. 2000).

Although the dune topography is very subtle, it has been measured remotely: large but shallow structures lend themselves to measurement by the precise range measurements from satellite laser altimeters, which yield measurements consistent with the field determinations (interferometric radar was also used with similar results) (Figs. 11.19 and 11.20). Field access can be challenging (for obvious reasons)—beyond the overall challenges of work in Antarctica, local mobility by snowmobile and sled can be restricted by sastrugi, low ridges of packed snow that form in a downwind direction (visible in Fig. 2.1). These form by a combination of erosion and accretion, somewhat like yardangs or the lee dunes that form in cohesive sands in China as described by Rubin and Hesp (2009).

## 11.5 The Role of Vegetation in Dune Motion

Vegetation is a significant component of nearly all desert areas on Earth. While deserts by definition receive limited amounts of rain, even minimal quantities of moisture seem to be sufficient to support a wide range of desert flora. The hardiness of desert plants is of less concern here than is the fact that the presence of almost any plant tends to have a strong impact on the wind flow over the surface. Plants alter the boundary layer of the wind velocity profile by

**Fig. 11.16** Timelapse image of a dune at Bruneau. The dark-bright patterning is the result of emerging transverse bedforms (dry sand over rain-produced wet sand) along the flanks of this somewhat linear dune. Streamers of sand are visible in the *center* of the image, while sand blowing off the dune is visible at *upper right*. Photo J. Zimbelman

**Fig. 11.17** Dunes formed in snow, on a 'blue ice' plain in Antarctica. Image courtesy Ted Scambos and Rob Bauer, National Snow and Ice Data Center

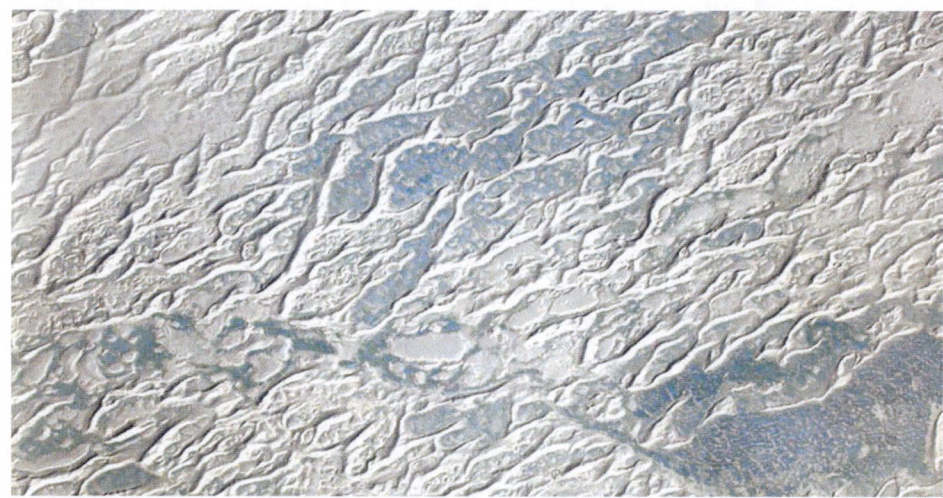

**Fig. 11.18** Aerial view of Antarctic megadunes. The irregular but repeating pattern is striking, as is the subtle shading. At *lower left*, the orthogonal downwind pattern of ridges (sastrugi) is evident. Image courtesy of Ted Scambos and Rob Bauer, National Snow and Ice Data Center

## 11.5 The Role of Vegetation in Dune Motion

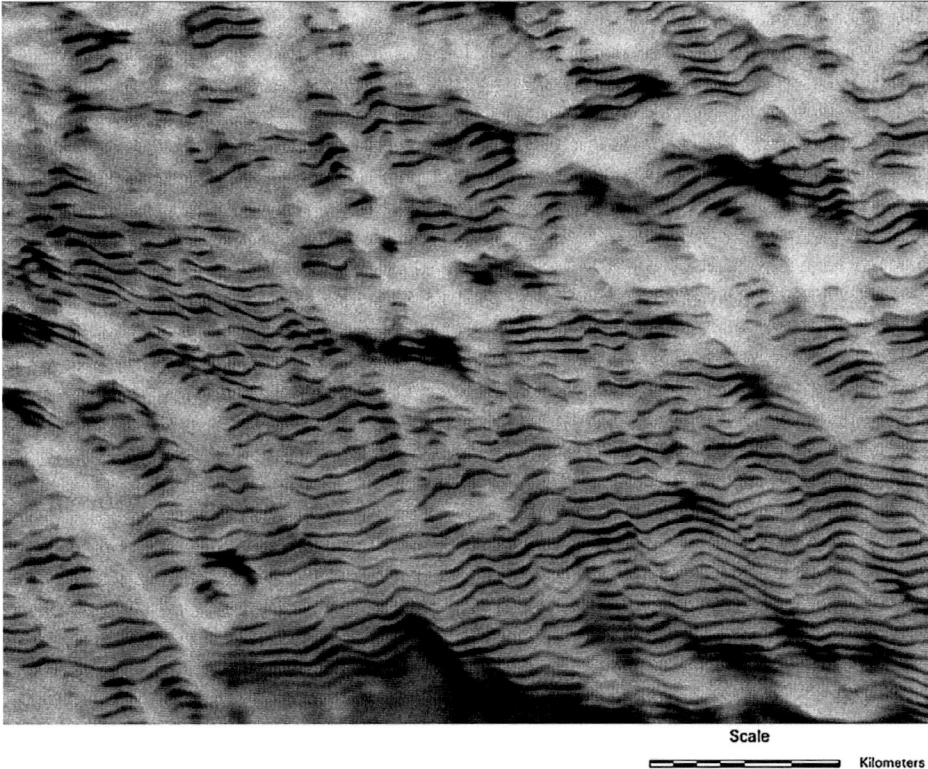

**Fig. 11.19** Synthetic Aperture Radar (RADARSAT) image of Antarctic megadunes. The contrast in the image derives from the different texture of the dunes and interdunes, not from topography. These megadunes were considered as a possible analog for radar-detected dunes on Titan, before taller and straighter examples of 'conventional' linear dunes were discovered. National Snow and Ice Data Center/RADARSAT Canada

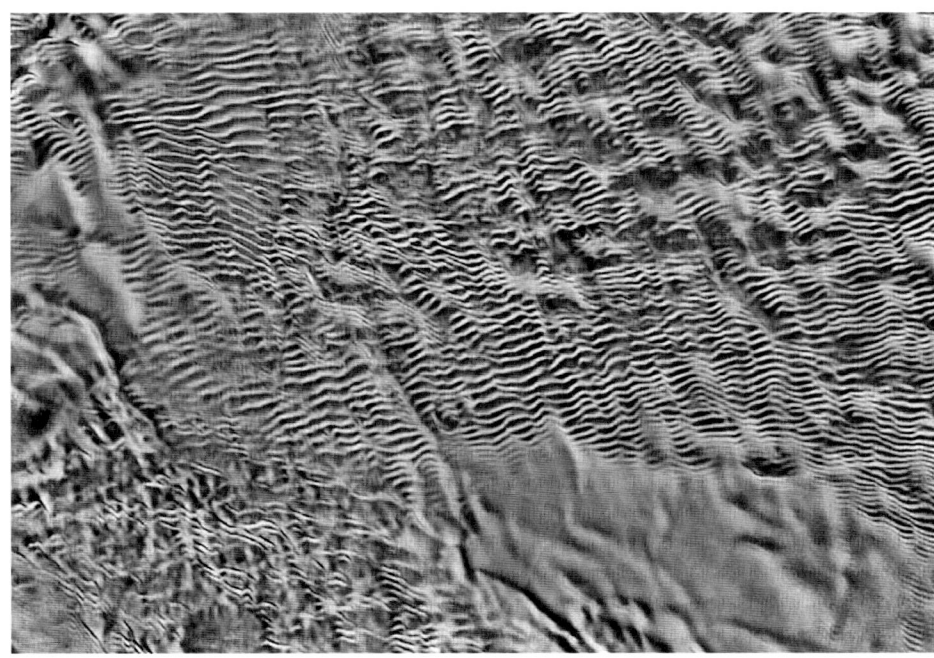

**Fig. 11.20** An optical image mosaic generated by stacking six dithered 250 m-resolution Moderate Resolution Imaging Spectroradiometer (MODIS) images and processing ('super-resolution') to generate a ∼100 m-resolution product. The formidable extent of the megadunefield is evident. This mosaic is part of the National Snow and Ice Data Center and the University of New Hampshire: image made available by NASA at earthobservatory.nasa.gov

increasing the roughness height for the surface, to the point that the height of zero wind velocity exceeds the diameter of the sand grains. Once this happens, the sand grains no longer experience the sensible force of the wind. As long as the plants are present, even after they have died, their impact on the wind flow can effectively 'stabilize' the sand to aeolian motion, thus stopping any further migration of dunes. Even if the plant spatial density is not sufficient to completely halt the motion of the sand by the wind, the plants can form local traps for sand on their downwind side, forming nebkhas (Fig. 7.6) There are many reports in the agricultural literature on the use of different plants to stabilize dunes found among or close to croplands or orchards, as well as to decrease the loss of topsoil to the wind. At present, the Earth is the only place within our solar system where vegetation plays an important role in the modification of aeolian transportation of sand.

## 11.6 Links Between Sand Dunes and Climate

The formation of dunes requires mobile (i.e., dry) sand, and wind. Clearly, changing climate can affect these quantities directly, but the influence of moisture on Earth is magnified by the role of vegetation. If sand is damp, not only is its cohesion higher (and thus the threshold wind stress higher), but it may allow plants to grow. Thus an increase in frequency of damp conditions can lead to further binding of the sand by plant roots themselves, and ultimately by organic matter from dead plants and animals that may feed upon it. To this is added the effect of plants on the aerodynamic roughness.

Changes in climate, particularly those over tens of thousands and hundreds of thousands of years, associated with the Croll-Milankovich cycles (periodic variations in orbital and rotational parameters of the Earth which influence the distribution of sunlight at high latitudes) are therefore expected to affect sand mobility. Notably, during the ice ages (when sea level was $\sim 100$ m lower than today because the water was locked up in continental ice sheets) the atmosphere overall may have been dryer. Furthermore, the shallow seabeds of the continental shelves would have been dry, making the sands there available for transport inland (e.g., much of the Arabian Gulf was dry, with the exposed carbonate-rich sand blowing into the Arabian deserts).

Although both dryer and wetter periods can prevail (notably, a few thousand years ago the Sahara was wetter and thus much more habitable than today) the climate of the last few centuries has been wetter and therefore less conducive to dune migration than, say, 20,000 years ago. Thus what were previously active dunes are now covered in soils and vegetation; there are many such dune systems in Europe, and in America the Nebraska sand hills serve as an example. However, the more recent warming of the climate may lead to dryer soils and thus reactivation of dune systems, notably in the Kalahari. Although dunes are likely never to rival lake sediment cores or ice cap profiles for paleoclimate reconstructions, some important information can be recovered from examination of dune sands with particular relevance to human survival, notably, measurement of the amount of organic matter, and age via carbon-dating or OSL (see Chap. 17).

Wind patterns can also change as the climate changes. For example, the annual monsoonal flow coming up from the Indian ocean may today not penetrate as far into the Arabian peninsula as it once did. This means that some areas (e.g., Fig. 7.11) which previously had seen a bidirectional wind regime now see a more unidirectional one, and the dune type (linear vs barchan) may be changing as a result. We discuss this morphodynamic age, and its application to other planetary bodies, in Chap. 21.

# Mars Dunes

## 12.1 Introduction

Mars has intrigued humans since this reddish object was first noted as one of the 'wanderers' (e.g., planets) that moved among the seemingly fixed stars. The red color is clearly visible to the unaided eye, particularly when Mars is closest to the Earth (roughly every 26 months), leading to the association of this planet with the god of war. As the positions of the planets were monitored systematically, careful observers noted that Mars occasionally reversed its direction of motion, only to revert to its normal progression among the stars after a few weeks. This perplexing motion eventually was explained once scientists accepted the fact that all of the planets orbited the sun, and the unusual motions of Mars results from its position as the next furthest planet out from the sun, after the Earth.

Telescopic observations revealed Mars as an ochre-colored disk displaying both dark and bright (but still reddish-hued) regions across its surface, with occasional white or rust-colored clouds (Fig. 12.1). The polar caps grow and shrink with the seasons, produced by a tilt of Mars' rotation axis that is very similar to tilt of the Earth. Telescopic measurements during the early 20th century began to suggest that the atmosphere of Mars, which was evident because of the documented movement of clouds and dust storms across the planet, was not very substantial in comparison to the atmosphere present around the Earth. Yet even the largest Earth-based telescopes failed to resolve individual landforms on the surface of Mars.

The advent of the space age provided the first opportunities to observe planetary surfaces up close using robotic spacecraft. The first spacecraft to return close-up images of another planet was Mariner 4, launched November 28, 1964; the spacecraft flew past Mars on July 14, 1965, and successfully transmitted 22 images back to Earth, taken by a television camera attached to a small telescope. While crude by the standards of spacecraft images of today, the Mariner 4 images shattered the hope that Mars might host the abundant life described in countless science fiction stories about the Red Planet. The surprising Mariner 4 images revealed that the surface of Mars was dominated by numerous large impact craters. While they covered <1 % of the Martian surface, these first close-up images were sufficient to reveal that Mars had a history that appeared to be more similar to that of Earth's Moon than to the Earth. The Martian crater morphology was subdued relative to comparable craters on the Moon, which was interpreted to suggest that winds on Mars may have modified the relief of the rugged Martian crater rims through either deposition or erosion, or both. As the spacecraft flew past Mars, the refraction of the radio signals transmitted by the spacecraft provided strong evidence that the Martian atmosphere was at most only 1–3 % the density of the atmosphere of Earth, confirming that the climate on Mars must be substantially different from that of anything on the surface of the Earth. The perception of an impact-crater-dominated surface was further strengthened by the Mariner 6 and 7 fly-bys on July 31 and August 5, 1969, only days after the Apollo 11 landing on the Moon. Hundreds of images with substantially better resolution (~300 m) than the Mariner 4 images were returned by both of these spacecraft, appearing to confirm the initial impression that impact craters dominated the geologic history of Mars, although both the south polar region and some isolated collapse features hinted that Mars still held some intriguing features that were distinctly non-lunar-like.

The prospects for a more diverse and interesting Martian history improved dramatically with the first spacecraft to be placed in orbit around another planet. The Mariner 9 spacecraft entered into orbit around Mars on November 14, 1971, but unfortunately it arrived at the height of one of the most intense dust storms of the last hundred years. But, unlike a flyby spacecraft, the orbiter could wait out the storm. In early 1972, the Martian atmosphere cleared sufficiently for Mariner 9 to begin its mapping mission, eventually returning more than 7300 images (with resolutions down to 100 m/pixel) before the mission concluded in October 1972. The wealth of orbiter images finally

**Fig. 12.1** Mars from Earth. In fact, many of the features visible in this Hubble Space Telescope image in 1997 could be discerned by 19th-century astronomers but can now be captured by amateur observers with fast digital cameras on modest telescopes. Some of the broad dark patches change shape over time due to removal and deposition of dust. It is possible to just make out a dark ring around the north polar cap, which forms the largest dunefield on the planet: Olympia Undae. *Image credit* Space Telescope Science Institute

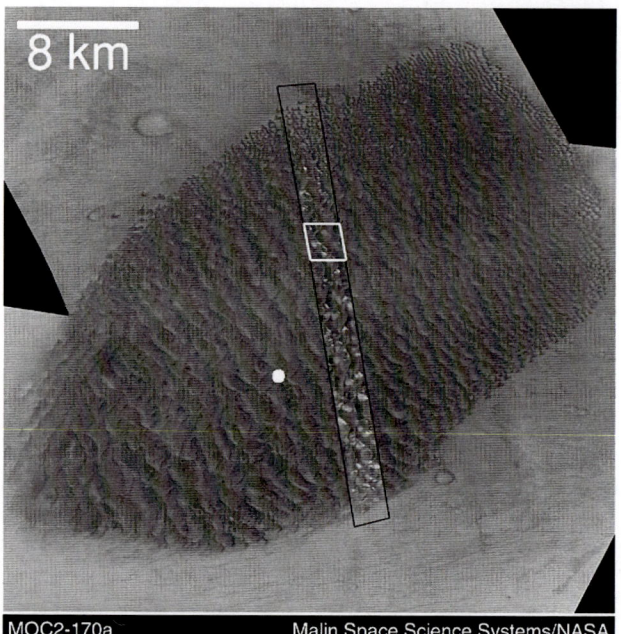

**Fig. 12.2** Proctor Dune Field, among the first dunes to be recognized on another planet. This is a processed mosaic of two Mariner 9 images with a resolution of ~60 m, acquired in March 1972 (for a raw Mariner 9 image, see Fig. 1.2). The rectangle indicates an area imaged by Mars Global Surveyor in 1999 (see Fig. 12.5); the dot indicates the location of a HiRISE image shown in Fig. 12.6. *Credit* Malin Space Science Systems/NASA. Image MOC2-170a

documented the entire surface of Mars, revealing the presence of not only impact craters that had dominated the terrains viewed by the first three fly-bys, but also the presence of enormous volcanoes (some hundreds of kilometers across), the largest canyon system in the solar system (appropriately named Valles Marineris, 'valleys of the Mariners'), evidence that a flowing liquid carved sinuous valleys across many portions of the surface, possible evidence of periglacial (ground ice) activity and, most important to the topic of this book, the first evidence of sand dunes on another planet (Figs. 12.2 and 1.2).

The dust storm itself demonstrated that dust can be blown around on Mars, and a number of streaks (see Figs. 7.1, and 13.3) were seen in the lee of craters that Sagan et al. (1972) inferred to be wind-generated deposits of dust or sand. However, the first actual dune fields were located on the floors of craters, notably the Hellespontus crater (Cutts and Smith 1973—see Fig. 1.2), now called Proctor crater (Figs. 12.2, 12.3 and 12.4). These first observations set the longest time interval for dune migration/evolution studies on Mars (see Chap. 9). Proctor has been re-observed many times since, e.g., Figs. 12.5, 12.6).

The stunning success of Mariner 9—and the evidence of past liquids—was strong validation of the NASA decision to send an armada of spacecraft to Mars, with the goal of assessing whether life was present. The Viking mission involved two orbiters and two landers; on July 20, 1976, the Viking 1 lander became the first spacecraft to return images from the surface of another planet, at a site in Chryse Planitia that provided startling evidence of wind-blown drift deposits (sand shadows, or lee dunes) around some of the nearby large blocks. Viking Lander 2 showed a few similar features (e.g., Fig. 16.14).

The >50,000 images returned by the two Viking orbiters, over several years of operation, revealed that dune fields within large craters were quite common on Mars, and the largest 'sand sea' (erg) on the planet formed a ring around the north polar cap (e.g., Cutts et al. 1976; Tsoar et al. 1979). The Viking data prompted a systematic comparison of Mars dunes with those on Earth (e.g., Ward et al. 1985; Breed et al. 1979), especially since satellite data on Earth dunes was becoming comprehensive at similar or better resolution.

Unfortunately, a gap of nearly 20 years separated the Vikings from the next successful mission to Mars, but the sophistication of the spacecraft missions then steadily increased:

- the Pathfinder landing (on July 4, 1997) included the first spacecraft (Sojourner) to roam across the surface of another planet;

## 12.1 Introduction

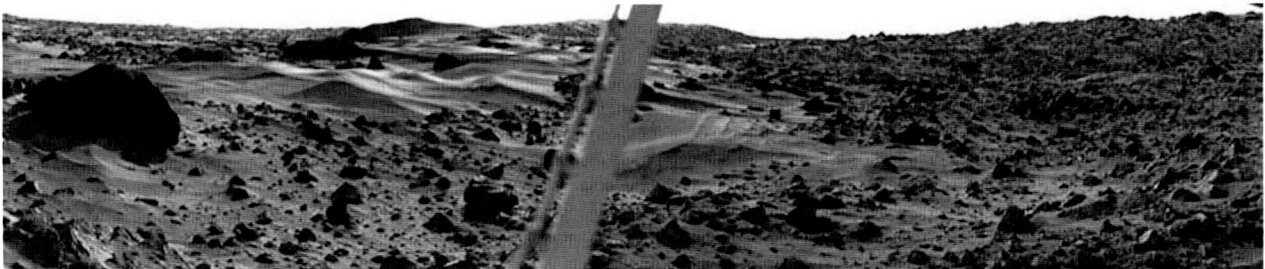

**Fig. 12.3** This spectacular picture of the Martian landscape by the Viking 1 lander shows aeolian features remarkably similar to many seen in the deserts of Earth. The dramatic early morning lighting—7:30 a.m. local Mars time, on August 3, 1976—reveals subtle details and shading. The picture covers 100° of azimuth, looking northeast at left (where the large rock was nicknamed 'Big Joe') and southeast at right. The sharp dune crests indicate the most recent winds capable of moving sand in the general direction from upper left to lower right. Small deposits downwind of rocks also indicate this wind direction. Large boulder at left is about 8 m from the spacecraft and measures about 1 × 3 m. The meteorology boom, which supports Viking's weather station, cuts through the picture's center. The sun rose two hours earlier and is about 30° above the horizon near the center of the picture. (For a similar view at Viking 2, showing changes over several years, see Fig. 16.14) *Credit* NASA PIA00393

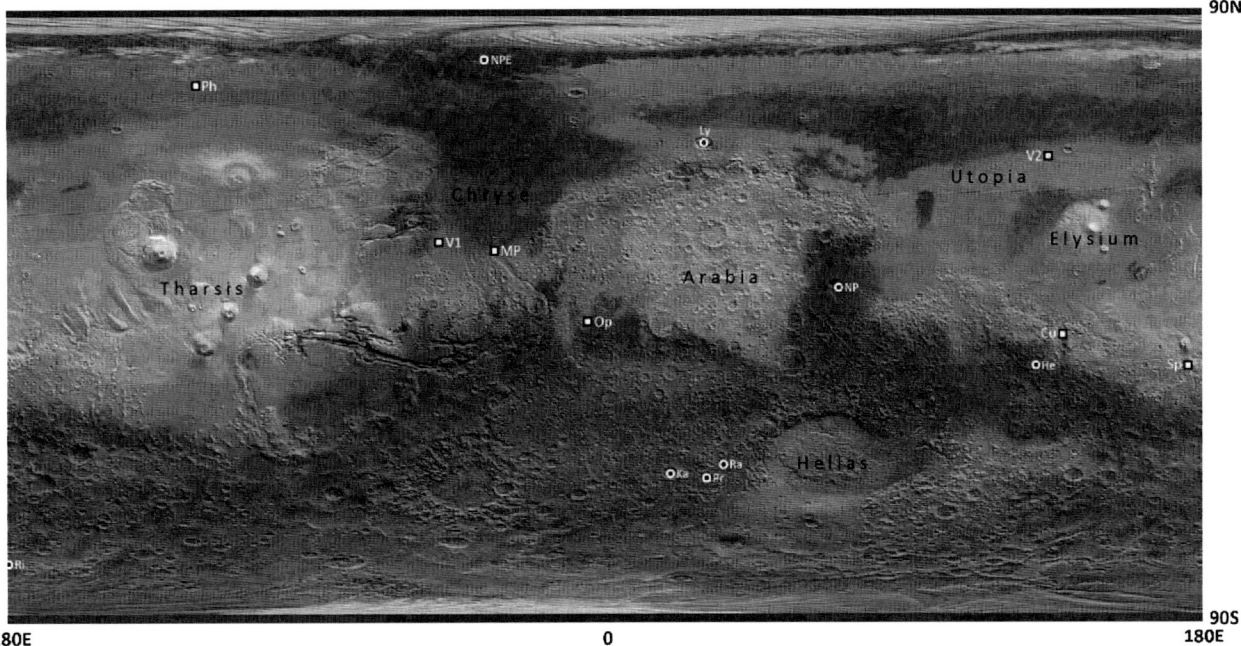

**Fig. 12.4** Locations of landers and rovers (squares) and example dune fields (circles), on TES albedo superposed on MOLA shaded relief. The spacecraft are V1,V2—Viking Landers 1 & 2; MP—Mars Pathfinder; Sp, Op—Mars Exploration Rovers Spirit and Opportunity; Ph—Mars Phoenix; Cu—The Mars Science Laboratory Curiosity. NASA/JPL/ASU, www.mars.asu.edu/data/tes_albedo, annotated by J. Zimbelman

- the Mars Global Surveyor observed the planet from orbit using multiple instruments for more than a decade;
- the Mars Odyssey and Mars Express orbiters brought even more powerful spectrometers and cameras to Mars;
- the twin Mars Exploration Rovers explored tens of kilometers across Mars, far exceeding their design requirements;
- the Phoenix lander sampled ground ice only a few centimeters below the surface at high northern polar latitudes;
- the Mars Reconnaissance Orbiter still continues to operate the best camera and spectrometer yet flown to Mars; and
- the Curiosity rover has only barely begun its mission to look for organic materials in sediments within the Gale crater (see a map of the various landing sites and major dune fields in Fig. 12.4).

Today there is a staggering wealth of spacecraft data available for Mars, with the in situ information at several locations (see also Chap. 16) supplemented by images from

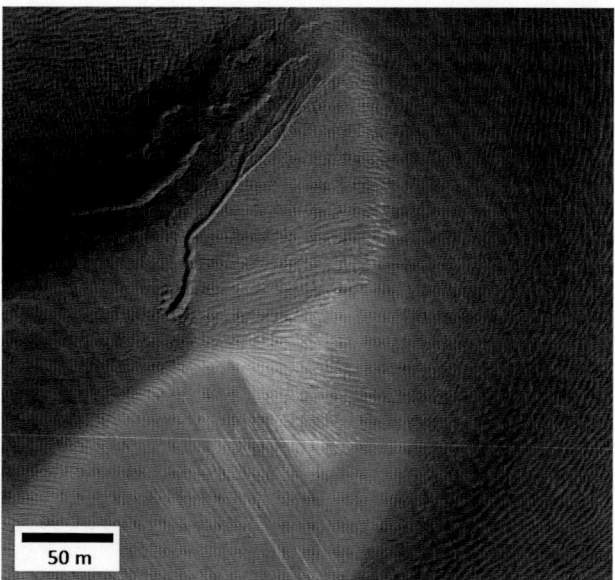

**Fig. 12.6** HiRISE image of part of Proctor crater (area indicated by dot in Fig. 12.2) showing avalanche streaks, and individual ripples. A gully seems to have formed. *Image credit* NASA/JPL/U.Arizona

**Fig. 12.5** Dunes in Proctor crater, imaged by Mars Global Surveyor in 1999 where the sand dunes are dark and patches of southern winter frost are bright. The sun illuminates the scene from the upper left. Dark streaks can be seen on frost-covered slopes, particularly just left of the center of the picture. These streaks result from recent avalanching of sand on the steep slip faces. Because the dark sand streaks are superposed upon the bright frost, these streaks can only be as old as the frost. This frost cannot be more than 11 months old, and was probably only a few months old at the time the picture was taken. Thus, the dunes must be active today in order to show such streaks. *Credit* Malin Space Science Systems/NASA

orbit (see also Chap. 18) with ever-higher spatial resolution, coverage, and spectral discrimination. As an example of the progression of Mars dune studies from orbit, compare the Mariner 9 image (Fig. 12.2) in 1972 with that from Mars Global Surveyor in 1999 and from the HiRISE camera on Mars Reconnaissance Orbiter in 2010 (Fig. 12.5): views of the same dunes are montaged together in Fig. 1.5.

In this section we will explore how these data show that the dunes on Mars are both similar to, and yet also different from, the sand dunes that are common on Earth. The story of the Martian dunes is intimately tied to the unique atmospheric conditions present on Mars today, including how the atmosphere may have varied dramatically in both the recent and distant past.

## 12.2 The Martian Atmosphere

The most significant aspect of the Martian atmosphere is that it is thin, as vaguely indicated telescopically, then more conclusively by radio measurements by the early Mariners, and then accurately demonstrated by the pressure measurements on the Viking landers which had to have huge parachutes for their safe descent. The surface pressure of about 6 mbar is over 100 times less that on Earth, such that the typical atmospheric pressure on Mars is comparable to the pressure at an elevation of 30 km (about 100,000 feet) above the surface of the Earth, three times higher than the cruising altitude for most commercial aircraft. This low pressure, and thus low density, means the wind stress for a given wind speed is much lower (see Chap. 4) and therefore sand should be rather difficult to set in motion. Once sand does move, however, the relevant length scales (see Chap. 5) are rather long, and so Mars ripples and elemental dunes are rather larger than their terrestrial counterparts.

The atmosphere of Mars is comprised primarily of carbon dioxide (>95 % $CO_2$), with minor amounts of $N_2$ and argon, plus trace amounts of other gases. The preponderance of carbon dioxide makes the Martian atmosphere chemically more like that of Venus than the Earth, but the atmospheric pressure is radically different between these three terrestrial planets. Perhaps the most important trace gas on Mars is water vapor, present in amounts far below what is present in the troposphere (lower atmosphere) on

**Fig. 12.7** Mars draws its own wind roses. Dark dust plumes indicate the prevailing winds as the underlying surface begins to defrost. Such features are seen widely in the Olympia Undae dunes and elsewhere. Defrosting dune surfaces are shown in Fig. 12.18

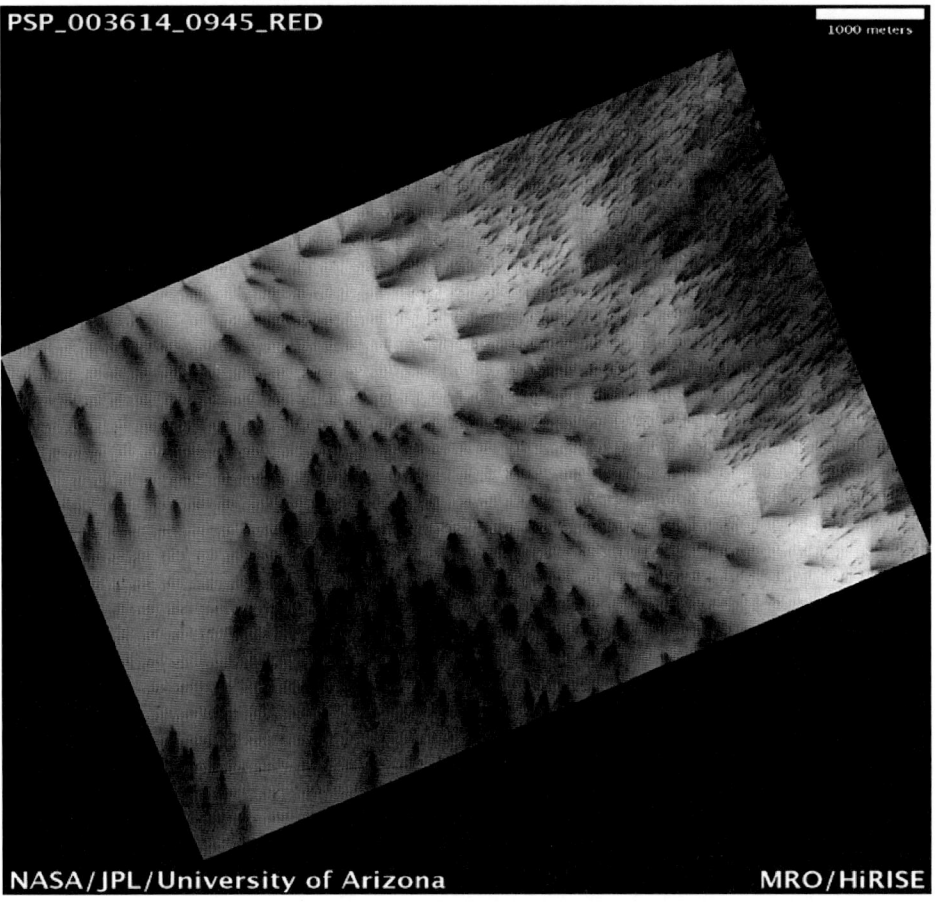

Earth, yet at those low abundances the thin atmosphere holds nearly all the water vapor that it can (that is, the Martian atmosphere is almost saturated with water vapor, making the relative humidity near 100 %, even though the total abundance remains incredibly small). Faint clouds of water ice sometimes form, and water ice is present in the near-subsurface (as dramatically shown by the Phoenix lander—see Fig. 16.16)—an important effect of the migration of water vapor is its possible deposition in the subsurface, indurating dunes and freezing them in place.

Polar temperatures during winter in both hemispheres get low enough to condense carbon dioxide to form seasonal polar caps on Mars (visible telescopically from Earth). The Viking landers measured variations in atmospheric pressure of ~30 % through a Martian year, indicating that almost one-third of the atmosphere freezes out at each pole in turn, only to be released again as the seasonal caps later shrink and eventually disappear completely. The deposition of $CO_2$ frost on polar dunes is evident when the dunes are observed closely from orbit—avalanches of dark sand are visible against the bright frost (and in fact sand and dust are sometimes sprayed out into ephemeral dark spidery patterns when $CO_2$ gas builds up under the frost cover in spring and escapes suddenly through cracks in it; see Fig. 12.7).

The low pressure of the Martian atmosphere causes it to respond quickly to temperature variations that take place throughout the day; several landers have documented day-night atmospheric temperature swings of up to 100 °C. One consequence of such dramatic temperature changes is that strong gradients can be produced quickly in the lower atmosphere, driven by rapid changes in the temperature of the Martian surface as the solar insolation varies through the course of a day—these gradients can drive strong slope winds. Another consequence of the low density is that the atmospheric boundary layer (see Chap. 3) can grow to be many kilometers thick. Thus Mars dunes, in many cases larger than Earth's, may be limited in size by how quickly they can grow rather than by how large they may ultimately grow to, capped by the boundary layer.

Insolation varies through the Martian year as well as through the day; the orbit of Mars is five times more elliptical than that of the Earth, causing large seasonal variations in the energy derived from sunlight (perihelion presently occurs during summer in the southern hemisphere of Mars, at which time the planet receives 45 % more solar energy per unit area than when the planet is at aphelion). This arrangement of the seasons changes dramatically over astronomical time, see below.

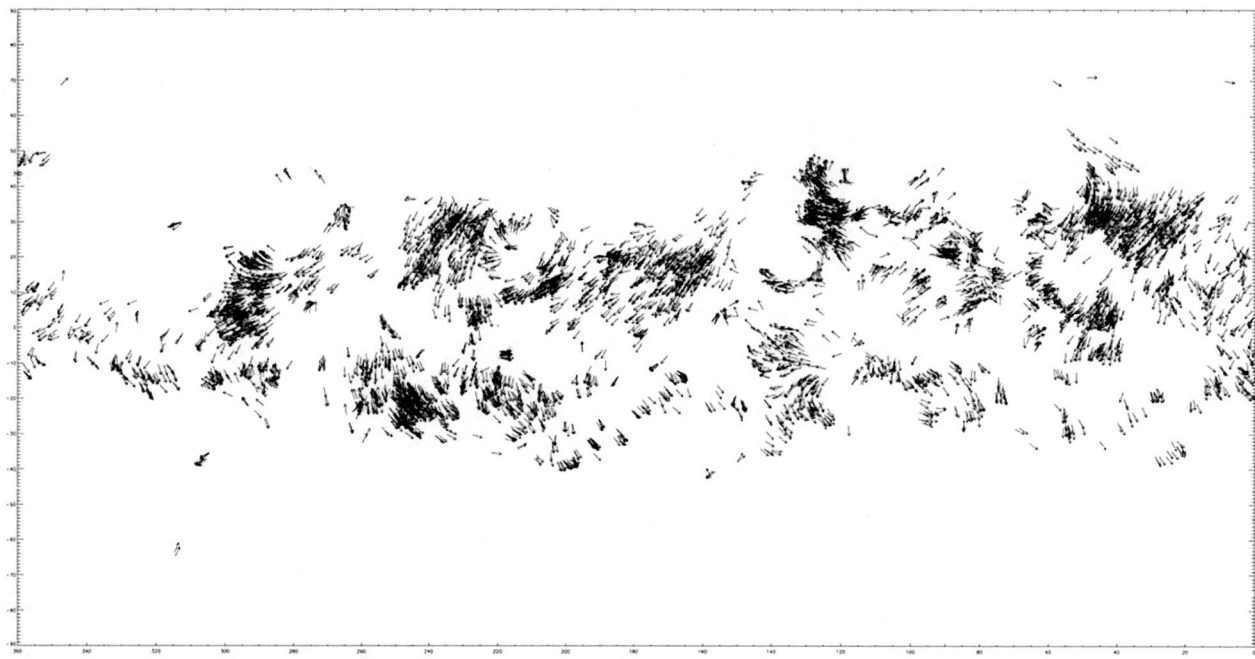

**Fig. 12.8** Dust streaks in the lee of craters (e.g., Fig. 8.2) provide another set of indications of Mars winds. This formidable compilation is courtesy of Peter Thomas of Cornell University

Both daily (diurnal) and yearly (seasonal) variations of several factors affect the winds on Mars, which essentially represent the global mechanism for redistributing sensible heat across the planet; it is the transfer of warm air from near the equator toward the poles, along with the transfer of cold air from the poles toward the equator, that drives global winds. Global Circulation Models (GCMs) are mathematical representations of this heat transfer process for a planet (see Chap. 19). GCMs indicate that the intensity and direction of global winds change through the course of a Martian year, and these patterns can be greatly affected by the strength and duration of seasonal dust storm activity (Fig. 12.8).

Recently, GCMs have been used to investigate how the Martian atmosphere and surface are influenced by changes in the orbital parameters induced by the gravitational effect of the other planets (primarily Jupiter and Saturn). The lack of a large moon at Mars allows such perturbations to alter greatly the conditions on the planet, primarily with regard to the tilt of the rotational axis relative to the orbital plane (called obliquity). While the obliquity of Earth ranges by about one degree around the present value of 23.5°, the obliquity of Mars can range from zero to near 80°, large deviations from the present value of 25°, which along with variations in orbital conditions (such as eccentricity) can cause enormous changes to the overall climate of the planet, all of which can affect wind patterns and their intensities (and thus dune locations, morphologies and orientations) as well as the stable distribution of ground ice (which may therefore cause dunes to be immobilized). For example, when the obliquity exceeds about 45°, the polar regions will on average over the year receive more sunlight than the equator!

## 12.3 Sediments on Mars

Mechanical weathering is the primary mode for the generation of sand-sized particles on Mars, mainly because of the dearth of water under present Martian conditions. The large temperature swings experienced daily and seasonally will act (albeit slowly) to break apart rocks exposed at the surface, perhaps enhanced by intermittent films of water molecules when conditions allow localized condensation of the trace amount of water vapor present in the atmosphere. Impact craters also provide an intense but very localized means to break up near-surface rocks into smaller constituent pieces. Still, when billions of years are available to carry out the mechanical weathering, it is not surprising that collections of sand-sized material can be found across much of Mars. When winds of sufficient intensity occur, the sand-sized material eventually tends to concentrate into protodunes or dune fields, usually within topographic depressions that act to trap the particles. The Viking, Pathfinder, and Phoenix landers, along with both MERs, all returned evidence of fine-grained material wherever these explorers looked (see Chap. 16), although only the last three spacecraft carried cameras that could resolve sand from the silt to clay-sized dust particles that settle ubiquitously out of the

Martian atmosphere. The two rovers have documented particles ranging from dust and sand to granules and cobbles everywhere along the tens of kilometers of the surface over which they have ranged.

Sophisticated remote sensing instruments on both landed and orbiting spacecraft have revealed much about the fine particles present on the Martian surface. Measurements of the thermal infrared energy radiated from the surface provide important clues to the physical size of the constituent surface materials. As discussed in Sect. 18.4, thermal inertia refers to how well (or how poorly) the surface maintains its temperature through the course of a Martian day. Materials with high thermal inertia (large blocks and rock outcrops) tend to stay relatively cool through the day and warm through the night, in contrast to low thermal inertia materials (such as a thick dust mantle). When multiple thermal wavelengths are measured, the distribution of high or low thermal inertia materials within a single pixel can be quantified, even when the individual particles are not resolved.

The thermal inertia of windblown deposits within impact craters on Mars indicates that these deposits respond to temperature changes as if they were comprised of a uniform layer of coarse sand (400–600 µm in diameter), which is larger than the fine sand that makes up most of the dunes on Earth (100–200 µm in diameter). Spectroscopic measurements of the dark deposits in craters also revealed that they consist primarily of a basaltic composition, some showing evidence of minerals called pyroxenes, which are common in basalt rocks. The Opportunity rover was able to investigate a dark sand deposit (named 'El Dorado') in the Columbia Hills, confirming the basaltic composition of the medium-to-coarse sand found there (discussed in more detail below). Both rovers have also obtained evidence of localized deposits of sand-sized aggregates that, when pressed upon by the rover arm, break up into dust; such aggregates may be similar to 'parna' on Earth, where sand-sized aggregates of clay particles form dunes around dried lakebeds. Aggregates on Mars likely could not survive the saltation required for transport over long distances, but they do appear to contribute to localized collections of sand-sized materials (e.g., see Figs. 16–20 of Sullivan et al. 2008).

The observed distribution of sand dunes is widespread around Mars, but not in a uniform or systematic manner. By far the largest accumulation of sand dunes occurs in the sand sea that surrounds the north pole within the 70–80 °N latitude band. Portions of the north polar erg were recently shown to include the spectral signature of the mineral gypsum, a material common around the margin of some saline lakes, leading to speculation that the dune field could incorporate materials derived from ancient lakebed deposits. Outside of the north polar erg, more than 900 sand dune patches larger than 1 km$^2$ have been mapped across the rest of Mars (Fig. 12.9), with the resulting collection of dune forms displaying a considerable diversity across the planet (Hayward et al. 2007). No obvious sources are recognized for the identified dune deposits, although the north polar sand may be derived from a specific dark basal unit within the polar layered deposits. Dunes within craters appear to be trapped within the topographic depression rather than derived from a localized source exposed within the crater rims. Mars also lacks obvious sinks for mobile sand; the lack of current oceans makes it difficult to remove sand from the system except through burial inside impact craters that became buried by some later event.

## 12.4 Types of Dunes on Mars

A variety of dune types have been recognized on Mars over the years, including barchan, barchanoid ridge, transverse, star, linear, dome, and complex dunes (the last displaying attributes of two or more dune types). Notably, almost one-third of the dune fields identified in a global map of sand dunes on Mars do not seem to fit easily into one of the dune categories used to describe dunes on Earth (Hayward et al. 2007, 2010, 2012). The formation mechanism for Martian dunes is thought to be essentially the same as the recognized aeolian processes that generate similar dune forms on Earth. However, Martian dunes tend to be up to an order of magnitude larger than their comparable terrestrial counterparts, similar to the large size displayed by many other Martian landforms. Here we will not necessarily discuss in detail each of the Martian dune types, but rather present examples of the more common types for comparison with dunes on the other planets.

Barchans (see Chap. 6) are one of the most readily identifiable dune types on any planet, due to the distinctive crescentic shape of the entire dune mass. Barchans and barchanoid ridges (Chap. 6) are quite common near the margins of the NPE, particularly where the sand supply from the erg becomes more widely distributed as the sand spreads onto the surrounding plains. Barchans are also found at lower latitudes where the supply of sand is relatively limited, but still sufficient to produce large and impressive crescent-shaped sand deposits (Fig. 12.10). Some Martian barchanoid ridges are seen in Figs. 6.13 and 6.15.

One of the most common dune types on Mars is the transverse ridge (Chap. 6), comprised of one prominent crest line that is inferred to be transverse to the orientation of the wind that blew the sand to its present location (Fig. 12.11). Martian transverse features are often quite symmetric with respect to the crest axis, making it difficult to decide from which of the two transverse directions the driving winds blew, but also suggestive that two nearly

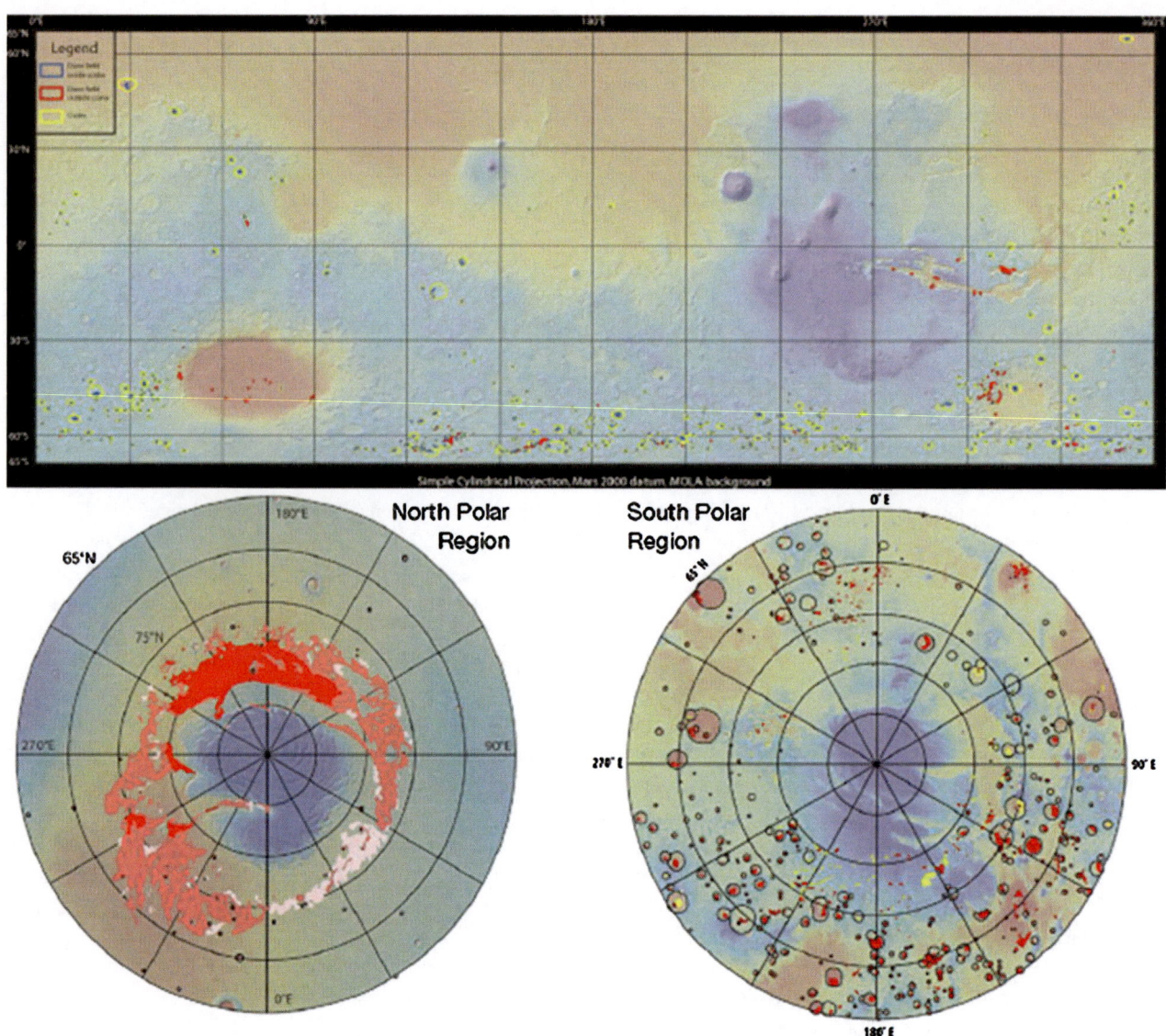

**Fig. 12.9** A map of dunefields on Mars assembled by the Mars Dune Consortium (credit Hayward et al. US Geological Survey Open File Report). Apart from the circumpolar erg in the north, most of the major dunefields are in craters

equal but opposite wind patterns may be present, perhaps driven by diurnal temperature and pressure variations, especially where winds are likely to be confined in valleys. For transverse features with wavelengths less than 100 m, the non-generic term 'transverse aeolian ridge' (TAR) has been applied, allowing for the smallest TARs to possibly be either small sand dunes or large ripples (see Chap. 5). Invariably, where TARs and dunes co-exist, the dunes are usually dark and the TARs bright, and the dunes are superposed on the TARs showing that they have been active more recently (e.g., Figs. 12.6 and 6.11).

True linear dunes (Sect. 6.6) are extremely rare on Mars (Lee and Thomas 1995), particularly since it is usually impossible to constrain local wind flow patterns to determine if the winds conform to what might be expected around terrestrial linear dune fields. Linear dunes are found near the margins of some intracrater dune fields, but these features generally lack an obvious crest line or the presence of a distinct slip face (Fig. 12.12). The linear forms often break up into individual dome dunes with increasing distance from the main portion of the dune field (again, Fig. 12.12). Dome dunes also lack a crest line or slip face but are small and are broadly elliptical to oval in planform.

Some dunes appear to be an amalgamation of dome and linear planforms (Figs. 6.10 and 6.11), producing a T-shaped or teardrop dune form that might historically have been simply dismissed as 'complex', but can now be interpreted (as a result of numerical models and flume experiments) as a barchan variant that results from an intermittently bidirectional wind regime (see Fig. 6.8). Both

## 12.4 Types of Dunes on Mars

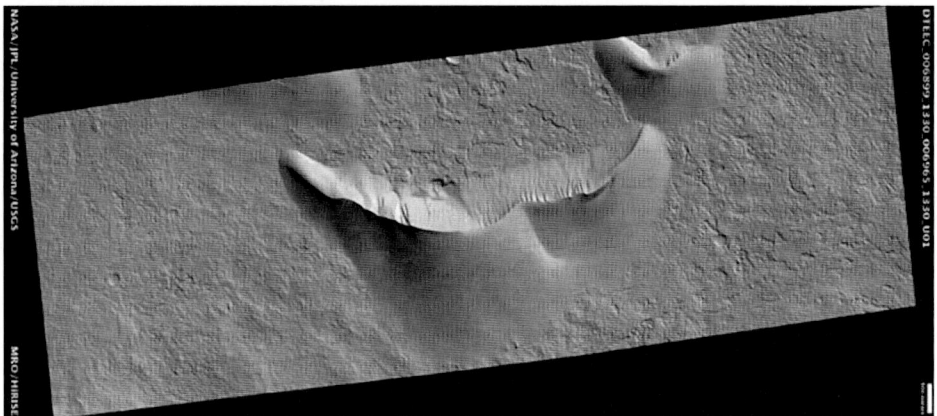

**Fig. 12.10** A shaded relief display of digital elevation data derived from a stereo pair of HiRISE images (PSP_006899_1330 and PSP_006965_1330) for a barchan dune on the floor of Kaiser crater. The elevation data were collected to investigate avalanche features on the slip face of the dune. The scale of this rather conventional-looking barchan is much larger than corresponding examples on Earth—the crest to the floor at the base of the slipface is a drop of some 300 m, over a magnitude larger than is typical for Earth barchans of the same shape. *Image credit* NASA/JPL/U of A/USGS

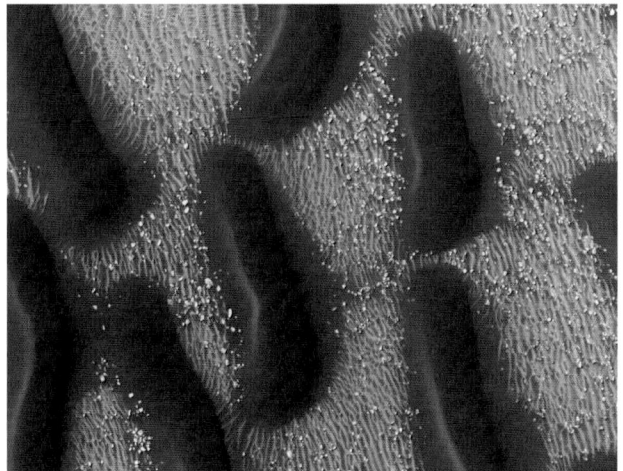

**Fig. 12.11** 'Wing-shaped' barchan dunes in the floor of Proctor crater, now observed in 2010 at a resolution of ~0.5 m by the HiRISE camera on Mars Reconnaissance Orbiter. Not only is the shape of the dunes clearly shown, but their superposition on bright transverse aeolian ridges (TARs) is evident (and close view would show individual ripples on the dunes). *Credit* NASA/JPL-Caltech/University of Arizona

the linear and dome features are usually much darker than their surroundings, including bright features that may consist of granule-coated sand ripples. Star dunes, or at least akle patterns, have formed on some sand deposits inside large impact craters, where the crater landform may contribute to the formation of complex wind patterns (Fig. 12.13). However, even the best stars typically have two arms that are better developed than the third (or fourth) arm, suggesting that either these features are evolving from a earlier transverse pattern, or that perhaps they may now be evolving toward a transverse pattern.

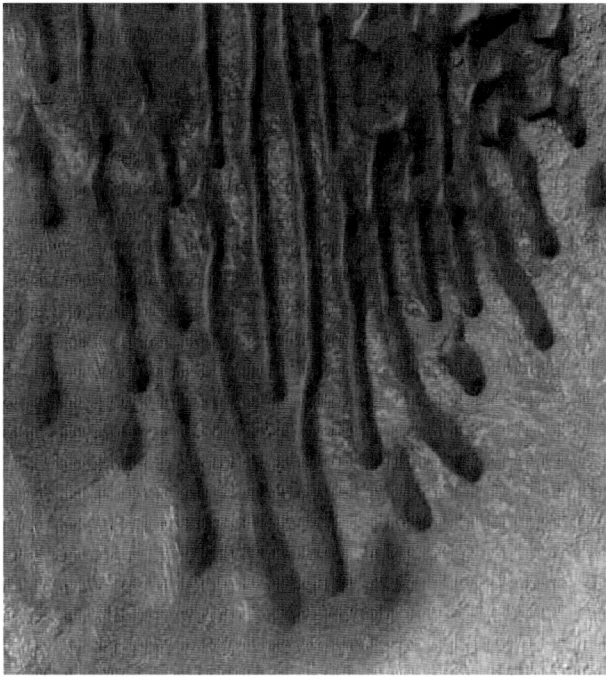

**Fig. 12.12** Transitional linear to dome forms. No sharp slip faces are obvious here. *Image credit* HiRISE browse image, NASA/JPL/U. Arizona

It is important to note that, after nearly 40 years of spacecraft imaging of Mars, only within the last several years have several cases of sand movement been documented on Mars. Starting with the disappearance of dome dunes in the NPE identified through the analysis of nearly a decade of MOC images, it has become readily apparent with HiRISE data that sand is indeed moving on Mars today. Observations made during the 4–6 years of operation of the

**Fig. 12.13** Interesting gradation between linear forms and 'teardrop' barchans across the floor of Bunge Crater in response to winds blowing from the direction at the top of the picture. The frame is about 14 kilometers wide. Mars Odyssey THEMIS. *Image* PIA13654

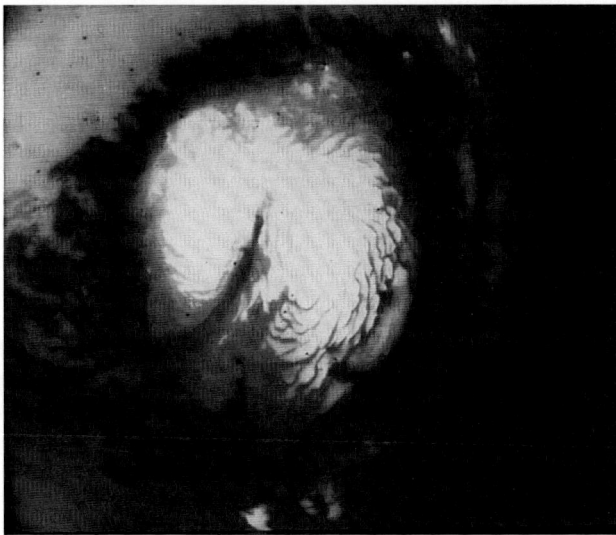

**Fig. 12.14** Mariner 9 image (DAS 13315770) showing the north polar cap with its spiral troughs and the deep gash Chasma Boreale. Some outlying patches of frost are seen, as well as the irregular dark ring around the 1000-km wide cap that make Mars' largest dune field, the polar erg Olympia Undae. *Credit* NASA/LPI

## 12.5 Examples of Dune Localities on Mars

Multiple orbiting spacecraft missions have now documented Mars at a variety of spatial scales, as discussed earlier. However, the highest resolution datasets do not have global coverage, since the data volumes would be prohibitive. The Thermal Emission Imaging System (THEMIS) on the Mars Odyssey spacecraft has provided a uniform image base with global coverage and adequate resolution to evaluate dune morphology. Thus the global inventory of dunes could be systematically mapped (Fig. 12.9), starting with the equatorial and mid-latitudes (65 S to 65 N; Hayward et al. 2007), followed by both the north (Hayward et al. 2010) and south polar regions (Hayward et al. 2012). This mapping effort identified dune fields down to a size of $\sim 1$ km$^2$; nearly 900 locations correspond to dune fields within impact craters of various sizes, and the North Polar Erg (NPE) is the only place where sand is abundant on a regional scale. We can't hope to treat even most of these widely dispersed sand deposits, but next we describe several examples that can serve to illustrate both the similarities and the differences of Martian dune fields as seen in both orbiter and lander imaging (Fig. 12.14).

*North Polar Erg* (72–82 N). A ring of sand called the North Polar Erg (NPE) surrounds the north polar cap on Mars, the one area on the planet that approaches the scale of sand seas (ergs) on Earth. The International Astronomical Union (IAU) has established guidelines for the nomenclature of features identified on all mapped planetary surfaces.

two Viking landers had suggested that at present the Martian winds only rarely exceed the threshold for initiation of saltation, but the sub-meter spatial resolution of the HiRISE camera has now documented that both sand ripples and sand dunes have moved perceptibly during periods of from months to years (see Chap. 9). Importantly, this sand movement has been detected at places widely spread across the planet, and the inferred movement rates on Mars are not unlike the rates of sand movement observed in some cold and dry environments on Earth (Bridges et al. 2012b). Below some of the sites of documented sand movement are discussed in relation to their setting, but these are likely to be only the first of a growing number of places where sand is observed to be in motion on Mars under present conditions, a situation that would have been very difficult to imagine only ten years ago.

## 12.5 Examples of Dune Localities on Mars

**Table 12.1** Major dune fields on Mars

| Name | Location | Width (km) |
|---|---|---|
| Abalos Undae | 78.5 N, 272.5 E | 440 |
| Hyperboreae Undae | 80.0 N, 310.5 E | 460 |
| Olympia Undae | 81.2 N, 178.5 E | 1500 |
| Siton Undae | 75.6 N, 297.3 E | 220 |
| Aspledon Undae | 73.1 N, 309.7 E | 220 |

The IAU Gazetteer of Planetary Nomenclature lists the term 'undae' to indicate 'a field of dunes', and the NPE includes the only five locations on Mars that have been approved so far to become named dune fields on Mars (Table 12.1).

The largest polar dune field has received considerable attention because compositional data from the Compact Reconnaissance Imaging Spectrometer for Mars (CRISM) instrument (see Sect. 18.5; Fig. 18.13) revealed that gypsum is a significant component of the Olympia Undae polar sands (Fishbaugh et al. 2007; Horgan et al. 2009). Imaging from the High Resolution Imaging Science Experiment (HiRISE) camera allows the dune sands to be traced back to eroded exposures of a thick bed within the basal component of the polar layered deposits, which would indicate that the origin of polar sand likely preceded the emplacement of the thick sequence of polar layered deposits (Fishbaugh and Head 2005). The NPE sand deposits can now be placed in context with other bedrock materials exposed throughout the north polar region (Tanaka and Fortezzo 2012), but this provides no additional insight into where the sand may have originated prior to its involvement in the deposition of the polar layered deposits. The dune morphology has been studied by Tsoar et al. (1979) and more recently subjected to a quantitative pattern analysis by Ewing et al. (2010). The high latitude of these dunes makes them particularly prominent sites for the frosting and defrosting processes that occur around winter (e.g., Hansen et al. 2011).

*Proctor crater* (48.0 S, 29.5 E). Proctor crater is 150 km in diameter, located in the southern cratered highlands to the west of the giant Hellas impact basin. The dune field on the floor of this crater was the first to be identified on Mars from Mariner 9 images (Fig. 12.2); the field covers an area approximately 35 by 65 km in area, and the dune field has been targeted often by subsequent orbiters. The resolution of the Mariner 9 image is not sufficient to allow potential long-term movement of the dunes to be detected; comparison of a higher-resolution image (Fig. 12.5) obtained in 1999 by the Mars Orbiter Camera (MOC) on the Mars Global Surveyor spacecraft to the Mariner 9 image showed that any movement must have been less than the size of a Mariner pixel (62 m/pixel) during the 14 Mars years (28 Earth years) that elapsed between the two images. On-going study of the Proctor dune field, notably with the HiRISE camera (e.g., Fig. 12.6) has yielded many insights into the emplacement and modification of dune fields within craters on Mars (see Fenton et al. 2003, 2005; Fenton 2006a).

*Kaiser crater* (46.6 S, 19.1 E). Kaiser crater is 210 km in diameter and is located in the southern cratered highlands to the west of Proctor crater. The dune field present on the crater floor was imaged late in the mission of Viking Orbiter 2 (Fig. 12.15), showing dune crests that are likely transverse to the winds that transported the sand into the crater. Much later, the Mars Orbiter Camera (MOC) on the Mars Global Surveyor spacecraft provided higher-resolution images of individual barchan sand dunes near the main dune field (Fig. 12.16), probably the result of sand that has escaped from the margin of the main dune field. The HiRISE camera obtained a stereo image pair from which very detailed elevation information is obtained for an individual dune in the Kaiser crater (Fig. 12.17). This dune showed avalanche features on its slip face in MOC image R06-00380, so HiRISE stereo data were collected to investigate the relief associated with avalanche modification of dune slip faces on Mars, in an effort to assess what mechanisms may have produced the avalanching of the sand.

*Lyot crater* (50.8 N, 28.8 E). Lyot crater is 240 km in diameter, located in the northern lowland plains of Vastitas Borealis. Lyot has the largest crater dune field found within the northern lowland plains of Mars. The Lyot dune field has been monitored regularly by orbiting cameras (e.g., THEMIS VIS image 6044, 9/10/12) to see whether there are distinguishable variations in how dunes in the northern hemisphere change with time, as compared to the more numerous dune fields in the southern hemisphere, but so far the dunes in Lyot do not display unique characteristics.

*Herschel basin* (4.5 S, 130.0 E). Herschel is a 300 km-diameter impact basin in the equatorial portion of the cratered southern highlands. MOC images of the dune field in Herschel showed a 'grooved' texture that was interpreted to suggest that the sand was moderately indurated and subject to erosion aeolian scour (see Fig. 42 of Malin and Edgett 2001). However, recently the Herschel dune field has shown 1 m of movement between two HiRISE images taken on March 3, 2007 and 1 December, 2010, and the grooved appearance is resolved to be a confluence of aeolian ripples on the dune surface (Fig. 12.17), so that induration of the sand is no longer required. This is merely one example of several locations where HiRISE repeat imaging is now documenting the planet-wide movement of sand under current climatic conditions (e.g., Bridges et al. 2012b).

*Rabe crater* (43.6 S, 34.8 E). Rabe crater is 108 km in diameter, and is located in the southern cratered highlands in what used to be called the Hellespontes region, the same general area where both Proctor and Kaiser craters are located. Monitoring of the Rabe dune field using MGS/MO data was unable to document any observable dune movement, leading to an estimate for the local sand migration

**Fig. 12.15** Viking Orbiter 2 image (575B60) of the floor of Kaiser crater, acquired in 1978 with a resolution of about 50 m per pixel. North is up, illumination is from the top right. To the right are large transverse ridges, while towards the bottom a flock of small barchans can be seen migrating towards the left. A few small barchans can just be made out in the box in upper left, an area imaged 20 years later (Fig. 12.16) by the Mars Orbiter Camera (MOC) on Mars Global Surveyor. *Image* NASA/MSSS

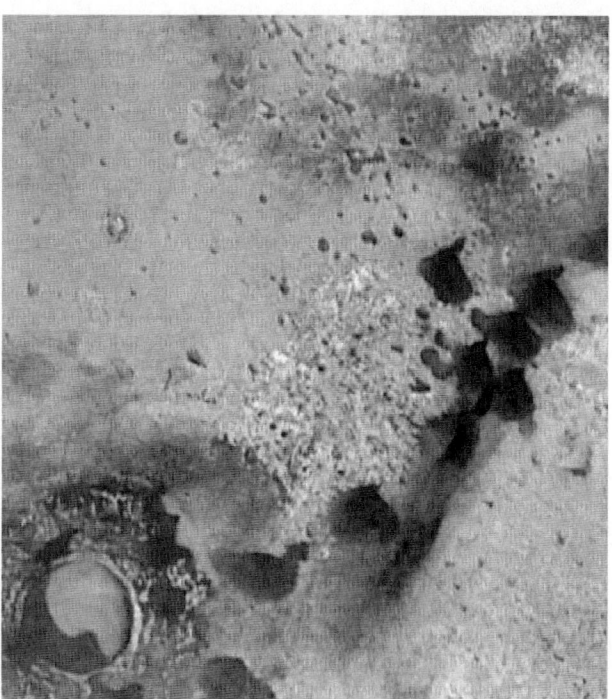

**Fig. 12.16** The MOC subframe, 10004, taken with a resolution of 20 m/pixel during the 100th orbit of Mars Global Surveyor on January 21, 1998, is centered at 47.1°S latitude, 341.3°W longitude and shows a close-up of some of the smaller, isolated 'fat teardrop' barchan dunes that occur west of the main dune field. The dune just below the center of the frame covers an area about 680 m by 480 m in size. The circular feature in the lower left corner is a small crater that formed by meteor impact some time before the dunes were present in this area—note the Yin-Yang accumulation of sand in the crater. This 'small' crater has a diameter of 1.2 km, the same diameter as the famous 'Meteor Crater' of northern Arizona, USA., on Earth and a little larger than Endeavour crater. *Image* MOC2-33C, NASA/JPL/MSSS

## 12.5 Examples of Dune Localities on Mars

**Fig. 12.17** Rippled (not deeply eroded, as had been thought from lower-resolution MOC images) dunes on the floor of Herschel crater. HiRISE PSP_002860_1650. NASA/JPL/ U of A. Inset, portion of browse image. Documented 1 m movement in the Herschel dune field, which occurred between two HiRISE images taken 1.5 Mars years apart. PIA14879

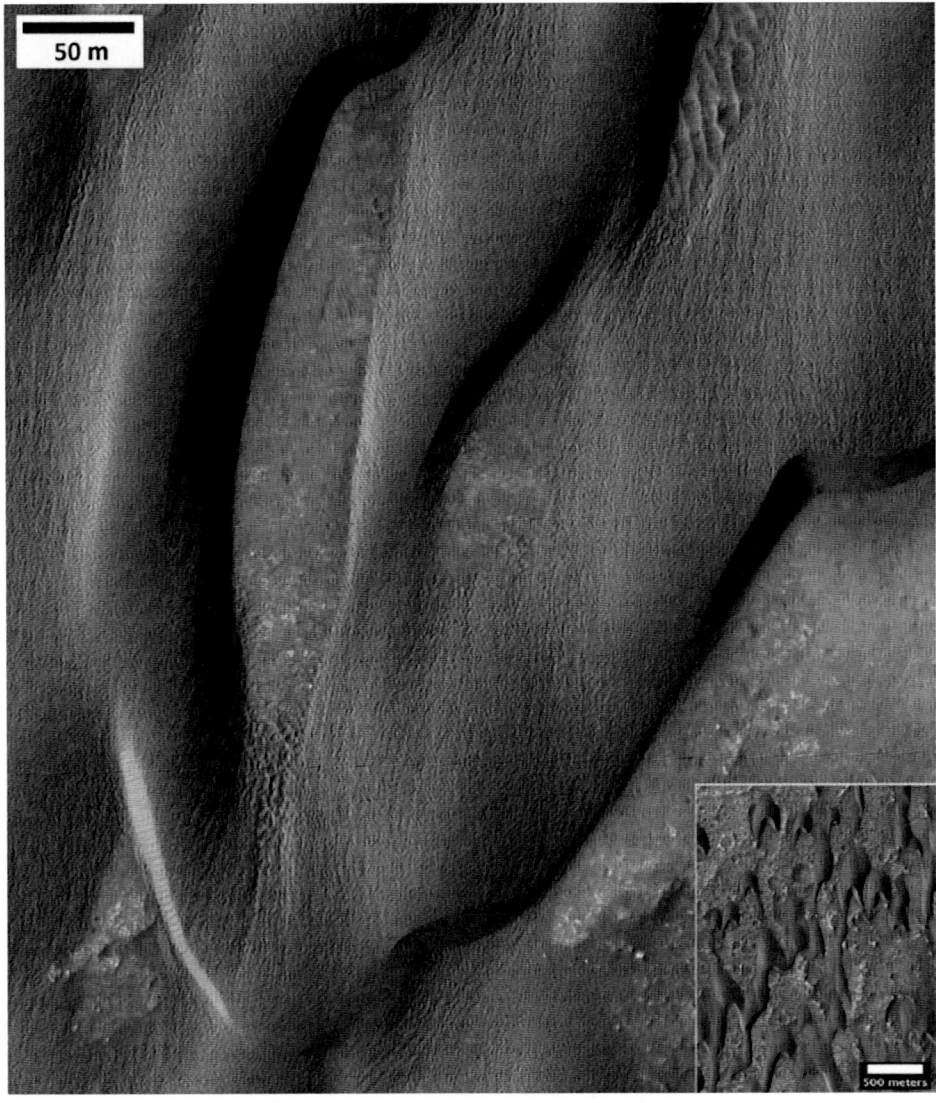

rate here of only 1–2 cm per Mars year (Fenton 2006a). The combination of THEMIS infrared and visible data (Fig. 12.18) provided interesting insights into the physical properties of this dune field (themis.asu.edu/feature/7). Color represents the nighttime temperatures, where blue is relatively cold and red is warm, corresponding to dusty and rocky exposures, respectively. The crests of the 150–200 m-tall dunes are consistently warmer (by ~6 °C) than the adjacent troughs between the dunes, suggesting that wind has swept most of the dust from the crest areas. Also, the red (rocky) areas are exposed layered material that may have provided a local source for at least some of the sand-sized materials.

*Richardson crater* (72.1 S, 179.5 E). Richardson crater is 55 km in diameter, and is located in the polar region of the southern cratered highlands. The dunes in this crater have received considerable attention from multiple spacecraft camera teams because they show distinctive 'spotted' patterns as the seasonal cap of carbon dioxide frost vaporizes from the sand surface during southern spring (Fig. 12.19). The defrosted 'spots' may expose sand that could then be blown or avalanched onto adjacent frost-covered areas, which in turn would enhance the removal of the frost from these areas.

*Nili Patera* (8.0 N, 72.0 E). Instead of occurring within an impact crater, this dune field (CRISM image) is located within the broad caldera present at the summit of the shallow-sloped Nili Patera volcano (Fig. 12.20). The barchan dunes of Nili Patera were the first place to provide documented evidence of ripple and dune movement on Mars of at least a meter, using repeat HiRISE images (Silvestro et al. 2010a). The rippled surfaces of the barchans and barchanoid ridges in the Nili Patera region have continued to be monitored with HiRISE as part of the continuing effort to document dune and ripple movement occurring today across the entire planet (Bridges et al. 2012a), as well as the first place on Mars to apply the sophisticated COSI-Corr software which was designed to

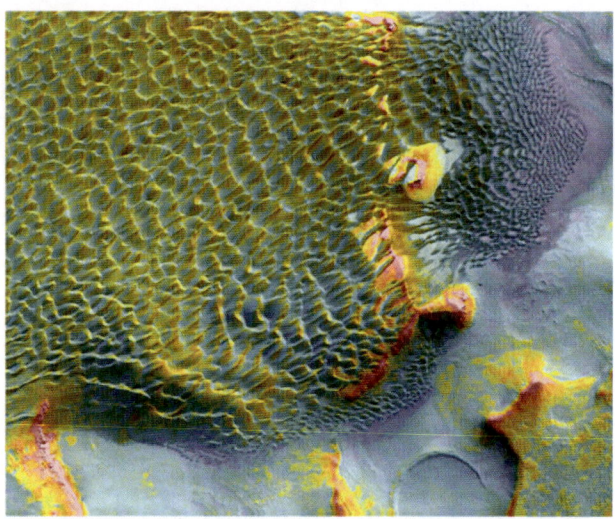

**Fig. 12.18** Rabe crater: Night-time temperature data overlain on a VIS image of the dunes in Rabe crater, both obtained by the THEMIS camera on Mars Odyssey. *Image* NASA/JPL/ASU

**Fig. 12.19** Defrosting carbon dioxide frost on the dunes on the floor of Richardson crater, southern spring. Dark jagged patches are dust blowouts, and white surrounds are frost. Portion of HiRISE ESP_012774_1080, credit NASA/JPL/U.Arizona

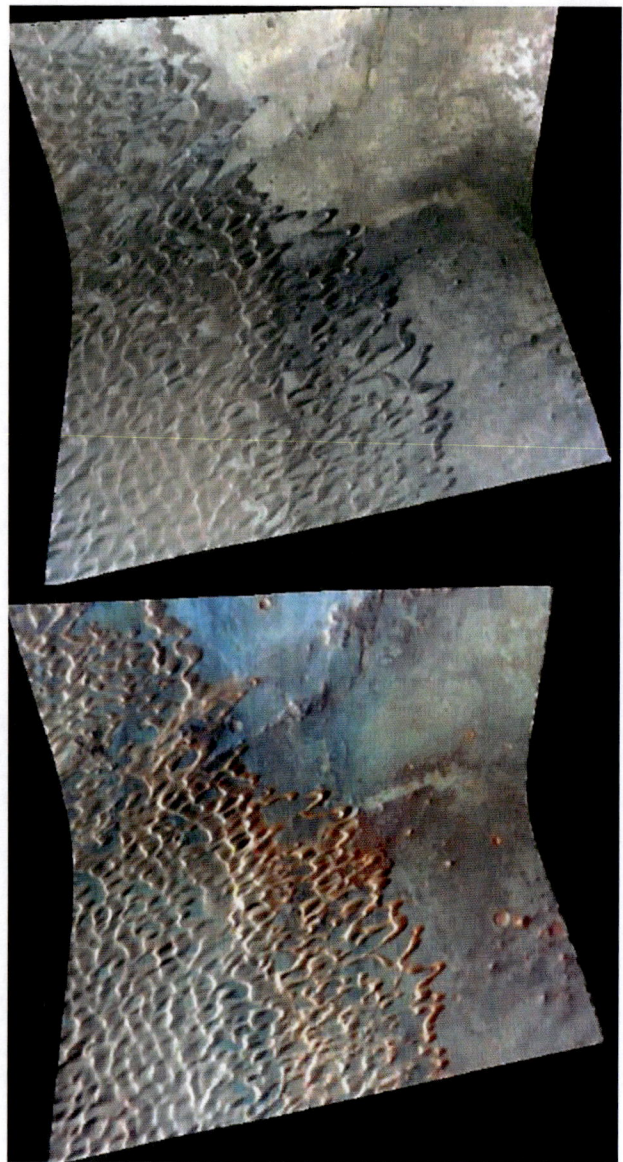

**Fig. 12.20** CRISM images of dunes in Nili Patera. Top, true color. Bottom, spectral bands chosen to highlight composition, with red dunes indicating concentrations of iron- and magnesium-rich igneous materials. *Image* PIA009347, NASA/JPL/JHUAPL

aid in the detection of sand dune movement on Earth (Bridges et al. 2012b; Fig. 12.21). Remarkably, given the nondetection of movement on Mars over the Mariner 9 to Mars Odyssey period of some 35 years, these latest results indicate that sand movement rates on Mars today are comparable to documented rates obtained in some places on Earth—see Chap. 9.

*Gale crater* (5.4 S, 137.8 E). Gale crater is 154 km in diameter, and is located at the transition from the southern cratered highlands to the northern lowland plains, south of the Elysium Mons volcanic center. A massive mound of material covers the central peak of the crater, rising 5.5 km above the level of the crater floor; layered clay-rich materials near the base of the central uplift (Fig. 12.22) are the primary target for the Curiosity rover, which landed in August 2012 to search for possible organic materials in sediments from the early history of Mars. The central mound in Gale crater is surrounded by an arc of dark sand dunes (orbital imaging of dune field reveals a rather complex wind flow pattern on the crater floor, likely strongly influenced both by the crater rim and the high central mound (Silvestro et al. 2013). During 2013–2014, Curiosity should get many images of a portion of this dune field as it drives toward a place to provide the rover a safe route to drive through the dune field. During testing of the Curiosity

## 12.5 Examples of Dune Localities on Mars

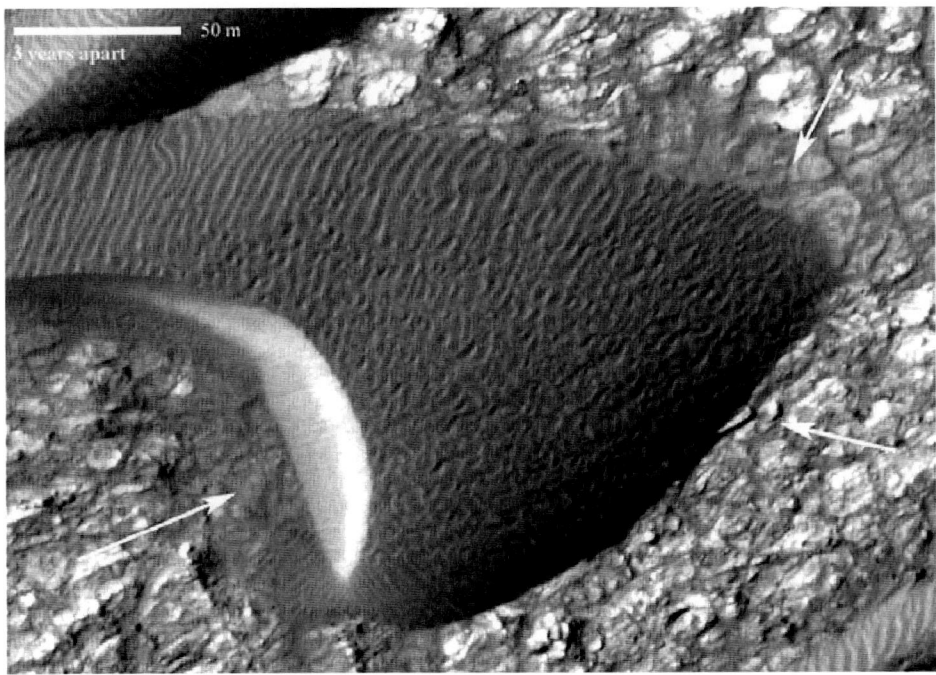

**Fig. 12.21** HiRISE subscene of Nili Patera barchans used to document ripple and dune movement. The two arrows at right point to the edge of the dune in an image acquired 3 years earlier. A closeup showing ripple movement is shown in Fig. 9.10

sampling instruments in November 2012, a sand deposit (called 'Rocknest') was scooped repeatedly; the final sieved portions of the sand were examined by precision chemistry and mineralogy instruments within the body of the rover. Olivine, plagioclase, and two pyroxenes were identified in the sand materials <150 μm in size using the Chemistry and Mineralogy (ChemMin) instrument, the first time that an X-ray diffraction pattern has been obtained on another planet (see Sect. 17.1), and the Sample Analysis on Mars (SAM) instrument confirmed the basaltic composition of the sand. The Mars Hand Lens Imager (MAHLI) on the movable arm of Curiosity obtained wonderful close-up images of the surface of the Rocknest deposit, revealing a surface coating of 1 mm-size large clasts overlying dark sand that comprises the bulk of the deposit (Fig. 12.23). This is only a 'first glimpse' of what the instruments on Curiosity will be able to tell us about the Martian surface materials; we are fortunate that, for engineering reasons, the first location selected for the scooping of samples was an aeolian sand deposit.

The small-car-sized Curiosity rover is the latest in a series of increasingly sophisticated robotic vehicles capable of driving across the Martian surface. In 1997, the microwave-oven-sized Sojourner rover became the first spacecraft to demonstrate the great advantages of having mobility for exploration spacecraft on Mars, as part of the Mars Pathfinder spacecraft that also demonstrated the practical use of airbags as a way to land instruments on Mars.

Sojourner provided images of large sand ripples that were not visible to the cameras on the Pathfinder lander (Fig. 1.6), as well as making the first Alpha Proton X-ray Spectrometer (APXS) measurements of Martian aeolian deposits, confirming their basaltic composition (Greeley et al. 1999). In 2004, the twin Mars Exploration Rover (MER) spacecraft Spirit and Opportunity brought golf-cart-sized robotic explorers to two sites on opposite sides of Mars. The Microscopic Imager (MI) and the Panoramic Camera (Pancam) provided numerous images of sand deposits at both landing sites, and the AXPS on the robotic arm of both rovers obtained dozens of chemistry measurements from sand deposits. Spirit obtained images of cm-scale sand ripples that visibly moved (see Fig. 16.15) during a 2-day period (Sullivan et al. 2008), and Opportunity investigated megaripples during its long drives across Meridiani Planum (Sullivan et al. 2005).

Next we mention a few of the MER investigation sites that have relevance to dunes:

*Endurance crater, Meridiani Planum* (~100 m diameter). Opportunity spent nine months exploring inside this impact crater, mostly intended to provide access to the rocks exposed in the crater wall. However, it was also able to drive to within a few meters of the small field of sand dunes or ripples on the crater floor. For reasons related to rover trafficability, Opportunity was not allowed to drive into the dune field in Endurance, mainly because other sand deposits had resulted in both MER rovers to become

**Fig. 12.22** Curiosity Mastcam panorama of the floor of Gale crater, showing layered mounds in Mount Sharp (aka Aeolus Mons) with dark dunes in the foreground. An aerial view of these dunes was obtained during the parachute descent of the rover Fig. (18.3)

temporarily stuck. Sullivan et al. (2005) review aeolian processes observed along the traverse to Endurance. Golombek et al. (2010) consider the interaction of ripples with small impact craters to constrain relative timescales of movement.

*Victoria crater, Meridiani Planum* (~800 m diameter). Victoria crater is substantially larger than Endurance crater, and thus the ripple field on its floor (Fig. 12.24) is correspondingly larger as well. Opportunity drove around roughly one-quarter of the rim of the crater, searching for the best path to gain access into it. During this drive, Opportunity crossed the surface of a broad dark wind streak that emanates from the northern side of the crater (Fig. 8.2), becoming the first rover to investigate a wind streak on Mars. Inspection of the crater walls (Figs. 5.21 and 5.22) at Cape St. Mary and Cape St. Vincent indicated crossbedding suggesting of an aeolian origin. Opportunity documented an increased quantity of basaltic sand on the surface of the streak (relative to outside of the streak), supporting the idea that the sand in the streak is derived from sand within the crater, blown out of the crater in a northerly direction (at present), and possibly eroding the rim in the process to form the distinctive scalloped appearance of the crater rim (Geissler et al. 2008). Opportunity eventually entered the crater on the northwest rim, but it only descended to the point where the bedrock became covered by fine detritus, beyond which posed a threat to the rover being able to leave the crater. The cross-bedding in the walls of Victoria crater has been studied by Hayes et al. (2011) to infer wind transport of sediments prior to the formation of the crater.

*Endeavour crater, Meridiani Planum* (22 km diameter). After nearly three years of driving south from Victoria crater, in 2012 Opportunity began to explore the rim deposits of by far the largest crater yet examined by a Mars rover. Several separate dune patches are present on the floor of this broad crater, some of which are visible in distant views from Opportunity, but orbiter imaging provides the best viewing option thus far. The dunes in Endeavour crater have been documented to move at a rate that is substantially greater than what has been observed in HiRISE images from elsewhere on the planet (Chojnacki et al. 2011). As long as Opportunity remains healthy, the rover might eventually drive closer to the dune fields.

*'Purgatory dune', Meridiani Planum.* Megaripples on Meridiani Planum steadily increased in size as Opportunity traveled south from Victoria crater. Finally the rover got stuck in one of the megaripples for about six weeks—we discuss the challenges of driving on sand and dust on Earth and Mars in Sect. 5.3. Termed 'Purgatory dune', the rover became trapped when several of its wheels broke through an indurated upper surface on the megaripple, and the fine-grained sediments inside the feature provided little traction to the wheels. Opportunity eventually extracted itself by repeated commands for drives that should have covered many meters but actually only moved the vehicle a few centimeters; engineers determined that the rover wheels were experiencing >99 % slippage during the extraction drives. Another ripple on which slippage was subsequently encountered (but detected onboard) is shown in Fig. 22.12. Jerolmack et al. (2006) have considered the sorting of sediments on ripples at Meridiani.

Spirit also got stuck in fine drifts that are visually quite different from Purgatory dune, so not all collections of fines must necessarily be associated with megaripples. Indeed, the demise of Spirit came about because it became stuck while attempting to drive to a site where it could orient its solar panels toward the sun during local winter; while it was stuck, the solar panels received insufficient sunlight to maintain heaters to warm the spacecraft instruments and

**Fig. 12.23** Close-up view of Martian soil with the Mars Hand Lens Imager MAHLI. At left is the surface of the 'Rocknest' sand shadow, while at right is a sample of sieved sand (<150 μm) from the interior of the feature. *Image* PIA16570. NASA/JPL-Caltech/MSSS

**Fig. 12.24** A mosaic of Opportunity images showing the view across Victoria crater, with the ripple pattern on the crater floor. An orbital view of the crater and the ripples is seen in Fig. (8.2), and a close-up of the rock exposures in the crater wall (which reveals bedding patterns suggesting the bedrock is an aeolian sandstone) are given in Figs. (5.21 and (5.22). *Credit* NASA/JPL PIA08779

computers through the cold Martian nights. The physical properties of Purgatory dune and the drifts in which Spirit became bogged down are dramatically different from what Spirit encountered at 'El Dorado', discussed next.

*'El Dorado', Gusev crater.* The only large patch of low albedo (dark) sand visited so far by a rover is the site called 'El Dorado', on the northern side of Husband Hill, the largest of the Columbia Hills within Gusev crater. During its descent from the summit of Husband Hill, Spirit drove well into the margin of the El Dorado sand deposit (Fig. 12.25). Wheel mobility was fine while driving over the well-sorted black basaltic sand that comprised the small dunes and ripples present throughout the deposit. Excellent microscopic imager data were obtained of sand (Fig. 2.5), as well as APXS measurements of a basaltic composition (Gellert et al. 2006; Sullivan et al. 2008; see also Sect. 4.2). Where the rover wheels dug into ripples, good assessments could be made of the subsurface, which was comprised of the same 200–300 μm, well rounded, coarse sand grains that were visible on the surface of the sand. Subsequent monitoring of the sand patch by Spirit, following its visit to the site, revealed that dust devil tracks were at times visible on the sand, but eventually all tracks were erased, presumably by a wind event that was strong enough to activate (even briefly) most of the surface sand. It appears that the mobility maintained by the sand in deposits like El Dorado allows the wind to move the grains sufficiently to disturb most dust that might settle onto the sand, thus maintaining the low reflectance of the entire deposit. El Dorado likely is the most representative

**Fig. 12.25** Portion of PanCam mosaic of El Dorado dunes, with Gusev crater rim visible in the background. Note the local slope. Image: NASA/JPL

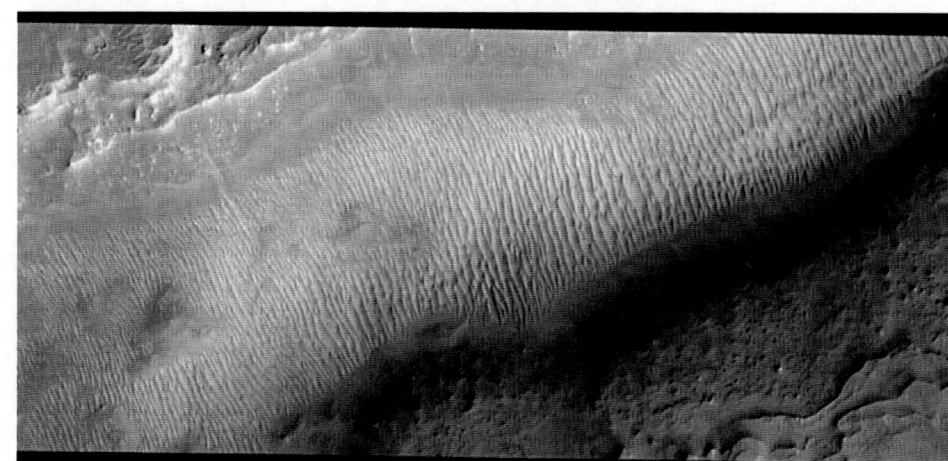

**Fig. 12.26** Mars Orbiter Camera (MOC) image of transverse aeolian ridges (TARs) in the floor of a trough. MOC image M12-00991 of Auqakuh Vallis. Credit: NASA/JPL/MSSS

'typical' sand deposit yet examined *in situ* by a rover, although eventually Curiosity may get a close examination of dark sand deposits as it crosses the dune field that surrounds the mound in Gale crater.

*Transverse aeolian ridges* (TARs). The generic term 'transverse aeolian ridge' has been applied to aeolian features that may possibly be either large ripples (megaripples) or small dunes. TARs have remarkably symmetric profiles perpendicular to their crest, and when compared to measured profiles of terrestrial ripples and dunes, TARs >1 m in height are most similar to reversing dunes while TARs <0.5 m in height are most likely granule-coated megaripples (Zimbelman 2010).

Among locations with nice examples of TARs is Nirgal Vallis (27.8 S, 316.7 E, see Fig. 5.13) with a field of TARs that cover the floor of this long sinuous valley; crater counts shows that the TARs on the floor of Nirgal Vallis are likely only 140–380 ka in age, with an upper limit of <1.4 Ma (Reiss et al. 2004). Another trough filled with TARs is Auqakuh Vallis (29.2 N, 60.4 E), Fig. 12.26.

## 12.6 Dunes in Relation to Topography

Dunes on Mars definitely do interact with the topography beneath or around them. As discussed above, most dune fields on Mars tend to be on the floors of impact craters, so that the crater topography represents a one-way trap; sand can move up the shallow slopes of the ejecta blanket, but once it passes over the crater rim and falls inside the main crater, it is unable to get out of the depression. For older, more subdued craters, the interaction between the dunes and the crater relief is less consistent, but the crater does still influence dune locations; this situation is best illustrated by dunes that have concentrated on and around the rim of one subdued crater (Fig. 6.17)—similar interactions can be seen with ripples (Fig. 5.8).

Perhaps the next most common occurrence of dunes in relationship to topography involves concentration of dune forms along the floor of sinuous channels (e.g., Fig. 5.13). Channels tend to have wall slopes that are often less steep than the slopes inside the rims of large impact craters, but the channel form still seems to be an effective concentrator of the mobile sand deposits. If the channel is large and oriented radial from a large topographic high (something like the Tharsis ridge), then it might funnel dense nighttime air somewhat analogous to katabatic winds emanating from over an ice sheet on Earth. Another example of the interaction between relief and mobile sand deposits is climbing or falling dunes (Fig. 7.4), where wind indicators around the dunes should confirm which way the dominant wind is blowing with regard to the cliff against which the dunes are banked. For example, in Fig. 7.4 the falling dunes break into barchans, where the barchans clearly indicate the direction of the sediment-driving wind. Dunes can also collect in the wind shadow behind a topographic obstacle, such as the elevated rim of a fresh impact crater, although this situation appears to be less common on Mars than it is in deserts on Earth. Interestingly, small (<100 m scale) dune features have been imaged everywhere from the summits of the Tharsis volcanoes to the deepest portion of the floor of the Hellas impact basin, covering a range in elevation of more than 30 km, so that there is no apparent elevation control on where dunes can form on Mars, although Lorenz et al. (2012) show that the bedform wavelength does appear to vary systematically with elevation (see Sect. 4.6 and Chap. 5).

## 12.7 Links Between Dunes and Climate?

It is very difficult to establish the relationship between dune occurrences or orientations and past climates on Mars. Unlike on Earth, where field studies can be used to constrain the timing for the emplacement of the dunes, it is difficult to obtain age constraints for dunes on Mars, except that they postdate the surfaces on which the dunes are found. One exception to this situation is the TARs that cover the floor of Nirgal Vallis (Fig. 5.13); at this location, enough impact craters could be identified superposed on the features to provide an upper limit of about 1.4 million years on the age of the dune field (Reiss et al. 2004). Recently, HiRISE repeat imaging of dunes is providing abundant evidence that Martian dunes are active under present wind regimes, so that the dunes should be very useful as local wind indicators.

The Hayward et al. (2007, 2009) global database of sand dunes was compared to surface winds predicted from GCM runs, but only limited success was achieved in relating observed dune patterns to the varying predictions from the GCMs. No general conclusions could be drawn from the few examples where dunes might possibly reflect paleo-winds rather than the current wind regime (see also Fenton and Hayward 2010). An earlier analysis of surface wind shear obtained from a GCM (Anderson et al. 1999) found a general correlation between the areas of the greatest predicted shear at the surface and regions of low albedo, but the global trends were not applied to individual dune fields. More localized (mesoscale) climate models, using a higher spatial density of grid points for the modeling, may have better success relating observed dune orientations to the more spatially constrained wind predictions that could come out of such regional scale modeling efforts (e.g., Fenton et al. 2005).

A more recent analysis by Gardin et al. (2012) examines the morphology of some 550 dune fields in some detail (they find 62 % are barchan and barchanoid, 18 % are transverse, 11 % are linear and the rest unclassified). They then use the slip face orientation and morphology to estimate the wind regime at a number of locations. They find that most are consistent with the global circulation estimated by present models, but some are not. This may be due to local topographic forcing unresolved in the models, by fossilized dunes recording a prior wind configuration, or to dunes having transient geometries not in equilibrium with the present regime (see Sect. 7.10). In this connection, it is estimated that the turnover time of Martian dunes (i.e., the timescale on which dunes should adapt to a new wind regime) is in the range of 10,000–100,000 years, which is the same timescale on which obliquity changes may cause climate variations, so some dunes might always be in transition, never quite reaching steady-state.

Sand dunes hold the potential to preserve evidence of past climates within their deposits. The interaction between active aeolian deposition and annual snow deposits means that some dunes might preserve internal ice deposits for extended periods of time, something that has been observed to be the case in some polar dunes on Earth. HiRISE holds the potential to look for possible ice exposures in the dunes that comprise the north polar erg, although great care would need to be exerted to demonstrate that the snow or ice was not the product of a recent winter. Whenever coring or detailed ground-penetrating radar sounding becomes widely available for Mars, the polar dunes would be a good place to apply such techniques.

## 12.8 Dune and Ripple Migration

An enduring puzzle for many years after the MGS/MOC images began to yield coverage with a fine enough resolution and a long enough time separation to detect Earth-like dune migration, was that no such migration was observed. This prompted several workers to speculate that perhaps most Martian dunes were immobilized somehow, via some sort of induration (with ice, salts or the action of volcanic acid fog, perhaps). In fact, as described in Chap. 9, slipface migration has now been seen in some locations on Mars, so

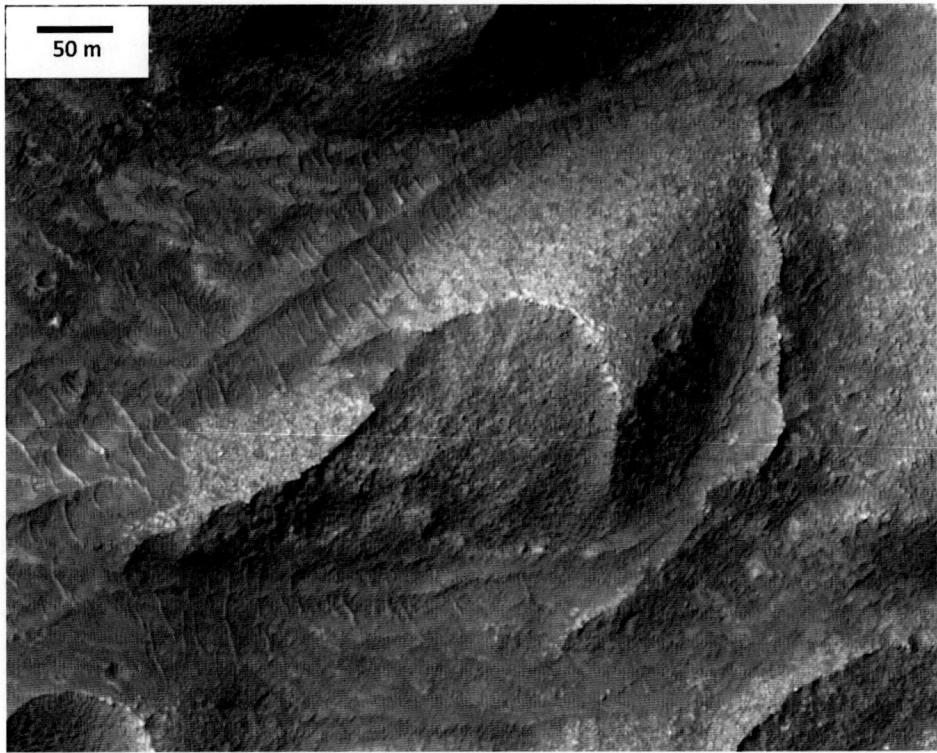

**Fig. 12.27** A Martian dune turned into stone. The pockmarked and blocky surface of this barchan dune indicates significant cohesion and thus evidence of induration. HiRISE image ESP_025389. *Credit* NASA/JPL/U.Arizona

not all dunes are immobilized. However, recent HiRISE imaging has also revealed convincing evidence that some Martian dunes are indurated (Fig. 12.27).

## 12.9 Future Possibilities

Future missions to Mars represent new opportunities to improve our understanding of aeolian sediments and landforms on Mars. The increasing scope and complexity of missions correlates with their increased costs, which (unfortunately) also then correlates with a decreasing prospect for selection of large missions in an era of tightly constrained budgets. We will point out some of the missions that might advance the state of knowledge of Martian dunes, from orbiters to specialized landers and rover and (eventually) to human exploration of Mars.

At present, the HiRISE camera continues to return a steady stream of highly detailed images from all over Mars, and it will likely remain the state of the art for orbital imaging for the foreseeable future. Alternate derivatives from the HiRISE design might sacrifice some of its magnificent spatial resolution (and thus the required size of the telescope) for improved spatial coverage and increased spectral bands, which would allow for more precise mapping of compositional variations among moderate-scale aeolian deposits than is available at present. Similarly, the CRISM spectrometer is likely to remain the state of the art for hyperspectral remove sensing of Mars for the foreseeable future, but perhaps someone will devise a very high spatial resolution spectrometer that could focus on details not resolvable today in CRISM data. An important addition from an orbital perspective would be an imaging radar at Mars, along the lines of the instrument that revealed >98 % of the surface of Venus during the Magellan mission. Correlation of radar return properties with existing imaging and spectral data sets would open up a new realm of understanding for the near-surface materials, including evaluation of the depth of some deposits.

A new approach for investigations near the surface may be either an airplane or balloon mission to Mars. Several such concepts already have been proposed to NASA, and technology may have advanced to the place where such missions are now more feasible than they may have been in the past. An airplane mission could allow for low-altitude remote sensing along flight lines of perhaps thousands of kilometers, while a balloon mission might cover a comparable distance over a much longer time frame, and even include multiple surface sampling opportunities as the altitude of the balloon decreased during each night. Geophysical investigations in particular may be more amenable to such low-altitude mission designs. Either mission type would need to address science objectives that are difficult or impossible to achieve using either orbiters or rovers.

It was recently announced that NASA plans to send another Curiosity-class rover to Mars in 2020, giving scientists the chance to apply its impressive suite of instruments at a site well removed from Gale crater. Future rovers may incorporate some of the emerging software developments termed 'shape from motion', where images obtained around a three-dimensional target can be merged to generate a full 3D digital model of the subject. Web-based versions of such software already allow anyone to make 3D models using their own images, and then even purchase a physical recreation from the digital model using 3D printers that are easily accessible via the web; imagine being able to generate a full 3D model of an actual Martian ripple! Age determinations are an important part of present-day aeolian studies on Earth, but it is unclear how difficult it may be to obtain meaningful results using something like optically stimulated luminescence (OSL) on Mars, where the radiation environment is currently totally unconstrained. Ground penetrating radar (GPR) on a rover holds great potential to reveal the subsurface layering and structure beneath aeolian (and other) landforms or deposits.

Ultimately, the advancement of science on a variety of questions about Mars will require the return of samples from documented locations to laboratories on Earth, which is also true for aeolian sediments. The need for documentation means that such samples will either be collected on specially outfitted rovers, or await the arrival of humans on Mars. Human missions to Mars remains a long-term goal of NASA, but such missions always appear to be at least '20 years out', and this 20-year extension has been the stated (or unstated) excuse for the last 40 years, since the end of the Apollo missions. Hopefully, the technical advances required to make a reality of either a sample return or a human mission to Mars can be achieved soon, leaving 'only' the securing of the funding required for such massive undertakings.

# Titan Dunes 13

## 13.1 Pre-Cassini Expectations

Titan is Saturn's largest moon, at 5150 km diameter is larger than the planet Mercury, and is unique among the satellites of the Solar System in having a dense atmosphere, four times denser than our own. The next-thickest atmosphere, that of Triton (see Chap. 7) is 100,000 times thinner. Titan's atmosphere was hinted at by the dark edges of its disk seen in telescopic observations in 1908, but gaseous methane was detected spectroscopically in 1944. This made Titan an object of interest, not only planetologically, but also from the astrobiological point of view, since the action of sunlight on methane was known to create solid (and liquid) organic compounds, which were expected perhaps to accumulate to a depth of 1 km on Titan's surface; accordingly Titan was made a target of the Voyager 1 spacecraft in 1980 (see Fig. 13.1).

This early study of Titan is summarized in Lorenz and Mitton (2008) and Coustenis and Taylor (1999, updated in 2008); a popular post-Cassini account is Lorenz and Mitton (2010). The most comprehensive research-level book is Brown et al. (2010). The Titan scientific literature has exploded in recent years—from a few dozen papers a year in the 1990s, to over 100 a year in the Cassini era. Recent reviews of saltation physics on Titan, and of the dunes overall, are by Lorenz (2013) and Radebaugh (2013), respectively.

Atmospheric conditions on Titan—and specifically the density of the atmosphere at ground level—were determined by passing Voyager 1's radio signal through the atmosphere and measuring how much the atmosphere bent it: a 'radio occultation' experiment. Although the surface was speculated to perhaps be wet with liquid methane (the surface temperature is 94 K), the atmospheric pressure is 1.5 times that of Earth, leading to an air density of 4 times that of Earth. It was realized that in such a dense atmosphere, and in Titan's low gravity, particles of material on the surface could be blown around with rather low windspeeds. In their book *Wind as a Geological Process* (1985), Greeley and Iversen calculated the windspeeds required for particle motion as a function of size, assuming that particle cohesion was the same as for sand on Earth. They found that the optimum size for saltation was 180 μm diameter, with a friction speed of about 0.04 m/s—note that this assumed a particle density of 1900 kg/m$^3$ (the average density of Titan) which may be a factor of almost 2 too high for most organic and ice surface materials.

As plans to explore Titan matured around 1990 with the start of the development of the Cassini mission, others began to consider the problem. In an analysis of likely winds on Titan in a 1991 conference on Titan, Allison (1992) suggested winds might be of the order of 0.3 m/s, and that this was comparable with that required for particle motion and thus 'wind-blown material might be an occasional feature of Titan meteorology'. Grier and Lunine, in an unrefereed abstract (DPS conference 1993) also noted the possibility of dune formation, in both subaerial and subaqueous environments.

Motivated by the possibility that Titan's atmosphere might have changed in density over time, and ease of aeolian transport would depend on the density, Lorenz et al. (1995) reconsidered aeolian transport. They used scaling arguments to assess likely near-surface windspeeds, based on estimated surface solar fluxes and the thickness of the atmosphere, and determined that, on average, winds would be lower than the threshold. However, the same was noted to be true for the Earth, prompting consideration of the statistical variation in windspeeds.

Specifically, expectations of thermally-driven winds near Titan's surface were extremely low, due to the low solar flux, the large column mass of the atmosphere and the small radius of Titan. This kind of energy-flux argument, which correctly predicts windspeeds of a few meters per second on Earth, suggests windspeeds on Titan of only about 1 cm/s. In contrast, despite Titan's thick atmosphere and low gravity, which both favor transport of material by wind, the threshold windspeeds required to move sand are of the order of 0.5–1 m/s.

**Fig. 13.1** A Cassini image of Titan, with Saturn and its rings (Titan orbits Saturn very close to the ring plane, so they are always seen edge-on). The shadow of the rings is cast on Saturn at lower right. This rather featureless view of Titan is comparable with (but prettier than) the Voyager view. The fuzzy edge of the day–night boundary is due to the same scattering/absorption effect (limb-darkening) that gave the first telescopic clues in 1907 that Titan had an atmosphere. *Credit* NASA/JPL/SSI

The second issue is of sand supply. Because of the weak sunlight, the hydrological cycle was expected to be weak overall, corresponding to about 1 cm of liquid methane per Earth year. Furthermore, raindrops would fall relatively slowly, making them only weakly erosive. The freeze–thaw action that serves to break down bedrock on Earth would not occur on Titan since the heat capacity of the atmosphere buffers the surface against large diurnal temperature changes. It had been further conjectured that seas of liquids on Titan's surface might act as traps for any sand that was formed (for example, the sand-sized fraction of ejecta from impact craters—this is believed to be the principal source of sand on Venus). Because of both the low expectation of sand abundance and the low expected windspeeds, Lorenz et al. (1995) were rather pessimistic about the prospects of detecting dunes on Titan, and little work was done on the topic thereafter while Cassini was en route.

It was noted in Lorenz et al. (1995) that measures of average windspeeds may be of only limited utility in assessing sand transport processes, in that winds above the transport threshold may in fact be quite rare, and thus the controlling factor in aeolian feature formation is the probability distribution of various windspeeds, and how long and fat is the high-speed 'tail' of the skewed distribution. Windspeeds are characterized by nonGaussian statistics, with log-normal or, more typically, Weibull functions being used to describe them, in the wind energy literature at least. Lorenz (1996) showed that Martian winds, measured by the Viking landers, could be well described by a Weibull distribution, and some studies on dust-lifting at Mars are now using this statistical approach.

However, Titan has the additional factor of a significant gravitational tide in its atmosphere, due to its eccentric orbit around the massive planet Saturn: while unique in this solar system, this may in fact be an important effect on many tidally-locked extrasolar planets. The tidal potential height difference associated with the tide is several hundred times larger than that exerted by the Moon on the Earth. Tokano and Neubauer (2002) explored the influence of atmospheric tides on Titan's windfield, and found that it could be a significant contributor, especially to near-surface winds, causing flows of the order of 0.5 m/s.

Little was known about Titan's surface prior to Cassini's arrival: the thick organic haze produced by the action of sunlight on methane made the atmosphere all but opaque to Voyager's visible-light camera. In the near-infrared, however, the atmosphere is clearer, and the first maps of Titan's surface were made from Hubble Space Telescope data in 1994, with a resolution of around 300 km. As 10 m telescopes with adaptive optics systems (like Keck and Gemini) became available around the year 2000, the maps improved by about a factor of 2, but were still about as detailed as naked-eye views of our own moon. Bright and dark features could be discerned, but their nature was unclear. It was tempting to speculate (as for our own moon, where the dark lava plains were named 'Mare' or seas) that the dark areas were seas of liquid hydrocarbons. Indeed, when a remarkable radar experiment using the giant 300 m radio telescope at Arecibo detected a mirror-like glint from Titan, suggesting patches of surface that were dead flat, it was a natural interpretation that the dark areas were lakes. Nobody suspected that they would turn out to be seas of sand.

## 13.2 Cassini Observations

The Cassini-Huygens mission to Saturn and Titan, conceived shortly after the Voyager encounter, began serious development in 1990 and was launched in 1997. It entailed the most massive planetary spacecraft built in the West, a 5-ton orbiter supplied by NASA, and a 2.7 m-diameter probe named Huygens, supplied by the European Space Agency. Scientific instruments for both orbiter and probe came from the USA and Europe, including several that would study Titan's surface. These included a camera (extending further into the near-infrared than Voyager's, allowing it to peer murkily through the haze), a mapping spectrometer ('VIMS') which would yield compositional

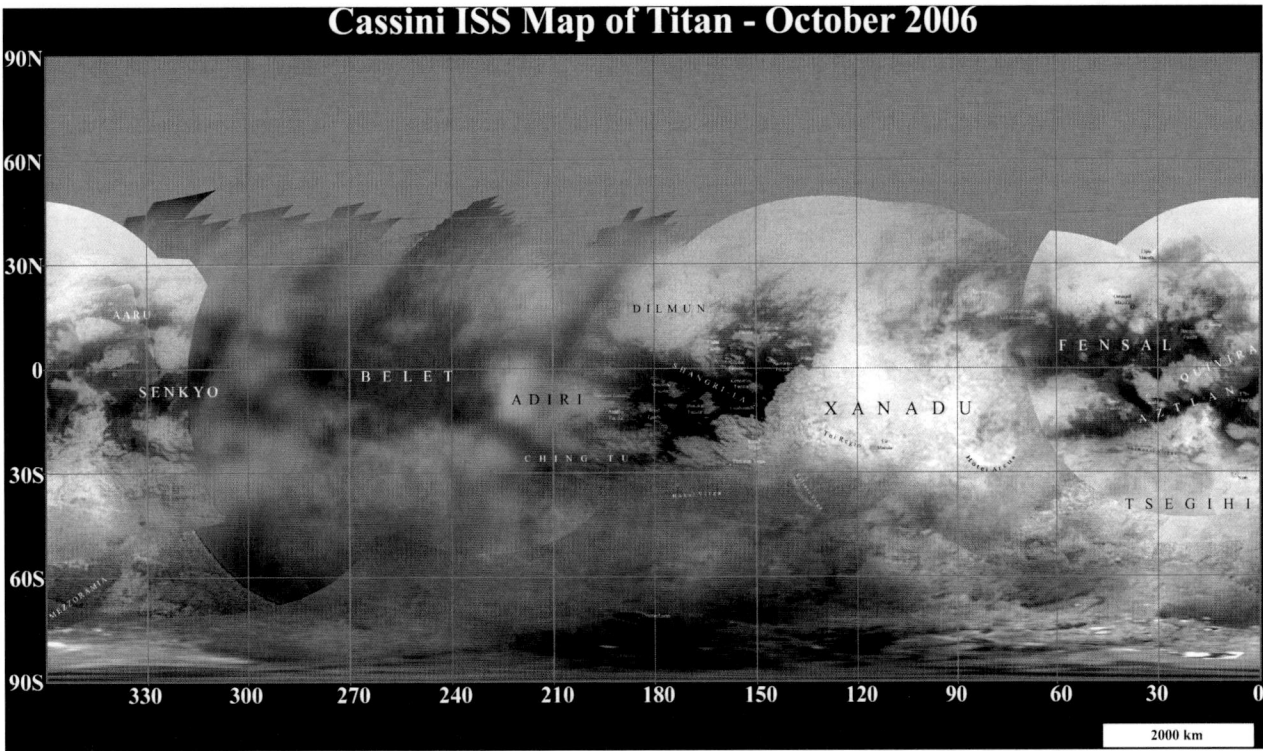

**Fig. 13.2** A near-infrared map of Titan, based on early Cassini camera observations. At this season (late southern summer) the northern high latitudes were in darkness and so are not mapped. Groundbased and HST observations since 1994 had detected the large dark areas (Fensal-Atzlan, Senkyo, Belet and Shangri-La) and in particular the large bright area Xanadu, but their nature was not known (and in fact names were only formally given in 2004). *Image credit* SSI

information and, specifically for studying Titan's surface, a Synthetic Aperture Radar (SAR) mapper. Additionally, the Huygens probe carried instruments to measure the mechanical properties of the surface if the probe survived landing (not guaranteed) and a camera.

Cassini arrived at Saturn in July 2004, and made some long-range observations of Titan a few days later. It would then spend its nominal four year 'Prime' mission orbiting Saturn, making 44 close encounters with Titan, and releasing the Huygens probe prior to the third encounter, to arrive in January 2005. (In fact, the spacecraft has performed well, and the mission was given an extension 'XM' or Equinox Mission through 2010, adding 26 further flybys (to 'T70') and has begun a further extension ('Solstice Mission') with the hope of operating through T126 near northern mid-summer in 2017.)

It was noted (Porco et al. 2005) in optical imaging (Fig. 13.2) from Cassini's first couple of flybys that there existed some 'streaky' boundaries between light and dark terrain. In particular, some streaks were identified and the eastern boundaries of several features were noted to be diffuse, while the western boundaries were sharp, suggesting possible surface transport. It could not be determined unambiguously whether these features were the result of aeolian or fluvial transport.

The radar observations from the first pass (TA) were of a rather inscrutable mid-latitude area, and no indications of aeolian features were seen (or, at this point, expected). However, the second SAR swath on Titan, T3, in February 2005, found many radar-dark subparallel features, nicknamed 'cat scratches' (Fig. 13.3), which were interpreted (Elachi et al. 2006) as being possibly aeolian in origin, but other processes were recognized as being impossible to preclude. The features were purely dark streaks (with no evidence of topographic shading) and some appeared to emerge and fork from a common point. At the time, longitudinal dunes were less familiar to the radar team and so other origins were still being entertained, even if in the context of this book, eight years later, their aeolian origin seems now obvious. It was only on T3, for example, that optically-dark areas were determined to be associated with these radar-dark features; this led to some speculation that perhaps the large dark areas like Shangri-La and Belet might also have them.

It was the near-equatorial flyby T8 (Figs. 13.4 and 13.5) in October 2005 covering Belet that made the interpretation clear (Lorenz et al. 2006). First, this flyby covered terrain that was in many places completely covered in the sand material; while on T3, the cat scratches were seen as some dark material deposited above a brighter substrate, on places

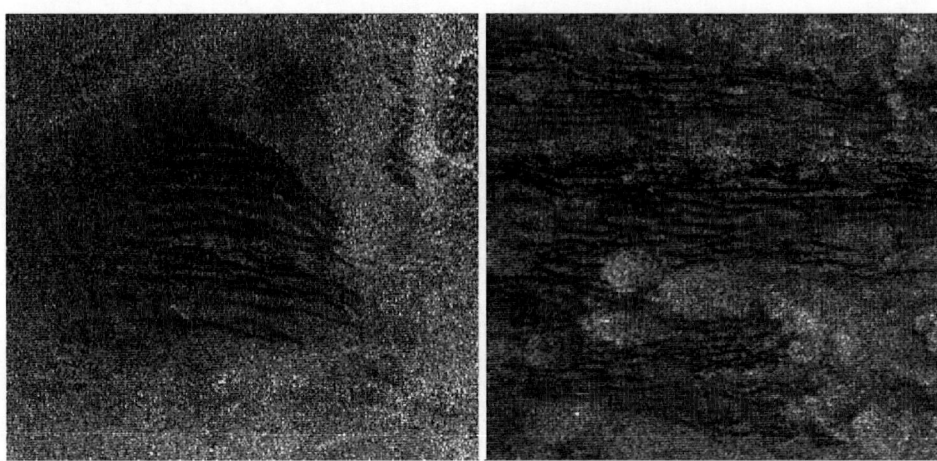

**Fig. 13.3** Two segments (about 100 km × 100 km) of the T3 Cassini radar image, showing dark features now known to be linear dunes ('cat scratches'). *Credit* NASA/JPL/R. Lorenz/Cassini Radar Team

**Fig. 13.4** The 'smoking gun'. This part of the radar image from T8, 100 km × 300 km, with radar illumination from above (north) shows very regular spaced dark features, with glints on their north faces, confirming their positive relief—their length, height and spacing (and appearance overall) were noted to be very similar to the linear dunes in the Namib desert. Their streamline-like arrangement with respect to the bright hills at the right supported an association with a wind or liquid flow, although some scientists maintained that an erosive origin should still be considered. *Credit* NASA/JPL/R. Lorenz/Cassini Radar Team

in T8 the topographic shape of the sand surface was the major control on the radar appearance. Second, the equatorial groundtrack of the spacecraft meant that the dunes, oriented E−W, were illuminated broadside-on, giving many opportunities to detect topographic glints (Fig. 13.4). These showed clearly the nature of the dunes as positive relief, and radarclinometry (Fig. 13.6) was used to show that the dunes attained 100–150 m in height: notably, the dunes were the same height, spacing and morphology as those in the Namib. Finally, in this dark region, the dunes themselves seem to have been larger (higher) than in T3, presumably as a result of a more abundant sand supply.

It was argued that ridges could be erosive in origin (i.e., yardangs), rather than depositional, but the presence of Y-junctions appeared to exclude that possibility. The arrangement of the dunes around bright and elevated obstacles suggested they followed a flow or streamline pattern (Fig. 13.7), making them longitudinal dunes (i.e., 'linear').

The dramatic implication of the observation (T8 covered about a million square kilometers, or 1 % of Titan's surface) was that, by analogy, all the large equatorial dark areas were covered with dunes—seas of sand! Radebaugh et al. (2008) document further study of the dunes with radar data, counting several thousand individual dunes. Some network dunes are noted, suggesting complex wind regimes in specific locations, but predominantly the dunes are linear.

The presence of dunes in the T8 swath, which also imaged the Huygens landing site, was in fact instrumental in being able to co-register the 2 cm radar image with the near-IR image taken by the Huygens probe's camera (e.g., Lunine et al. 2008). These different data did not always correlate well, especially at the small scale, but two dunes (seen only in the distance as horizontal dark streaks in the DISR side-looking images, as well as in the radar) were just visible, and provided the main tie-points to match up the radar and near-IR data.

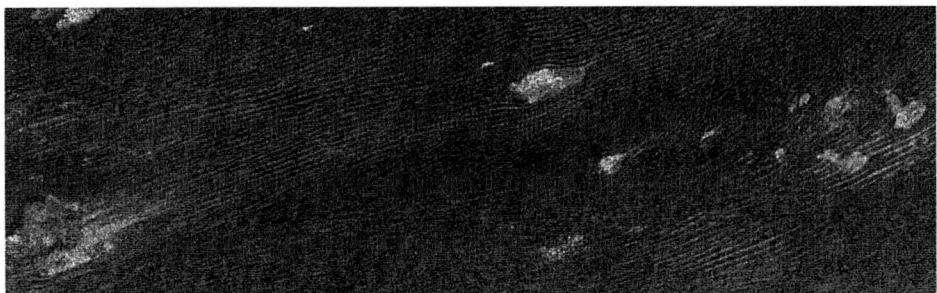

**Fig. 13.5** Elsewhere in the T8 image, the topographic glints were not obvious, but the same features appeared as dark streaks above a brighter substrate. In this image (again, 100 km high) the bright interdune areas are visible, as is a streamlined inselberg at center. This showed the features to be longitudinal in character and suggested a depositional origin. The case for dunes was generally convincing. *Credit* NASA/JPL/R. Lorenz/Cassini Radar Team

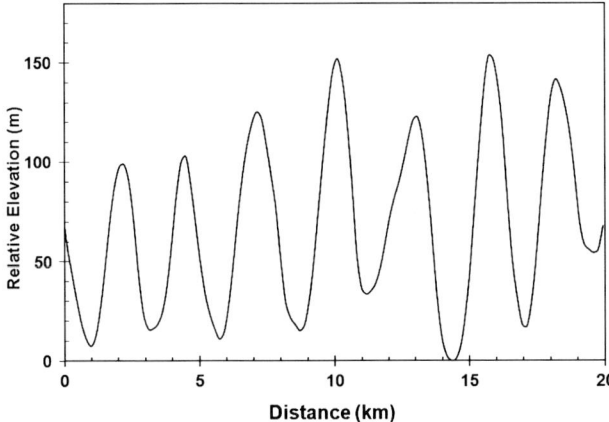

**Fig. 13.6** Radarclinometric profile of the Titan dunes in Belet, shown in Fig. 13.4. This profile shows well the typical ~3.3 km crest-crest dune spacing on Titan, and the remarkable >100 m height

It may be noted also that the Doppler tracking (e.g., Bird et al. 2005) of the Huygens probe during its parachute descent (given independent support from some optical and thermal measurements) indicated near-surface winds of the order of 0.3 m/s. This is of a comparable magnitude to the saltation threshold estimated in pre-Cassini work.

Dunes were detected in the T20 VIMS multispectral image strip (or 'noodle') in Fensal (Barnes et al. 2008). Up to this point, VIMS data had been of too low spatial resolution to measure individual dunes. In this instance, however, the observations were made close enough to Titan that the resolution was about 250 m. A challenge in interpretation is that, while long, the image is only 64 pixels wide (Fig. 13.8), but analysis here was facilitated by context from a high-resolution radar image of this location (~50 W, 10 N) acquired a few months earlier on T17. observation by radar. The dune material spectroscopically appears to contain less water ice than other units on Titan, and various organic materials would be consistent with the data. Barnes et al. (2008) measured a typical dune spacing of 2 km, and used photoclinometry to determine a dune height in this area of 30–70 m. They also noted that the strong contrast between dunes and interdunes suggested the latter were completely clear of dark sand (whereas radar might 'see through' some centimeters of sand, the near-IR light would not) and this might suggest that the dunes were still actively saltating.

Lorenz and Radebaugh (2009) mapped the direction of all the dunes observed with radar through the nominal or Prime mission of Cassini (Fig. 13.9). By this time, coverage of Titan's surface by radar imaging was around 30 % (and distributed enough to exclude large sampling biases). The dunes are typically aligned close to the E−W direction, and while in some cases the direction along the dune axis is ambiguous, the overall pattern is overwhelmingly one of net sand transport from west to east: dunes terminate abruptly at the western edge of obstacles (Fig. 13.10), and pick up gradually thereafter. Some obstacles have 'tails' in this downstream direction—some of these morphological clues are discussed in Radebaugh et al. (2010).

The growing Cassini coverage (radar imaging with resolution adequate to resolve dunes now covers about 40 % of the surface) permits some global trends to emerge (Fig. 13.11). Savage et al. (2013) have documented in detail the widths and spacings of the dunes and their variation with latitude: a pattern analysis shows that there is really only one major population. One notable variation is the general increase in radar backscatter from north to south, and a corresponding decrease in microwave brightness temperature. This has been interpreted (Le Gall et al. 2011, 2012) as a progressive increase in interdune area from south to north (see Fig. 13.12). This may be connected with the general configuration of Titan's seasons; in the present epoch (as coincidentally is the case for both Mars and the Earth, although the effect on Earth is very small), the perihelion of Saturn's eccentric (e = 0.09) orbit around the sun occurs close to the southern summer solstice, which has the effect of making southern summers more intense, but shorter, than those in the north. Although the total solar energy delivered to each hemisphere is the same, nonlinear effects can cause

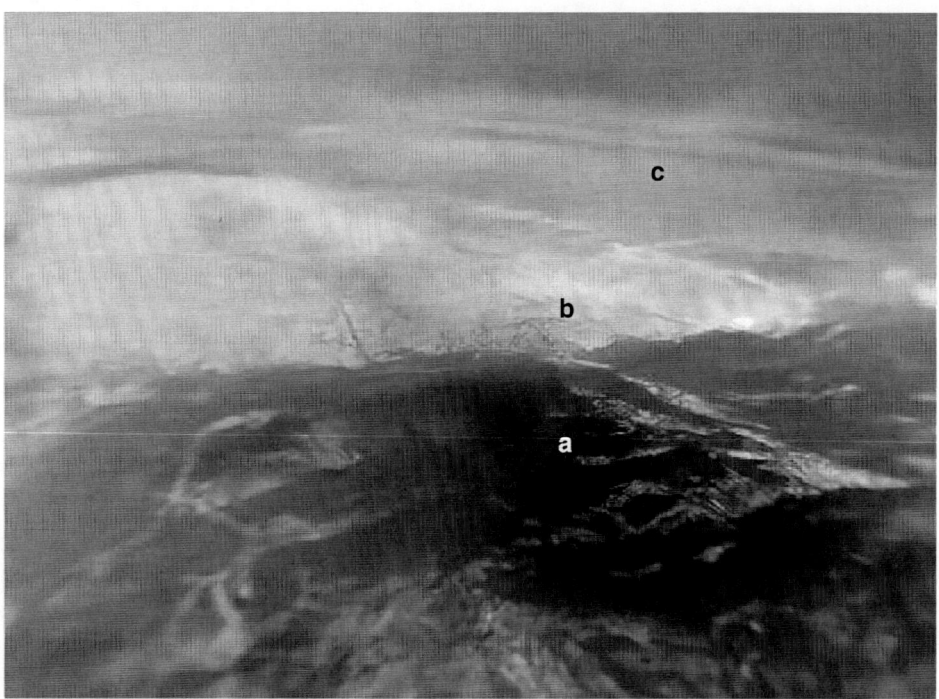

**Fig. 13.7** The closest thing we have to an aerial photo on Titan: a mosaic of images (assembled by Erich Karkoschka of the University of Arizona) taken during the parachute descent of the Huygens probe. This scene looks north–west from a simulated altitude of about 25 km. The probe landed in the dark foreground (**a**), revealing a riverbed scene with rounded cobbles on a damp sandy substrate. The most dramatic pictures were of a network of small rivervalleys in a bright highland (**b**)—the distance **a**–**b** is about 7 km. Not noticed at the time, but instrumental in locating the landing site in radar image data, were two dark streaks—dunes—at (**c**), about 30 km north of the landing site. *Image credit* NASA/ESA/U.Arizona

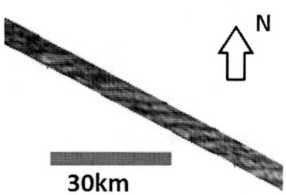

**Fig. 13.8** Titan's dunes (and sand-free interdunes) at about 13 N, 56 W seen at 2 microns wavelength by the Cassini Visual and Infrared Mapping Spectrometer (VIMS). The sand is dark at this wavelength and at optical wavelengths and in radar, providing an important compositional constraint. Image courtesy of J. Barnes

this different forcing to have important consequences, notably favoring a transport of moisture to the northern hemisphere. This appears to be a compelling explanation (Aharonson et al. 2009) for the fact that the vast majority of Titan's hydrocarbon lakes are found around the north pole. Recent work suggests that the net interhemispheric transport of moisture is manifested as a longer rainy season (summer) in the North (Schneider et al. 2012). Clearly, moisture availability can affect sand transport, and thus an effect of the asymmetric seasons on the dunefields is not surprising. An interesting side-note to this astronomical climate forcing is that, much like the Croll-Milankovich cycles that pace the Earth's ice ages and influence the polar layered terrain on Mars, the hemispheric arrangement would be reversed only about 30000 years ago—perhaps during our last ice age, there were more seas in Titan's south than in the north.

Although the dunes appear almost invariably to be linear (longitudinal), there have been a handful of counterexamples identified, albeit at the threshold of the radar resolution. Lorenz et al. (2006) noticed a 'lattice' pattern where linear dunes south of a bright topographic obstacle appeared to blend in with some superposed orthogonal (i.e., transverse) forms. It was suggested that perhaps the obstacle 'straightened' the fluctuating flow that is associated with the linear form. Hayes et al. (2012) have examined sharpened radar images of dunes and found some evidence of barchan forms, possibly undergoing a transition in orientation. The possibility that such dramatic rearrangements of the climate system may occur on such short periods raises the question of whether the aeolian geomorphology can respond fast enough—the dune pattern observed may represent the result of winds over tens of thousands of years, and thus perhaps not only the currently (poorly) observed climate.

An additional class of feature (Fig. 13.13), that may be dunes, are what appear to be ridges that Radebaugh et al. has found in a number of mid-latitude locations. These are notable in that they resemble the dunes in their size and shape, but apparently have a different composition or surface texture

## 13.2 Cassini Observations

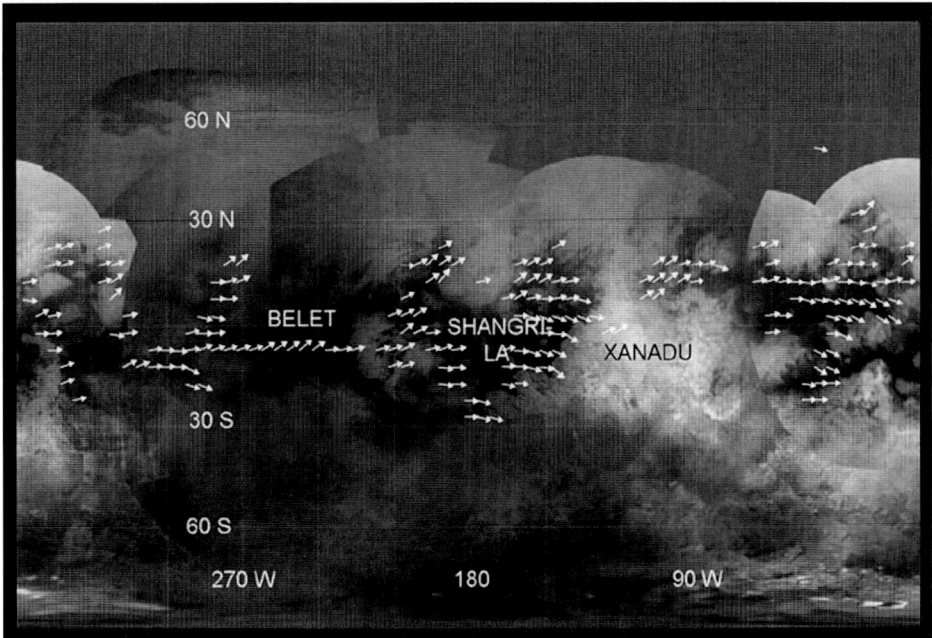

**Fig. 13.9** The average orientation of Titan's dunes observed in the prime mission in 5 × 5 degree boxes indicated with arrows, from Lorenz and Radebaugh (2009). There is some hint of a poleward divergence, and there are local deviations around bright areas. Note that large dark areas still remain unobserved by radar

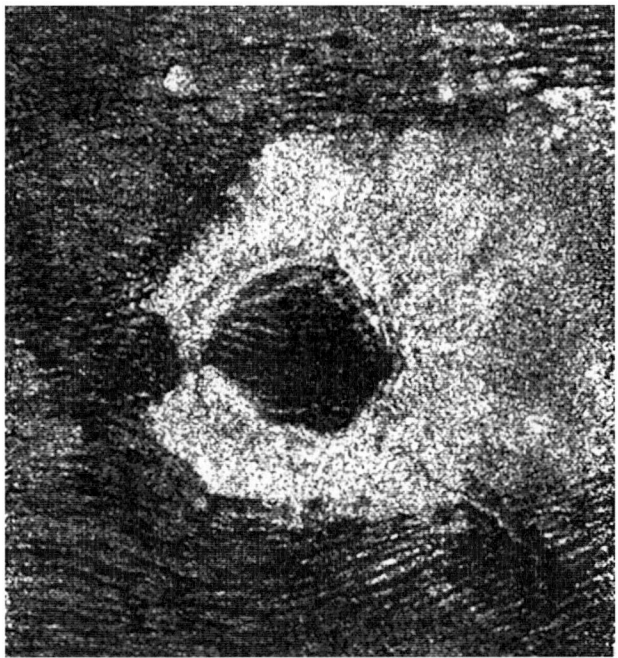

**Fig. 13.10** A segment (120 km$^2$) of the radar image acquired on flyby T56, showing the dunes breaching into a presumed impact crater. Towards the bottom, the dunes appear to be diverted by the bright ejecta blanket of the crater. A similar interaction occurs on Earth at the Roter Kamm structure in Namibia (Fig. 18.5) *Credit* NASA/JPL/Cassini Radar Team/R. Lorenz

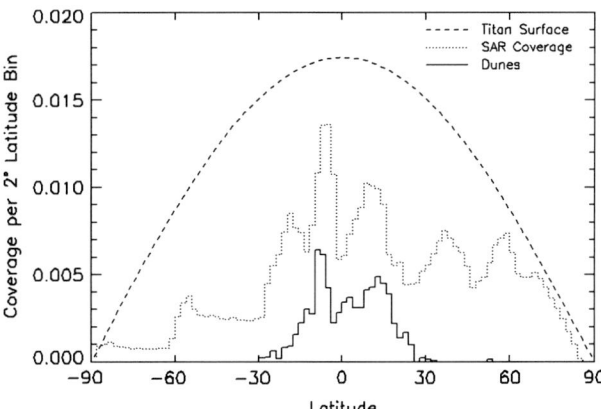

**Fig. 13.11** A histogram (*solid bars*) of the observed dunes, together with the radar coverage (*dotted line*). The relatively uniform radar coverage with latitude shows that the dune distribution is not a sampling effect—the dunes are genuinely confined to equatorial latitudes

in that they are radar-bright. They also have consistently more north–south orientations than are typical for the 'classic' dunes. A favored interpretation is that these are relic dunes from a previous climate epoch wherein dry conditions persisted further poleward than they do now, and where the wind regime favored a different orientation than seen today.

## 13.3 Sand Source, Amount and Composition

The optimum particle size for saltation in Titan's atmosphere, assuming interparticle cohesion similar to terrestrial materials, is about 250 microns in diameter. The likely means for creating such material is by breakdown of coarser

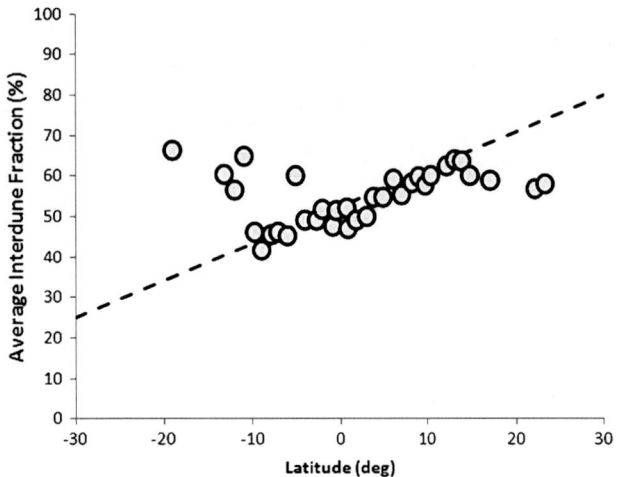

**Fig. 13.12** Radar measurements of the interdune area fraction in dune-covered areas as a function of latitude by Le Gall et al. (2011). Work by others has substantiated this trend, which is presumably a result of varying sand mobility due to the asymmetry of Titan's seasons in the present epoch

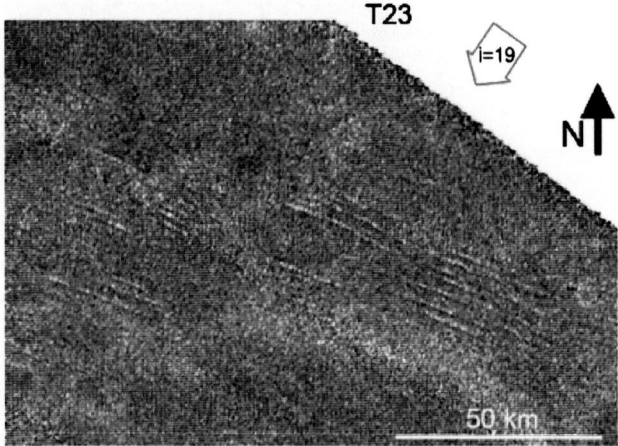

**Fig. 13.13** Possible stabilized or 'fossil' dunes, at a latitude of some 53°N. Not only does their orientation deviate from the typical, but these features (well north of the sand seas) have a generally radar-brighter appearance than their equatorial counterparts, suggesting a somewhat different surface texture. *Image* NASA/JPL/Cassini Radar Team/J. Radebaugh

materials such as impact ejecta or fluvial sediments, or by agglomerating finer material such as the atmospheric haze.

Lorenz et al. (2008) have estimated the total sand volume, noting that about 40 % of the low-latitude half of Titan (i.e., about 20 % of the total) appears covered in dunes, and used radarclinometric, radiometric, and similarity arguments to estimate the average depth to be between 200,000 and 800,000 km$^3$ of material, corresponding to a thickness of several meters over the whole planet. That estimate has been revised with the benefit of further radar coverage and more careful analysis by Le Gall et al. (2011) who find 12.5 % of the surface (i.e., a total of ~10 millions km$^2$) covered in dunes with an estimated total volume of 250,000 km$^3$, although this of course relies on the assumed dune height.

Neish et al. (2010) used radarclinometry to study the height H of a variety of dunes and validated the accuracy of this method with the relatively low Cassini resolution (see Fig. 18.24) against SAR images and topographic data on the Namib desert on Earth. They found that the dunes were fit with a relationship of the form $H = cD^n$, where D is the spacing, and c and n are constants. They suggest that the values for Titan (n = 0.78 for the large Belet dunefield) can be interepreted to imply the dunes are active, noting that n ~ 0.54 at the margins of the Namib, but n = 1.72 at the center.

An initial estimate of the volume in river channels (Lorenz et al. 2006) was that there is not enough volume in the channels to account for the volume of sand needed to construct the dunes. Since more heavily-eroded areas have since been found, it may be that this calculation has to be revisited. The observed impact crater distribution—originally thought to be a likely source for sand-sized material (it is believed to be the dominant sand source on Venus)—is unable to provide the required volume, unless some other process has broken down larger ejecta.

Thus, at present, a photochemical origin appears to be favored: observations (see next paragraph) favor an organic composition, and estimates (Lorenz et al. 2006, 2008, 2010) of organic production in the atmosphere seem to be consistent with the dune volume, suggesting that the sand formed from what were once haze particles. Conversion of <1 micron haze particles into 250 micron sand grains could perhaps occur by sintering over long timescales (although some details need to be worked out, like why the process would stop…), or perhaps, more likely, it may involve cycles of wetting and drying in Titan's lakes. The latter scenario would require that the sand move from the lakes at the poles to the equatorial regions where the dunes are found. The sand source remains an outstanding problem in Titan science.

As for its composition, there are a few constraints. The microwave properties of the sand seas (Le Gall et al. 2011, 2012) support a porous, organic composition. Water ice (once suspected to be the dominant bedrock on Titan, but apparently largely covered by an organic veneer in most places) appears to be ruled out. The material is dark at visible and near-infrared wavelengths, and the infrared

## 13.2 Cassini Observations

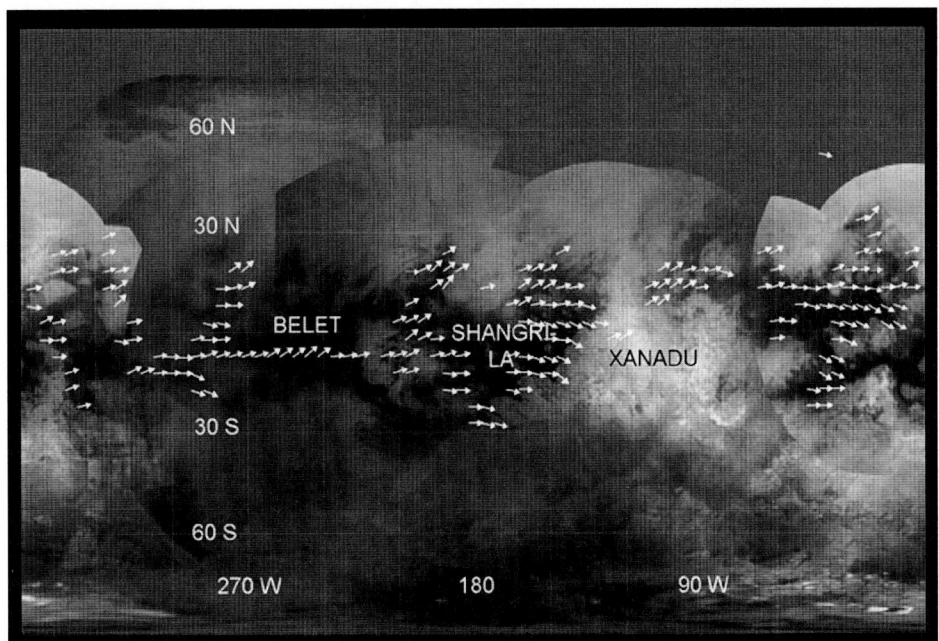

**Fig. 13.9** The average orientation of Titan's dunes observed in the prime mission in 5 × 5 degree boxes indicated with arrows, from Lorenz and Radebaugh (2009). There is some hint of a poleward divergence, and there are local deviations around bright areas. Note that large dark areas still remain unobserved by radar

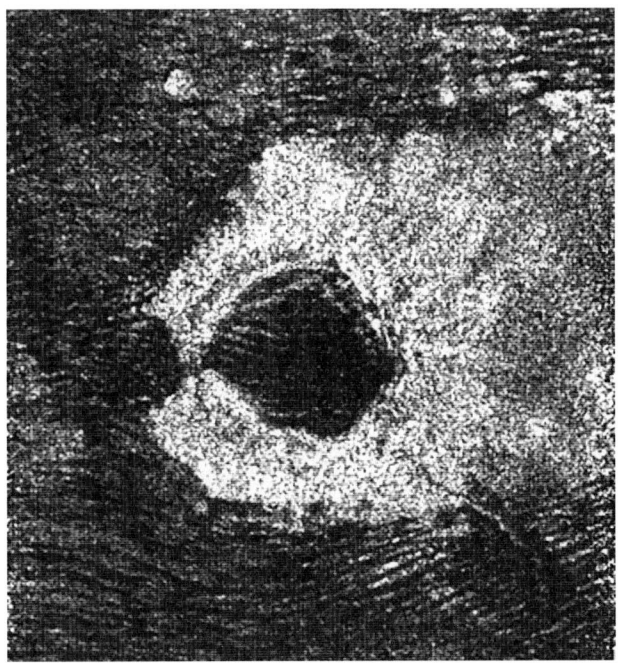

**Fig. 13.10** A segment (120 km$^2$) of the radar image acquired on flyby T56, showing the dunes breaching into a presumed impact crater. Towards the bottom, the dunes appear to be diverted by the bright ejecta blanket of the crater. A similar interaction occurs on Earth at the Roter Kamm structure in Namibia (Fig. 18.5) *Credit* NASA/JPL/ Cassini Radar Team/R. Lorenz

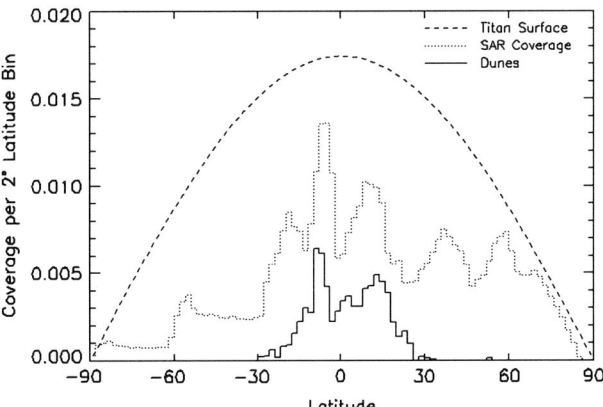

**Fig. 13.11** A histogram (*solid bars*) of the observed dunes, together with the radar coverage (*dotted line*). The relatively uniform radar coverage with latitude shows that the dune distribution is not a sampling effect—the dunes are genuinely confined to equatorial latitudes

in that they are radar-bright. They also have consistently more north–south orientations than are typical for the 'classic' dunes. A favored interpretation is that these are relic dunes from a previous climate epoch wherein dry conditions persisted further poleward than they do now, and where the wind regime favored a different orientation than seen today.

## 13.3 Sand Source, Amount and Composition

The optimum particle size for saltation in Titan's atmosphere, assuming interparticle cohesion similar to terrestrial materials, is about 250 microns in diameter. The likely means for creating such material is by breakdown of coarser

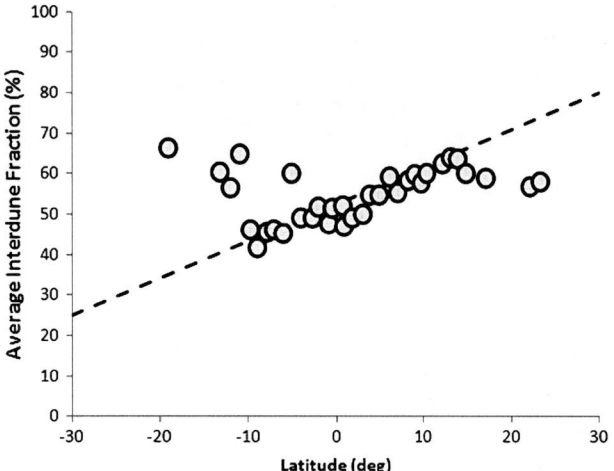

**Fig. 13.12** Radar measurements of the interdune area fraction in dune-covered areas as a function of latitude by Le Gall et al. (2011). Work by others has substantiated this trend, which is presumably a result of varying sand mobility due to the asymmetry of Titan's seasons in the present epoch

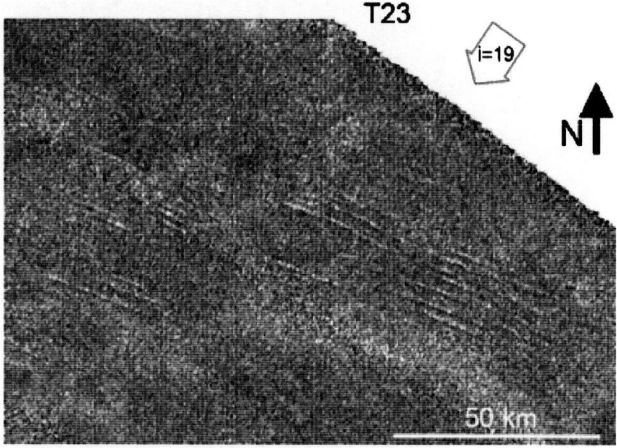

**Fig. 13.13** Possible stabilized or 'fossil' dunes, at a latitude of some 53°N. Not only does their orientation deviate from the typical, but these features (well north of the sand seas) have a generally radar-brighter appearance than their equatorial counterparts, suggesting a somewhat different surface texture. *Image* NASA/JPL/Cassini Radar Team/J. Radebaugh

materials such as impact ejecta or fluvial sediments, or by agglomerating finer material such as the atmospheric haze.

Lorenz et al. (2008) have estimated the total sand volume, noting that about 40 % of the low-latitude half of Titan (i.e., about 20 % of the total) appears covered in dunes, and used radarclinometric, radiometric, and similarity arguments to estimate the average depth to be between 200,000 and 800,000 km³ of material, corresponding to a thickness of several meters over the whole planet. That estimate has been revised with the benefit of further radar coverage and more careful analysis by Le Gall et al. (2011) who find 12.5 % of the surface (i.e., a total of ~10 millions km²) covered in dunes with an estimated total volume of 250,000 km³, although this of course relies on the assumed dune height.

Neish et al. (2010) used radarclinometry to study the height H of a variety of dunes and validated the accuracy of this method with the relatively low Cassini resolution (see Fig. 18.24) against SAR images and topographic data on the Namib desert on Earth. They found that the dunes were fit with a relationship of the form $H = cD^n$, where D is the spacing, and c and n are constants. They suggest that the values for Titan (n = 0.78 for the large Belet dunefield) can be interepreted to imply the dunes are active, noting that n ~ 0.54 at the margins of the Namib, but n = 1.72 at the center.

An initial estimate of the volume in river channels (Lorenz et al. 2006) was that there is not enough volume in the channels to account for the volume of sand needed to construct the dunes. Since more heavily-eroded areas have since been found, it may be that this calculation has to be revisited. The observed impact crater distribution—originally thought to be a likely source for sand-sized material (it is believed to be the dominant sand source on Venus)—is unable to provide the required volume, unless some other process has broken down larger ejecta.

Thus, at present, a photochemical origin appears to be favored: observations (see next paragraph) favor an organic composition, and estimates (Lorenz et al. 2006, 2008, 2010) of organic production in the atmosphere seem to be consistent with the dune volume, suggesting that the sand formed from what were once haze particles. Conversion of <1 micron haze particles into 250 micron sand grains could perhaps occur by sintering over long timescales (although some details need to be worked out, like why the process would stop...), or perhaps, more likely, it may involve cycles of wetting and drying in Titan's lakes. The latter scenario would require that the sand move from the lakes at the poles to the equatorial regions where the dunes are found. The sand source remains an outstanding problem in Titan science.

As for its composition, there are a few constraints. The microwave properties of the sand seas (Le Gall et al. 2011, 2012) support a porous, organic composition. Water ice (once suspected to be the dominant bedrock on Titan, but apparently largely covered by an organic veneer in most places) appears to be ruled out. The material is dark at visible and near-infrared wavelengths, and the infrared spectral signature of the dune fields ('brown' in one widely-used color mapping) is distinct from that of water ice exposures. A final clue comes from the presence of a spectral band associated with the sand seas (Clark et al. 2010). Although absorption of light by methane in Titan's atmosphere only permits spectroscopy in a few isolated windows, making much less of the wavelength range of the instrument (1–5 microns) available for composition identification, a band at 5.05 microns does appear to be spatially correlated with the 'brown' sand unit, and is suggestive of aromatic organics like benzene (though the actual composition is likely more complicated than that).

## 13.4 Implications for Meteorology

The dunes represent an important set of constraints on Titan's meteorology in three respects: their extent, their form, and their orientation. They further give implications for photochemistry, in that the dunes represent the largest known reservoir of organic material (Lorenz et al. 2008) as described above.

First, their distribution, confined to the tropics, defines the latitudes equatorward of 30° N and S as having at least sometimes the conditions required for dune formation (namely available and transportable (i.e., dry) sediment, and winds strong enough to move the material), whereas other latitudes appear not to satisfy these requirements. Since the sand source is not yet unambiguously identified, this constraint pertains to winds and humidity. With regard to humidity, models had already (e.g., Rannou et al. 2006) suggested low latitudes on Titan should eventually dry out unless resupplied by a surface methane source.

Mitchell (2008) explored this question further and estimate some 1–2 m of liquid methane per year could be removed from low latitudes. He found that the latitudinal extent of the dry region depends on the total methane inventory, with between 7 and 20 m agreeing best with observations (such as dune extent, and the humidity recorded by Huygens).

Second, the predominance of the longitudinal (linear) dune form requires a modestly-changing (typically bidirectional) wind regime (e.g., Lancaster 1995). Sources of such a variation include seasonal change (the usual reason for this wind regime on Earth) and possibly the gravitational tide in the atmosphere.

Finally, the dune orientation pattern represents an important diagnostic on the tropospheric winds, for which there are few clouds to act as tracers. Tokano (2008) has explored the winds in a global circulation model (GCM) and finds that indeed surface winds should not infrequently exceed the saltation threshold of 0.5–1 m/s and, furthermore, that bidirectional winds are encountered over the course of a Titan year, due to a seasonal change in the hemisphere-to-hemisphere Hadley circulation. However, this model (indeed, all models of Titan's general circulation) predicts that the near-surface winds at low latitudes are predominantly easterlies (i.e., blowing westwards), in contradiction to the appearance of the dunes. Tokano (2008) reports GCM experiments with introducing Xanadu as a large positive relief feature (i.e., a hill) and changing its albedo in an attempt to modify the winds, but was unable to form low-latitude westerlies.

This apparent paradox (Wald 2009) may have been resolved (Lorenz 2010) at least in part by studying the windspeed history with finer time resolution. Tokano (2010) found in a refined version of his model (incorporating a crude topography, although this is not believed to be a principal factor) that while the equatorial near-surface winds were indeed easterlies, for a brief period around the equinoxes, as the intertropical convergence zone (ITCZ) crosses the equator, stronger vertical mixing leads to strong westerlies. Even though these winds persist for only a small fraction of the Titan year, if the saltation threshold is high enough (i.e., $> \sim 1$ m/s), the sand motion may reflect only those stronger westerlies and not the weaker (albeit more prevalent) easterlies and thus the dune pattern indicates a west-to-east sand transport. Of course, one model showing that a mechanism is possible does not prove that the mechanism is responsible, and the problem remains an active area of research.

Whilst the dunes almost exclusively indicate eastwards sand transport, there are regional deviations of up to 45° (see Fig. 13.9). The pattern remains to be fully interpreted, although one immediate impression is one of deviation around the 2500 km continental-scale feature Xanadu—much as dunes deviate around bright obstacles only a few kilometers across. There may also be some convergence of flow towards the center of large dune fields like Belet which might be a 'sea-breeze' effect of daytime solar heating of the dark dunefield causing an updraft and convergence.

A further interesting question is: why, given that Titan's gravity is so different from Earth, the atmosphere 4× thicker, and the sand presumably made of rather different stuff, should the resultant dune forms resemble the Earth's so much in morphology and size? The morphology is, of course, a rather basic effect—the dune-forming process is the same everywhere, and the morphology depends on the sand supply and the variability of the wind vector as discussed in Chaps. 6 and 9. But why the size should be the same in Belet and in the Namib is less obvious.

**Fig. 13.14** Artistic impression (by Mike Carroll) of Titan's dunes, with an airplane landing on the flanks. In reality, the rings will be invisibly edge-on, and scattering in Titan's atmosphere will not allow Saturn to be seen as sharply, but such artistic license is usually employed to highlight Titan's exotic setting. If one could see the rings, they would be oriented more vertically, since Saturn is fixed over Titan's equator, and the dune fields are found only at latitudes less than 30°. *Image courtesy* Michael Carroll

The answer appears to be (Lorenz et al. 2010) that dunes grow up to a limiting height (and thus spacing) which is determined by the thickness of the planetary boundary layer, the layer of air that is well-mixed near the surface. The top of this layer acts, according to the theory of Andreotti et al. (2009) as a 'capping surface', much like the free surface of a liquid layer and resists upward deformation. Thus the airflow diverted by a dune becomes compressed as the dune grows up to the PBL top, and the resultant increased shear stress prevents further sand accumulation, and so growth stops. The geometry of the compressed streamlines and the duneform is such that the dune spacing roughly equals the PBL thickness, and the maximum height is about a factor of 12 smaller.

In the coastal regions of the Namib, the thermal inertia of the sea prevents the air warming strongly, and the PBL can grow over the course of a day to only about 300 m. On the other hand, some tens to hundreds of kilometers inland, the dry ground can heat up much more and the PBL can grow to $\sim 2$ km. Indeed, the dunes at the coast are generally smaller, and the dunes inland can grow to $\sim 150$ m high and $\sim 2$ km apart. The dunes on Titan are well distant from the long-lived polar seas (though transient lakes may exist in Titan's overall dry tropics, as on Earth) and so there is little variation of thermal inertia, and thus little variation of boundary layer height. The Huygens probe descent profile suggested a PBL thickness of $\sim 3$ km (although an earlier assessment suggested $\sim 300$ m, this appears to have been the new growing PBL during the morning of the Huygens descent, distracting attention from the remains of the fully-grown 2 km PBL from the previous day). Recent modeling has explored the behavior of Titan's PBL (Charnay et al. 2012).

## 13.5 Future Exploration of Titan

Cassini has revealed Titan to be a world of great geomorphological diversity, with river channels, polar seas of hydrocarbon liquids, and dunefields. It is therefore of considerable interest for future exploration, and Titan's unique thick atmosphere opens many possibilities: not only is it fairly straightforward to deliver landers by heat-shield and parachute, but the atmosphere can be used to decelerate vehicles into an orbit by 'aerocapture', a much more fuel-efficient approach than using retrorockets. A future Titan orbiter could use a radar or near-infrared imager to obtain global mapping with a resolution 5–10 times better than

Cassini, dramatically improving our understanding of the dune morphology.

Higher-resolution imaging, given Titan's thick atmosphere, likely requires study from closer to the surface. But here Titan offers exciting possibilities too: flight in Titan's low gravity and thick atmosphere by both heavier-than-air and lighter-than-air means (Lorenz 2009) is easier than at Earth, and rather serious proposals for airships, hot-air balloons, and most recently airplanes (Barnes et al. 2012) have been made (Fig. 13.14).

In the meantime, the Cassini mission is presently ongoing. While the spatial resolution of its instruments will not increase, the geographical coverage (including the highest-resolution VIMS data, able to resolve dunes, and opportunities for radar stereo to give topography of dunes and their settings) will improve over coming years.

# Venus Dunes

## 14.1 Introduction

Since Venus has a thick atmosphere (Fig. 14.1), it seems a natural place to contemplate aeolian transport. Although speculations about Venus' surface ranged from a global ocean world (of water), to a tarry swamp, other speculations were of a desert world. In many respects, this last perspective has been borne out, although Venus seems largely to be a rocky rather than a sandy world, dominated by volcanic and tectonic processes, rather than erosion and deposition.

Authoritative research reviews of Venus science as a whole are given in two books in the University of Arizona planetary science series, namely *Venus* (edited by Hunten et al. 1983, written in the wake of Russian probes and Pioneer Venus) and *Venus II* (Bougher et al. 1997, a mere 1362 pages compiled in the wake of Magellan) and an AGU volume (Esposito et al. 2007). An excellent and readable text is that by Marov and Grinspoon (1998). The best comprehensive review of Venusian aeolian processes specifically is the chapter in *Venus II* by Greeley et al. (1997a).

## 14.2 History of Venus Exploration

Even before space probes revealed the details of surface conditions on Venus, it was expected to be a windswept place. Indeed, one early theory (Öpik 1961) even advanced friction by wind-blown dust as being the reason that Venus' surface might be hot, although this mechanism doesn't stand up to thermodynamic scrutiny. As early spacecraft data came in, it became obvious that the greenhouse effect was responsible for elevating Venus' surface temperatures, but until probes reached the surface in the 1970s, it was not known how dense and hot the atmosphere really was, nor which greenhouse gases might be responsible. One theory—which formed the Ph.D. thesis of the later-prominent climate scientist James Hansen—was that airborne dust would provide greenhouse warming. In fact, we now know that the lower atmosphere of Venus is relatively clear, and the greenhouse effect is due primarily to carbon dioxide and water vapor.

The Venus probes revealed the atmospheric pressure at the surface to be some 90 bar: this translates, given the molecular weight of the predominantly $CO_2$ atmosphere and the high temperatures, to an air density of about 64 kg/m$^3$, or about 50 times that of sea-level air on Earth. This high density might lead one to expect aeolian transport to be rather easy. However, in contrast to the abundant sand drifts seen in images from the surface of Mars by the Viking landers in 1976, images returned from the torrid Venusian surface by several Soviet Venera probes[1] showed (Fig. 14.2) relatively little fine-grained material (compared with the Moon or Mars), although there were some indications that the landings kicked up a cloud of dust. Since the Venusian surface is hidden from view at optical wavelengths, there was no camera survey of Venus from orbit comparable with the Mariner 9 and Viking orbiter surveys of Mars which had revealed ample evidence of aeolian activity on that world.

Wider reconnaissance of the Venusian surface was accomplished with radar measurements using large radio telescopes on Earth, and later by radar mapping of the northern polar regions of Venus by the Venera 15 and 16 spacecraft in the early 1980s. However, these techniques had spatial resolutions (1 ~ 2 km) too poor to resolve dunes.

The next (and for the moment, latest) step was the near-global radar mapping at ~120 m resolution by the NASA Magellan spacecraft in the early 1990s (Fig. 14.3). Despite the thick atmosphere which might a priori suggest aeolian features might be widespread, only a couple of areas of resolvable dunes were discovered on Venus in Magellan radar imaging (e.g., Greeley et al. 1992; Weitz et al. 1994;

---
[1] Florensky et al. (1977) suggest on the basis of Venera 9 and 10 imaging that fines have been moved in the atmosphere, but there are no obvious bedforms.

**Fig. 14.1** Venus in ultraviolet light, observed in 1972 by Mariner 10. At visible wavelengths (like Titan) Venus appears nearly featureless because of its thick, cloudy atmosphere. *Credit* NASA

Greeley et al. 1995), although wind streaks and downwind dispersal of impact ejecta were observed in many locations, and some possible yardangs were also identified.

## 14.3 Venus Dunes

The two prominent dune fields identified on Venus are Algaonice (Fig. 14.4) and Fortuna-Meshkenet (Greeley et al. 1992) (Fig. 14.5). The Algaonice dunes at 25°S, 340°E cover some 1300 km² (about the same size as the Lencois Maranhenses dunefield in Brazil) at the end of the ejecta outflow channel from the impact crater of that name (the dunes themselves were subsequently formally named Menat Undae, although the Algaonice name seems to have been more widely used). The dunes are 0.5–5 km in length and are quite bright, likely because there are slip faces oriented towards the radar illumination, which was at an incidence of 35°.

The more northern dune field, Fortuna-Meshkenet lies at 67°N, 91°E in a valley between Ishtar Terra and Meshkenet Tessera (the dunes are formally named Al-Uzza Undae). The dunes are 0.5–10 km long, 0.2–0.5 km wide and spaced by an average of 0.5 km. They appear to be transverse dunes, in that there are several bright wind streaks visible in the region, which seem generally orthogonal to the dunes. Glints are not observed strongly on these dunes, although here the incidence angle of the radar observation was 22–25°.

Small dunes, however, might be much more widespread than the fields of resolved dunes above. In many areas where no resolvable dunes were detected, it was noted that the radar echo had a substantially different strength when the area was observed from one side versus another. One explanation of this radar asymmetry is that unresolved 'microdunes' may be present (Weitz et al. 1994; Kreslavsky and Vdovichenko 1998), wherein the asymmetry is due to shallow stoss slopes being prominent from one direction against the steeper slip faces seen from the other (Fig. 14.6). A possible additional factor in shaping the radar reflectivity of microdunes is that wind tunnel experiments (see later) show that in a sand of mixed composition, dense (and therefore radar-reflective) minerals such as pyrite or chromite become concentrated on the windward (stoss) side of the microdune (Greeley et al. 1991a).

A significant reason for the lack of observed dunes may be a planetwide paucity of sediment. The Venusian surface may have been completely resurfaced by lava flows about 500 Myr ago, and the kinetics of breakdown of basalt by the Venusian atmosphere is uncertain (and of course, other sand-generating processes such as freeze–thaw, glacial action or fluvial erosion do not occur on present-day Venus). Indeed, the dominant source of sand-sized sediment may be the ejecta from impact craters. Garvin (1990) calculated that impacts could produce enough fine-grained materials (<1 mm) to form at most a globally-averaged layer 1 m thick. Note that the Fortuna and Algaonice dunefields likely account for only about 1000 km³ of sediment in total.

## 14.4 Aeolian Transport Under Venus Conditions: Experiments

The conditions on the surface of Venus are far from common terrestrial experience (although somewhat similar conditions are generated in certain industrial processes). The three major factors (surface gravitational acceleration at 8.79 ms$^{-2}$ being not too different from the terrestrial 9.81 ms$^{-2}$) are the 90 bar pressure, the 750 K temperature, and the principal gas being $CO_2$. One reason these features were interpreted to be microdunes rather than ripples is that they have slip faces and form bedding planes (see Chap. 5) (Figs. 14.7 and 14.8).

Just above the saltation threshold, at 0.63 m/s, ~18 cm long dunes with slip faces developed. The wavelength became shorter as windspeed was increased, reaching 8 cm for a windspeed of 1.07 m/s. The dunes became more degraded at this point, with the slip faces tending to

**Fig. 14.2** Surface image from Venera 14 (note the distortion—the image was scanned cylindrically, looking down at an oblique angle). Most of the surface here is platy lava, probably somewhat representative of the surface as a whole. Some other landing site images show more blocky terrain, and a few more patches of regolith, some of which may be fine enough to saltate. *Image* NASA NSSDC

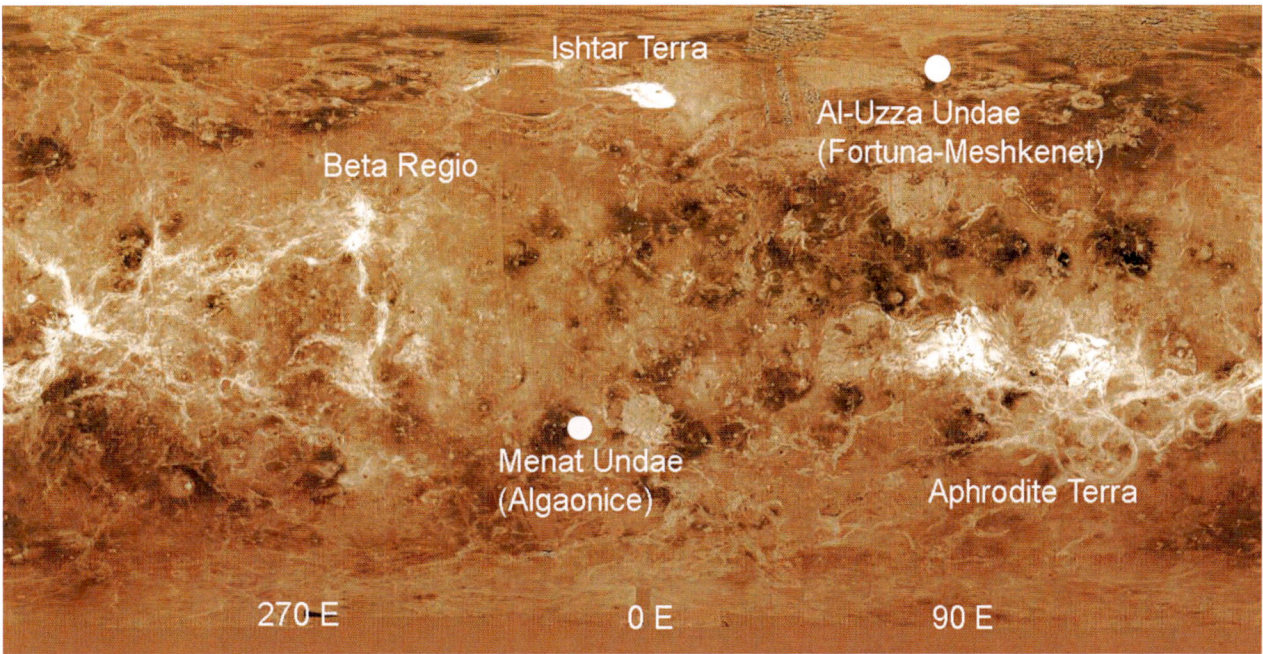

**Fig. 14.3** A global mosaic of radar reflectivity measured by the Magellan spacecraft (cylindrical projection). The two major dunefields are marked with *circles*. *Image* NASA/JPL

disappear. Further increases in windspeed caused the 'waves' (they may not meet the definition of dunes any more) to grow longer, to 27 cm for a windspeed of 1.35 m/s. At 1.5 m/s and higher no dunes or waves were formed.

Greeley et al. suggest that at the high wind speeds, particles are able to jump across the lee of a wave onto another, such that there is no longer a clear separation between zones of deposition and removal of sand, and thus the slip faces are lost. The transfer of material from one structure to another resembles that in ripples, but the length of the saltation path may be rather higher (compared to the wavelength of the structures) than is the case for terrestrial ripples.

Greeley et al. also found that a run of 90 s of wind at 1.25 m/s removed as much material as a 5 h run at 0.6 m/s, near the threshold. Thus occasional 'high' winds (and 1 m/s winds had been measured at the Venusian surface by the Venera landers) could quickly obliterate small dune forms.

Although the Venus conditions replicated in the tunnel are of most interest in understanding bedforms on Venus itself, the range of pressures (and thus atmospheric density) that can be explored in the tunnel is useful in understanding aeolian processes more generally. Marshall and Greeley (1992) mapped out the parameter space of windspeed, atmospheric density and particle size, and found distinct regions of behavior.

**Fig. 14.4** Magellan radar mosaic of the Algaonice dunefield. The dunes form a roughly V-arrangement in the center of the image, just above the circular pts. More obvious than the small, bright dunes themselves are some background volcanic features and wind streaks that permeate the region. *Credit* NASA/JPL

## 14.5 Aeolian Transport Under Venus Conditions: Theory

The formulae for predicting the windspeed threshold of motion of particles can of course be applied to Venus conditions and, unsurprisingly, given the much denser atmosphere than any other world, the required threshold windspeed is low (see Chap. 4). Experiments in the Venus wind tunnel largely bear out the predictions (see Fig. 14.9).

The thick atmosphere essentially forces the momentum of a grain to rapidly couple to that of the freestream airflow. White (1981) predicted the shape of Venusian saltation trajectories using relatively simple equations of motion, and found that for the likely 1–2 m/s surface winds, saltation trajectories would be likely to be only 2–8 mm long, with similar height. Under the assumption (common at the time) that ripple length scales might relate to the saltation length, White (1981) predicted that ripple wavelengths would be small—likely too small to affect radar remote sensing by Bragg scattering.

Perhaps counterintuitively, the saltation flux on Venus under dynamically similar conditions (same ratio of friction speed to threshold) would be 10 times less than on Earth (White 1981). This is largely because the threshold speeds are so low on Venus, and flux scales roughly as the third power of friction speed. Measurements in the Venus wind tunnel appear to support these predictions (Greeley et al. 1984a).

As discussed in Sect. 4.6, Claudin and Andreotti (2006) have advanced a scaling theory for the scale of an 'elementary' duneform, the wavelength at which a flat bed of sediment will destabilize most quickly. This scale $\lambda$ is found to be $\lambda \sim 53(\rho_s/\rho_a)d$ wherein a particle of size d, density $\rho_s$ is accelerated to a velocity approaching that of the wind in air of density $\rho_a$. In the Venus atmosphere, where $\rho_a$ is quite large, this scale is rather small. Specifically, for particles of $\rho_s \sim 3000$ kgm$^{-3}$ and d $\sim 0.1$ mm $\lambda$ is predicted to be $\sim 22$ cm. This is in rather good accord[2] with the experiments by Greeley in the Venus wind tunnel which ranged from 8 to 27 cm, depending on windspeed.

At present, little is known about the Venusian boundary layer and so the success of the theory that the size of dunes is limited by the boundary layer thickness cannot be assessed. However, if it is to hold, the $\sim 1$ km wavelength of the Fortuna-Meshkenet dunes is the benchmark to be met.

## 14.6 Venus Winds

Venus has a thick, torrid atmosphere. Although winds at altitude sweep the cloudtops rather swiftly in the sense of the planet's rotation (much as there is a superrotating flow

---

[2] One must be careful to avoid circular reasoning—the Venus wind tunnel data was one of the datapoints used to compute this relationship.

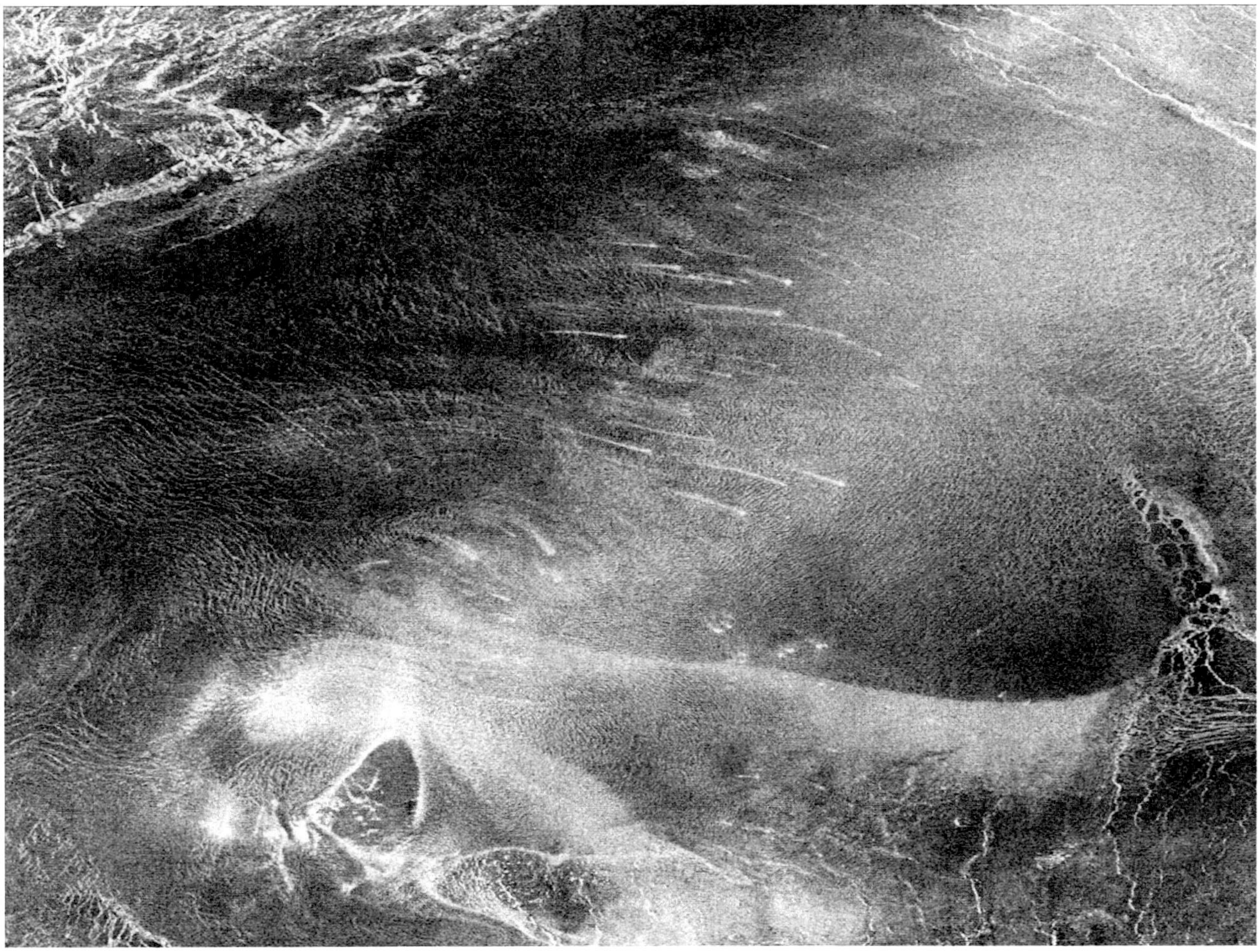

14.6 Venus Winds

**Fig. 14.5** Magellan radar mosaic of the Fortuna-Mesknet dunefield. These appear rather more extensive than the Algaonice dunes. Similar to those, the dunes appear to be transverse, orthogonal to the prominent windstreaks. *Credit* NASA/JPL

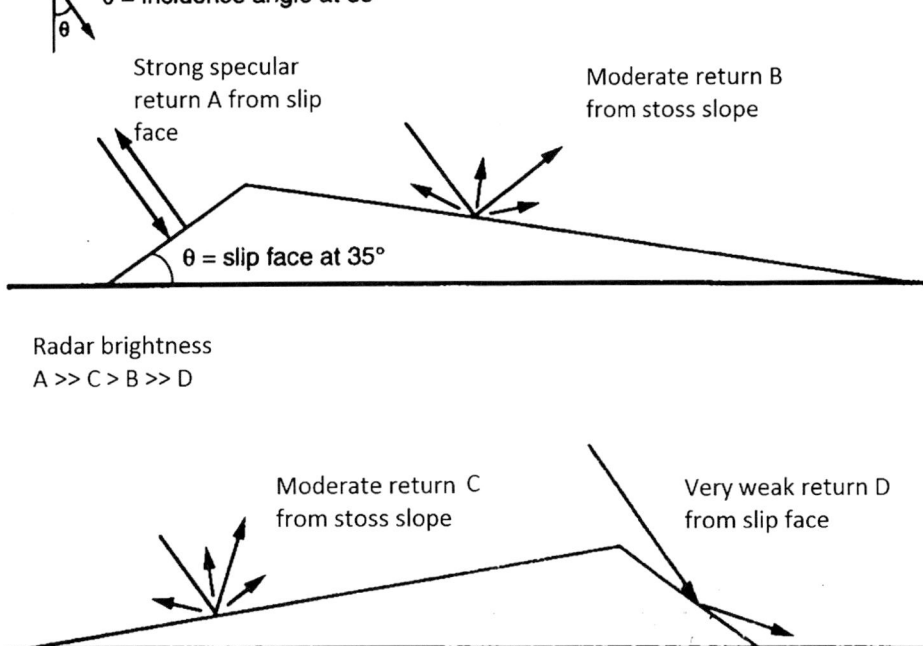

**Fig. 14.6** The asymmetric slopes of microdunes lead to enhanced brightness when viewed from the slipface side than when viewed from the stoss side, an effect that is apparent even when the dunes themselves are too small to be resolved. See also Fig. 18.22

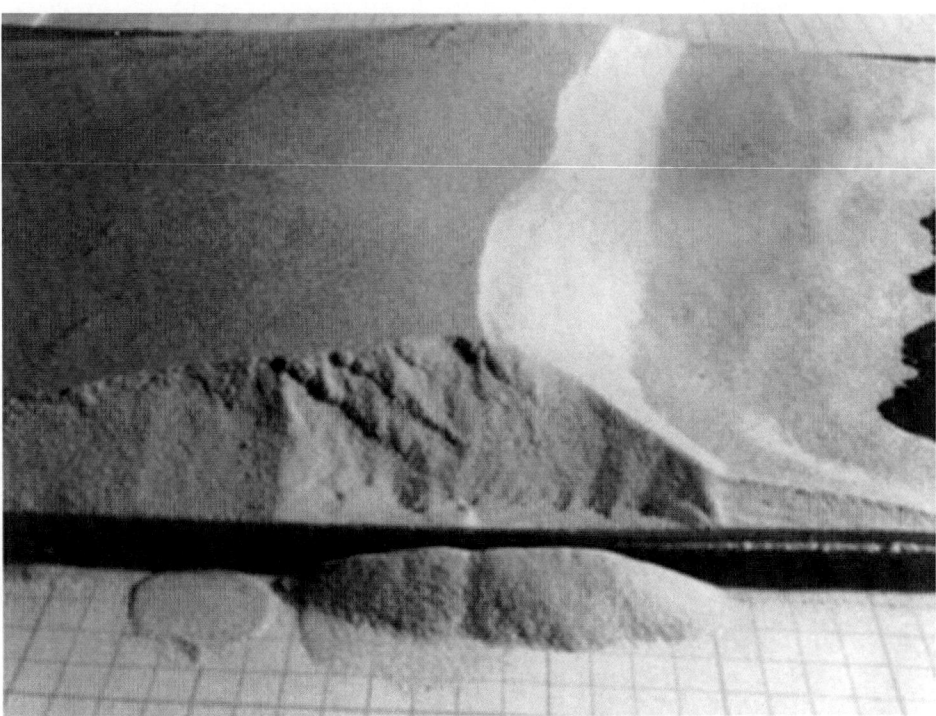

**Fig. 14.7** Close-up view of a microdune formed in the Venus wind tunnel (see also Sect. 4.6). The bedding planes in the sand are visible, as is the steep slipface. *Image courtesy* of Ron Greeley

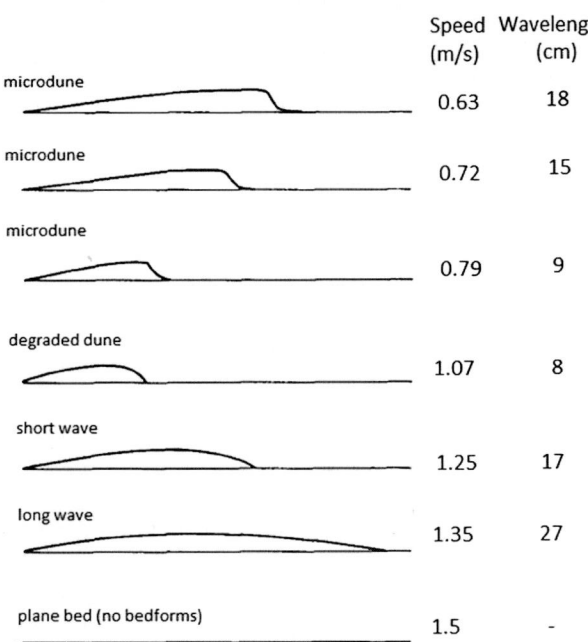

**Fig. 14.8** Schematic (adapted from Greeley et al. 1984b) showing the progression of bedforms in the Venus wind tunnel at Venus densities as a function of speed (see also Fig. 4.18). At the lower windspeeds, sand is pushed up the stoss slopes and avalanches down to form a slipface. At higher speeds, the slipface disappears and a wave or ridge morphology appears, lengthening as windspeed increases. At higher winds yet, the waves become shallow and a plane surface results

in Titan's upper atmosphere), the near-surface winds are almost unknown except for a handful of radio tracking measurements of descent probes, some brief wind measurements on the surface, and indirect indication of winds such as the orientation of wind streaks.

Venus orbits the Sun in 224.7 terrestrial days, but because of the planet's slow retrograde rotation, a solar day (i.e., the time between successive solar noons, and thus the period with which atmospheric motions are forced) is 116.8 terrestrial days. The planet's orbital eccentricity is only 0.007, and the obliquity is 177.4° (i.e., a tilt of 2.6° with retrograde rotation), so seasons are effectively nonexistent. The cloud-tops can be tracked, especially at ultraviolet and infrared wavelengths, and the predominant motion is zonal, as noted above. The uneven deposition of sunlight drives a slow meridional (Hadley) circulation, and the angular momentum balance of this circulation is a key problem in Venus meteorology. Because the atmosphere is so massive, it has a large inertia or 'memory', which makes it rather challenging to simulate numerically.

Dobrovolskis (1993) suggests that slope winds on Venus may be significant, as on Mars, and that such winds may be strong enough to transport sand. This picture does seem to be borne out in part by the Magellan mapping of some 5700 wind streaks (Greeley et al. 1995) which show a preference in the downhill direction where slopes are observed (about 20 % of the total being biased downhill), even though the

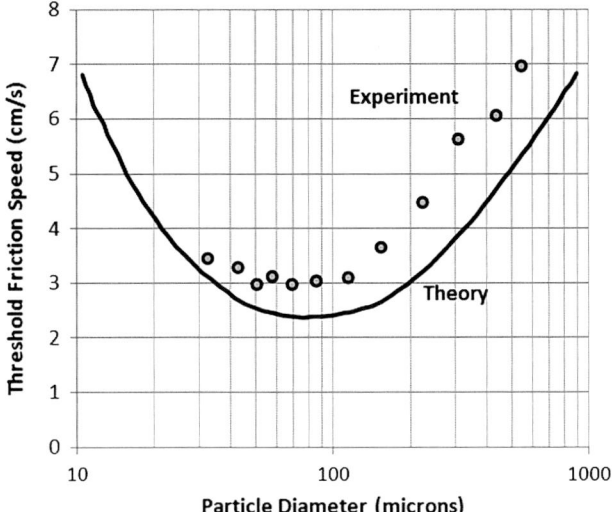

**Fig. 14.9** Wind tunnel data for Venus conditions (gravity is 'faked' by using lower density rock, although the effect is in any case small) showing the optimum size for saltation. The experiments suggest a slightly higher threshold than theory would predict. This friction speed is an order of magnitude lower than that required on Earth. Data from Greeley and Iverson (1987)

slopes themselves are typically only ~0.3°. They also argue that a Hadley-type flow may control many of the wind directions inferred from streaks, although their maps of streak orientation do not make this obvious. Most likely local terrain may modify directions substantially.

There are sadly very few surface windspeed measurements, only those by Venera 9 and 10 in 1976. These carried cup anemometers able to operate at 500 °C. The Venera 9 wind record spans 49 min (Avduevsky et al., 1977) and that of Venera 10 only 90 s, sampled at a 0.4 Hz frequency with a resolution of the order of ~0.2 m/s. The Venera 9 and 10 records have been characterized as having a means of 0.4 and 0.9 m/s respectively, with standard deviations of 0.1 and 0.15 m/s (Golitsyn 1978). Surface winds were also estimated from the intensity of sounds recorded by the Groza microphone on Venera 13 and 14 to be in a similar range (Ksanfomality et al. 1983). These data do seem to show winds within a factor of 2 or so of the saltation threshold, even for a small number of observations, so aeolian transport may be common. However, more data are sorely needed.

## 14.7 Venus Sand

The expectation, given that Venus' relatively low population of craters suggests it has been resurfaced—principally by volcanism—is to give an effective surface age of ~500 Myr. Whether that resurfacing was a global cataclysm ~500 Myr ago, or was a more protracted process that yields this effective age is still under debate, but there seems little doubt that the Venusian surface is dominated by basalt. Without liquid water on the surface, more evolved igneous rocks are not produced (although perhaps outcrops of such, formed early in Venus' history, may be present).

In the absence of fluvial processes (or glacial abrasion, or freeze–thaw), it is hard to imagine how to make much sand. Similarly, the high atmospheric pressure on Venus prevents gas in magmas from expanding dramatically to produce Strombolian eruptions that erupt a lot of fine ash—most of the lava will have simply emerged from the ground and flowed. Some chemical weathering processes may produce some fines, and impact ejecta also produces fine particles (dark parabolic haloes around some impact craters may be radar-dark because they are made of smooth deposits of fine-grained material—these appear around the apparently youngest craters).

Regardless of how sand may be produced, there are questions about how materials behave under unfamiliar conditions (questions much like those which confront Titan). Specifically, is basalt at 500 °C hard enough to behave as sand?

Marshall et al. (1991) studied the adhesion of basalt particles at a range of temperatures, noting that rocks may be appreciably softer at Venus surface temperatures. A small jet of air, pulsed by a valve at roughly once a second, was used to propel ~3 mm angular particles of rock at a polished target slab at known speeds inside a chamber.

They found that at temperatures below 440 K, abrasion occurred as expected. Results were highly variable between 440 and 570 K: above 570 K particles almost always accreted (i.e., welded themselves onto the target). This adhesion threshold of about 500 K is roughly 40 % of the melting temperature and is higher than temperatures even on Venus' highest coolest spot (Maxwell Montes, ~650 K) and so grain adhesion may well be an important aspect of aeolian processes on Venus.

## 14.8 Future Exploration of Venus

Post-Magellan exploration of Venus has been dominated by atmospheric studies from two missions, Venus Express and Akatsuki. Venus Express is a mission launched by the European Space Agency in 2005, and has been operating at Venus since 2006. However, its instrumentation (with the exception of a handful of isolated radio science and infrared observations) does not address the surface. Akatsuki (formerly 'Venus Climate Orbiter' or 'Planet-C') was launched successfully in 2010, but suffered an engine failure during its attempted Venus orbit insertion (which may be re-attempted in 2015).

Future Venus exploration (Bullock et al. 2009) is likely to include platforms able to return image data on spatial scales 1–2 orders of magnitude better than Magellan, employing either higher-resolution (possibly interferometric) radar from orbit, or near-infrared imaging from a balloon near the surface. Such future missions will likely reveal many more duneforms than are presently known, and would likely reinvigorate aeolian studies on Venus.

# Other Dune Worlds

15

The last decade has seen a transformation in our understanding of the abundance of planets in our galaxy. From a handful of known exoplanets in the 1990s, the number of confirmed candidates (albeit mostly gas giants) is approaching 1000. Some of these planets have rocky surfaces, and it seems certain that our galaxy, and indeed the universe, is teeming with worlds on which dunes may exist. However, for the forseeable future, we can learn little about these planets and their moons and must content ourselves with our own solar system.

Beyond the terrestrial planets and Titan, which we have already covered, two moons in our solar system are presently known to have tenuous (and in fact rather variable) atmospheres that may be thick enough to transport particulate material: Triton and Io. We will discuss them briefly in this section in order of decreasing pressure; Triton also serves as a prototype for Pluto and perhaps other similar worlds in the chilly Kuiper belt of bodies at the edge of the solar system. However, as we discuss, it is at best marginal to think of these as 'Dune Worlds'. We may note that the acceleration of solid particles by gas is a known (and spectacular) process on comets, and on Saturn's moon Enceladus, but no details of horizontal gas flows are known and thus, even if these fell under the definition of aeolian transport (and they may not), there is nothing at present to say.

It is conceivable that other worlds (such as the Galilean moons) may have had atmospheres soon after they formed but which are no longer present. If this is the case, aeolian transport might have occurred, but there seems little hope that traces of such processes are preserved to an observable extent. Similarly, aeolian processes can occur in transient volcanic 'atmospheres' (witness, say, the bedforms generated in ashfall deposits during maar eruptions on the Earth) in which case perhaps some ripples or related features may have formed even on our own moon. Again, however, there is little more that can be said than to note the possibility.

## 15.1 Triton

Triton is the largest moon (radius 1353 km) of Neptune, and revolves around its planet in just under 6 days. Its density suggests it has a rock or rock/iron core, but substantial amounts of water and carbon dioxide ice. Unusually, its orbit is retrograde, suggesting perhaps that Triton is a 'captured' moon, rather than having formed in a protoplanetary disk. The orbital arrangement and Triton's rotation lead to very large changes in the amount of sunlight on different parts of Triton (its situation resembles the planet Uranus, where during the course of a year the sun passes nearly overhead each pole), with the result that there may be substantial seasonal migration of volatiles.

Triton has an atmosphere that is quite thin, predominantly of nitrogen. The surface pressure determined by the Voyager 1 radio occultation experiment was a mere 14 microbar: the gas appears to be in vapor pressure equilibrium with nitrogen frost on the surface; this frost makes Triton's surface very bright and thus it has an exceptionally low surface temperature—about 40 K.

Aeolian transport on Triton was considered by Sagan and Chyba (1990). They found, given the pressure conditions and the low Triton gravity of 0.78 m/s$^2$, that even an atmosphere as tenuous as Triton's 16 microbars could lift 5 micron or smaller particles into suspension if cohesion between particles is small. They noted that, even though the Triton surface pressure is 400× smaller than Mars', the low temperature on Triton (6× less) mitigates the atmospheric density by that factor; also the Triton dust would be likely organic (as on Titan) and thus less dense than Martian rock. These factors, together with the low gravity, mean that the threshold friction speed on Triton would be only ~2.5 higher than on Mars.

Measurements of the atmosphere by stellar occultation in 1997 showed that the atmosphere had in fact thickened somewhat (at these conditions, even the observed small rise in surface temperature of ~2 K can increase the vapor

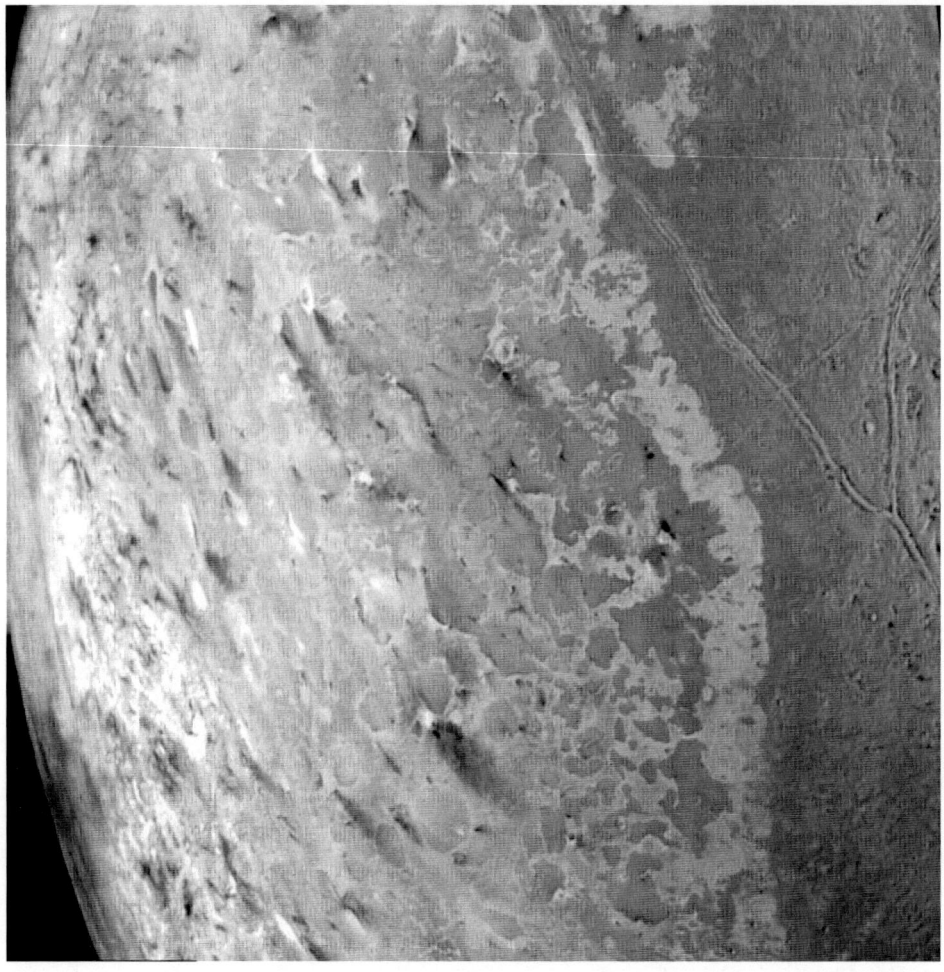

**Fig. 15.1** Voyager 2 image of the surface of Triton. Many dark streaks can be seen, all pointing at around 5-o'clock, suggesting transport by a regional wind pattern. The extent to which the material merely drifts downwind while settling out from a vertical plume, versus being transported by saltation along the surface, is not known. *Image* NASA

pressure substantially, and only a small change in heat flux can change the surface temperature). The changes are likely caused by the elliptical orbit of Neptune around the sun, and thus further change (unobserved at present) has likely occurred.

Groundbased spectroscopy (Grundy et al. 2002) suggests that nonvolatile materials are widely distributed on Triton: the fact that such materials are evident even though sublimation and condensation of nitrogen and methane should seasonally deposit 'clean' frosts may imply that winds can redistribute finely-grained material quite quickly. The material should not be produced fast enough by photochemical processes to be detectable in seasonal frosts (Fig. 15.1).

The particles appear to derive from point sources, likely associated with eruptive plumes (e.g., Hansen et al. 1990; these probably solar-driven plumes have been sometimes referred to, almost certainly incorrectly, as 'geysers') in which case the streaks are more likely just the downwind fallout of dust blasted into the sky by the plumes, rather than being saltated along the ground and forming dunes or ripples. There is, however, no observational evidence for or against the existence of bedforms.

While no near-term exploration of Triton is expected the New Horizons spacecraft, launched in 2006, is set to fly past Pluto and its moon Charon in 2015. Pluto may in some respects resemble Triton, although the best imaging resolution of Pluto will likely be of the order of 50 m/pixel, so dunes—if present at all—would have to be large to be detectable. On the other hand, the saturation length in a thin atmosphere, and thus the expected bedform length scale, is large. Another moon, Io, offers a tentative example.

## 15.2 Io

Jupiter's innermost large moon, Io, has a constitution rather similar to the terrestrial planets, with iron, sulphur and silicates. It is in an orbital resonance with Europa and Ganymede, that maintains a substantial eccentricity to its orbit

**Fig. 15.2** The margin of the lava flow field associated with the Prometheus volcano on Jupiter's moon Io is seen in this 12 m/pixel image, acquired by the Galileo spacecraft on February 22, 2000. The dark lava has margins similar to those formed by fluid lava flows on Earth. This entire area is under the active plume of Prometheus, which is constantly raining bright material. The older plains (*upper right*) are covered by ridges with an *east–west* trend. These ridges may have formed by the folding of a surface layer or by deposition or erosion. Bright streaks across the ridged plains emanate from the lava flow margins, perhaps where the hot lava vaporizes sulphur dioxide. The bright material must be ejected at a low angle because it only coats the lava-facing sides of the ridges. North is slightly to the *right* of straight up. *Image credit* NASA/JPL/University of Arizona, Photojournal image PIA02557

around Jupiter. This eccentricity leads to strong tidal forces which deform the world slightly and this tidal kneading generates appreciable heating of Io's interior, with the result that Io is volcanically very active. This activity was discovered by Voyager 1 in 1979, when a large fountain-like plume was observed on Io's limb. Further Voyager and Galileo observations have found many active volcanic centers, some with exposed silicate lavas with temperatures in excess of 1400 K, others merely sulphur dioxide plumes jetting into space before falling back down (as sulphur dioxide frost).

Io's atmosphere is predominantly of sulphur dioxide, but is very tenuous indeed. Spectroscopic measurements from Earth show that the pressure on the sub-Jovian hemisphere is only about a tenth of a nanobar, although the same data show that the antiJovian hemisphere has a pressure of several nanobar. This large discrepancy can be sustained because the gas can condense out onto the surface before it is transported around the moon. The discrepancy suggests some uneven supply of gas—perhaps a greater supply from volcanic plumes, which are also more abundant on the antiJovian hemisphere, or perhaps more abundant or warmer frost deposits. This spatial variability might similarly imply that temporal variability can occur too.

Aeolian transport in such a tenuous atmosphere seems unlikely. However, the 'atmosphere' in the immediate vicinity of volcanic plumes might be substantially thicker, and while particle transport here may severely stretch the definition of 'aeolian', it cannot be ruled out entirely and is instructive to consider.

Figure 15.2 shows sulphur dioxide frost that appears to have jetted horizontally (perhaps from a frost deposit overrun by silicate lava) that highlights the rippled texture of the ground. This rippled texture may simply be tectonic, but the possibility of aeolian deposits cannot be ruled out. The ridge spacing is 100–200 m: it seems possible to contrive scenarios (using the drag length or saltation path arguments discussed in Chaps. 4 and 5) wherein such wavelengths are consistent with particle movement in a thin atmosphere.

Future missions to Io are being considered in NASA's planning, although would not arrive until the mid 2020s at the earliest. We may hope that such missions have the capability to explore these features in more detail.

# Part IV
# Dune Studies

In this section we review the various methods and approaches used to study dunes. A staggering range of scientific techniques have been brought to bear in the study of dunes and the sand and wind that create them. We have arranged these in four broad categories: field investigations, laboratory investigations, remote sensing, and models. In all these areas there has been remarkable progress has been made in the last couple of decades.

Inevitably, there are some inconvenient borderline cases where particular methodologies may transcend this somewhat arbitrary classification; for example, is a ground-penetrating radar better discussed in the context of other radars (remote sensing techniques) or in the context of other field methods? Similarly, wind tunnel measurements are most usually laboratory techniques, but portable field wind tunnels have also been used. We crave the readers' indulgence in recognizing that everything has to go somewhere, and we hope they can find what they seek.

# Field Studies

# 16

Sand dunes are an outdoor phenomenon, and one of the joys of dune research is to go out into the field, tramp around and over dunes, and study them closely. Here we review some of the field research techniques that have been and are employed. Crudely speaking, these techniques may be gathered in four categories: the study of sand, the study of shape and structure, observation of wind, and the study of aeolian processes.

We close with some discussion about the special challenges of making field observations with robot proxies on other worlds. As discussed by Clancey (2012), this practice challenges our somewhat arbitrary distinction in Chaps. 16 and 17 of field and laboratory work. While Mars rovers have acted as surrogate field geologists, moving around, observing, and occasionally manipulating the environment, they also carry sophisticated instrumentation that is more traditionally retained in a laboratory on Earth. Nonetheless, in the spirit of exploration, we discuss them in this section.

## 16.1 Sand Measurements in the Field

Although the more elaborate investigations of sand composition require a laboratory (see Chap. 17), some basic properties are easy to evaluate in the field. The use of a set of sieves can determine the size distribution, which can also be crudely gauged by close examination with a microscope or even a hand lens. Particle shape, and the heterogeneity of sand (how much quartz vs dark minerals vs shell fragments) can be similarly estimated by eye, or more quantitatively by spectroscopy or color imaging (see Chap. 18). As noted in Fig. 5.10, the presence of magnetic minerals is readily indicated by using a permanent magnet. Some laboratory techniques have become sufficiently miniaturized to employ in the field—hand-held X-ray fluorescence instruments, much like those used on spacecraft, are now available, for example.

Even where sand is to be merely sampled in the field, in order to permit closer study in the laboratory, elaborate protocols may be required. For example, samples for OSL dating require that samples from depth be acquired without exposure to sunlight, so stainless steel tubes which can be sealed in a light-tight manner are driven into the dune.

A variety of field instruments exist for measuring the cohesion of soils. A common one is the cone penetrometer, a device driven or hammered into the ground and the force needed to do so is measured (or, equivalently, the number of calibrated hammer blows to drive a cone a given distance is recorded). The Apollo astronauts used such an instrument on the moon. Another measurement is the shear vane, a cruciform set of plates which are turned in the ground.

A property of interest is the saltation threshold—the windspeed or shear stress at which material can be lofted from the ground (this is perhaps studied more in the context of dust than sand). This can be measured in the field with a portable wind tunnel, an approach pioneered by Bagnold (see Chap. 17).

A more convenient apparatus is the PI-SWERL (Portable In Situ Wind Erosion Lab) developed at the Desert Research Institute (e.g., Etyemezian et al. 2007). Essentially, this is a chamber placed over the ground, with a spinning blade to develop shear (not unlike a lawnmower). The wind shear applied to the surface can be related to the blade speed, which is controllable, and air is sucked out of the chamber through an optical dust counter, so the dust emission as a function of shear stress can be quantified. The chamber is easily moved from spot to spot to assess the heterogeneity of the dust emission threshold.

Some interestingly ad-hoc methods of estimating soil strength or saltation threshold have been tried. Li et al. (2010) report a correlation of friction velocity with measurements of the size of crater produced when shooting pellets into the ground with an airgun.

**Fig. 16.1** Examining the internal structure of a dune (to expose bedding—see Chap. 25) the brute-force way. A bulldozer digs a trench across a dune at White Sands National Monument. USGS photo

## 16.2 Dune Shape and Structure

It is its shape that defines a sand dune, and therefore quantitative measurement of the shape is instrumental in the study of dunes and their evolution. In the early days, measurement of shape was performed using the traditional methods of surveying with a plane table or theodolite to measure positions by triangulation, and chains to measure length. It is sobering in this satellite age to recall that in the 1800s continental-scale areas of our planet were surveyed by teams of determined men battling the terrain and disease: despite their primitive equipment, their impressive precision not only permitted the accurate mapping of, for example, India, but also the realization that our planet is not spherical, but is bulged at the equator. At a more local scale, these geometric methods have still been used well into the late 20th century, A slightly less accurate device for measuring angles in the field, but widely used because of its convenient portability, is the Brunton Pocket Transit (often referred to as a Brunton compass), much beloved of geologists. New electronic sensors are beginning to surpass this traditional sort of instrument.

While remote sensing (see Chap. 18) now allows wide areas to be uniformly surveyed, shape measurements of higher resolution and/or precision are still needed in many cases and can be performed by GPS, field imaging and laser methods. Additionally, in the field it is possible to probe the internal structure of dunes, by digging (Fig. 16.1) and by ground penetrating radar and other geophysical methods.

### 16.2.1 GPS

The modern miniaturization and wide availability of GPS equipment have made the terms GPS and 'satnav' more or less synonymous with navigation itself. Satellite navigation more generally was developed (notably the US Navy Transit satellites built by the Applied Physics Laboratory, using purely Doppler methods) soon after Sputnik, but the particular implementation of the Global Positioning System using satellites in high orbits with very precise clocks began in the 1970s, becoming fully operational in 1994 (although it had been an important facility in desert navigation in the 1991 Gulf War). The arrival time of signals received by a

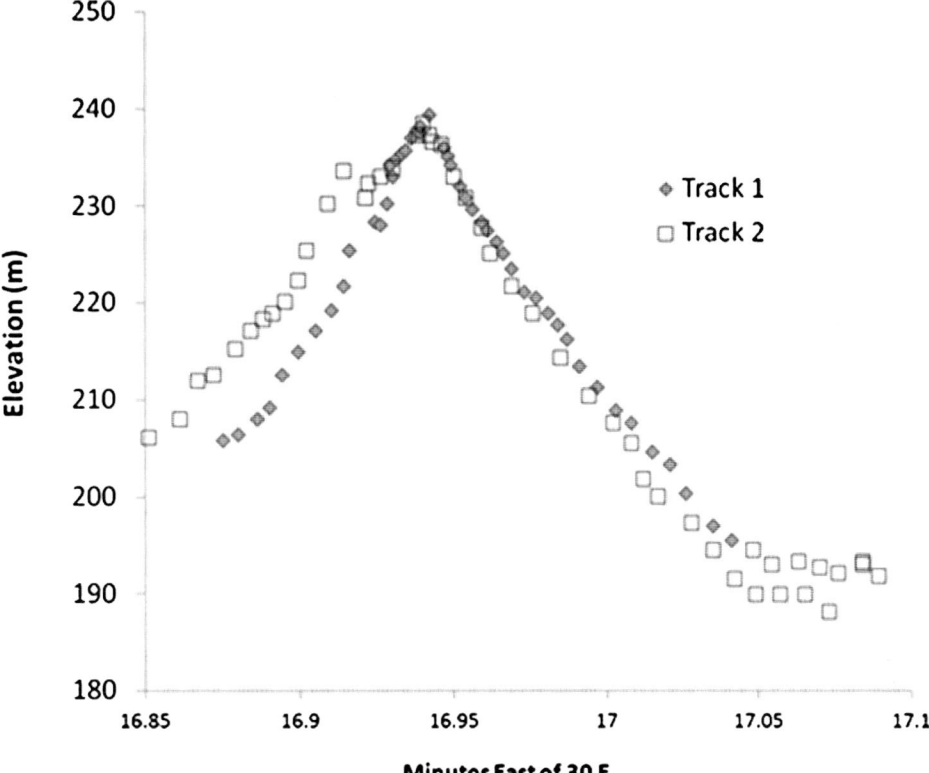

**Fig. 16.2** Two adjacent GPS traverses across a linear dune in Quattaniya, Egypt (see also Figs. 6.21 and 16.10) made using a simple handheld GPS receiver. The traverse is about 300 m long; the dune grades into a wide shallow plinth, not shown here, to the west (*left*) and is about 50 m high

user depends on the distance between the receiver and each satellite, so each time measurement defines a spherical shell in space where the receiver must lie. Having fixes from several satellites defines several shells, and the intersection of these shells defines a small volume (usually only a few meters across) in which the receiver must lie. All this elaborate calculation was performed by computer in back-pack-sized units in the 1980s, which progressively shrank (more or less in parallel with the comparable technology in mobile telephones) to handheld units in the 1990 and 2000, and then mere thumbnail-sized devices incorporated in phones and other appliances in the 2010s.

In the 1990 and 2000s, when GPS units became available and affordable for scientific use, the out-of-the-box accuracy was often poorer than some tens of meters, since the 'native' precision of the signals was restricted to US military users, a policy termed 'Selective Availability'. However, in 2000, the SA policy was revoked, allowing accuracies of a few meters to be obtained with only a single handheld receiver. This has made mapping the boundaries of geological features like dunes, or measuring their profiles by marching or driving across them, almost trivial.

Rather effective shape models of dunes even only a few meters high can now be constructed with simple handheld GPS units (which can be set to record coordinates at short intervals, and can then be downloaded to a computer— e.g., Fig. 16.2.). The accuracy of these may vary, but can be just a few meters (and except for the measurement of dune migration, it is generally the case that the relative positions of the slipface, crest, etc. are more important than the absolute position in some reference frame). Convenient tools now exist (notably Google Earth) to overlay GPS tracks on satellite imaging (see Fig. 9.5).

For higher precision (centimeters) a technique called differential GPS must still be used (it had already become popular to circumvent the limitations of the SA policy), wherein a fixed base station and a mobile survey unit simultaneously record fixes. Since many of the errors (such as those induced by the propagation of the signal through the ionosphere) are common to both receivers, they cancel out when the position of the survey unit relative to the base station is calculated. Because the accuracy is so high, the antenna of the survey unit is often mounted on a staff so that its distance from the ground can be precisely controlled as the surveyor walks around; a backpack (Fig. 16.3) is a more convenient and often adequately accurate solution.

### 16.2.2 Imaging

Dunes are naturally photogenic targets. G.K. Gilbert's photographs from the 1890s (a number of which appear in this book) are striking were largely enabled by the innovation of photographic film, which let the geologist take his own pictures (rather than needing a dedicated photographer along, as had been the case only a decade or two before

**Fig. 16.3** The second author, with a precision GPS (antenna sticking vertically out of backpack). Some ripples are visible at left. Note the gloves—it is in fact quite cold. Great Sand Dunes National Park and Preserve. Photo by Cheryl Zimbelman

when photography required glass plates). Much of Bagnold's field work relied on photography as a means of data acquisition—not only of dunes, but also of the paths of saltating grains, and the instantaneous measurement of wind pressures by taking photography of manometer columns.

We have recently seen a new revolution in imaging—the digital one. Storage of data in image form is, of course, made vastly easier by modern digital cameras, which are not only inexpensive but can store thousands of images or hours of video on small memory cards. The marginal cost of acquiring an image, both in time (point-and-shoot cameras take only a couple of seconds to pull out of a pocket) and in media costs is so close to zero that the field researcher can essentially take an unlimited number of pictures. Modern image processing tools allow images to be compared, or mosaiced together, or features measured precisely.

That said, taking some care can be rewarded in the field. Near-overhead illumination tends to wash out topographic shading but may emphasize any compositional sorting (like patches of red spherules in the Rub' Al Khali, or black magnetite-rich patches in Death Valley) whereas sensuous shadows develop near dawn and dusk, emphasizing the rounded curves of a dune.

Dunefields often lack an intrinsic scale, so inclusion of reference objects for scale is useful—a backpack, a colleague or a vehicle. For studying ripple topography, a handy trick is to mount a horizontal edge like a ruler just above the ripple pattern, and measure the shape of the shadow of the edge cast onto the ripple surface.

It is often useful to structure the illumination of a scene, either in space or time or both, in order to study saltation, although this is more usually done in the laboratory than in the field. A long exposure of a closeup of a saltating dune may just show a blur where the sand is in motion, yet a flash of suitable length will show the streaks of individual sand grains.

A new avenue for photography in aeolian studies is the use of photogrammetry to construct 3-dimensional models of a natural surface; software packages are now available via the internet (sometimes for no charge) where a series of individual digital photographs can generate surprisingly robust 3D models, as long as the surface photographed includes sufficient textural information to allow the software to identify abundant tie points. As an example, a series of nine photographs taken while walking around a megaripple field near Grand Falls in central Arizona was able to result in a good 3D model of the ripple (Fig. 16.4) when submitted to the 123D Catch web site (Autodesk 2012); no geometric information was required to be logged in the field (other than including something for scale in the photographs), since the software automatically reconstructs all camera positions while generating the surface model. Utility of this new photogrammetric technique, however, may be restricted to sites where the photographed surface has either sufficient textural or albedo diversity to provide abundant tie point opportunities, which is not always the case on many natural dune surfaces.

Another new capability is afforded by inexpensive digital timelapse cameras that can take thousands of images over a period from a few hours to many months. These may be

## 16.2 Dune Shape and Structure

**Fig. 16.4** Screenshot of an auto-photogrammetry program (Autodesk's 123DCatch) showing the image texture draped over a digital model of the shape of some ripples; the model can be spun around to be viewed in different directions, or exported for numerical analysis. The positions of the camera that have taken the images are automatically estimated and are shown at the *top*. A Brunton compass, in homage to earlier methods, is used to show the scale of the ripple. *Photos* J. Zimbelman

used to document changes in dune shape, or in the position of ripples. So far, no movies of dunes in motion have been made, but it is likely just a matter of time.

Ripple migration has been observed with timelapse imaging in Egypt by Lorenz (2011) and at Great Sand Dunes National Park and Preserve, e.g., Fig. 16.5, by Lorenz and Valdez (2011). These papers describe some image processing techniques that can help sift through the prodigious amount of data that movie sequences can generate. These observations were used to estimate ripple migration rates shown in Chap. 9 from about 1 to 15 mm/min. Sullivan et al. (2008) report measuring the migration of ripple crests on Mars by 2 cm towards the Spirit rover at Gusev crater between sols 1260 and 1265 during a particularly windy period in 2007 (the first time movement of whole bedforms was documented by a surface vehicle—see Fig. 16.15).

Finally, although this verges on aerial photography (see Chap. 18), it is possible—indeed, now rather easy, given the availability of very small digital cameras with timelapse or video settings— for scientists in the field to obtain images from a vantage point tens or hundreds of meters up by using kites (Fig. 16.6; see also Figs. 6.12, 6.21, 6.25 and 11.15) or remote-controlled planes or helicopters. These are rather fun to use, and make a useful bridging scale between the 1–2 m height of field photographs and the distant perspective of true aerial or orbital imaging. Aber et al. (2010) discuss at length the possibilities and techniques of small-scale aerial photography from kites, drones and balloons. An exciting possibility is the combination of kite imaging with the new photogrammetry tools mentioned above to build shape models of dunes quickly in the field.

### 16.2.3 Thermal Imaging

We discuss thermal imaging in more detail in Chap. 18, but note here an effect seen in field imaging with hand-held thermal cameras. Avalanching flows can bring sand to the

**Fig. 16.5** Images a couple of days apart of the same spot (an anemometer and several reference stakes 20 cm apart are fixed in the ground). Not only have the ripples coarsened and thinned, but they have become somewhat steeper and more sinuous. These images were taken by a timelapse camera on a fencepost, the shadow of which is visible at *left*. *Photo* R. Lorenz

**Fig. 16.6** An image taken from a kite at about 200 m altitude, showing Sand Mountain, Nevada at *right* (a few dark spots are vehicles). A maze of narrow linear dunes snake away from it to the edge of the sand field; the profiles of these dunes are somewhat rounded due to recreational vehicle activity (see also Fig. 7.2). Four persons (including the kite 'pilot') are just visible on the dune crest at bottom left. *Photo* R. Lorenz

surface from depths of a few centimeters, where (depending on the time of day) it may be warmer or cooler than the undisturbed surface. Thus even when the optical contrast between undisturbed and avalanching sand is tiny, their temperatures are quite different and so avalanching lobes show up well in thermal images (e.g., Fig. 16.7).

### 16.2.4 Laser Scanning

Shape measurements of dunes and ripples can be made with field laser scanners (Fig. 16.8) which measure the distance (to ∼mm precision) to the ground over a specified range of angles from a fixed point. For cm-scale measurements

## 16.2 Dune Shape and Structure

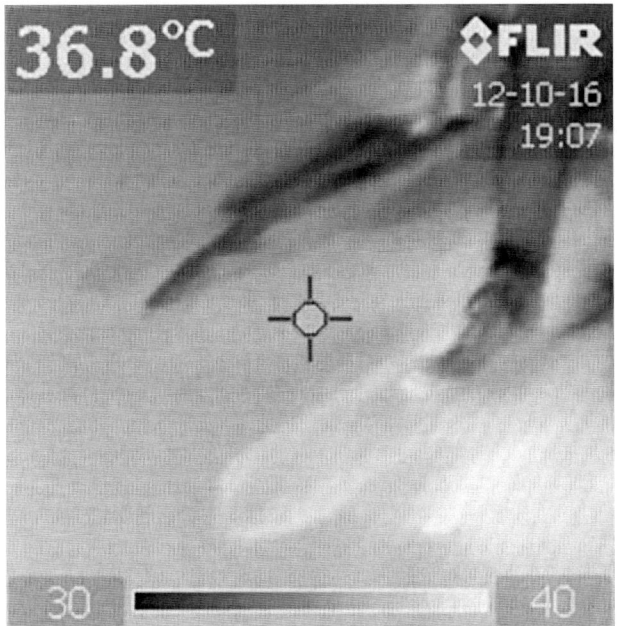

**Fig. 16.7** Warmer and cooler avalanche lobes visible in a thermal image at Sand Mountain, Nevada. Note the feet of two experimenters. *Photo* R. Lorenz

(see Fig. 16.9), this is probably the most accurate technique (similar scanning has been used to quantitatively measure the shape of ventifacts to measure erosion by saltating sand), but involves a certain amount of bulky hardware and an electrical generator. It seems that, for all but a few applications, this scale of measurement may in the future be performed mostly by photogrammetry.

### 16.2.5 Ground Penetrating Radar

Radar methods can measure distances with high precision—this is after all the origin of the acronym Radio Detection and Ranging. Thus the time-domain information of an echo can reveal structure along the propagation direction of the radar waves. This can be exploited to sound the structure within a dune, if there are variations in dielectric constant due to moisture, minerals or porosity. This approach is termed ground-penetrating radar or GPR, and typically involves an antenna on a sled (Fig. 16.10) or cart, or sometimes carried by hand (Fig. 10.11). Radar pulses are sent out directly into the ground and echoes received, the echo time relating to the depth in the target via the speed of light in sand. By traversing the antenna across the surface, an image can be built up, with the horizontal coordinate corresponding to the position of the antenna (estimated from GPS, or sometimes from an odometer wheel on the antenna), and the vertical to the depth. The brightness at each point is estimated from the echo intensity.

Because the beam from the antenna is not typically pencil-narrow, a buried point reflector like a rock or a landmine will provide a reflection even when not in the (usually-vertical) boresight of the antenna. However, the distance will be slightly larger when the reflector is off-axis, and thus the reflection will map to a slightly greater depth. As a result, point reflectors often appear as hyperbolic traces in an echogram.

As with remote sensing radars, the choice of wavelength is important. Longer wavelengths will typically penetrate deeper before being attenuated, but demand physically larger antennas for a given beamwidth. Long wavelengths will be less sensitive to small reflectors like pebbles. Thus the prudent researcher will visit a new site with a selection of antennas, so that whatever is best for the project at hand, and the conditions (notably, moisture, e.g., Bano and Girard 2001) can be chosen.

Probably the most significant findings from GPR studies of dunes were those from one of the first applications of the method—to the structure of the linear dunes in the Namib (Bristow et al. 2000, 2005). These showed reflecting layers consistent with the dune having migrated somewhat sideways, west-to-east, whereas of course the general sand transport is along the south–north direction along the dune.

GPR studies have been made of booming dunes such as Dumont in California (see Chap. 10) to identify layering suggested to be important in the booming process. Some studies of Antarctic (Fig. 16.11) and Alaskan dunes have been reported, where snow, ice and water are important factors in generating reflecting surfaces. Other studies include those of dune/fluvial interaction in Australia (Hollands et al. 2006) and of the internal structure of the Algodones dunes. The internal structure of a barchan is nicely illustrated in Fig. 1.9.

The reader will likely wonder whether the antenna needs to be in contact with the ground. In principle, it need not (and vehicle-mounted GPR systems a few tens of centimeters above the ground are used to detect landmines, for example), but the signal-to-noise is vastly improved if it is, and this need not preclude vehicular operation (Fig. 16.12). The analogy would be looking vertically into water on a bright day with the sun overhead—it is much easier to see the bottom by putting your face underwater than being confronted with the glare from the air:water interface. The air:sand gap similarly produces a major reflection which challenges the dynamic range of the receiver. Airborne radars are able to produce profiles of structures in terrestrial ice sheets and glaciers (which if sufficiently cold and dry are somewhat radar-transparent, like dry sand) but this relies on the layers being detected having a separation rather larger than the amplitude of the surface roughness within the beam footprint. Such measurements have also been made by

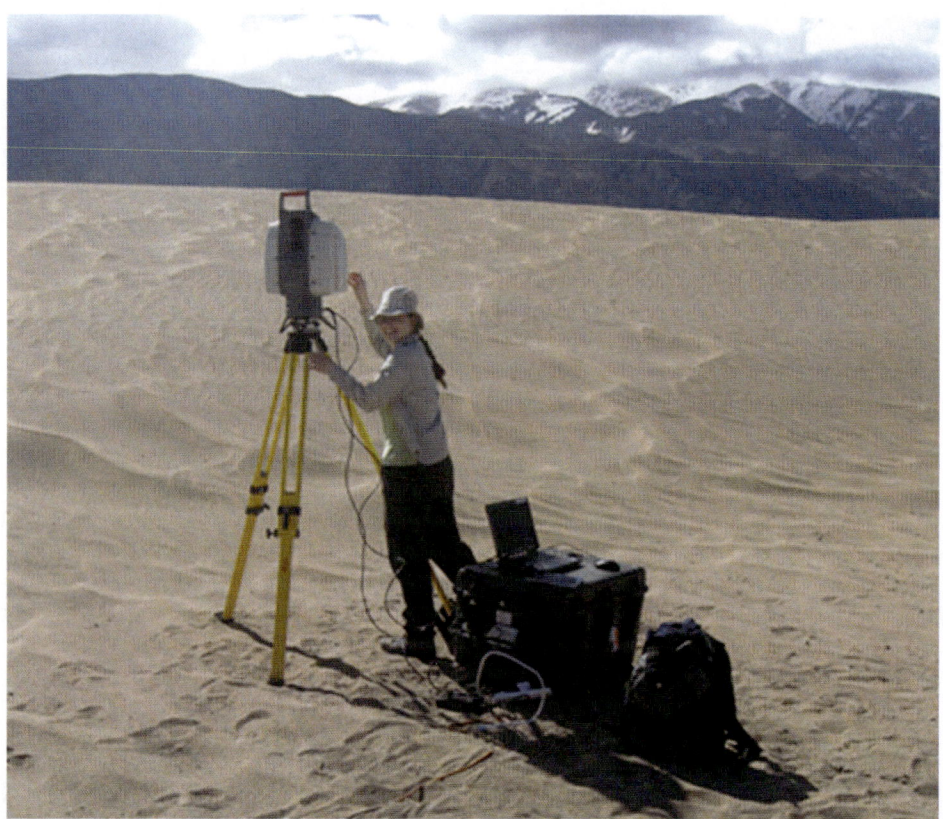

**Fig. 16.8** Jo Nield and a laser scanner being used at Great Sand Dunes National Park in May 2010 to measure the microtopography of aeolian ripples on the inside of a parabolic dune (this same ripple site was observed later that year by timelapse imaging—see Fig. 16.5). Note the laptop for data acquisition; a cable runs off the bottom of the photograph to a generator which also had to be lugged out to the site. *Photo* R. Lorenz

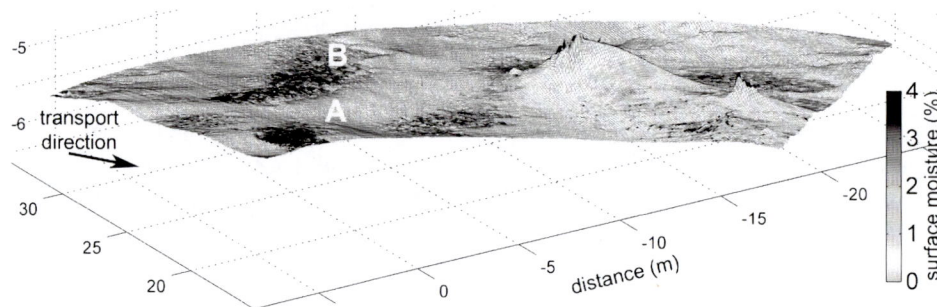

**Fig. 16.9** Surface topography field of coastal dunes measured with a laser scanner. The amplitude of the laser reflection can be processed to estimate the surface moisture. Image courtesy of Jo Nield

satellite radars of the layered ice caps on Mars. While such methods could correspondingly measure the thickness of regolith layers or sand sheets on the large scale, measuring structures within an individual dune from an aircraft would be very challenging.

## 16.2.6 Gravity and Other Geophysical Methods

Another geophysical measurement that occasionally contributes to dune studies is gravity. Insofar as sand has a lower bulk density from solid rock, the surface gravitational

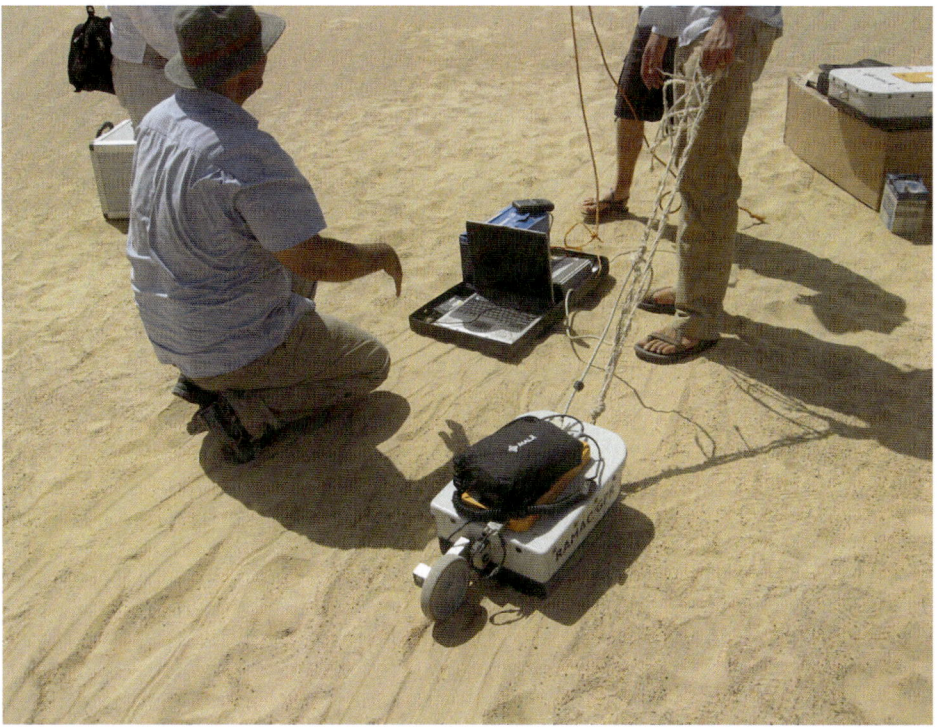

**Fig. 16.10** Ground-penetrating radar equipment. The antenna sled is in the foreground, with the radar electronics in a package on top, and a laptop is being used to record and display the data (were it windy, placing a laptop so close to the ground would invite disaster). Note the wheel behind the sled—this measures the distance travelled along the ground so that the time history of radar pulses can be converted into a spatial profile. *Photo* R. Lorenz

acceleration is subtly lower over a thick layer of sand than over a thin layer. Yang et al. (2011) used gravity data, combined with SRTM topography, to study the highest dunes on Earth (commonly 200–300 m high, with some as high as 450 m), in the Badain Jaran desert. They were able to infer that even though one dune stood some 150 m high above its surrounds, in fact the sand at the crest was only ~100 m thick, the dune standing on a local rise in the bedrock topography.

Seismic methods also can be used to study the internal structure of dunes (e.g., see Chap. 10) although GPR has become far more common. Other geophysical methods like well logging could of course also be applied, but the compositional variations within dunes are typically very small so may be difficult to detect.

## 16.3 Measuring Wind

It was Bagnold who shaped the field by bringing systematic measurements of wind and its effects to bear, and field measurements of wind remain an important aspect of aeolian studies. A good all-round reference on the challenges of making measurements in the environment is that by Strangeways (2003), Brock and Richardson (2001) discuss meteorological sensors in more detail. Various organizations (such as national meteorological services) also issue standards documents on how certain measurements should be made. Note that instrumentation technology develops rapidly, so texts such as those above may not include the most recent innovations (e.g., the use of flash memory in dataloggers).

Although wind and its effects are of principal interest, other meteorological variables such as temperature and humidity are sometimes relevant in that they may influence the saltation threshold. There is some evidence, for example, that more saltation occurs in the afternoon than in the morning, even when allowance is made for the stronger winds in the afternoon. A likely possibility is that dew in the morning provides stronger cohesion for the sand. Note that in some aeolian applications, the windspeed at a given height is of less interest than is the wind shear. The two are of course related, but often not simply (as discussed in Chap. 3). A variety of different anemometers can be used.

### 16.3.1 Cup and Propeller Anemometers

The most familiar kind of anemometer has a set of cups mounted on spokes from a central bearing, such that the wind pushes them in a horizontal circle. Assuming the bearing friction is adequately small compared to the drag forces on the cups, the intrinsic calibration of these sensors is quite straightforward, in that the tip speed of the cups will relate closely to the wind speed, except below some low speed threshold where the bearing friction is too strong for the wind force to overcome. Because the cups rotate horizontally, the sensor operates independently of the wind direction.

The cup rotation is most usually measured with a magnetic reed switch, by optical means (e.g., a light-emitting

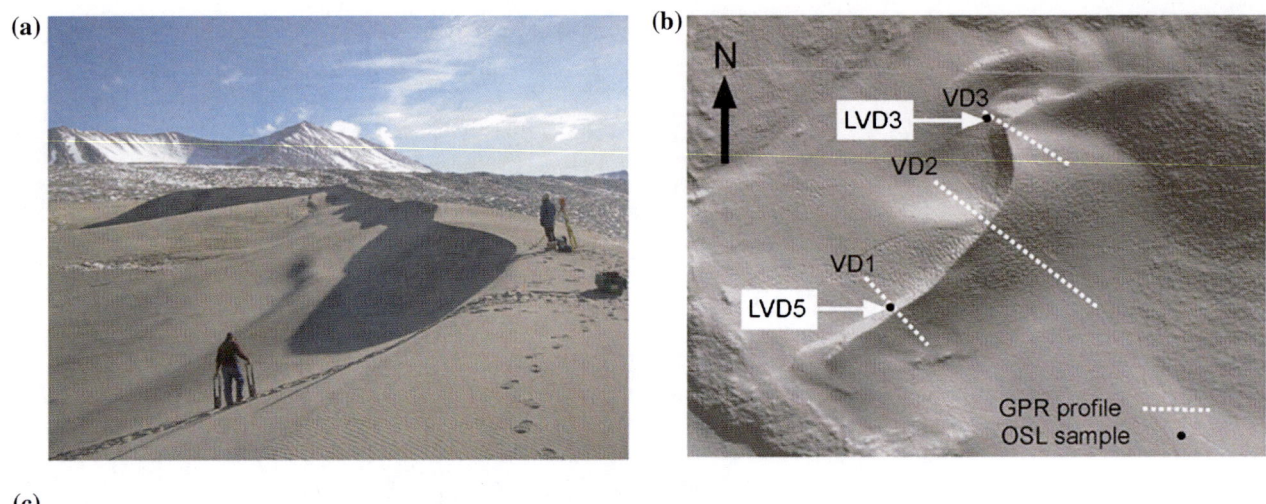

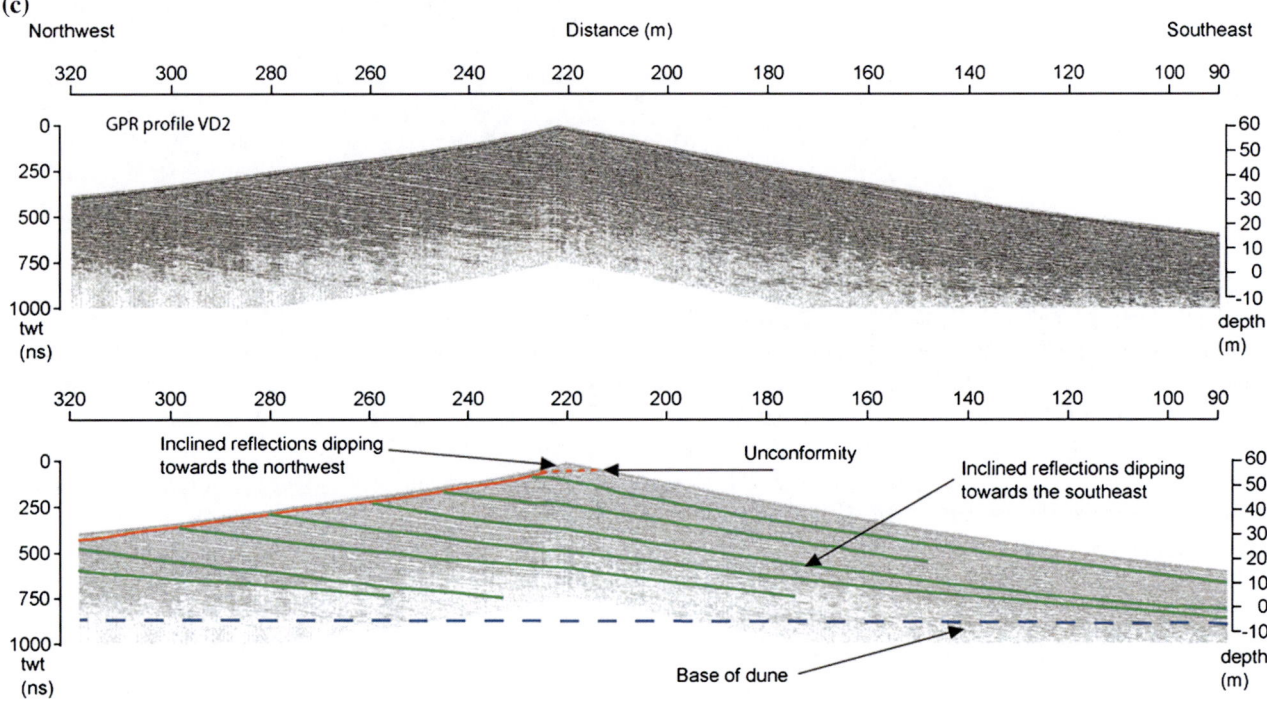

**Fig. 16.11** Ground penetrating radar survey of a dune in the Victoria (dry) Valley in Antarctica. The low temperatures minimize the absorbing effect of any water that may be present in the sand, so this remarkable observation sees through some 60 m of sand. The dipping layers are seen in the radargram, and are interpreted in the lower panel. Image courtesy of Charlie Bristow

diode and phototransistor pair, with a reflective patch or chopper mounted on the cup shaft)—in both instances, the data recording system must count pulses. Another approach, less often used, is to drive a small dynamo or generator (i.e., an electrical motor in reverse) whose output voltage will relate to the rotation speed.

These sensors are relatively inexpensive, and are often used for measuring the wind gradient in the boundary layer with, e.g., three anemometers at different heights supplemented by a single wind vane for direction. The major issue with this type of sensor is that the cups have some inertia which makes it difficult to measure fluctuations in wind speed of less than a second or so. Cup anemometers were used by the Russian Venera landers on Venus. Another approach is to use a propeller, mounted on a horizontal axis. This in turn is mounted on a swivel with a wind vane, such that it is free to turn into the wind. Again the shaft rotation is typically measured optically.

### 16.3.2 Hot Wire and Hot Film

The electrical resistance of most metals is temperature dependent. A wire subject to electrical heating will reach a

**Fig. 16.12** Using a vehicle to drag a GPR system across a linear dune at Quattaniya, Egypt. Note this GPR is a longer wavelength antenna than that shown in Fig. 16.8 and thus is physically rather larger. *Photo* R. Lorenz

temperature that depends on how much heat is applied, and how efficiently that heat is removed by the medium around the wire—and thus on wind speed. Thus, for example, when a wire is driven by a constant current, the voltage across it will depend on wind speed, albeit in a nonlinear way. Because the sensing element can simply be a short length of narrow wire, these sensors can be very compact and are often used in arrays in boundary layer studies in wind tunnels. Furthermore, because the thin wire (often just a few tens of microns across) can change temperature rapidly, these sensors are very good for characterizing fluctuations in windspeed and thus for measuring turbulence. An important consideration in using these otherwise attractive devices in the field is their modest robustness—the wire may be easily broken.

There are variants of the principle, using metallic films deposited on a surface rather than wires, and with different heating and sensing elements. One approach is to have a central heater, whose temperature gives the wind speed, and a set of temperature sensors around it, such that one will be in the plume of warm air from the heater and can thus indicate the direction. Usually hot wires are mounted such that they respond to wind from some range of directions and comparison between differently-mounted sensors is needed to recover direction.

Hot wire anemometers have been used on the Viking and Pathfinder Mars landers (although results from the latter are in some doubt, fluctuations in air temperature in the afternoon defeating the calibration assumptions). A hot film anemometer was developed for the Beagle 2 lander, and is on Mars today on the Mars Science Laboratory Curiosity.

### 16.3.3 Pitot Tubes

Measuring the 'ram' pressure of air is a convenient way to infer windspeed. This method, wherein the difference between the static pressure and the total pressure (dynamic plus static) is measured by pointing a 'Pitot' tube upwind, is the basis of airspeed indicators on aircraft. While modern systems use semiconductor pressure transducers, pitot tubes allow 'analog' measurement, in that the ram pressure can be balanced by the weight of a column of liquid in a glass tube (i.e., a manometer), and thus the length of the liquid column (usually colored with a dye) can simply be read off a scale next to the tube. Such systems are readily scaled up to display several pressures simultaneously, such as those corresponding to winds at different heights. Before the advent of electronic data acquisition, an effective way to record wind data was to take a photograph of the manometer array. It is apt that the cover photograph of Bagnold's autobiography, *Sand, Wind and War* (1991) is of the man doing exactly this while a sand trap is measuring the transport flux. The remarks by Bagnold (1941, p. 77) on the design of his field manometer array (with which he made the first quantitative measurements of wind velocity and sand transport in the field, in the Libyan desert in 1938) are worth recalling, since they underscore some of the general challenges of field data acquisition:

> The design of the manometer, which was home-made, had to embody several special features. It had to register both velocity and static pressures at least six pitots simultaneously, to be capable of carriage by car across rough country without requiring readjustment, to be set up, leveled and zero-ed rapidly, and to be read easily under the unfavorable conditions of an open dune during a violent sand storm.

### 16.3.4 Ultrasound

The transit time of a pulse of sound through a flowing medium will depend on the intrinsic propagation speed of sound through it, and the speed and direction of the flow. The sound speed of air is temperature-dependent, but this can usually be corrected in real time, e.g., by measuring the pulse transit time in both directions. This measurement can be performed with audible sound, but more usually is performed with ultrasound, in particular to facilitate measurement of short pulses. The wider the physical separation, the more accurately can the transit time be measured, but the weaker the pulse and the larger the volume of air over which the measurement is averaged: 10–20 cm is typical, crossed in about 500 microseconds.

Most frequently, a set of transducers is mounted in a tetrahedral or other arrangement such that the transit time in several different directions and the three components of the wind direction can be calculated (see the discussion on turbulent structure in Sect. 3.6). Ultrasonic anemometers are capable of high sample rates and thus for characterizing turbulent flows.

Because of the more complex signal processing required compared with other sensors, ultrasonic anemometers are typically rather expensive. Such anemometers have been proposed for planetary measurements (an application for which their lack of moving parts makes them ideal), but have not yet flown. A challenge for Mars application is that coupling transducers to the thin atmosphere is poor, so adequate signal levels are difficult to obtain.

### 16.3.5 Optical Tracer Methods

The flow of a medium can be estimated if there are visible tracer particles in it. Even small amounts of dust or smoke are enough to permit speed measurements via laser Doppler anemometry, although this is a fairly expensive apparatus. Similar systems have been used at a distance in the field (Doppler lidar) to measure the dust loading and wind speed in dust devils.

A rather neat sensor has been developed, notionally for Mars applications, by Merrison et al. (2006) using light structured by a holographic optical element and a photodiode to detect the sequence of pulses of reflected light as a dust particle crosses the light pattern. This yields the speed of the particle—of course, if the particle is a saltating sand grain rather than a dust particle, its speed may not be the same as that of the wind.

### 16.3.6 Large-Scale Tracer Methods

In order to measure the wind flow pattern over a dune or dunefield, a variety of approaches can be applied. One method is to install an array of wind sensors as described above. Other approaches include release of tracers such as smoke or bubbles, or the use of kites or streamers (e.g., Fig. 5.1) to visualize the flow. A promising modern technique in boundary layer meteorology is to use a small Unmanned Aerial Vehicle (UAV) to measure the wind field; comparison of its air speed and its ground speed allows recovery of the wind.

## 16.4 Measuring Saltation

The simplest way of measuring saltation, or specifically the sand flux, is a sand trap. At its crudest this might be a bucket set flush into the ground, so that moving sand accumulates in it. A rather higher-fidelity approach is a rectangular box, usually with a slightly flared opening (see Fig. 21.1). For unattended operation over long periods, this might be equipped with a door that closes after a finite period, or closes and stops a clock after a given amount of sand has accumulated. Another design (Jackson 1996) operates like a rain gauge, emptying a bucket after 1 g of sand has accumulated. Often several traps are used together, e.g., at different heights. When the mass of sand can be measured before any one of the traps becomes full, then both the vertical and the temporal distribution of the moving sand can be determined. Many diverse designs have been used for the sand traps, but each has its own set of issues with regard to influencing the wind conditions present at the entrance to the trap (e.g., all traps produce some amount of 'back pressure' opposed to the wind at their entrances, hence box traps tend to be somewhat trapezoidal in cross section, with a narrow opening upwind, flaring wider downwind). Sand traps have proved most useful when they can be calibrated in the controlled conditions found inside a wind tunnel prior to their deployment in the field. Thus far, sand trap results have proved to be generally consistent with predictions of sand flux based on both wind tunnel experience and on saltation theory.

Sand traps are integral sensors, in that they measure the amount of sediment that has been transported over a long period. It is often desirable to have information on shorter timescales. One approach is to use a sand trap with a balance that can read out the weight of sand at any given moment, and thus the sand flux can be computed by differentiating these data.

Another class of sensors measure saltation in real time. The most common modern sensors in this category are the safire ('Saltation Flux Impact Responder', e.g., Baas 2004) and SENSIT. The safire uses a metal tube attached to a piezoelectric disk. When a sand particle strikes the cylinder, a vibration is communicated to the piezoelectric disk, which generates a small electrical pulse. Electronics in the sensor filter amplify the signal to generate a digital pulse for each impact above a threshold. A convenient feature of the electronics is that they also provide an analog voltage between 0 and 5 V which provides a count of the number of pulses in a 50 ms period. Since most field datalogging equipment records analog voltages, this greatly simplifies data acquisition, especially if a large array of sensors is to be monitored, since it avoids the need to have separate pulse-counting systems.

The operation of the SENSIT (Stockton and Gillette 1990) is broadly similar, and it records information on an internal data logger that can subsequently be downloaded to a computer. Early versions of the sensor recorded both impact kinetic energy and particle counts as a function of time, but more recent sensors can store the information as energy per individual impact event. Yet another device is the Saltiphone (Spaan and Van den Abele 1991). These various devices have been successfully used in field investigations (Arens 1996; Gillette et al. 1997; Sterk et al. 1999; Helm and Breed 1999). A piezoelectric impact sensor was carried to Mars on the Beagle 2 lander in order to detect saltating particles. Even a simple microphone can detect saltation, which often manifests as a recognizable hiss even when particles do not impact the microphone itself. Acoustic emissions from avalanching is an entire research field of its own ('booming dunes')—see Chap. 10.

Optical sensors can measure the passage of saltating or suspended dust. An optical mass flux sensor (Butterfield 1999) has been employed in wind tunnel studies, and an optical device has been proposed for Mars use (Merrison et al. 2012).

## 16.5 Data Acquisition

There is more to measuring something than just a sensor; the data must be recorded, and when time-resolved measurements are needed—especially at turbulent timescales (see Chap. 3) with many sensors (Fig. 16.13)—the data management challenge can be formidable.

Early dune studies relied either on manual recording of measurements, or occasionally on photographic recording (e.g., of the heights of the liquid columns in an array of manometers). Chart recorders, or magnetic tape, were a later solution for recording instrument data, but for the last 20 years or so, electronic datalogging has become by far a more convenient solution.

Dataloggers take two principal forms: a standalone unit (which can be as small as a box of cigarettes) or as a peripheral to a personal computer. The latter arrangement is usually cheaper for a given capability, and generally allows larger capability in terms of the number of channels to be recorded, how often they are sampled, and so on. However, the need for a computer makes them more suitable for laboratory installation, or (usually using a laptop computer) for only brief, attended, operation. On the other hand, the number of measurements is limited only by the disk capacity of the computer, which can be huge.

Standalone dataloggers perform the same function, but record the data in their own memory for subsequent download via a serial cable to a computer (or, more recently, record the data on removable storage like SD memory cards which can be physically transferred to a computer for analysis). They are generally smaller, more robust, and with lower power requirements than a PC solution. This means they can be left for long-term operation using battery or solar power, although the memory capacity may be a factor in how long they might be left unattended. A final solution is for data to be digitized locally and transmitted via radio to a base station equipped with a computer.

## 16.6 Aeolian Abrasion

One basic 'instrument' type that has been used successfully in a variety of aeolian settings are diverse targets used to document the amount of abrasion produced by wind-blown particles. Wooden stakes and fence posts provided clear visual evidence of the abrasive power of wind-blown sand long before Bagnold initiated quantitative aeolian investigations. Indeed, both natural and man-made materials still provide graphic visualization of the erosive power of sand and wind; for example, the distinctive 'butterfly' pattern that typically forms on metal sheets attached to the base of telephone poles in windy desert environments (e.g., Fig. 1.5 of Greeley and Iversen 1985), illustrates that the wind and particle density profiles above a surface combine to produce zones of maximized abrasion. Robert Sharp elaborated on what could be learned from the abrasion of wooden stakes; he placed carefully weighed rods and blocks of diverse materials (such as Lucite, gypsum, and common building bricks) in the sand-and-wind-rich environment of the

**Fig. 16.13** A modern field experiment can be an orgy of instrumentation. Here we see an array of sand traps (*boxes*), safire detectors (narrow cylinders in the ground), cup anemometers and wind vanes. The data acquisition equipment is prudently protected in a tent (the shade in which makes laptop screens easier to see). This array, at Windy Point in California, was used to study turbulent structure in saltation (Baas 2008). Note the pickup truck, with its left wheels somewhat sunk into the sand, and the ripples in the foreground. Photo courtesy of Andreas Baas

Coachella Valley in southern California, from which he was able to obtain direct field measurements of the rates of abrasion of different materials caused by saltating sand (Sharp 1964, 1980). One of Sharp's students, Michael Malin, took this approach to a new level; he placed weighed samples of sawed rock chips at multiple locations throughout the dry valleys of Antarctica in order to quantify the present abrasion rates in this harsh (but Mars-like) environment. Ronald Greeley brought the study of sand abrasion into the laboratory by constructing a device that could repeatedly throw individual sand grains at prepared targets, under constrained velocities and angles of incidence, providing data on the rate of sand abrasion which was then extended to the conditions currently present on Mars (Greeley et al. 1982; Greeley and Iversen 1985, pp. 112–118). Such investigations have proved to be important foundations for understanding the aeolian abrasion of individual rocks (producing ventifacts) or large bedrock exposures (producing yardangs).

Yardangs (a Turkic word) are streamlined erosional wind forms, similar in form to inverted boat hulls, that in terrestrial deserts range from meters to kilometers in length. They require very erodible rocks (more typically, consolidated sediment) in which to form. They have been observed on Mars (e.g., Ward 1979).

Ventifacts are clasts of rock that have been shaped (often into triangular 'trikanter' forms) or textured (with pitting or fluting) by windblown sand. These too have been observed on Mars (e.g., Bridges et al. 1999; e.g., Fig. 16.14). A somewhat analogous process, although with some deposition added to complicate matters, forms sastrugi in snow surfaces.

An entire book could be written on aeolian erosion. However, a full discussion of these topics is beyond the scope of this book, but there is some good treatment in Greeley and Iverson (1985), and in Laity and Bridges (2009). A great (free) resource on rock breakdown in planetary environments more generally is the photographic atlas by Bourke and Viles (2007).

**Fig. 16.14** The fluted texture and pyramidal shape of this rock (named Jake Matijevic, after the architect of the Sojourner rover) is believed to have been caused by abrasion by saltating grains. The rock is 25 cm tall and 40 cm wide, imaged by the rover Curiosity in late 2012. *Image* NASA/JPL

## 16.7 In Situ Spacecraft Observations

We live in an age when robot explorers are traversing ripples (and, at the time of writing, declining to risk traversing dunes!) on another world, Mars. Field imaging of various types is performed by landers and rovers on other worlds—panoramas, stereo pairs, and timelapse sequences (e.g., Fig. 16.15, also Fig. 5.4) just as discussed above. In this situation, however, there is a much higher cost per image (if nothing else, an opportunity cost) in that, unlike the memory card of a modern camera, the data downlink capacity to the Earth is limited, so only a finite number of pictures can be taken.

The act of landing on a planetary surface can itself provide some information on the properties of the surface. The landing loads on legs or the underside of the vehicle can be inferred from accelerometer measurements and can be influenced by the bearing strength and cohesion of the surface material. The penetration of feet, or even of ejected items such as restraint pins or covers, can be assessed by imagery to allow similar insights. Loose granular material can also be mobilized at landing, either by the turbulent wake of a probe in an atmosphere (e.g., landers on Venus kicked up clouds of dust that could be detected by their blocking sunlight for many seconds after landing), or by erosion by the exhaust from rocket motors used to control the descent on Mars—an extreme example being the Phoenix lander (Fig. 16.16) in the polar regions, which cleaned off ~10 cm of regolith to expose a layer of solid ice underneath. (Thruster erosion has also been documented on the moon, too, but that body is of little interest in aeolian studies.)

Some landers (e.g., Surveyor, Viking, Phoenix) have been equipped with arms; in addition to acquiring samples for on-board instruments, these arms can be used for soil mechanics investigations,(e.g., Fig. 16.17). Moore et al. (1987) give a very detailed report on the mechanical properties of the Martian regolith from Viking measurements. The arm can be used to dig a trench—whether the walls of the trench collapse will indicate the degree of cohesion in the material. Soil may be dug and drizzled into a pile (e.g., Fig. 5.3), the shape of the pile indicating the angle of repose of the material (and thus indicating the friction between grains). Figure 2.5 shows a microscopic imager picture of basaltic sand, with evidence of very low cohesion.

Spacecraft surfaces can become coated with windblown material, and that material can be detected by imaging. Although fine airborne dust can be deposited on lander surfaces, and is responsible for the typically steady decline of the output of solar panels (e.g., the solar panel on the Sojourner rover lost about 0.3 % of power per day), the size of some observed particles on the MER panels (Fig. 16.18)

**Fig. 16.15** Rapid aeolian change on Mars: a timelapse pair of images taken by the Spirit rover on Mars, five days apart. These images are analyzed in Sullivan et al. (2008) to determine a ~2 cm ripple movement. The movement is easy to observe by blinking images back and forth, but takes close scrutiny on the printed page, so we have added an *arrow*. *Image* NASA/JPL

**Fig. 16.16** The quick and easy way to measure how much regolith overlies an ice layer in the frozen northern plains of Mars. The dirt has been blown away by the exhaust plumes from the landing rockets (the nozzles of which are just visible at *top*) of the Phoenix lander in 2008. NASA/JPL/U.Arizona

**Fig. 16.17** Operation of the surface sampler in obtaining Martian soil for Viking 2's molecular analysis experiment in an area named 'Bonneville Salt Flats'; the exposure of thin crust appeared unique in contrast with surrounding materials and became a prime target for organic analysis in spite of potential hazards. The large rock in the foreground is 8 in. high. **a** The sampler scoop has touched the surface, missing the rock at upper left by a comfortable 6 in., and the backhoe has penetrated the surface about 0.5 of an inch. The scoop was then pulled back to sample the desired point and **b** the backhoe furrowed the surface pulling a piece of thin crust toward the spacecraft. **c** This picture was taken 8 min after the scoop touched the surface and shows that the collector head has acquired a quantity of soil. With surface sampler withdrawn (**d**), the foot-long trench is seen between the rocks. The trench is 3 in. wide and about 1.5–2 in. deep. Penetration appears to have left a cavernous opening roofed by the crust and only about 1 in. of undisturbed crust separates the deformed surface and the rock. http://photojournal.jpl.nasa.gov/catalog/PIA00145

**Fig. 16.18** 150-micron sand grains (*arrowed*) can be seen next to the wires (a little over 1 mm wide) on this close-up of a Mars Exploration Rover solar panel. These grains are far too large to have simply settled out of the atmosphere—they must have saltated onto the rover. NASA/JPL

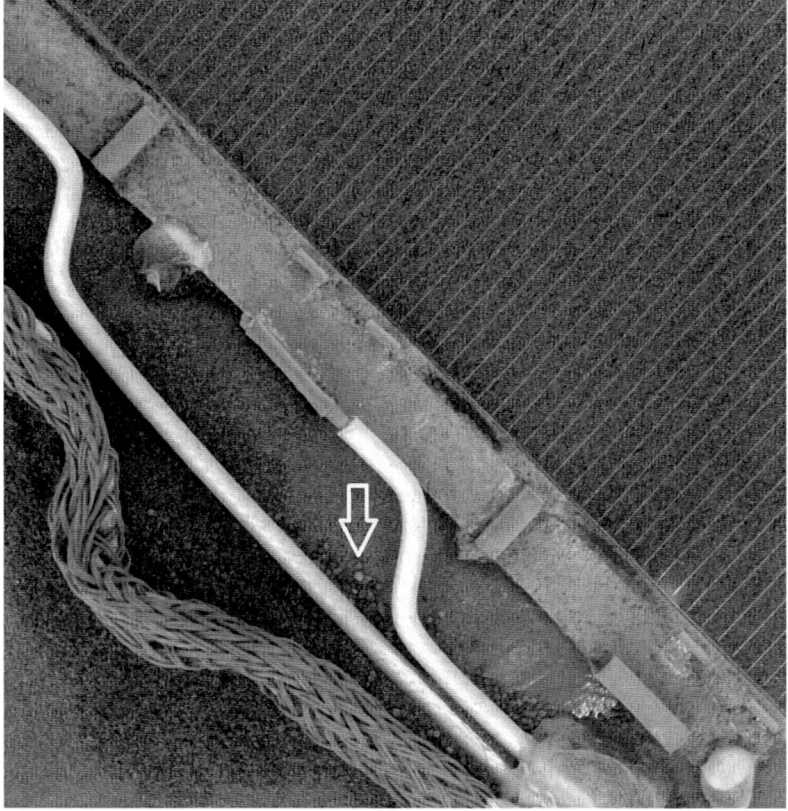

**Fig. 16.19** The capture magnet on the Spirit rover, showing a range of particle sizes adhering to the rings of different magnetic strengths (rather stronger than a typical refrigerator magnet). NASA/JPL

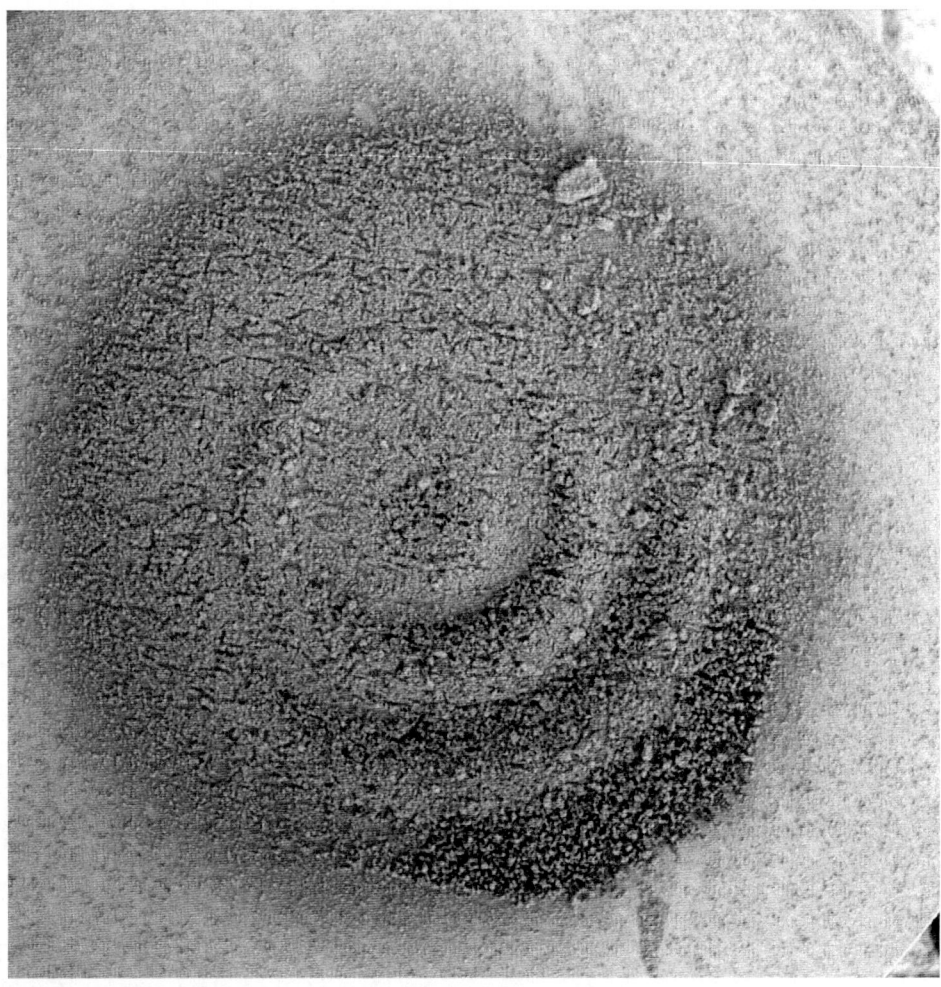

indicates they must have been saltated. Some surfaces may be specifically designed to observe the accumulation of material—among these are surfaces with magnets of various strength (Fig. 16.19) mounted behind them so that the amounts of certain minerals such as magnetite can be assessed.

In addition to manipulation with an arm, another interaction of a vehicle with the surface can be studied, namely that of the wheels on a rover. Usually, the motor current required to drive the wheel is measured, as well as the number of turns a wheel makes. This in turn can be compared with the forward motion measured from camera imagery—usually the wheel rotation corresponds to rather more than the distance moved, owing to slippage of the wheel in soft ground (see Chap. 22). This slippage and the energy required to move a given distance are diagnostic of the surface properties. The textures in the track (Fig. 16.20) also reveal the cohesive properties and friction angles of the regolith; a detailed study of those from the Mars Exploration Rovers is given in Sullivan et al. (2011).

Because tracks or other marks made by rovers are created at known times, their subsequent erasure can be used to estimate rates of processes. Spirit and Opportunity tracks have been given close follow-up observations by orbital imaging (Geissler et al. 2010) to determine that erasure by erosion and deposition by surface winds takes about one Martian year.

While the manipulations of the Martian environment by rovers and landers over long-planned hours and days are hopelessly clumsy in comparison with what a human could perform in minutes, these vehicles are often equipped with formidable analytical instrumentation. The most sophisticated of this is devoted to looking for evidence of life, but many mineralogical and other investigations can be performed. Sometimes these are contact measurements made with instruments on an arm; more generally, material must be ingested into an instrument, a relatively straightforward process for sand but requiring grinding for rocks. A common technique has been APXS (Alpha Particle X-Ray Spectroscopy), wherein a small radioactive source induces

## 16.7 In Situ Spacecraft Observations

**Fig. 16.20** Mars exploration rover wheel tread on a ripple. Note the 'bulldozed' lobe of material, the crisp imprint of the rover wheel tread pattern, the steep walls of the 'trench' and the radial fractures in the material indicating some cohesion. NASA/JPL PIA 1620543

**Fig. 16.21** The Sojourner rover imaged by the Pathfinder lander in 1997. The rover has stopped atop a shallow ripple, nicknamed 'Mermaid', where it used an APXS instrument to measure the sand composition. NASA/JPL/U.Arizona

**Fig. 16.22** An image from the robotic arm camera on the Phoenix lander, having dumped Martian regolith onto the sample hopper of a Thermal and Evolved Gas Analyzer (TEGA) instrument. One of the two doors (about 2 cm wide and 8 cm long) is only partially open, and much of the regolith has proven too cohesive to fall through a wire mesh screen with 1 mm holes (the first author worked on ground testing of the soil loading system some years previously—such cohesive soils were not anticipated). NASA/JPL/U.Arizona

X-ray emission which is diagnostic of the elemental composition of surface rocks or sand. Such an instrument was mounted on the Sojourner rover (Fig. 16.21) and rovers since have also carried this. (In the terrestrial atmosphere, this technique does not work, but similar field instruments using X-ray fluorescence instead have recently become available.) X-ray diffraction is a mineralogical technique (see Chap. 17) that measures the spacing of atoms in the mineral lattice and is carried on the Curiosity rover. 'Cooking' material in an oven to decompose it is another technique, to measure organics or to detect hydrated minerals or carbonates; the TEGA instrument on Mars Phoenix, despite some operational challenges (Fig. 16.22) used this approach and measured carbonates. Wet chemistry experiments have also been performed, e.g., on the Phoenix lander, detecting perchlorate minerals. Microscopic imaging, and even Atomic Force Microscopy, has been applied to observe the texture of grains.

For all these techniques, the analogy with terrestrial measurements and materials is straightforward. In-situ operations at Venus face tremendous challenges due to the high temperature and corrosive conditions. At Titan, conditions (albeit cold) are more benign for instrument function, but designers must grapple with much less familiar materials and properties. Given that Titan's sand seas may be one of the easiest places on that world to land (its liquid seas being the other), it may be that in a couple of decades we see interaction of landers or rovers with the widespread dark sands of that world.

# Laboratory Studies

# 17

Although dunes themselves are planetary phenomena, sand and the processes that make it and shape it can be studied in the laboratory. Laboratory settings allow more elaborate analyses of sand than are possible in the field, enabling its elemental composition to be measured with great sensitivity and precision. Mineralogical and age analyses can also be performed. Sand transport can be studied under much better-controlled conditions in laboratory wind tunnels than in the field, and detailed experiments on rates and thresholds, and deposition patterns including interactions with scale model topography or even model vehicles can be explored. Remarkable recent experiments in water tanks expose the dependence of bedform morphologies on wind direction variations. A special challenge in the laboratory is to reproduce in wind tunnels the atmospheric conditions on other planets.

## 17.1 Sand Studies

While detailed investigations of sand texture and composition can be performed in the field, it is often more convenient to simply and quickly acquire dozens of samples in bags or bottles during a field expedition, label them with their sampling locations, and bring them back to the laboratory. There they can be efficiently and systematically examined and documented without the risk of notes blowing away in the wind or sand getting under a laptop's keyboard. Measurements like sieving can also be at least semi-automated in lab conditions, and access to samples from other locations for comparisons, and other institutions (e.g., libraries) for information is easier.

In addition to sieving (or now, more commonly, optical size classification) and examination of shape and texture (see Chap. 2), more detail can be derived by studying sands through a microscope with crossed polarizing filters which highlight mineralogical differences between the grains. Even closer textural studies can be made with scanning electron microscopes (SEMs) (Krinsley and Smalley 1972; Krinsley and Doorkamp 1973; Smalley and Krinsley 1979) and today grains can be studied at the atomic level by Atomic Force Microscopy (AFM), a technique that has even been applied on Mars by the Phoenix lander, yielding a resolution of close to 10 nm, a hundred times better than optical images.

Various methods are employed to assess the chemical composition of sands. Traditional approaches using Bunsen burners[1] and wet chemistry are progressively being supplanted by newer methods, although some simple wet analyses are still useful and efficient (e.g., separating heavy mineral grains via settling in a dense solution; salinity measurements by washing the sand and measuring the electrical conductivity, etc.) A proliferation of modern techniques (perhaps stimulated by the commercial need to discriminate them for patent purposes), is characterized by a blizzard of acronyms: we introduce at least a few in the following paragraphs.

A revolutionary technique to obtain elemental composition information is the so-called microprobe (or sometimes EMPA—Electron Micro Probe Analyzer) which uses an electron beam to stimulate the sample to produce a spectrum of X-rays; the energy of each line in the X-ray spectrum is characteristic of an element, and thus by calibrating the strength of each line against known materials, the elemental abundance of the sample (e.g., the relative amounts of magnesium, iron, etc.) can be measured. Since the electron beam can be aimed (and is usually conveniently integrated as part of an electron microscope) to a spot just a micron or so across, individual grains in a rock, or the matrix between them, can be separately analyzed.

The same X-ray spectroscopy technique can be used by using an X-ray source to stimulate the sample instead of an electron beam (i.e., X-ray fluorescence, XRF), and handheld XRF analyzers are now available which could be used in the

---

[1] E.g. a grain of sand with even a trace of sodium placed into a Bunsen burner flame will briefly turn the flame a characteristic yellow.

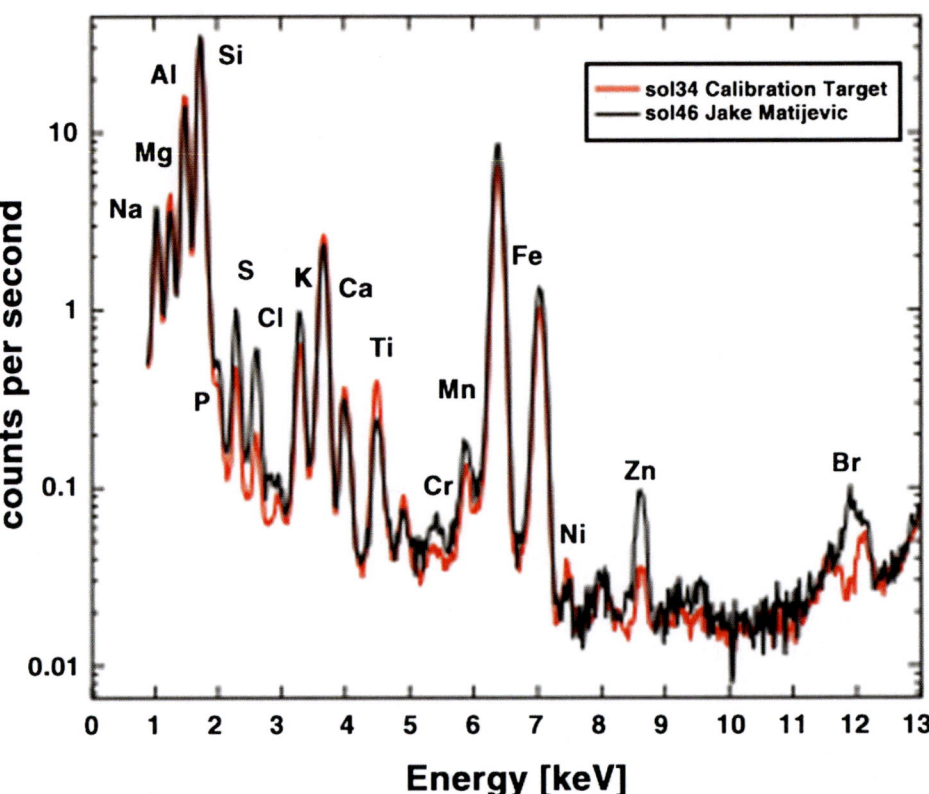

**Fig. 17.1** An X-ray spectrum from the APXS instrument on the Curiosity rover. A calibration target is compared with the rock Jake Matijevic (which may be a ventifact—see Fig. 16.13). Compared to previously found rocks on Mars, the Jake rock is low in magnesium and iron, and high in elements like sodium, aluminum, silicon and potassium which often found in feldspar minerals. It has very low nickel and zinc. The salt-forming elements sulfur, chlorine and bromine are likely in soil or dust grains visible on the surface of the rock. These results point to an igneous origin. Sharper spectra could be obtained with laboratory analyzers, while the Pathfinder APXS had a rather lower resolution. *Image*: NASA/JPL-Caltech/University of Guelph/CSA

field. This approach has been used on spacecraft too, although most in situ elemental composition measurements on Mars have used an alpha-particle source to stimulate the sample. In addition to the X-rays thereby produced by particle induced X-ray emission (PIXE), abundances of some lighter elements can be indicated by the energies of protons that are also produced, and by the energies of alpha particles that are scattered by the samples, the combined technique being named Alpha-Proton-Xray Spectroscopy (APXS). This was the main science instrument on the Pathfinder rover Sojourner (where it can be seen in action on the Mermaid ripple in Fig. 16.21), and has also been carried on the two Mars Exploration Rovers and the Curiosity rover. An example spectrum is shown in Fig. 17.1.

Somewhat related is the technique of neutron activation analysis (NAA) wherein a neutron source is used to form radioactive elements in a sample, which can be detected by their characteristic emissions of gamma rays. This technique is particularly effective for detecting very small amounts of rare earth elements.

Elemental composition can also be indicated by vaporizing mineral samples with extreme heat, most commonly with an inductively-coupled plasma. Elements can give off light with characteristic wavelengths (ICP-AES, atomic emission spectroscopy), essentially a high-tech version of the bunsen burner investigation. Even greater sensitivity is obtained by analyzing the vaporizing sample in a mass spectrometer (ICP-MS) which allows detection of even parts-per-trillion (ppt, one part in $10^{12}$). For example, Zimbelman and Williams (2002) applied ICP MS to sands from Mojave and the Colorado River, distinguishing two source regions.

The generation of a plasma from a mineral sample can also be accomplished with a powerful pulsed laser, and the Curiosity rover carries an instrument (Chemcam) which analyzes the light from the plasma (Laser Induced Breakdown Spectroscopy) produced on a sample up to 7 m away from the rover. Successive laser pulses can be aimed at the same spot, producing a profile with depth into the target (thereby differentiating the underlying sample from the dust that tends to appear everywhere to some extent). Chemcam observations (they are discussed in this chapter because of the AES connection, although obviously this is also a field measurement and in some respects could be considered a remote measurement, too) have been used to profile the chemistry in the 'Rocknest' dune deposit.

The methods described above are all elemental abundance techniques. While useful in assessing the provenance of sands (e.g., one source may be more calcium-rich than another), they do not indicate the mineralogical composition, e.g., whether a sample with calcium, silicon and carbon in it is calcium carbonate plus silica, or a calcium silicate with diamonds. Some of these possibilities can be discriminated by microscopic examination, but the determination of mineral composition uses a laboratory technique called X-ray diffraction (XRD). An X-ray beam is applied to a

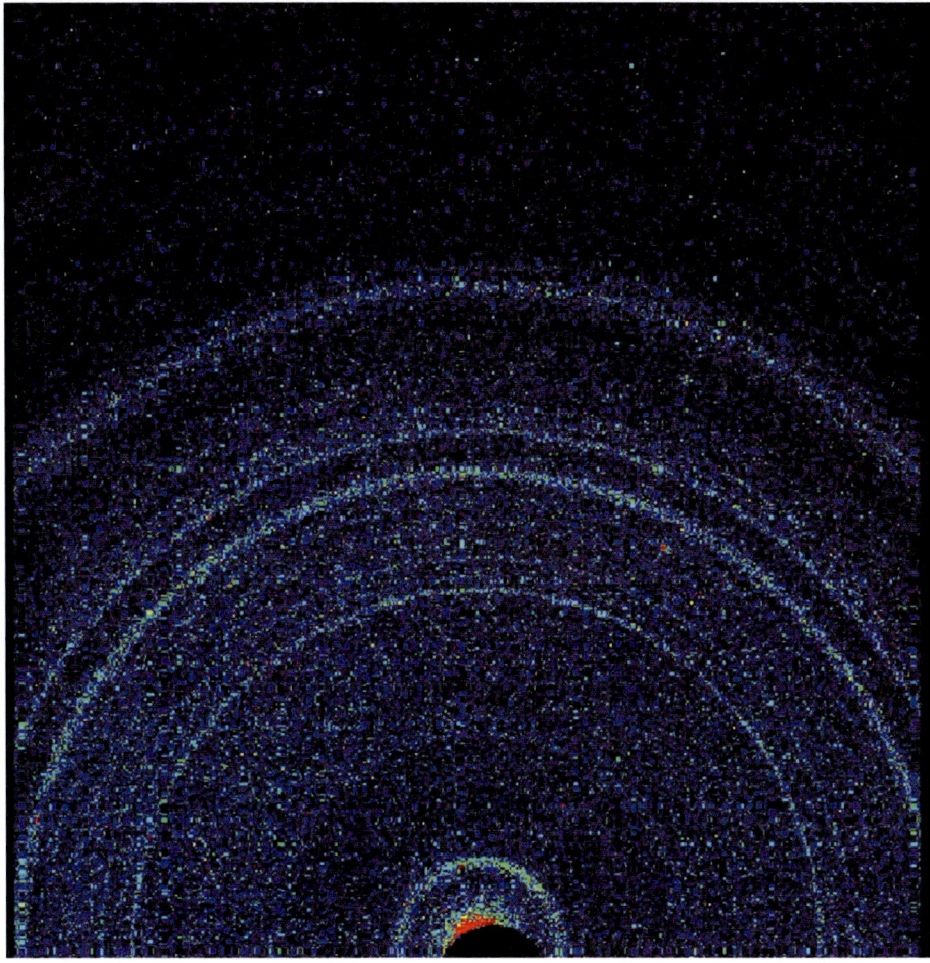

**Fig. 17.2** An X-ray diffraction image of the Rocknest sand shadow on Mars, taken by the Chemin instrument on Curiosity. The arcs correspond to different diffraction angles and thus lattice spacings in the minerals in the soil; the overall pattern is similar to that of volcanic rocks in Hawaii. *Image credit*: NASA

sample, and the intensity with which the beam is scattered away at different angles relates to the spacing of atoms in the crystal lattice, which is characteristic of the mineral composition. (Essentially, the crystal structure is acting as a diffraction grating, which relies on the spacing of the atomic layers in the lattice being of the order of the wavelength of electromagnetic radiation being diffracted by Bragg scattering. For minerals, this spacing corresponds to the wavelength of X-rays although, as we discuss in Chap. 18, aeolian ripples can act as Bragg gratings for microwaves in radar remote sensing.) Laboratory instruments have beamed X-rays at a sample, and swept an X-ray detector on a goniometer through a range of angles. A more compact implementation (the Chemistry and Mineralogy Instrument, Chemin) has reached Mars for the first time on the Curiosity rover—the X-rays scattered by a sample are imaged by a CCD camera, with the various scattering angles forming circular arcs (Fig. 17.2) on the detector whose radius relates to the lattice spacing in the constituent minerals of the sample.

Finally, a prominent goal in aeolian studies is age dating—this is particularly important in efforts to relate changing sand mobility and dune morphology to paleoclimate studies. In addition to carbon-dating of buried organic materials like plants, the principal laboratory techniques here are OSL (Optically-Stimulated Luminescence) and Thermoluminescence, which are ways of estimating the time since sand grains were exposed to sunlight. In essence, energetic particles (cosmic rays, and emission from radioactive elements in the sand itself) cause defects in the crystal lattice of sand-forming minerals like quartz, kicking electrons out of the lattice essentially into metastable storage sites. If these electrons are nudged out of these storage traps and fall back into holes in the electronic structure of the lattice, they may emit a photon of light, which can be measured in sensitive instruments. The amount of light emitted is a measure of the total dose of radiation seen by the sand sample since some 'zeroing event' which cleared the material of trapped electrons. Sunlight acts as a zeroing process (so obviously the sand samples from a given depth should not be exposed to light; elaborate protocols for sampling, using light-proof steel tubes, must be used).

One way of nudging the electrons out of their traps in an instrument is by heat (hence thermoluminescence). Another is by exposing the sample to infrared light (the emitted light is usually in the ultraviolet, and so the stimulating light and

**Fig. 17.3** The aeolian processes wind tunnel at ASU, with a plexiglass working section. Note the lights and cameras. Photo courtesy Rob Sullivan

the emitted light are easily discriminated). These techniques rely on somewhat empirical calibrations of how many trapped electrons are generated per unit time, for a given material (like quartz or feldspar) in a given environment (the cosmic ray dose depends on altitude, and the amount of radioactive elements like thorium or uranium in the sand). OSL especially has become a widely-used technique in paleoclimate and dune evolution studies.

## 17.2 Process: Terrestrial Wind Tunnels

Any discussion of the instrumented study of aeolian processes typically must begin with Bagnold's now classic measurements of wind conditions and their effect on the movement of sand, as described in his seminal 1941 book. Bagnold built a wind tunnel at his home where different segments of the tunnel floor could be weighed individually during a simulation run. This allowed him to generate the first quantitative measurements of sand flux under diverse but controlled wind conditions, which he could compare with his measurements in the field (see Chap. 16). Even though these represent the first quantitative aeolian measurements, nearly all of his results have stood the test of time when compared with modern measurements, a testament to the great care Bagnold devoted to the various measurements he made.

Bagnold devotes three pages of his book just to the illumination he used in his tunnel in order to photograph the trajectories of saltating grains. As he notes, although visible to the naked eye, the streaking trajectory of a grain gives only a short effective exposure time which makes photography challenging, so he had to install a specially intense arc light to photograph the grain trajectories (shown schematically in Fig. 4.1). Modern digital cameras, xenon lamps (e.g., Fig. 17.3) and laser light sheets now make this sort of observation (e.g., Fig. 17.4) much more straightforward, and remarkable close-up pictures showing the rotation of individual grains (see Fig. 4.11) can now be taken.

Today, many wind tunnel facilities can be found in universities and research laboratories around the world. Notable aeolian facilities include the wind tunnels at Iowa State University (Fig. 3.7a of Greeley and Iversen 1985) and at Arizona State University (Figs. 17.3 and 17.5), which have been used by Iversen and Greeley (respectively) to conduct a wide array of experiments; another example is the sediment transport tunnels at the University of Guelph and nearby Trent University in Canada. A commercially operated tunnel is also operated in Guelph to study wind loads on buildings and also to estimate where blown snow deposition may occur, which can be a structural danger.

While the instrumentation has changed dramatically, the basic measurements obtained from modern wind tunnels remain quite similar to those obtained by Bagnold. Sensors for measuring the wind velocity as a function of height above the test surface remain the most essential requirement for monitoring any wind tunnel run, augmented by specialized instruments for whatever the specific question is being investigated. Recently, it has become quite common for portable wind tunnels to be deployed in field settings,

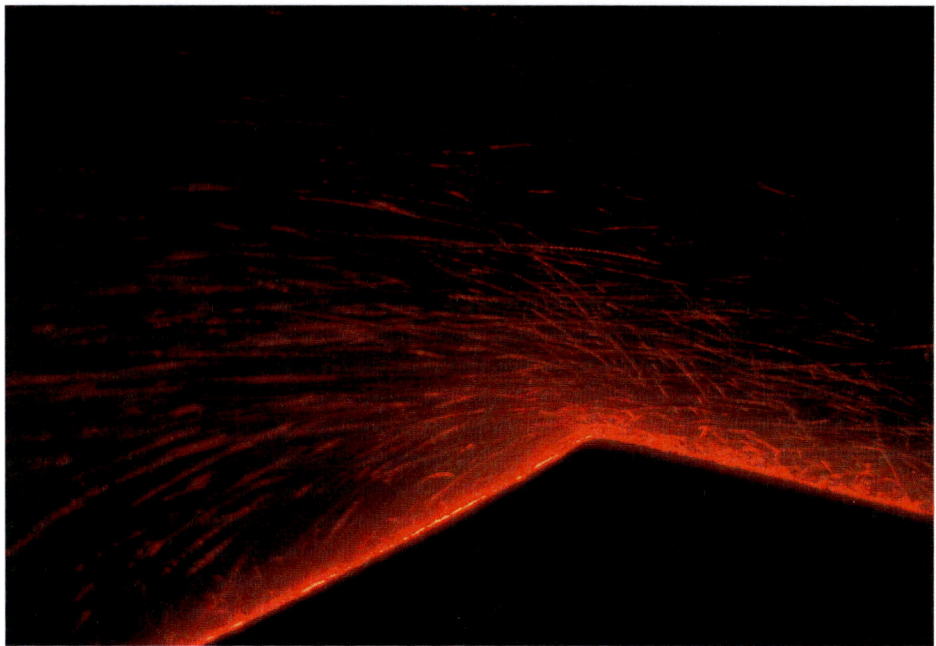

**Fig. 17.4** As an example of the close examination of transport processes in wind pioneered by Bagnold, this modern laser-illuminated example shows grains flying over the crest of a ripple in the Trent University tunnel. Note the slightly rounded crest of the ripple. Image courtesy of Cheryl McKenna-Neumann

**Fig. 17.5** Bedforms in a wind tunnel at ASU. Here the propagation of ripples up a slope is being investigated (slope alters the saltation threshold slightly, as well as the saltation path relative to the unrippled surface). The ripples here are not linear, due to the boundary layer of the plexiglass walls of the tunnel. Photo courtesy of Rob Sullivan

allowing investigators to run controlled experiments over naturally-occurring desert surfaces. Such portable wind tunnels of necessity tend to be considerably smaller than most laboratory wind tunnels, so that great care must be exerted to isolate the test section from any significant wall effects (i.e., the build-up of a boundary layer associated with each wall of the wind tunnel; e.g., see Fig. 17.5), but this has proved to be possible with many portable facilities.

Purpose-built wind tunnels (Fig. 17.6) may incorporate specific structures to control the boundary layer turbulence, e.g., Fig. 17.7.

In addition to the usual aerodynamic considerations for wind tunnels, sediment transport tunnels must also grapple with the challenges of the sediment itself, which obviously moves. Thus sand may be introduced by some kind of hopper or drizzling mechanism, and may accumulate in

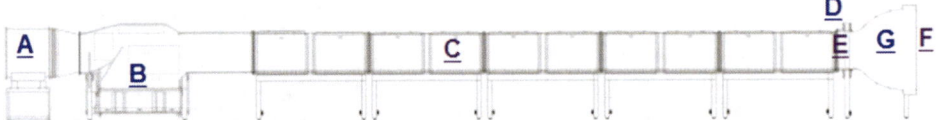

**Fig. 17.6** Cross-section of a purpose-built wind tunnel at Trent University in Canada used to study sediment transport. **a** Drive section, housing the motor and fan. **b** Settling chamber. **c** The test section, where measurements are made. **d** A hopper for drizzling sand into the flow. **e** A boundary layer trip. **f** Honeycomb flow straightener. **g** Contraction zone. The overall arrangement is similar to open-circuit wind tunnels used in aeronautics, although these typically have a shorter working section, and have no need for elements **b**, **d** and **e**

**Fig. 17.7** An arrangement of roughness elements designed to control the thickness of the boundary layer (see Sect. 4.2) in the Trent tunnel

some kind of sump. Closed-circuit wind tunnels in particular have to address this issue, in that sand may cause wear or damage on motors or fans.

As well as the transport by sand and bedform behavior, terrestrial tunnels have also been used to study ventifact formation with application to Mars. Another recent study is the scour or deposition expected on Mars rovers, notably Spirit and Opportunity which are solar powered. Figure 17.8 shows an example of a small rover model, with removal of material in the wake of the wheels.

## 17.3 Process: Planetary Wind Tunnels

In an ideal universe, we might set up portable wind tunnels on other planets and observe the transport processes just as we do here, so that the effects of the different gravity, atmosphere and so on can be accurately measured. The best we can do is attempt to recreate aspects of these environments in chambers or wind tunnels on Earth. This usually entails some compromises—while air density is not too hard to reduce, gravity is difficult to modify!

The MARSWIT (Mars Surface Wind Tunnel) is a facility intended for studying particle motion at atmospheric conditions representative of current Martian surface conditions (Greeley et al. 1974a; Greeley and Iversen 1985, pp. 79–80). The MARSWIT is located at the NASA Ames Research Center at Moffet Field, California. The building was originally designed to conduct pressure tests on Atlas rockets (Fig. 17.9); a vertical chamber large enough to hold an entire Atlas missile could be pumped down to very low pressures in order to simulate flight conditions encountered at very high altitudes.

Once the rocket-testing use for the facility was no longer required, the floor space was used to house a 14 m-long wind tunnel with a test section cross-sectional area of 1.1 m². For ambient and near-ambient pressure tests, a simple fan is used to drive air through the tunnel.

The entire chamber can be pumped down to pressures representative of the atmospheric pressure conditions on the surface of Mars. Pumping down such a large volume of space

## 17.3 Process: Planetary Wind Tunnels

**Fig. 17.8** Mars rover model sitting on a layer of *white sand* on a *dark red* substrate. The dark substrate is visible (*arrows*) where wind scour in the lee of the rover wheels has removed sand. Photo courtesy of R. Sullivan

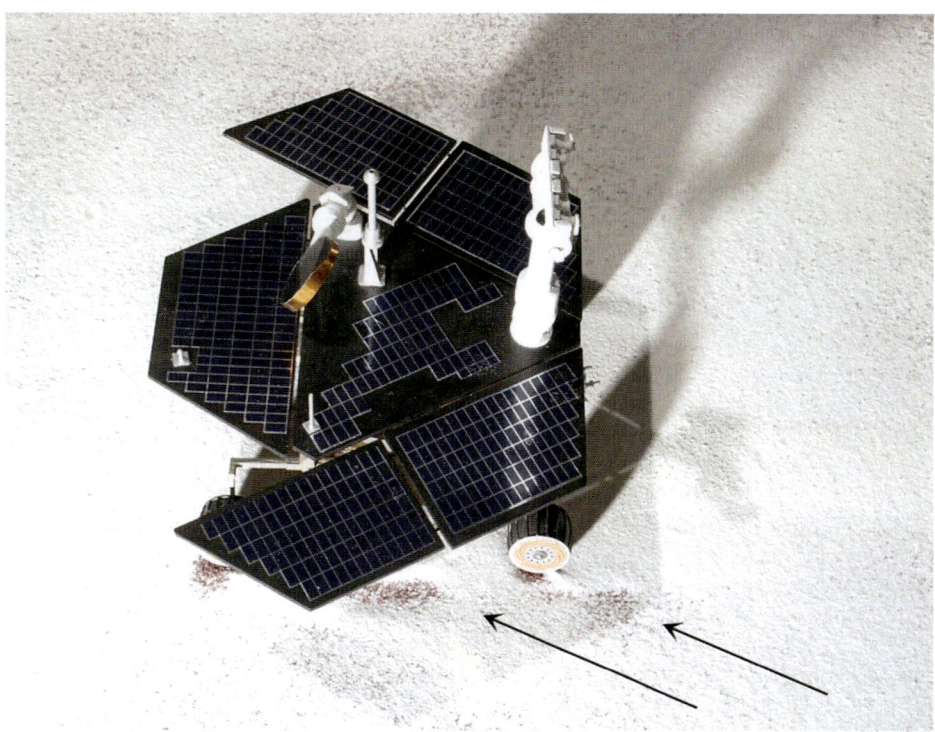

**Fig. 17.9** An Atlas booster is introduced through the massive steel door (*slid to the right*) into the large airtight building which can be pumped down to vacuum or Mars pressures. The Marswit facility (Fig. 17.10) was installed in this building. *Image*: NASA

**Fig. 17.10** The fan at the exhaust end of the MARSWIT wind tunnel: the fan is roughly where the Atlas launcher is in Fig. 17.9 (in which the *vertical pipe at the left* can also be seen). The rings of bolts of the Venus wind tunnel are just visible behind some bins at *right*. Here the tunnel operator is removing boards that cover the rails for the steel doors in preparation for closing up and pumping down the facility. *Photo*: R. Lorenz

to low pressure relies on a nearby steam ejector plant whose prime use is maintaining low pressure during arcjet testing of heat shield materials—through it is only the availability of this infrastructure that the Marswit facility exists.

At ~5 mb pressure, a typical bladed fan would not provide sufficient wind energy to approach the speeds necessary to initiate or maintain saltation of sand, so an innovative pressurized nozzle-injector system was devised to shoot the chamber atmosphere through the tunnel. The injector system is capable of producing a maximum speed of 13 m/s at one atmosphere pressure, increasing to 180 m/s at 5 mb pressure (Greeley and Iversen 1985, p. 79), which are more than adequate to explore saltation conditions on both Earth and Mars. Threshold measurements were collected under Martian conditions (Greeley et al. 1980), but when pumping to the building was discontinued, threshold speed measurements were also conducted as the chamber pressure slowly increased, allowing the exploration of threshold conditions at pressure levels in between that of Mars and Earth. In order to simulate the effects of the reduced gravity level on Mars (~38 % that on Earth), experiments were also run using ground-up walnut shells (sieved to sand sizes) in place of typical terrestrial quartz sand (Greeley et al. 1974a, b). These ground-breaking measurements provided the first experimental data on saltation under Martian conditions, as well as supporting numerous follow-up experiments to investigate a host of aeolian processes at pressures relevant to Mars conditions. Note, however, that while substituting a lower-density sediment correctly influences the force balance for 'fluid threshold' measurements (see Chap. 4), the momentum in saltating particles is less than it should be, so these experiments may overestimate the 'impact' threshold for saltation.

The advent of European interest in performing Mars lander or rover missions has stimulated wind tunnel work there too. Here, lacking the massive infrastructure at NASA

**Fig. 17.11** The Aarhus Mars Wind tunnel is a large chamber in which low-pressure gas is recirculated. A large number of portholes and cable feed-throughs are needed to access the interior with instrumentation, although this is a small price to pay for not having to pump down the entire building. Note the tunnel sits on rollers, so sections can be separated to adjust equipment inside (or clean it of dust!). Photo courtesy of Jon Merrison

**Fig. 17.12** The inside of the Aarhus tunnel (when it was unusually clean). The large fans needed to drive the low-density air are visible at *top* and *bottom*. Photo courtesy of Jon Merrison

Ames, various university groups have developed smaller-scale facilities to test instrumentation in Mars conditions. The University of Oxford has a transient low-pressure tunnel for anemometer testing, although the authors are aware of no aeolian investigations with it.

At the University of Aarhus in Denmark, a facility (Fig. 17.11) has been built in a large vacuum chamber to simulate Mars conditions with winds. The wind is generated by a set of large recirculating fans (Fig. 17.12). A range of studies on dust-lifting, including triboelectric effects, has been performed, and instrumentation for sand- and dust-transport and wind measurements, often using optical means (Fig. 17.13) has been tested.

In order to replicate Venus conditions (or at least come close), very serious hardware is needed. Whilst replicating Mars conditions on a large scale in Marswit was enabled by the presence of an unused facility which was endowed with the capability to pump down to low pressures, the generation of high pressure and high temperature characteristic of Venus cannot be done on such a large scale (Fig. 7.14).

**Fig. 17.13** Lasers are a feature of much modern instrumentation for studying sand and dust transport. They also make for some appealing photographs. Image courtesy of Jon Merrison

For a start, there is a great structural difference between the pressure differential for Mars simulation (1 bar minus 6 mbar is ∼1 bar) and that for Venus (90 bar minus 1 bar is 89 bar, the pressure at about 1 km depth in the Earth's oceans), so a Venus facility must have engineering reminiscent of a submarine, and a sturdy one at that.

Venus conditions were somewhat replicated in a small wind tunnel (the Venus Wind Tunnel, VWT) set up in the early 1980s at NASA Ames (Greeley et al. 1984a). (The replication was 'somewhat' in that the high pressures and gas composition was reproduced, but not the high temperature.) The high pressures necessitate some very heavy-duty pipework (see Fig. 17.13). The overall dimensions of the tunnel are 6 m × 2.3 m, although the working section is quite small, and the gas flow is driven by a 1 horsepower (750 W) DC electrical motor which can drive the flow at up to 4 m/s. The test section where particle motion and bedform formation (see Figs. 4.16 and 5.7) was observed was 20 cm in diameter and 122 cm long.

The tunnel had been largely unused through the 1990s, but following the interest in Titan aeolian processes, has been recently refurbished with new instrumentation and is being used for Titan conditions (again, replicating atmospheric density but not temperature). Several other Titan chambers exist in various facilities, but none have yet been used for aeolian studies. (As an aside, the first author conducted some experiments on the generation of waves in water at different atmospheric pressures and, as a Titan analog, kerosene, in the Marswit facility (Lorenz et al. 2005)). Note that it is generally less easy to set up a Titan chamber than a Mars chamber. First, the chamber must be rated for positive pressure (which is more hazardous since a failing window will then blow

**Fig. 17.14** The Venus Wind Tunnel at NASA Ames. Note the obviously heavy engineering. A cart with instrumentation is in the foreground. Image courtesy of Ron Greeley

## 17.3 Process: Planetary Wind Tunnels

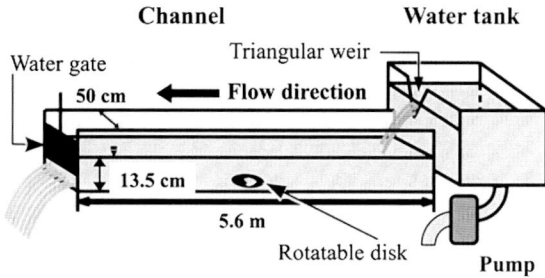

**Fig. 17.15** Schematic of a steady-flow water flume, with a circular platform that can be rotated to expose the bedform to varying flow directions. Graphic courtesy of K. Taniguchi

outwards) and, second, Titan involves much colder temperatures which adds to the hazard, as well as the cooling demands. (For example, if the power to the building failed, a 1.6 bar Titan chamber at 94 K would start warming up, becoming a 5 bar chamber at room temperature.)

At the time of writing, NASA is constructing a large Venus simulation chamber with a volume of a little over 1 m$^3$ at the Glenn Research Center in Cleveland. This chamber, the Extreme Environment Chamber, will be able to generate not only the pressure and gas composition of Venus, but also its temperature. While the principal purpose of the chamber is to test instruments and equipment for future Venus missions, perhaps some aeolian experiments can be conducted by introducing some kind of blower or fan. This facility might also be capable of simulating Titan conditions in a similar way.

## 17.4 Water Tank Experiments

A range of remarkable experiments has been conducted with sand in flumes (essentially a water version of a wind tunnel) and in water tanks where a platform loaded with sand can be moved back and forth. Clearly, the saltation threshold is far off what it will be in air. But such studies are of interest for studying the morphology of bedforms, because another scaling parameter (see Sect. 5.5) falls in a useful range. This is the saturation length, proportional to the diameter of the particle divided by the fluid density. Dunes in air on Earth may have length scales of the order of a few meters and larger; by substituting water for air (with a factor of 800 difference in density) and perhaps using especially fine sand, the saturation length and thus the bedform wavelength shrink accordingly, and therefore dunes form with scales of just a centimeter or two. At this scale, the bedforms (Fig. 7.18) and their arrangement can be conveniently studied.

The interaction between migrating barchans has been studied (Endo et al. 2004) in a continuous water flume (Fig. 17.15, using different colors of sand to track the material during their merger) and the variety of barchan forms under different flow direction variations has been studied (Taniguchi et al. 2012; see also Fig. 17.16).

Perhaps the most spectacular examples of this type of study is that by the Paris group. They used a platform immersed in a water tank (Fig. 17.17), where the platform could be translated some distance (thereby setting up a quasi-steady flow for a brief interval). The platform had a rotating table on which the sediment bed is deposited (incidentally, a rotating table was used on a beach, with natural wind, in Rubin and Ikeda's original experiments on bedform orientation). Thus to simulate reversing winds, the platform is just moved back and forth. To simulate unidirectional winds, the platform is rotated 180° between each cycle. And, of course, a range of wind angles can be simulated by different angles; Reffet et al. (2010) show the transition from transverse to longitudinal bedforms (e.g., Fig. 17.16).

## 17.5 Abrasion Experiments

The susceptibility of materials to aeolian abrasion can be expressed as the ratio of the mass of material eroded to either the mass of the impacting particles or to the number of impacting particles; the susceptibility to abrasion is *not* equivalent to a coefficient of abrasion because the susceptibility can vary with parameters such as impacting particle velocity and the angle of impact with the target (Greeley and Iversen 1985, p. 112). How does one go about measuring the many different physical aspects that can contribute to determining the susceptibility of aeolian erosion for diverse natural materials, since very careful control of both the impacting particles and the target materials is required? The solution to this question was a rather unique experiment where individual particles could be directed at targets within a chamber, which provided careful control of both the environmental conditions in the test chamber and the geometry of the individual sand grain impacts on the target surface.

The experimental apparatus used to determine the susceptibility to abrasion for various rocks ended up being a circular chamber in which individual particles could be flung at controlled speeds against a prepared rock target (Figure 4.5 of Greeley and Iversen 1985). Individual sand grains were dropped into a rotating tube that accelerated the grain to the desired velocity at the point of impact on the target plate. The mass of the individual impacting grains was carefully controlled (through sieving), and the mass of the target plates before and after a run provided a direct measurement of the amount of material removed by the accumulated effect of the impacting sand grains. The angle of the target surface relative to the direction at which the

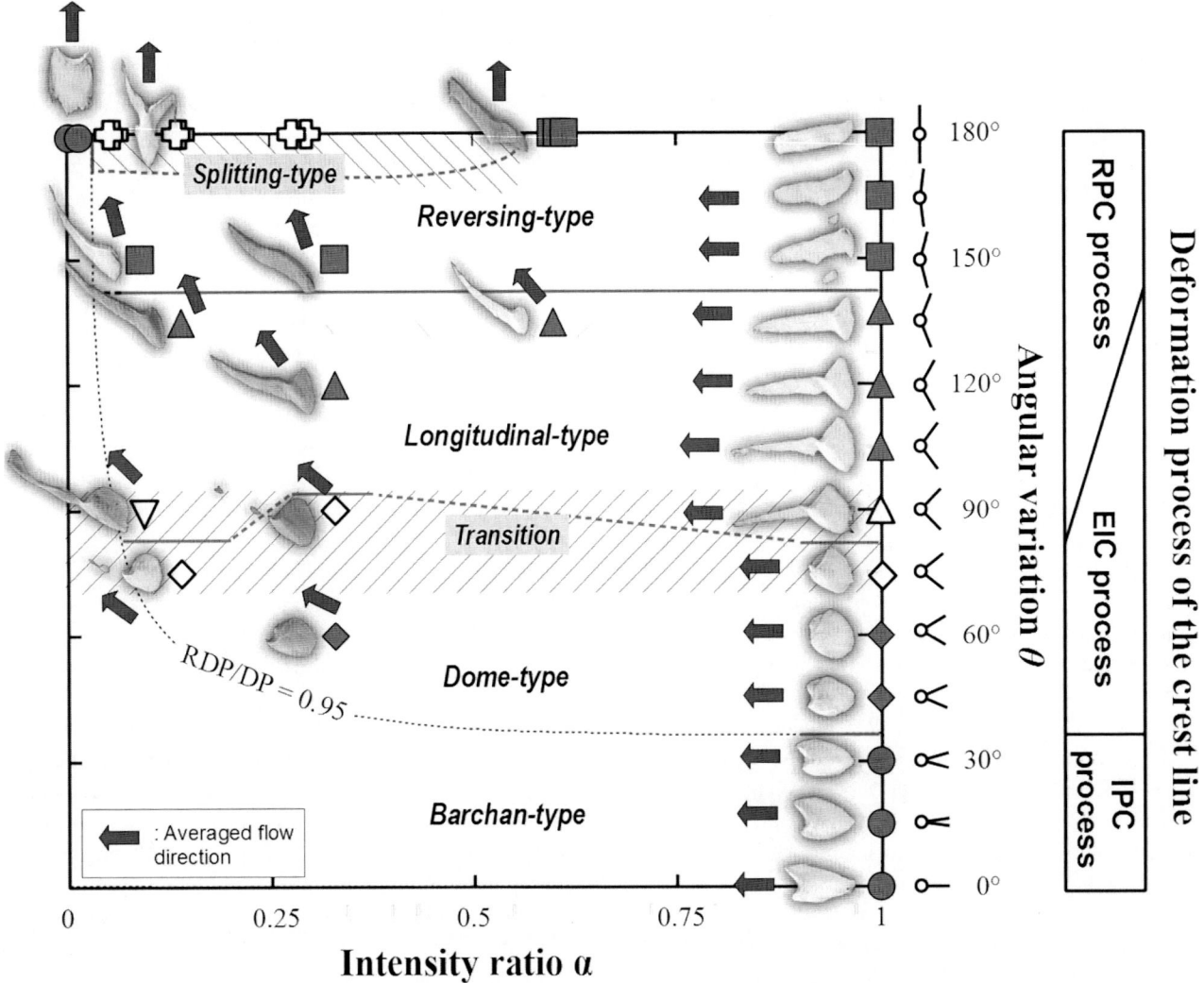

**Fig. 17.16** Results of the flume experiments on barchan morphology, showing the grading from the classic barchans at the *bottom*, up through more teardrop and *triangle/linear* forms as the wind angle diverges more, until the wing shapes (sometimes with two slip faces dash see also Fig. 6.25) appear in reversing winds. Hooked barchans at left form when one wind prevails much more than another. Image courtesy of K. Taniguchi

sand grain hit the surface was also varied while holding other variables constant. All of the experiments could be reproduced under either terrestrial or planetary atmospheric pressure conditions. The experimental runs produced a large matrix of results that represented the first well controlled documentation of both impacting grain and target conditions for a variety of atmospheric conditions (Greeley et al. 1982; Greeley and Iversen 1985, pp. 112–118).

The results of the experiments demonstrated that susceptibility is directly proportional to both the cube of the sand grain diameter and to the square of the velocity for the impacting particles, which in turn means that the mass loss from the target is directly proportional to the kinetic energy of the impacting particle (Greeley et al. 1982; Greeley and Iversen 1985, p. 115). This conclusion was in agreement with earlier experiments by Dietrich (1977) that explored the abrasion of natural materials by windblown dust. The damage done to the target is thus primarily a function of the energy that the impacting particle can impart at the point of impact, but the resulting damage is also governed by the strength of bonds present between the atoms that comprise the target minerals (Dietrich 1977). The sand-slinging experiment also demonstrated that the angle of impact played a significant role in the amount of material abraded from the target, consistent with inferences about the abrasion of rocks to form wind-shaped features called 'ventifacts'.

Some related experiments were performed by Marshall et al. (1991) to explore the strength and adhesion of basalt particles at Venus-like temperatures (see Sect. 14.6). Similar experiments might usefully be performed on Titan analog materials.

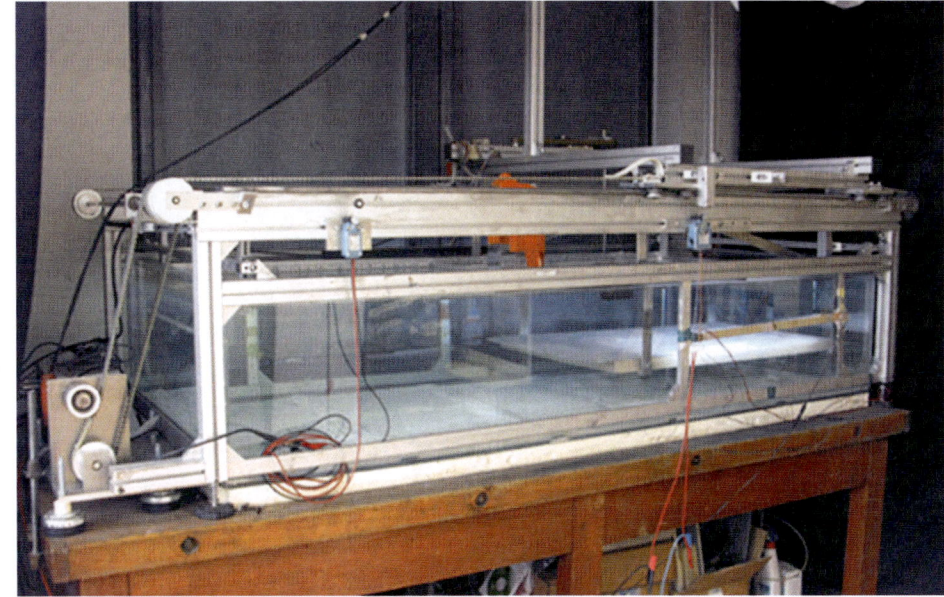

**Fig. 17.17** The water tank used to perform the experiments reported in Reffet et al. (2010). The platform can be seen at right, pulled along rails by a chain that drapes from a pulley at upper left down to the computer-controlled motor. A circular sediment table on the platform can be rotated between each back-and-forth cycle. Photo courtesy of Sylvain Courrech du Pont

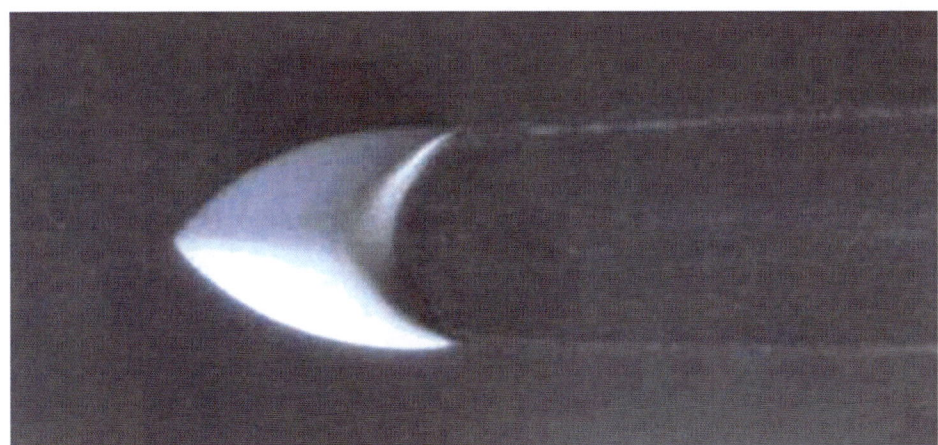

**Fig. 17.18** A barchan in the water tank, showing beautiful crescent slipface and sand loss from the arms. Photo courtesy of Sylvain Courrech du Pont

## 17.6 Avalanching and Granular Flow

In addition to sand-blowing experiments in wind tunnels and abrasion experiments, there is an entire field of study in granular flows (with industrial as well as geological applications). Experiments that study the sorting of different materials in avalanching flows, the mobilization of material by vibration (and—as in the singing sand of Chap. 10—the emission of sound by flowing sand) have been common investigations. Triboelectricity is another field of study in which a range of laboratory investigations may bear fruit.

# Remote Sensing 18

In this section, we review the various remote sensing methods used to study dunes. The quality, extent and, most especially, the accessability of satellite remote sensing data have transformed studies of dunes on Earth. And, of course, it is remote sensing that has for the most part made studies of dunes on other worlds possible. We devote some particular attention to radar remote sensing, in part because this is less familiar to most readers than are optical pictures, but also because it is the means by which dunes on two of the four dune worlds in the solar system were discovered. We begin the discussion with satellite imaging's predecessor, aerial photography.

## 18.1 Aerial Photography

A fundamental limitation with field work is the limited perspective afforded by a viewpoint only a couple of meters above the ground. Sometimes there is a convenient mountain nearby from which a good regional view of a dune or dunefield can be obtained, but usually there is not. Thus a wider view needs a higher viewpoint. This can be obtained from kites, balloons or aircraft (we discuss kite imaging in the field in Chap. 16). Jet airliners, cruising at what the pilot will call 30,000 ft, are 10 km above the ground, so a typical unzoomed point-and-shoot camera through the window will show an area on the ground about 20–30 km across; although the smallest dunes may be hard to see, nonetheless some impressive views can be had by the alert airline passenger (Fig. 18.1; also other examples are of the Algodones dunes, Fig. 24.3; barchans in Egypt, Fig. 6.5; the Moenkopi dunes, Fig. 7.3; and the Badain Jaran, Fig. 18.1). Usually the best pictures are taken on the side of the plane away from the sun (for most readers, this will mean the north side), to avoid glare and scattering by dirty or scratched windows. A good zoom lens is useful, as may a near-infrared filter (see below). Sometimes, some contrast enhancement in post-processing (e.g., Photoshop) can help substantially.

More impressive views may be had from lower altitude, typically from a light plane (e.g., Fig. 18.2); usually images are clearer as well as of larger scale. Some spectacular commercial photography has been obtained from paramotors (powered paragliders) which fly both low and slow, and can be conveyed by truck into the middle of sand seas for sustained operations.

Due to the same factors (gravity vs. atmospheric density) worlds without dunes are worlds on which one can fly, and in terms of ease of flying near the surface are Titan, Earth, Mars and Venus. From an aeronautics standpoint Titan is easy, albeit cold and distant. Mars' thin atmosphere presents aerodynamic and thus structural challenges, although viable aircraft and balloons have been proposed. While the Venus lower atmosphere is dense, it is so hot as to be difficult to operate in.

Vehicles such as landers or rovers on their way down to the surface sometimes have the opportunity to image the surface from the air: descent cameras of one sort or another have been carried on all the Mars landers and rovers in the last decade, and a spectacular recent example, which happens to show some dunes, is Fig. 18.3. Interestingly, although it was not realized at the time (see Chap. 13) the first images showing Titan dunes were those taken by the Huygens probe's camera during its parachute descent (see Fig. 13.7).

A higher altitude and wider view inevitably means longer paths through the atmosphere, which in dusty environments can lead to poor contrast in images. This can be ameliorated somewhat by using longer wavelengths, such as by using an infrared filter. Aerial images are often attained at a range of times of day so, unlike most orbital images, shadows can often be present. These can be dramatic and useful in showing the shape of dunes, which are usually covered in sand of a uniform color and are thus 'washed out' near noon.

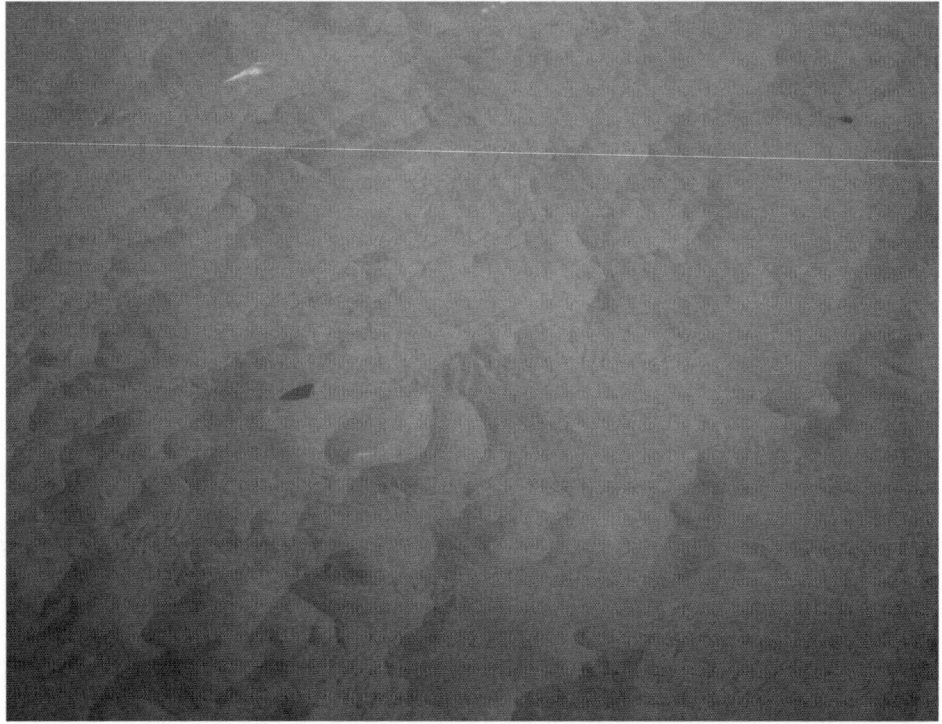

**Fig. 18.1** A view of the somewhat crescentic star dunes in the Badain Jaran desert in China, seen through a rather murky airplane window (but contrast-stretched after the fact to compensate). These dunes are seen from space in Fig. 7.15, and in the field in Fig. 11.8. Note the dark triangular lake to the *left* of the dune in the *center*. *Photo* R. Lorenz

**Fig. 18.2** An aerial view, taken in 1953, of rather sensuous shadows on barchanoid dunes in the Ubari sand sea in Libya. While we may hope to see aerial photos from Mars or Titan, it may be a long time before we see aerial photographs showing other aircraft! *Photo* US Geological Survey

## 18.2 Orbital Imaging

Imaging of planetary surfaces began soon after the dawn of the space age: experimental cameras on captured V-2 rockets showed the wide views of terrain and weather that would be possible from altitudes of hundreds of km (e.g., Lowman 1964) but, ironically, rarely showed the White Sands dunes near where the rockets were launched from. Within 5 years of Sputnik, Corona spy satellites were returning capsules of film from orbit, and Luna 3 had

18.2 Orbital Imaging

**Fig. 18.3** The closest thing we have to an aerial photograph of dunes on Mars. This image was taken (as this book was being completed) by the down-looking MARDI camera on the Mars Science laboratory Curiosity, shortly after its parachute deployed. The circular object at *left* is MSL's heatshield falling away: in the background are dark sand dunes on the floor of the Gale crater. MSL landed a couple of minutes after this image was taken. *Photo* Malin Space Science Systems/NASA/JPL

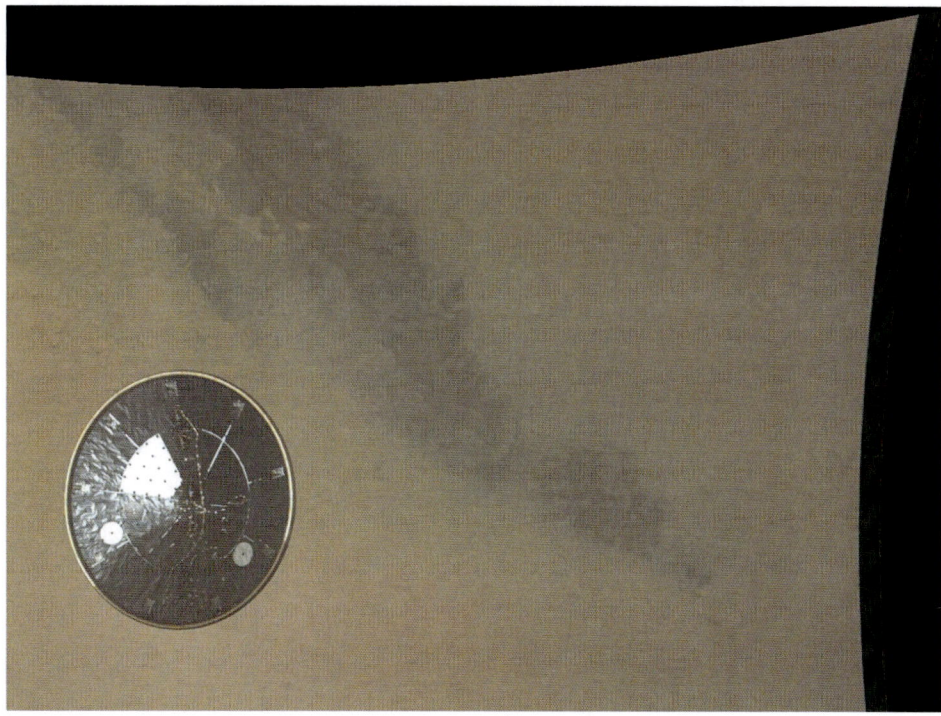

**Fig. 18.4** Gemini 4 handheld film camera image of remarkably straight and regular linear dunes in the Rub' Al Khali in Saudi Arabia. NASA image s65-34765. Image Science and Analysis Laboratory, NASA-Johnson Space Center http://eol.jsc.nasa.gov

**Fig. 18.5** The 2.5 km-diameter Roter Kamm impact crater in Namibia is seen here in this ASTER image of the Namibian sand sea. Compound linear dunes are seen to cross the crater floor, much as seen at Titan (Fig. 13.10). *Image* NASA/GSFC/METI/ERSDAC/JAROS, and U.S./Japan ASTER Science Team

imaged the far side of the moon. Within another 3 years, Mariner 4 would be the first spacecraft to fly by Mars, and the first Ranger close-ups of the Moon would take place.

An early documented report is of dunes in north-west Africa, recorded by automatic cameras on an unmanned test flight (MA-4) of the Mercury series (Morrison and Chown 1964); doubtless there may be similar reports in the Soviet literature. The first high-quality pictures of dunes from space were, however, taken by hand. While the first manned missions concentrated on learning how to function in orbit, as mission durations increased it became possible to contemplate making scientific observations. The Gemini 4 mission in June 1965, generally more famous for Ed White's spacewalk—America's first—lasted a record 4 days, and featured a dedicated experiment to perform high-quality scientific imaging of the Earth using a Hasselblad film camera. Among the features captured and documented (Lowman et al. 1967) were large linear dunes in Egypt (see Fig. 18.4; see also the section on fictional dune worlds).

Unlike manned missions (usually launched into more energetically-favourable orbits), systematic imaging of the Earth's surface is usually performed from satellites in near-polar orbits such that they can observe the entire surface of the Earth. These orbits are usually tuned to be sun-synchronous, such that they observe sites at the same time of day, which facilitates change detection. Generally, the time of day chosen is close to noon, when shadows are minimized.

The first dunes on another world were seen by Mariner 9 at Mars in 1972. Although modest in quality and resolution ($\sim 100$ m) by modern standards, these first images are still being used as a basis for comparison to detect dune changes. Mars has been imaged progressively better by the armada of spacecraft sent there since, with the spatial resolution improving by about an order of magnitude each quarter-century (through $\sim 1.5$ m on Mars Global Surveyor in 1997, and to the $\sim 0.5$ m on Mars Reconnaissance Orbiter in 2006).

Similar improvement had been seen at Earth only a decade or so earlier than at Mars. Remote sensing for geological/geomorphological applications really took a step forward with the US Landsat program (e.g., Fig. 18.7). Since 1972, this has provided various imaging data as technology has improved but, most significantly, allowed study of sites worldwide with resolutions of $\sim 20$ m. The availability of that dataset made possible one of the most

## 18.2 Orbital Imaging

**Fig. 18.6** Landsat 7 image of part of the Namib sand sea acquired in 2000, showing a staggering diversity of dune forms, and the interaction of the sand sea with the ephemeral Tschaub river. Image provided by the USGS EROS Data Center Satellite Systems Branch as part of their Earth as Art series

significant global surveys of sand dunes on Earth, the formidable *Global Sand Seas* publication by the US Geological Survey, edited by McKee (1979a). Considerable effort has been made in maintaining a continuous capability and in robustly archiving the data, permitting long-term studies of changes in land use and geomorphology (notably, tracking dune movement) and these data, once accessible only by a few, can be readily obtained for free by anyone on the internet.

Outside the USA, a major player in remote sensing was the French SPOT program, whose first satellite was launched in 1986. It too provided imaging of 10–20 m resolution in a range of wavebands. Since most commercial remote sensing is geared towards characterizing vegetation, the data are often presented in a somewhat false color (a key giveaway in an image being when vegetated areas look 'red', as healthy plants reflect strongly in the near-infrared). In such situations, the evaporite pans on which many dunes

**Fig. 18.7** A spectacular sidelong view from the International Space Station of dunes migrating past and across a lake in Mongolia. *Image Science and Analysis Laboratory, NASA-Johnson Space Center http://eol.jsc.nasa.gov*

**Table 18.1** Thermophysical properties of selected geological materials[a]

| Material | k<br>Thermal conductivity<br>(J m$^{-1}$ s$^{-1}$ K$^{-1}$) | $\rho$<br>Density<br>(kg m$^{-3}$) | c<br>Heat capacity<br>(J kg$^{-1}$ K$^{-1}$) | I<br>Thermal inertia<br>(J m$^{-2}$ s$^{-1/2}$ K$^{-1}$) |
|---|---|---|---|---|
| Granite | 1.78–5.02 | 2650 | 900 | 2060–3460 |
| Basalt | 1.12–2.38 | 2900 | 830–900 | 1642–2492 |
| Sandstone | 2.1–3.80 | 2310 | 850 | 2054–2731 |
| Water | 0.56 | 999.8 | 4217 | 1536 |
| Dry sand | 0.29 | 1650 | 850 | 637 |
| Mars sand[b] | 0.044 | 1650 | 850 | 248 |
| Mars dust[c] | 0.004 | 1100 | 850 | 61 |

[a] Values from Table 18.1 of Mellon et al. (2008), p. 401
[b] 200 μm sand at 6 mb atmospheric pressure, typical of the Martian surface
[c] 1 μm dust at 6 mb atmospheric pressure, typical of the Martian surface

## 18.2 Orbital Imaging

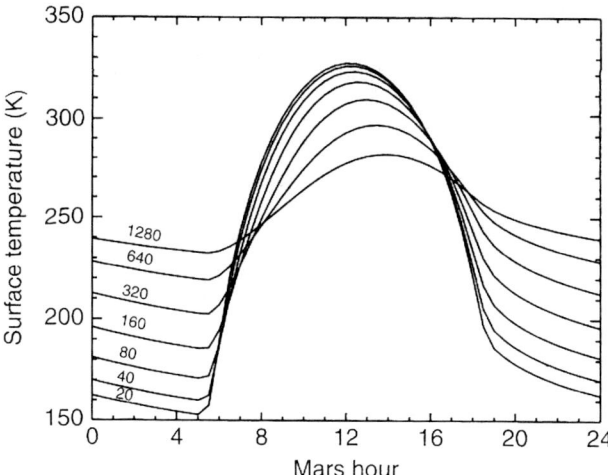

**Fig. 18.8** Surface temperature behavior of Mars as a function of thermal inertia and albedo. *Top*, diurnal temperatures for a range of thermal inertia during the summer, assuming an albedo of 0.10. Generally, thermal inertia controls the amplitude and the phase of the diurnal temperature cycle and albedo controls the mean. Mars hour represents 1/24th of a Mars solar day (88775 s). After Fig. 18.1 of Mellon et al. (2008)

are found can look blueish, whereas to the human eye they would be simply white (the same effect is seen on some Mars images).

Until a decade or so ago, resolutions much better than ~20 m for scientific (as opposed to military surveillance) applications meant using aerial photography. However, electronic cameras have dramatically improved, as well as the capability for relatively affordable spacecraft to achieve the pointing accuracy needed to image targets on-demand, and (importantly) the connection of customers with providers made not just possible, but easy, by the internet. As a result, a variety of commercial enterprises worldwide now provide satellite imaging down to less than 0.5 m resolution, which can be explored easily with tools like Google Earth.

Geomorphologically, the Earth's dunes are overall well-characterized globally at the 'free' 20 m level (e.g., Figs. 18.5 and 18.6, from ASTER and Landsat). However, the higher spatial resolution now available commercially means that areas of particular interest, notably those where dunes are migrating, can be studied much more closely without mounting an expensive or hazardous expedition (both a good and a bad thing!). This higher resolution permits changes to be detected over much shorter timescales than was the case in the past (e.g., only the fastest-moving barchans were clocked by Landsat data). We report our own analysis of barchan migration using commercial imaging of the Star Wars film set in Tunisia in Chap. 24.

Originally, such change detection measurements required careful matching of images (perhaps via wet photographic processing) and many tedious manual measurements. Now computational tools such as Geographical Information System (GIS) products, Google Earth and so on make such measurements easier. Furthermore, automated and semiautomated processes can now extract differences in position of features by a fraction of a pixel, by computationally cross-correlating images. One, but by no means the only, implementation of this technique is called COSI-CORR (Co-registration of Optically Sensed Images and Correlation), developed at Caltech. Similar automated correlation methods are used (with different viewing geometries) to make digital elevation models from stereo imaging.

It may be remarked that the hand-held image is not dead: high-quality digital stills and video are routinely obtained by astronauts on the International Space Station (there is even a big high-quality windowed cupola for Earth observations), and NASA makes these often striking data available too (and indeed some of the illustrations in this book come from this site). The cameras used here are typically true-color and, of course, because someone thought the image was worth pressing the shutter button for, are often well-composed and appealing. Notably, many images are off-nadir (e.g., Fig. 18.7), which makes a pleasing change in perspective from the flat view normally used in Earth observation.

## 18.3 Thermal Imaging

In addition to imaging reflected light, it is possible to sense thermal infrared light emitted from surfaces. This gives indications of surface temperature, as well as being somewhat influenced by composition. The surface temperature of course varies by latitude, season and time of day, as well as factors such as the particle size, and whether volatiles are present.

A prominent technique in planetary remote sensing is to compare daytime and nighttime thermal images. Rocks and hard ground warm up and cool down more slowly (an effect due partly to density $\rho$ and heat capacity c, and in part due to thermal conductivity k) and so are cooler in the day and warmer at night than are porous materials like sand and dust. Table 18.1 lists the thermal conductivity for a few rock and sand examples, values that are up to two orders of magnitude lower than the thermal conductivity of metals. Density represents the mass per unit volume of the material. The table illustrates that density does not vary greatly for most natural materials; water is the least dense common material. Heat capacity (also called the specific heat

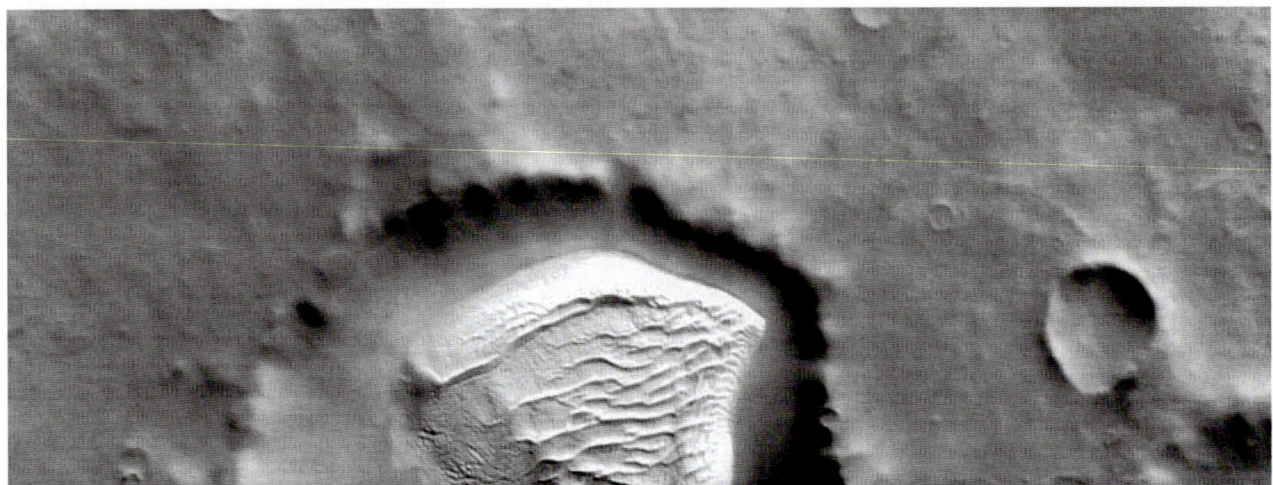

**Fig. 18.9** A daytime thermal image of the surface of Mars (a crop of THEMIS image I25817004) around 70°S. The image is 3.2 km from *top* to *bottom*. Illumination is from the *top right*, and cold shadows are apparent in the crater walls. The large sand deposit in the *center* of the crater, with superposed dune forms is *white*, indicating a high temperature due to the low thermal inertia of the sand. *Credit* ASU/NASA

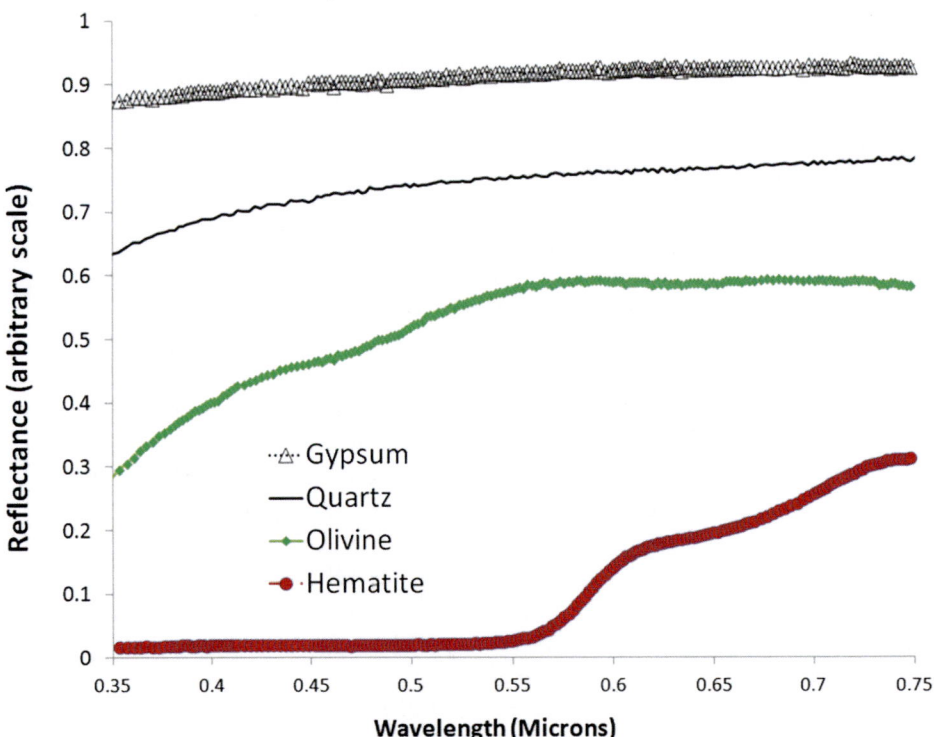

**Fig. 18.10** Visible wavelengths have some ability to discriminate mineral composition. Olivine is of course greenish (*green* corresponds to ∼0.55 microns), and hematite *red* (*red* light is ∼0.65 microns). Pure quartz is nearly *pure white*, although is slightly more *yellow* (i.e., less *blue*) than gypsum

capacity) provides a measure of the ability of a material to store heat. The higher the heat capacity, the larger the quantity of heat needed to raise the temperature of a unit volume; water has a heat capacity 4–5 times that of common rocks (Table 18.1). For most geologic materials, the product $\rho c$ can vary by a factor of 2–4, while k can vary by orders of magnitude. Calculated values for sand and dust on Mars illustrate how the thin Martian atmosphere greatly reduces the thermal conductivity of granular materials (compare Mars values to those of dry sand).

Usually a model (e.g., Fig. 18.8, taking into account the illumination geometry and other factors) is tuned to fit the observed temperatures, and the free parameter in the model that is diagnostic of the surface material is the 'thermal

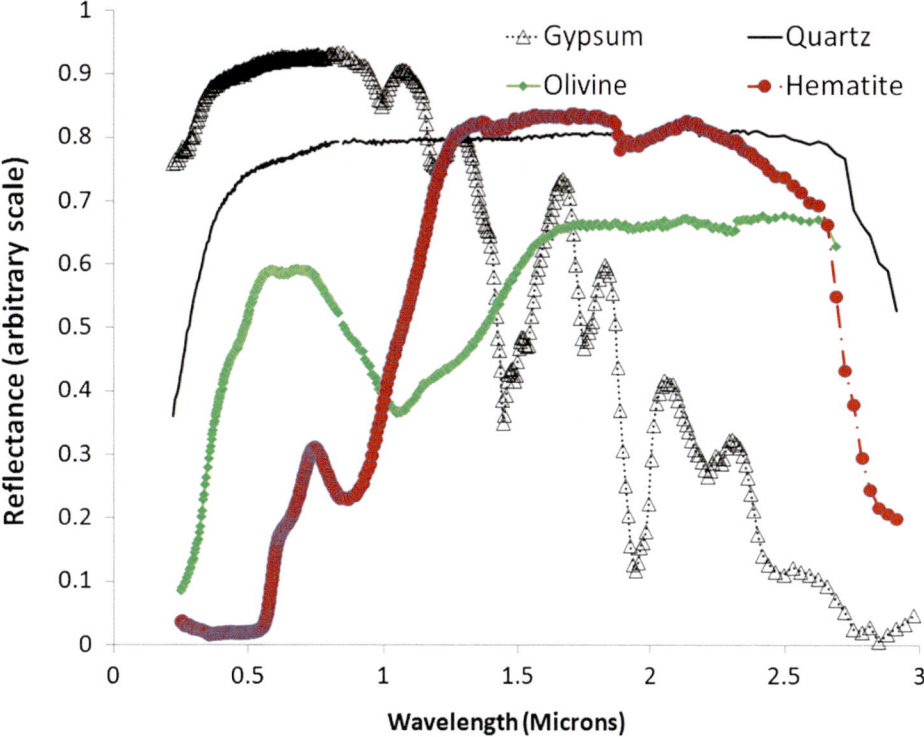

**Fig. 18.11** The near-infrared provides many more diagnostic spectral features. Olivine has a wide band at just above 1 micron, and gypsum has a very deep band at 1.9 microns. These data (and those in Fig. 18.10) are from the USGS Spectral Library

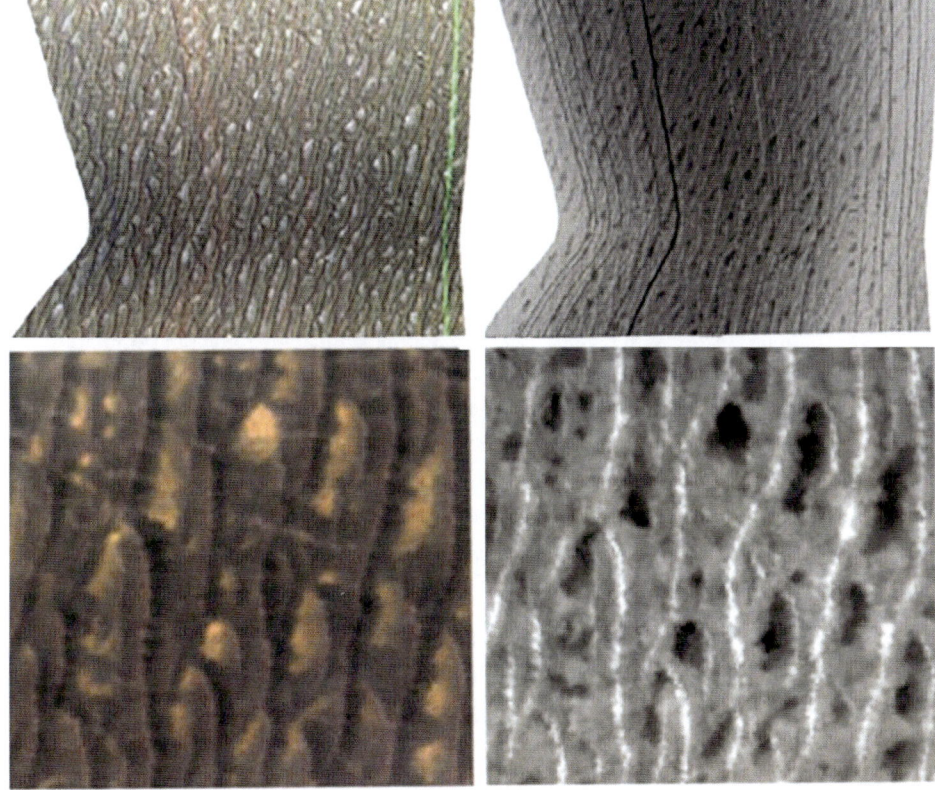

**Fig. 18.12** A CRISM image region of sand dunes surrounding the Martian north polar cap acquired on October 1, 2006 near 80.0°N latitude, 240.7°E longitude. It covers an area about 12 km (7.5 miles) square. At the *center* of the image, the spatial resolution is as good as 20 m (65 ft)/pixel. The image was taken in 544 colors covering 0.36–3.92 μm: the image at *right* shows the strength of an absorption band at 1900 nm wavelength, which indicates the relative abundance of the sulfate mineral gypsum

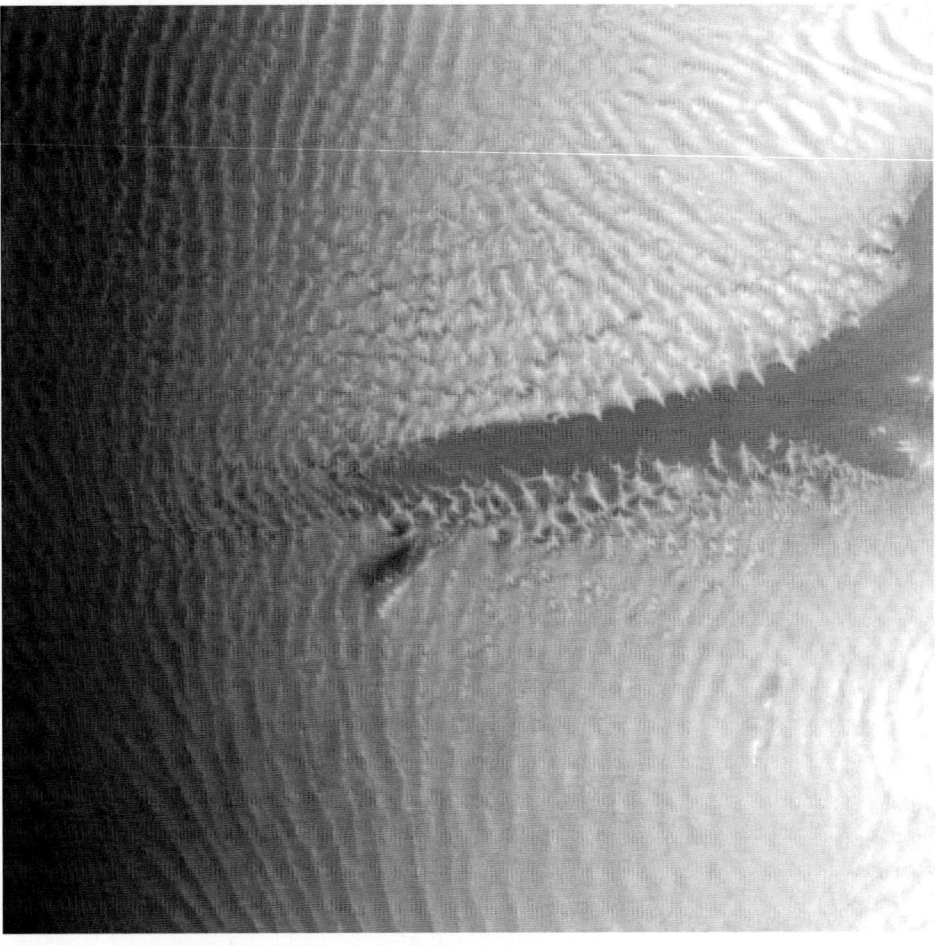

**Fig. 18.13** ASTER DEM of the Namib from stereo optical images, with 30 m postings. Compare this with the Landsat image (Fig. 18.6) and the DEM generated by radar interferometry during the SRTM mission (Fig. 18.26). Profiles 1 and 2 in Fig. 18.14 were extracted from horizontal (E–W) lines about 25 % from the *top* and *bottom* of this image, respectively

inertia', I, which is defined as $I = (k\rho c)^{0.5}$. This is measured in the rather ungainly aggregate units of $Jm^{-2} K^{-1} s^{-0.5}$, which are mercifully abbreviated as 'SI units' or 'tiu— Thermal Inertia Units'. Rocky surfaces typically have thermal inertias of several hundred. A value of 200 tiu is typical for sand, although measurements of I for the north polar erg on Mars are lower, 75 tiu, perhaps indicating a smaller grain size. Fine-grained sand can have a high tiu value if the grains are cemented together ('indurated') and thus the mobility of a dune may be connected to its observed thermal inertia. Studies of several Mars regions have been made, e.g., the dunes in Proctor crater (Fenton and Mellon 2006) and the north polar erg (Putzig et al. 2008).

Plots of surface temperature through the course of a day clearly show that there are times in both the early morning and the late afternoon when curves for different thermal inertias tend to cross (Fig. 18.8). Near these daily crossover points, it is the most difficult time to determine the thermal properties of geologic materials, so measurements obtained during both day and night are best for constraining the thermal properties. The minimal variation in $\rho c$ among geologic materials means that most thermal inertia variations on Mars can be attributed to changes in the thermal conductivity; for Earth, the effects of the thick atmosphere and the potential presence of surface water make such an assertion far more problematic. Albedo differences on the surface can shift the daily temperature curves up or down without greatly altering their basic shape.

On Mars, many areas (and especially the north polar erg) are covered for part of the year in a seasonal $CO_2$ frost which pins the surface at a uniform value (day and night) of about 150 K. It goes without saying that sand sealed under such a frost layer cannot saltate.

On Earth, the Landsat Thematic Mapper has a thermal channel—often dunes may appear to have a quite different contrast in the thermal images compared with the optical reflectivity. In orbit around Mars, a series of instruments has flown starting with a radiometer on Mariner 9 with varying spectral and spatial resolutions. The highest spatial resolution instrument (and thus arguably the most useful for dune studies—e.g., Fig. 18.9) is THEMIS, whose thermal imaging resolution is about 100 m.

At Titan, Cassini's Composite Infrared Spectrometer (CIRS) has very modest spatial resolution, that nonetheless has been able to detect that the large dark sand seas on Titan are about $\sim 1$ K warmer than their surrounds, and that there

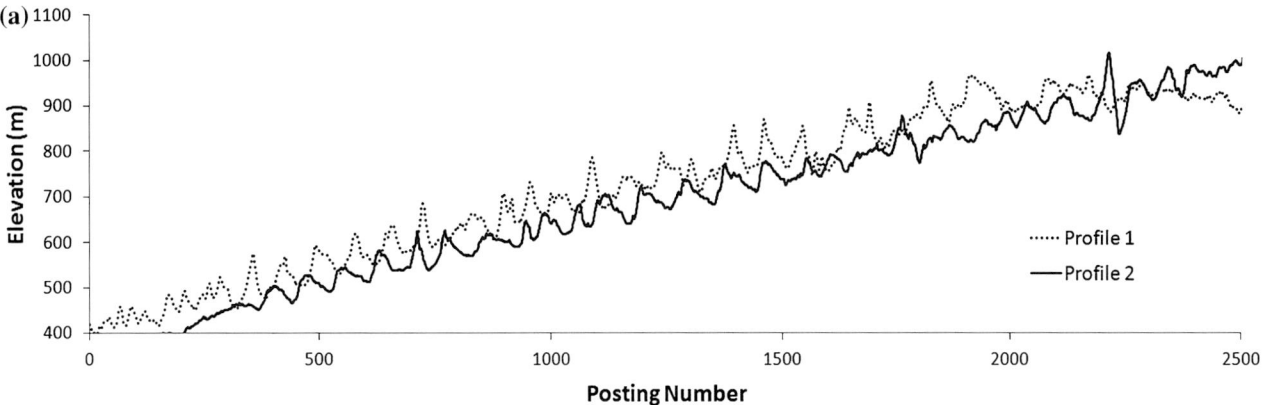

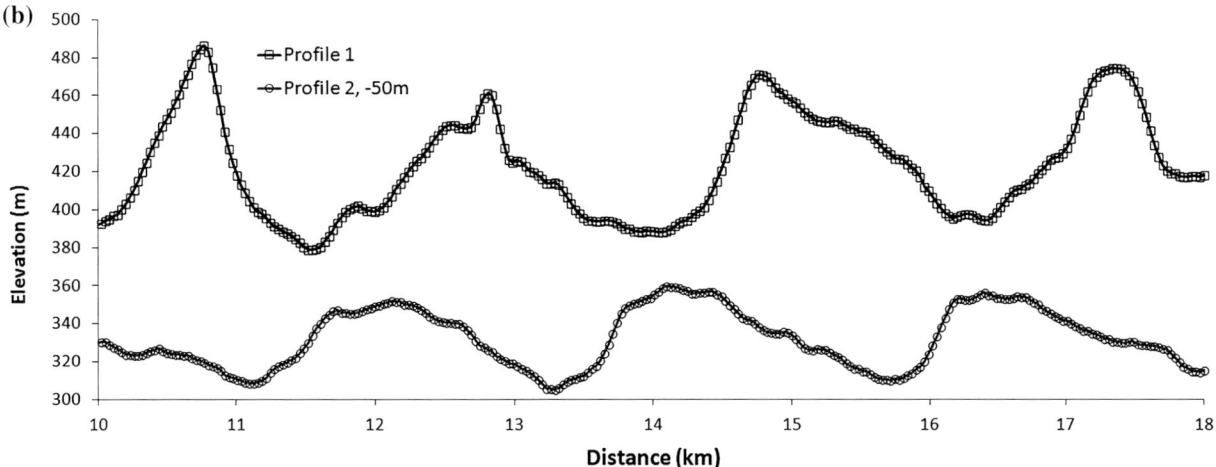

**Fig. 18.14** **a** Profiles across the DEM in Fig. 18.13. Note the overall slope, dropping about one dune-height every five dune wavelengths towards the Atlantic to the west (*left*). **b** Zoom-in on parts of the profile to make individual postings visible (at 30 m intervals), with a linear trend removed—the regularity of the profiles is seen. Compare these profiles with GPS profiles

is some indication of diurnal temperature change. These findings are consistent with overall dry, porous sands with a low albedo, although Titan's thick atmosphere will probably preclude the more quantitative analysis of thermal inertia that has been possible at Mars. On Venus, sadly, the atmosphere is so thick and hot that thermal observations from orbit are impossible.

## 18.4 Hyperspectral Imaging

The reflectivity of a mineral can be a complex function of wavelength (Figs. 18.10 and 18.11); this is after all why some minerals—especially ores of transition metals—are pretty colors. Thus color imaging can give insights into the composition of surface minerals, including those segregated into dunes. The ability to discriminate different compositions can be enhanced by looking at a large number of narrow wavelengths across a wide range of wavelengths into the infrared, rather than just the three wide bands of primary color our eyes have evolved to exploit (which after all serve principally to discriminate what is good to eat).

A plethora of instrumental techniques exist to develop this sort of information. One possibility is narrowband imaging through a number of filters—usually giving a high-quality image (high spatial resolution, but poor spectral resolution). Another approach is a point spectrometer wherein the light (or emitted thermal radiation in some cases) from a patch of terrain is passed through a grating or scanned with an interferometer, giving a high-quality spectrum. In between there are many variants: imaging spectrometers, hyperspectral imagers, with all kinds of variations of scanning and sweeping. A common approach for orbiters has a slit with a grating projected onto a 2D detector, with one dimension corresponding to a spatial position along the slit (and thus along the ground) and the other dimension corresponding to wavelength. If the slit is

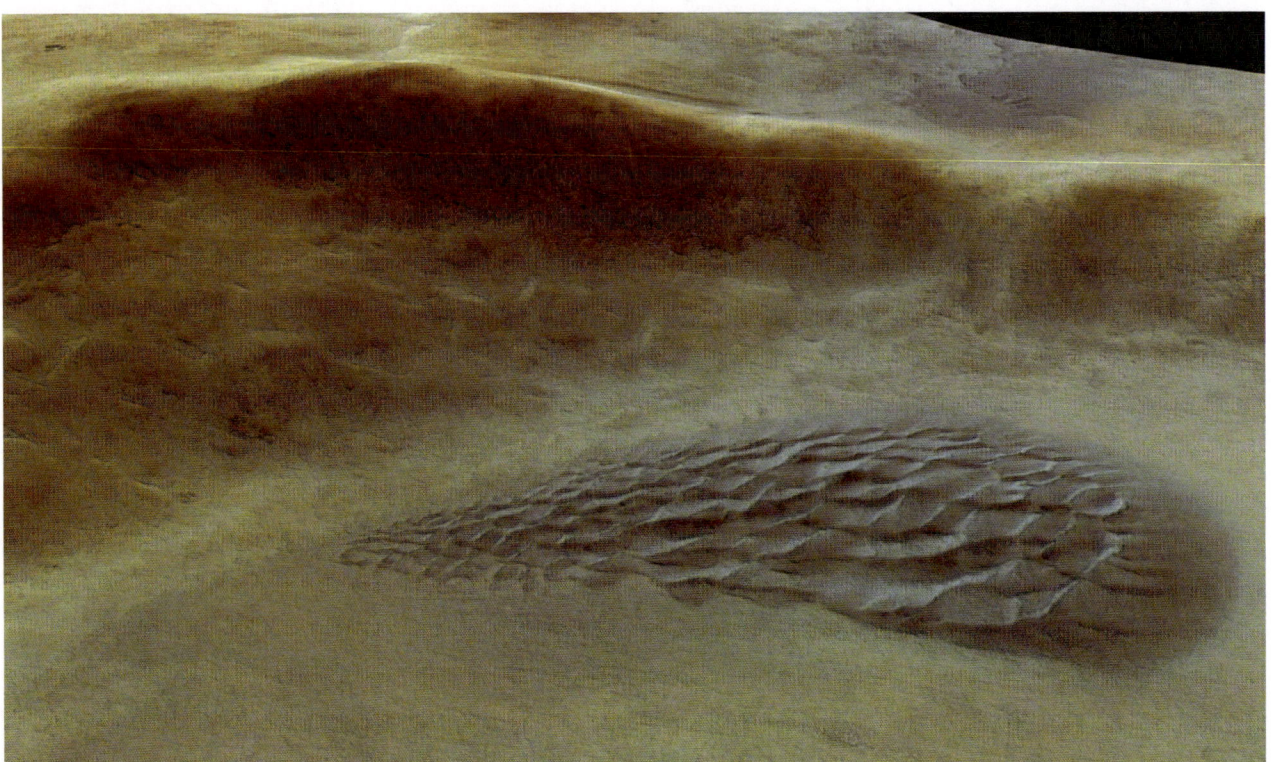

**Fig. 18.15** A view generated by draping color nadir imaging from the Mars Express HRSC over a DEM generated from stereo images acquired on May 27, 2004, and shows the crater with a dune field located in the north-western part of the Argyre Planitia crater basin. The image is centred at Mars longitude 303°E and latitude 43°S. The image resolution is approximately 16.2 m/pixel. *Credit* ESA/DLR/FU Berlin (G. Neukum)

aligned orthogonal to the orbital motion of the spacecraft, then a long noodle of ground, hundreds of pixels wide, is covered, with a spectrum for each pixel.

Data of this type are acquired by several instruments flying today at dune worlds. Cassini has a Visual and Infrared Mapping Spectrometer (VIMS) which has provided information on the composition and geometry of Titan's dunes. At Mars, prominent instruments have been CRISM (Compact Reconnaissance Imaging Spectrometer for Mars, built at the first author's present institution) and OMEGA (Observatoire pour la Minéralogie, l'Eau, les Glaces et l'Activité). CRISM (Fig. 18.12) obtains spectral images over a $\sim 10$ km-wide swath with a spatial resolution of $\sim 40$ m, analyzing light from 0.36 to 3.92 μm with a spectral resolution of 6.5 nm; OMEGA's performance is spectrally similar but with spatial resolution of only $\sim 300$ m (CRISM is a much more modern design, reaching Mars on the Mars Reconnaissance Orbiter in 2006, OMEGA originally being developed for the Russian Mars-96 mission in the mid-1990s, being reflown on Mars Express in 2003). Data from these instruments suggest that the sand in Mars' north polar erg (see Chap. 12) contains gypsum (e.g., Horgan et al. 2009).

## 18.5 Topography From Imaging

We are evolved to instantly derive the shape of things by reconciling the slightly different views from our two separated eyes: stereoscopic vision. The same geometric operation can also be performed with image pairs or sets taken from orbit, and is now done rather systematically on both Earth and Mars.

A near-global topography dataset for Earth, spanning 83°N–83°S, has been generated using stereo-pair images by the ASTER (Advanced Spaceborne Thermal Emission and Reflection Radiometer) instrument onboard the Terra satellite, as a result of a joint effort between NASA and the Japanese Ministry of Economy, Trade, and Industry (METI). The instrument has shortwave- and thermal-infrared cameras, as well as a 15 m-resolution visible and near-IR camera which features a nadir-looking and a canted backward-looking imager. The nadir and canted images form a stereo pair from which a 60 km × 60 km digital elevation model (DEM) can be extracted (Figs. 18.13, 18.14). Hundreds of thousands of these local DEMs have been assembled into a global product covering 99 % of the Earth's landmass with 30 m postings. This product is a

## 18.5 Topography From Imaging

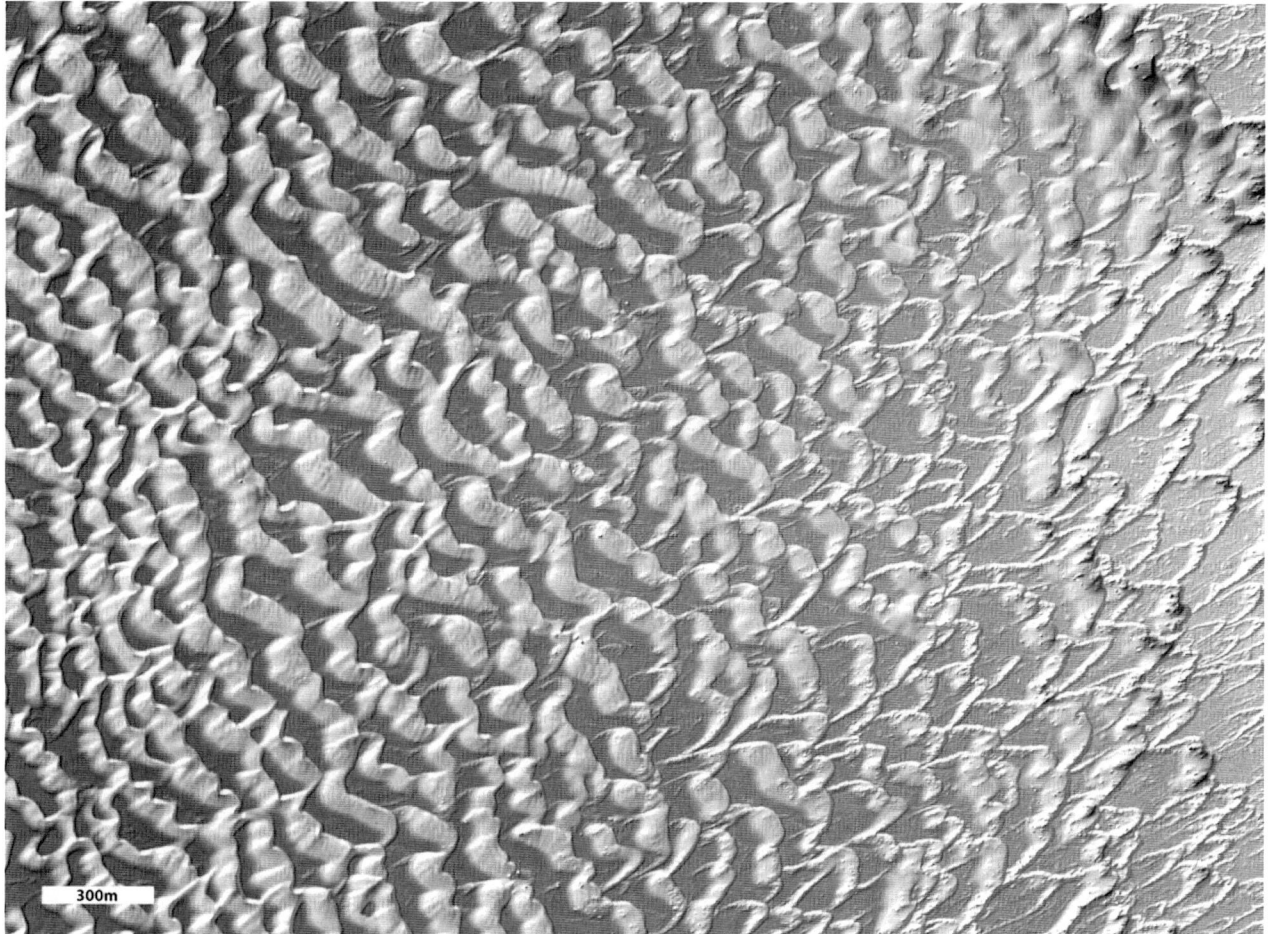

**Fig. 18.16** Shaded relief LiDAR topography of dunes at White Sands National Monument. Sand migration is from *left* to *right*, and over the ~4 km from one side to the other vegetation becomes progressively more abundant and the dunes grade from barchanoid to parabolic (Note the inversion of the crescent.) Uniform data of this sort are excellent for morphometric analyses. Data acquired in January 2009 by the National Center for Airborne Laser Mapping (NCALM) *Image courtesy* Ryan Ewing

convenient one for dune studies (e.g., Hugenholtz and Barchyn 2010).

In many respects, this DEM (made with years of images) is quite comparable with that made with just a few days of data from SRTM (see Radar imaging section). An optical DEM relies on daylight, and also on cloud-free conditions, although these are usually the norm for many areas with dunes. The optical DEM does not suffer from the decoherence gaps that the radar interferometry approach can suffer from in radar-dark areas (compare Figs. 18.13 and 18.26). Different applications may favour one dataset over another.

The same imaging approach has been carried out at Mars by the German-built HRSC (High Resolution Stereo Camera) on the European Mars Express spacecraft, in orbit since 2003. In fact, HRSC should have got to Mars some 6 years earlier, on the Russian Mars-96 mission, but this spacecraft was lost soon after launch; the Mars Express instrument is a refurbished spare. This camera performs simultaneous forward and backward (18.9°) as well as nadir imaging with 10 m pixels (and a 2.5 m high resolution channel). This instrument has yielded many spectacular color images and DEMs (e.g., Fig. 18.15) including some of dune features. More recently, very high resolution DEMs have become available from the HiRISE imager on Mars Reconnaissance Orbiter (e.g. Fig. 12.10).

In the absence of stereo imaging, topography can be estimated by radar or laser altimetry, and stereo image products are often validated by comparison with altimetry (which is usually better-calibrated at an absolute level). The global-scale altimetry products from this type of instrument—typically with footprints spaced hundreds of meters apart—are only just able to resolve large individual dunes. At Mars, the MOLA instrument did detect the North Polar Erg (Neumann et al. 2003), although morphometric studies at Mars have tended to use stereo DEMs. On Earth, data

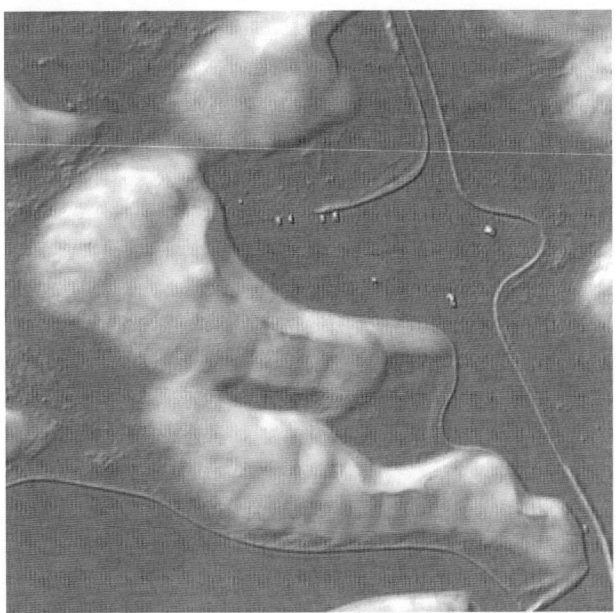

**Fig. 18.17** Shaded relief LiDAR map of White Sands: Zoom into an area just ~200 m across showing the shape of individual dunes, which are up to about 7 m high. The looping lines a berm about 0.5 m high, bulldozed to maintain clear parking areas and roads; the small rectangular features are cars and visitor facilities. Compare with kite camera image (Fig. 11.15). Note the wave texture on the dunes; these presumably correspond to the destabilization wavelength (see Chap. 5). Data acquired in January 2009 by the National Center for Airborne Laser Mapping (NCALM) through Seed Grant program to Ryan C. Ewing. *Image courtesy* Ryan Ewing

from the ICESAT laser altimeter have been used to study Antarctic megadunes of snow (on both Earth and Mars, these altimeters are on satellites in near-polar orbits, so the density of measurement points is maximized where the groundtracks converge at high latitude) as well as to document dune advance in the Arabian Peninsula (of 6 and 25 m over a 6-month timeframe, Dhabboor 2008).

In recent years, computational tools to perform topographic modeling from 'random' images (as opposed to the well-posed pairs in the satellite systems above) have become widely available, and it seems that taking a series of images with a simple digital camera may be the preferred way of measuring topography in the field at scales smaller than the orbital datasets. One caution should be noted: stereo matching relies on finding tiepoints in the images so in a surface covered with a material that is uniform in brightness and color, such methods may fail. This may cause problems for studies of some dunes with very uniform sand properties, especially on overcast days where the lighting is isotropic. Thus laser ranging does have application at smaller scales, e.g., from aircraft and at ground level (the latter was discussed in Chap. 16).

Airborne laser scanning does an excellent job of obtaining precision topography over areas up to several kilometers across. Such topography is good for defining subtle (e.g., partly vegetated) dunes such as the parabolic dunes on the Canadian prairies (e.g., Wolfe and Hugenholtz 2009) and on particularly uniform-colored dunes such as those at White Sands (see Chap. 11). Some examples of the airborne lidar there are shown in Figs. 18.16 and 18.17.

## 18.6 Radar Studies of Dunes

RADAR (RAdio Detection And Ranging) has evolved from its origins as a means of detecting aircraft and ships to a wide range of sensing techniques, both imaging and nonimaging. We discuss certain nonimaging techniques (GPR) later in this section, but of most interest to geomorphologists is imaging radar, which is in fact our principal source of knowledge about dunes on Venus and Titan. In part for this reason, we devote some space in this book to describing radar remote sensing.

Most imaging radar uses a technique called Synthetic Aperture Radar, or SAR. This relies on the radar platform moving relative to the scene, and the radar echoes are processed into an image by isolating echoes from different parts of the scene by their different echo time and different Doppler shift. This computationally-intensive processing allows an image with much higher spatial resolution (i.e., smaller pixels) to be synthesized than the footprint of the antenna (which is determined by the wavelength and the size or 'real aperture'). This magical transformation of the signal into an image was originally performed with curved mirrors, but is now rather straightforward to perform with modern computers. The first civilian SAR imaging of the Earth was accomplished in 1978 by NASA's Seasat, which imaged land surfaces as well as the sea (though the fact that a radar reconnaissance satellite experiment, 'Quill', was flown some 14 years earlier was only declassified as we write this in 2012!). An example Seasat image, of the Algodones dunes, is shown in Fig. 18.18.

After Seasat, several brief but productive experiments were made with the space shuttle (Shuttle Imaging Radar, SIR-A in 1982; SIR-B in 1984; and SIR-C, augmented with a German X-band radar X-SAR, flew twice in 1994). In the meantime, after a number of false starts, Venus mapping radars were developed in the 1980s, with the Soviet Venera-15 and -16 flying in 1983, and NASA's Magellan in 1989, and several other countries developed Earth-observing radar satellites. These include the European Space Agency ESA's ERS-1 (European Remote Sensing Satellite) in 1991, ERS-2 in 1995 and Envisat in 2002; the Japanese JERS-1 in 1992, and Canada's Radarsat-1 and -2 (1995 and 2002); several countries since have launched radar imaging satellites, largely for military purposes. Since SIR-C, the only imaging radar NASA has flown has been Cassini (in 1997). Many of the satellites mentioned above also conduct nonimaging

## 18.6 Radar Studies of Dunes

**Fig. 18.18** One of the first spaceborne radar observations of dunes. The Algodones Field. Mountains are seen at *right*, while the checkerboard pattern at *left* shows the different reflectivity of different crops, ploughing and irrigation on fields. The dunes in this image (*center* band of irregular strips—compare with optical image in Fig. 7.14) are bright due to topographic shading on a relatively deep (and thus dark) sand substrate

radar studies (altimetry and scatterometry), and NASA has flown a number of scatterometers since Seasat—e.g., NSCAT and Quickscat—and an NSCAT view of the Earth showing the major sand seas as 'stealth' areas with poor radar reflection is shown in Fig. 18.19.

The spatial resolution achieved by Earth-orbiting SARs is usually a few tens of meters; that of Magellan was about 100 m, while Cassini achieves about 300 m. The footprint of scatterometer measurements is often several tens of km. The first dune observations from airborne SAR as well as Seasat and the shuttle experiments were reported by Blom and Elachi (1981, 1984, 1987).

The interpretation of a radar image requires some familiarity with how radar is reflected from the scene. Crudely, radar is a radio equivalent of a flash photograph—the source of illumination and reception of the reflected energy are co-located. This means that any facets of a reflective surface that are normal to the illumination will reflect strongly, whereas more generally, very smooth surfaces that are not normal to the illumination will look dark. For typical scenes of planetary surfaces observed at incidence angles well away from vertical, all else being equal, radar brightness is often a measure of roughness—rougher surfaces tending to look brighter (see Fig. 18.20). It should be understood also that here 'roughness' is defined on a scale comparable with the wavelength of the radar: thus a surface that is 'smooth' at a long wavelength can be 'rough' at a shorter wavelength, and will look dark in one image and bright in the other. There is a correlation between the roughness one might infer from a radar image, and the 'aerodynamic roughness' (see Chap. 3) related to the wind drag on the ground (e.g., Greeley et al. 1997).

For historical reasons, radar wavelengths are often referred to by bands designated by letters, associated with the introduction of different kinds of radar in World War 2. Specifically, most terrestrial radar observations have been at L-band (23.5 cm: Seasat, SIR-A, B, C, JERS-1) and C-band (5.8 cm: SIR-C, Radarsat, ERS-1) with a few shorter-wavelength observations at X-band ($\sim 3$ cm, XSAR on SIR-C). Ku-band (2 cm) is often used in nonimaging radars such as altimeters and scatterometers, but is also used by the Cassini radar system used to observe Titan.

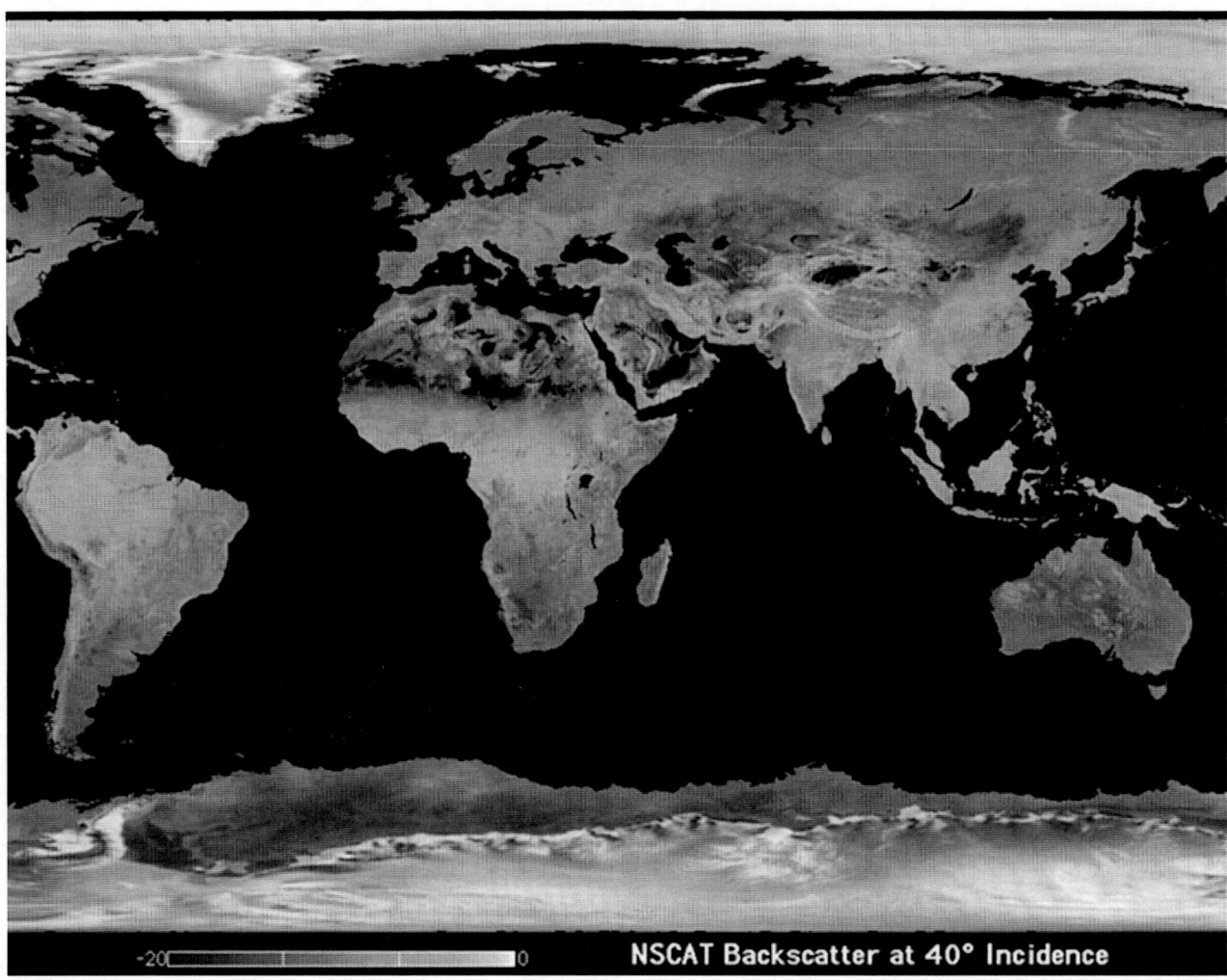

**Fig. 18.19** A near-global low-resolution radar backscatter (Ku band—2 cm) map of the Earth's land surface. Ice-covered terrain is exceptionally bright, while forested areas (Brazil, Equatorial Africa and S.E. Asia) are somewhat bright and uniform. The darkest areas are the Sahara, Arabian and Gobi deserts, compare with the optical appearance of Earth in Fig. 11.2. *Map courtesy* David Long, Brigham Young University

Shorter wavelengths allow more compact antennas to be used and generally achieve higher spatial resolution. The Venus atmosphere is in fact thick and absorbing enough that Ku-band radiation does not penetrate (Cassini did not receive an echo when it tested its radar during a Venus flyby en route to Saturn and Titan); for this reason, Magellan at Venus used a 12.6 cm wavelength. The information from different wavelengths can be combined as different primary colors (e.g., Fig. 18.21).

Beyond the choice of wavelength, there are other important aspects (of a radar) that influence the appearance of surfaces in an image (to the point of making some bright surfaces dark and vice versa). These include the incidence angle, and the polarization (e.g., Blumberg 1998). Further, because the energy returned from a scene depends so much on the orientation of surface facets at the instant of the observation, a radar image can often look 'speckled', which can appear noisy (even though the formal signal-to-noise can be high). Yet a fraction of a second later, when the imaging spacecraft has moved on some hundreds of meters and the geometry has changed, a different (but still speckled) image would result. A more interpretable image often results when many of these instantaneous images ('looks') are combined together. Thus the number of looks in an image is often an important quality metric. There are other, geometric, aspects of radar imaging that can be challenging to those unfamiliar with them, like 'layover' (the apparent tilt of mountains towards the observer due to the way the range to the target is mapped onto the image). Perhaps due to the complexity of these factors which may seem daunting to people other than electronics engineers, and perhaps due to the poorer accessibility of data in the pre-world-wide-web days when imaging radar first became available, the full potential for radar in dune studies has not yet been penetration of the radar energy into a material tends to increase with longer wavelength (and 'ground penetrating radar' (GPR) used in field measurements tends to use longer wavelengths than imaging radars—often about 1 m).

Generally, sand dunes look somewhat dark to radar because dunes are usually smooth on the scale of the radar wavelength and thus specularly reflect most of the energy away from the transmitter, and the dune has little internal structure to scatter any energy that does enter the dune. Thus apart from topographic glints, dunes generally appear dark against what is usually a rougher and/or denser interdune, and indeed, on the global scale, the sand seas are the darkest land areas (see Fig. 18.19). However, brightness on the dunes themselves can be highly variable, with strong glints from slip faces that are oriented towards the radar (see Fig. 18.23).

Whether truly specular glints are present or not, the backscatter curve (e.g., Fig. 18.20) can be exploited to gauge the average orientation of the surface within the pixel. This approach, termed radarclinometry as an analog to photoclinometry, works best on surfaces that have a uniform composition. With a known viewing geometry and a backscatter curve assumed to be uniform, the slope and thus the height of resolved dunes can be calculated. This was done on Titan by Lorenz et al. (2006), determining heights of over 100 m: Neish et al. (2010) performed further measurements on Titan, and also showed that the method works on terrestrial dunes (in the Namib—Fig. 18.24) and is not significantly degraded by the lower resolution of the Cassini radar compared with terrestrial radars.

Measuring dune size, spacing and orientation is of course as easy with a radar image as with an optical one. Because radar engineers are familiar with working in the frequency domain, it has been natural to apply spectral methods (Fourier transforms) on the image (e.g., Qong 2000; such approaches are more routinely applied to SAR images of the sea surface, to determine wavelength and orientation of waves, e.g., associated with hurricanes, so it is a straightforward step to apply the same algorithm to dunes).

However, unlike (e.g., polar terrain) dunes tend to be found in dry deserts, where optical visibility is good and thus at the spatial scale afforded by radar imaging, the terrestrial sand seas had already been mapped optically (e.g., McKee et al. 1979). While radar has been the principal tool in discovering and characterizing dunes on Titan and Venus (which are optically very challenging), most of the terrestrial literature has been devoted to understanding how dunes appear to radar (i.e., assessing the role of different contributions to the echo) rather than to actually discovering something new about the terrestrial dunes. Some likely potential applications are to estimate the small-scale slope distributions over wide areas, or to estimate the thickness of interdune sand deposits.

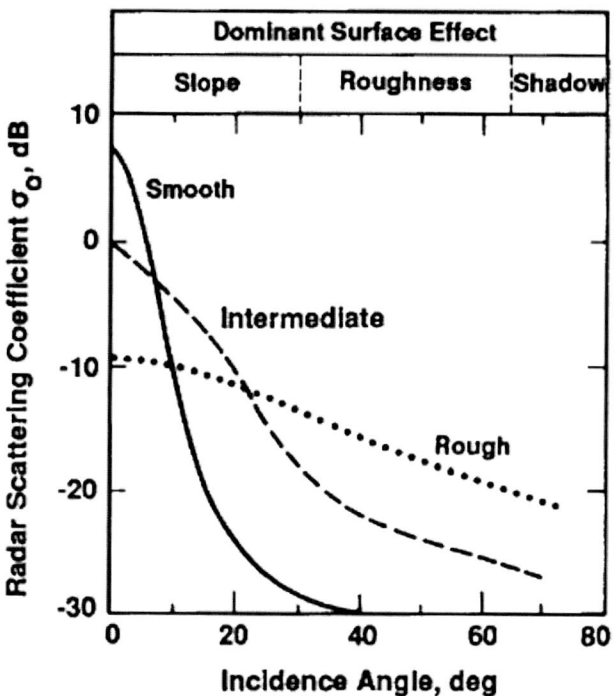

**Fig. 18.20** Brightness of a horizontal surface illuminated by radar as a function of roughness and incidence angle (0 incidence is vertical). The effective incidence of an arbitrary surface will also depend on its slope relative to the illumination direction—this sensitivity is particularly strong at low incidence angles, making radarclinometry possible and leading to glints. At shallower incidence angles, the wavelength scale roughness is usually the dominant factor. (Optical reflectivity shows similar behavior, but most geological surfaces are optically rough.)

realized. There are a number of good texts on radar for planetary surface observation, notably Elachi (1988); the three-volume epic by Ulaby et al. (1982) is widely regarded as the classic text for serious radar scholars. A short but useful guide was produced for the Magellan mission (Ford et al. 1993) and is available online.

For now, we will discuss radar reflectivity and the interpretation of images in general terms (and Fig. 18.22 summarizes this). The relation of reflectivity to slope and roughness is adequate description for many terrestrial surfaces where the radio energy is simply reflected at the surface, especially where water is present. However, radar backscatter can be complicated somewhat by the penetration of radar energy into a target and its subsequent reflection within the material, termed 'volume scattering'. This can occur in materials that are free of moisture (which includes very cold ice), and in fact is quite common for porous, dry materials like desert sands, and for the lunar regolith. Radar can often see 'through' sand sheets with thicknesses of tens of centimeters, and the SIR experiments revealed, for example, river channels in bedrock that had been buried by more recent aeolian deposits. The

**Fig. 18.21** Multi-wavelength Space Shuttle Imaging Radar image of the Namib near Luderitz. The Atlantic Ocean is at *left* and the town itself. In the *middle* are transverse dunes, and linears inland at *right*. The colors indicate the relative contribution of different wavelengths, which respond to different length scales of roughness. NASA/JPL image PIA01856

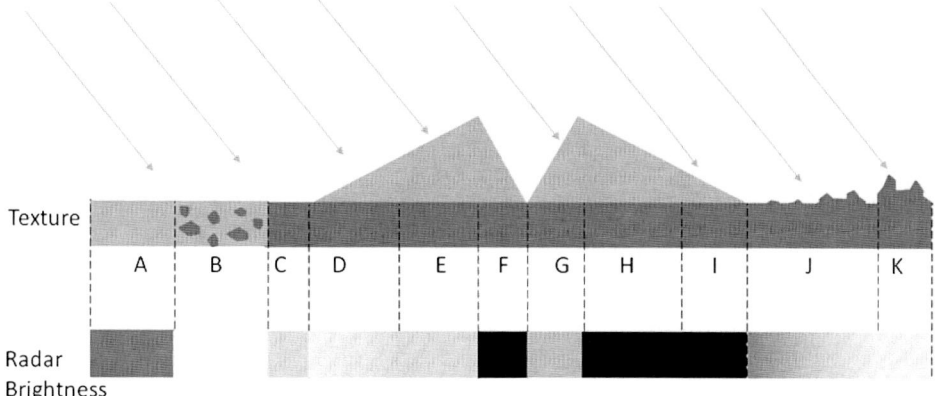

**Fig. 18.22** A schematic of the radar-brightness (*bottom bar*: *white* indicates a strong reflection, *black* a weak one) as a function of surface texture. In the surface texture schematic, *light grey* indicates low dielectric constant material like loose, dry sand, whereas packed clay or rock is shown as *dark grey*. Radar energy illuminates the surface from *upper left* as shown by *arrows*. A deep loose sand surface (*A*) is somewhat dark, whereas if the sand contains many boulders larger than the wavelength scale (*B*) the reflection can be bright due to volume scattering—remarkably bright in some cases where the reflections are coherent. A rock or other dense substrate (*C*) gives a brighter reflection (all else being equal) than a sand substrate (*A*). Radar energy can penetrate a thin layer of sand, and in fact the sand can focus the energy leading to a strong reflection (*D*). If a smooth dune facet (*D, E*) is almost normal to the incident radar energy, a strong reflection will result, often seen as a glint in a radar image. In contrast, facets inclined away from the beam will reflect very little energy back to the radar and so will be dark (*F, H*). The brightness is a function of angle (see Fig. 18.20), thus the facet at *G* is inclined further from the beam than at E and so is slightly darker; this angular dependence is the basis for radarclinometry. Note that the thin sand-on-substrate at I is still dark to a monostatic radar; even though the surface reflects energy, most is going away from the sensor. A given rough horizontal surface illuminated by an inclined beam will generally be brighter the larger the roughness (*J–K*). Note that dunes with different stoss and lee slopes in the plane containing the radar beam will generate an anisotropy even if the individual features cannot be resolved—the brightness averaged over *D–E–F* is substantially higher than *G–H–I*. Such an anisotropy has been used to infer small dune textures in Magellan images of Venus (see Sect. 18.3). Different surfaces with the same notional brightness but different scattering mechanisms (e.g., *B* vs. *D* vs. *K*) can sometimes be discriminated by radar with different polarizations. The scale of roughness elements (*B, J*) can often be estimated by comparing reflectivity in different wavelengths

**Fig. 18.23** Shuttle Imaging Radar (SIR-C) scene of the Namib. The contrast between slopes toward and away from the radar is dramatic, much more than is typical in optical images of the same scene; the fine-scale information in such radar images has not been fully exploited. Note that the glints here are strong enough that the radar energy bleeds into adjacent pixels (especially in the Doppler direction) giving the *dotted lines* flaring away from the brightest points

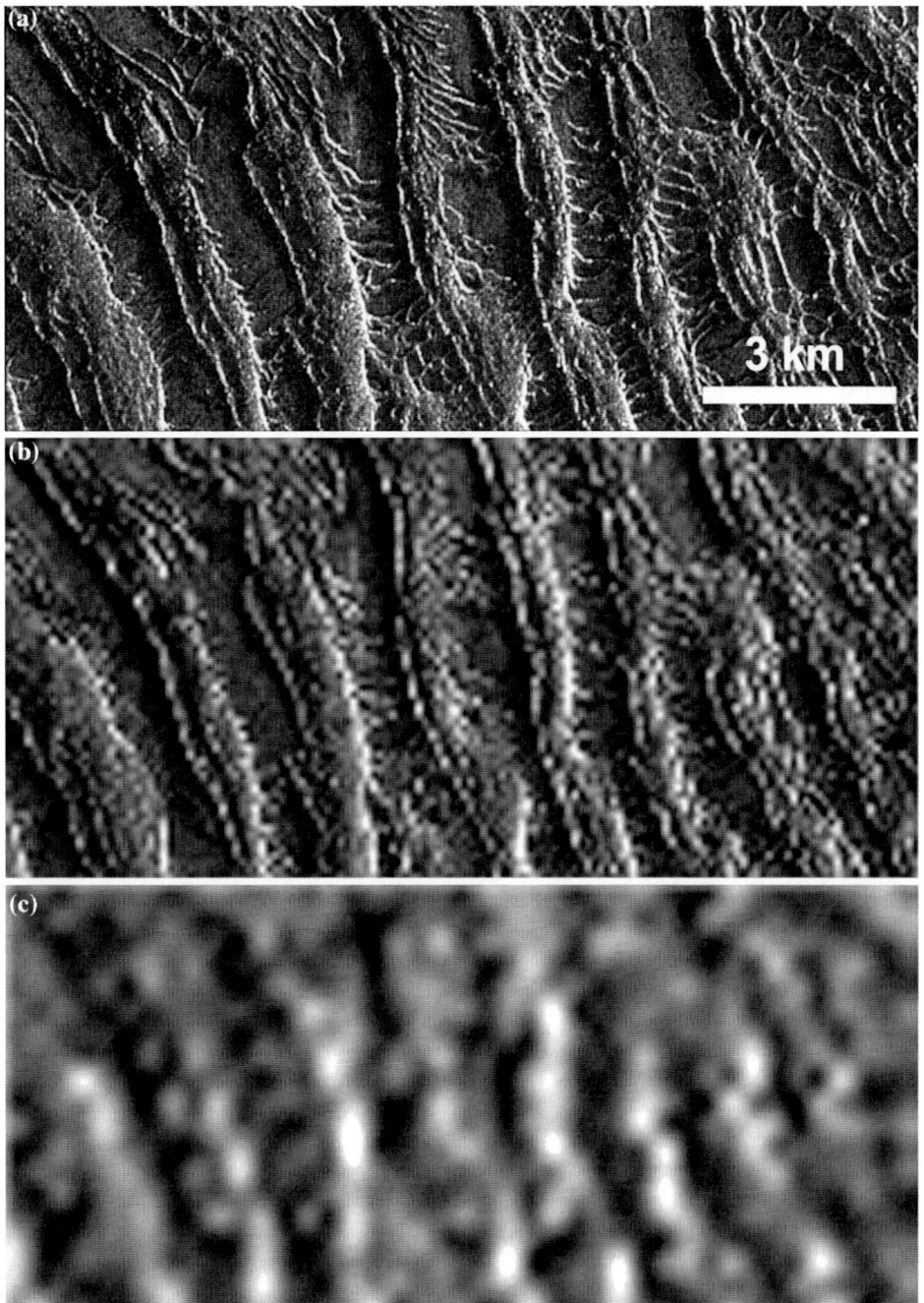

**Fig. 18.24** Space Shuttle Imaging Radar SIR-C of the Namib. The image is shown at **a** full resolution (25 m/pixel); **b** degraded to 100 m, comparable with Magellan imaging of Venus; and **c** degraded to 300 m/pixel, comparable with Cassini imaging of Titan. Notice that the interdune areas in this image are not especially bright. Note also the small duneforms visible in the full-resolution image orthogonal to the main north–south dunes

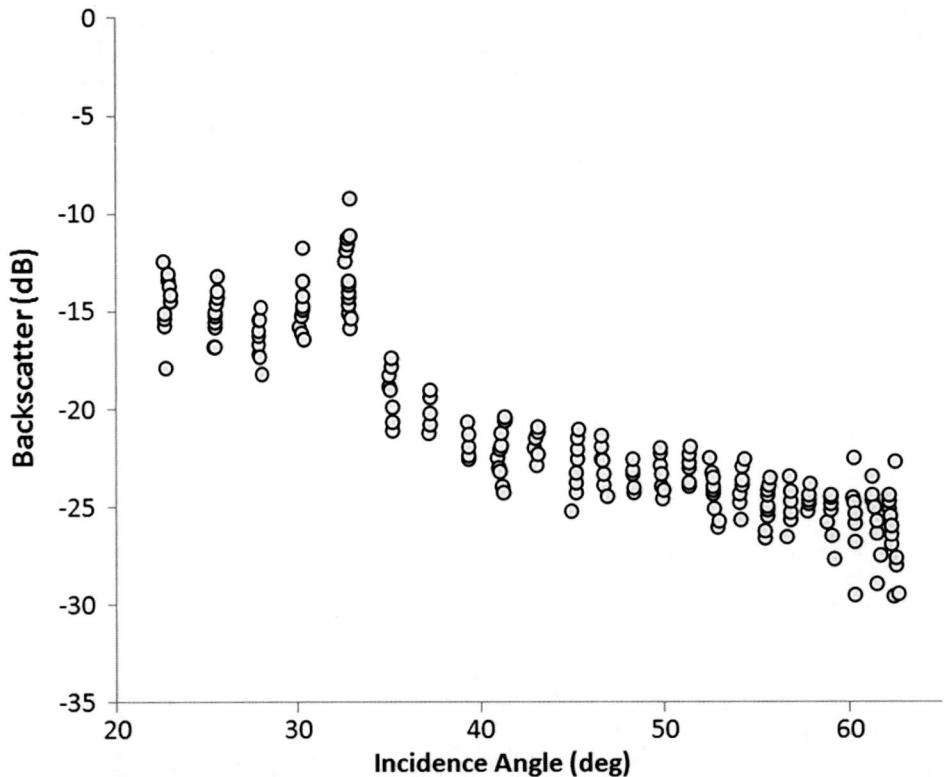

**Fig. 18.25** Scatterometer observations of part of the Sahara covered in linear dunes. The observation does not resolve individual dunes, but the systematic alignment of the dune faces leads to a peak in backscatter close to the angle of repose (compare with Fig. 14.6). Data from the NSCAT scatterometer, from Stephen and Long (2005). A similar backscatter function at Venus has been used to infer the presence of unresolved microdunes. Analysis of this type has been attempted at Titan but was confounded by the footprint being dominated by scatter from bright interdune areas

Even when a radar does not resolve the shape of a dune, the presence of dunes (or ripples) can be inferred from an azimuthal variation in radar backscatter of a region. Specifically, the presence of facets oriented towards the radar can give an overall brightness increase. It was thought that Bragg scattering (a coherent reflection from periodically corrugated surfaces—in essence, the surface acts like a diffraction grating) from ripples might be significant, as it is for ripples on the sea, but this generally seems not to be the case.

On the other hand, where (unresolved) dunes are present, there may be an anisotropy in the slope distribution (e.g., all the slip faces in the downwind direction) leading to a strong variation in the total brightness as a function of azimuth and/or elevation. The azimuth variation, in particular, has been used to infer the presence of small dunes in some areas on the surface of Venus (see Chap. 14). Similar azimuth variations can be measured by scatterometers in some sand seas; these also (Stephen and Long 2005) can show non-monotonic variation of backscatter with incidence angle (see Fig. 18.25). The backscatter function of sand seas on Titan has been studied by Le Gall et al. (2011, 2012) with some important results on the dune/interdune ratio and its variation with latitude.

A particular technique for detecting very small (cm-scale) shifts in surfaces is Interferometric SAR. This relies on the orbit of the platform being well-controlled, such that the illumination conditions between two successive images are identical. Changes of a fraction of a wavelength in the pathlength between the platform and the scene can be detected as a change in the phase of the returned echo; this technique has been used to detect, for example, the deformation of volcanoes such as Kilauea and the relative movement of terrain after earthquakes. The technique has been applied to sand dunes (e.g., Bodart et al. 2005); among the examples in their study are image pairs acquired 24 h apart (with about 100 m separation) and it is clear that the biggest phase differences (and thus the largest surface changes) in the scene took place at the slip faces of dunes.

Exploitation of the same effect was used by a radar system flown on the Space Shuttle on an 11-day mission in 2000. This Shuttle Radar Topography Mission (SRTM—e.g., Farr and

## 18.6 Radar Studies of Dunes

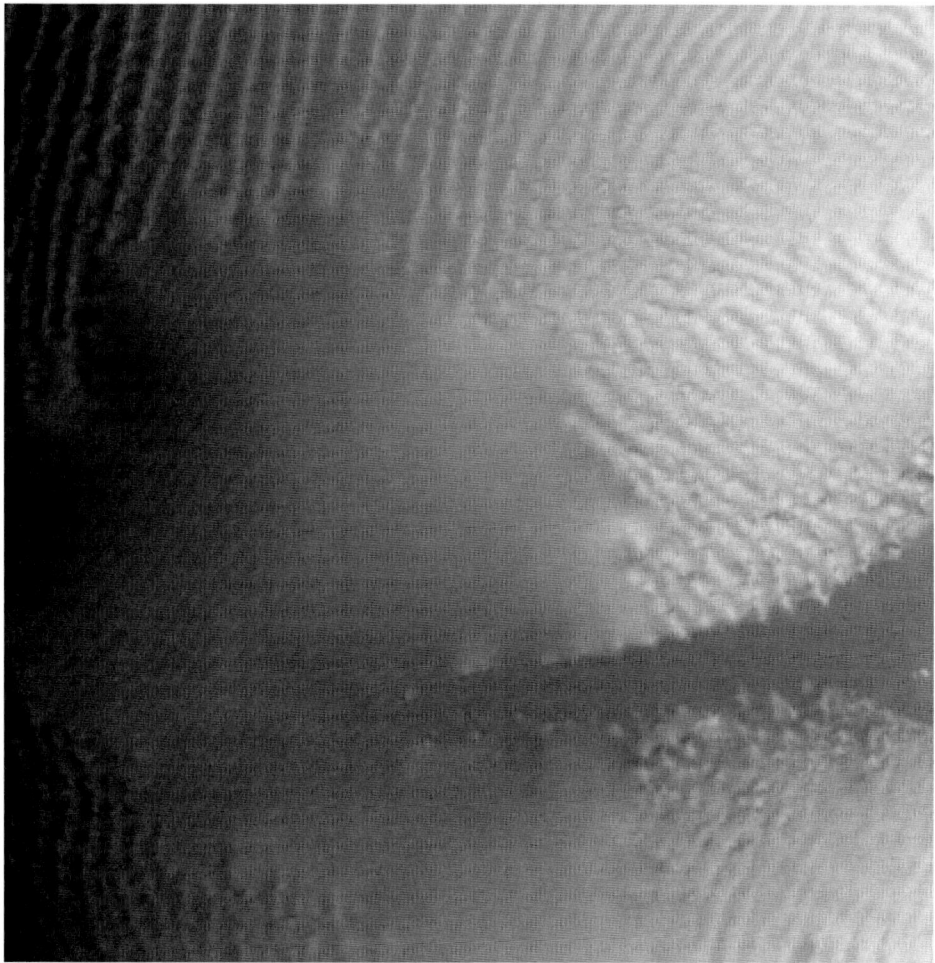

**Fig. 18.26** Topography generated from SRTM data, showing part of the Namib. The uniform-quality topography with 90 m spacing over 80 % of the land area of the Earth opened new vistas in geomorphological studies. This tiny example part of a scene is actually one of the worst-quality parts of the whole product: the patch at the *right* where data are missing is because the reflection from the deep sand was too weak to permit the interferometric reconstruction of the surface height. Neish et al. (2010) used this data as 'ground truth' to compare with topography generated from radarclinometry on Titan. Compare this topography with that generated by stereo optical imaging in Fig. 18.13

Kobrick 2000) used a radar with an additional receive antenna at the end of a boom that was some 60 m long: the phase difference between the echo received at the main antenna onboard and this outboard antenna was used to generate a precise topographic map of the Earth's surface between 60°N and 54°S latitude, covering about 80 % of the Earth's land area (this was limited by the Shuttle's orbit). This dataset, with 16 m vertical accuracy and measurements at 30 m spacing (though the generally-available data products have a 90 m spacing), is a powerful tool in the study of large dunes, although some patches in sand seas are actually too radar-dark for the technique to work correctly (see Fig. 18.26). A couple of papers point out the applicability of the DEM for dune studies (e.g., Blumberg 2005; Potts et al. 2008) using examples in the Namib and Taklimakan deserts.

One advantage of SAR in the study of dunes is that the imagery is independent of solar illumination, and thus can take place at night or through cloud. The presence of water can strongly affect the appearance of a scene to radar, however, so some care is needed if an image of a dry scene is to be compared with one after rainfall.

SAR is of course an important technique on planets where an optically thick atmosphere inhibits visual or near-infrared observations—in particular Venus and Titan.

# Numerical Models

# 19

The last 1–2 decades have seen a transformation in dune studies due to our growing ability to simulate using numerical models on computers the wind-driven processes that shape dunes. The transformative aspect is that, due to computer power, these simulations can now study the processes in a time- and spatially-resolved way (and not just in the bulk-averaged manner considered by Bagnold and others). These models are able to reproduce the wide range of observed morphologies and behaviors, bringing a new level of understanding to dune studies. A deep hierarchy of modeling approaches now exist, pertaining to dunes themselves as entities, to the transport of individual particles, or to the wind, or some combination of these. We will discuss these in approximate order of maturity.

## 19.1 Modeling the Wind

Applying numerical models of global or regional airflow to dune studies is perhaps more highly developed for other planets than for Earth, in that measurements of wind are very limited and thus to put the presence or pattern of dunes in a meteorological context requires numerical simulation of the wind environment.

At the global level, this is simulated by a Global Circulation Model (GCM, sometimes also a General Circulation Model), which is the same sort of computer program used to forecast our weather. Essentially, the atmosphere is divided up into a 3-dimensional grid (which on a sphere entails some tricks) and the equations of force and motion (the Navier-Stokes equations) are solved for each cell in the grid to work out any pressure changes and how fast the flows are in and out in each direction. Where global scale flows are considered (as in Global Circulation Models), some major factors are the variation of pressure with altitude and the effect of planetary rotation which affects dynamics via the Coriolis effect. GCMs typically have grid cells of a few degrees, such that the total number of cells, tens or hundreds of km across (the model resolution), may be of the order of $10^4-10^5$, depending on the number of layers in the model vertically. These equations must be solved in short enough timesteps that the solution is numerically stable, such that thousands of iterations a day may be needed. Thus GCMs can be very computationally demanding, and indeed are a major application of supercomputers and large computer clusters.

While the Navier-Stokes equations are universal, the way in which other physical processes are represented in GCMs is not. In particular, how friction with the ground is handled, how sunlight and thermal infrared radiation is absorbed, scattered and emitted in the atmosphere (the radiative transfer scheme), and how convection, clouds and precipitation are modeled, are aspects that see a wide variety of approaches. It is simply not feasible to model all the features of these processes at a fine enough level explicitly—consider the spatial variability of clouds on Earth—and so at the chosen grid scale, these processes must be 'parameterized' (i.e., approximated by some computationally convenient fudge). Each planet has its own peculiarities: Mars has dust storms and sublimating polar caps, Titan a thick haze and a methane hydrology, etc. There is also an intrinsic difficulty in modeling the dynamics of a more massive atmosphere: not only does the model take longer (in computational time as well as simulated time) to 'spin up' to a steady state, but the larger inertial effects in a thicker atmosphere give it more 'hidden variables'—in essence, more potential for chaotic unpredictability. In this respect, the Martian atmosphere (Fig. 19.1) is rather easy to predict, in that it responds quickly and directly to solar heating (although the distribution of surface dust provides the system with some year-to-year 'memory').

Global models are useful to estimate regional environments, for example the overall ('trade wind') patterns of airflow, which latitudes may be systematically dry and which may see much rainfall, and so on. An extensive comparison between the pattern of dunes and measured winds were made in the landmark volume by McKee et al. (1979a), but only a few studies have explored simulated

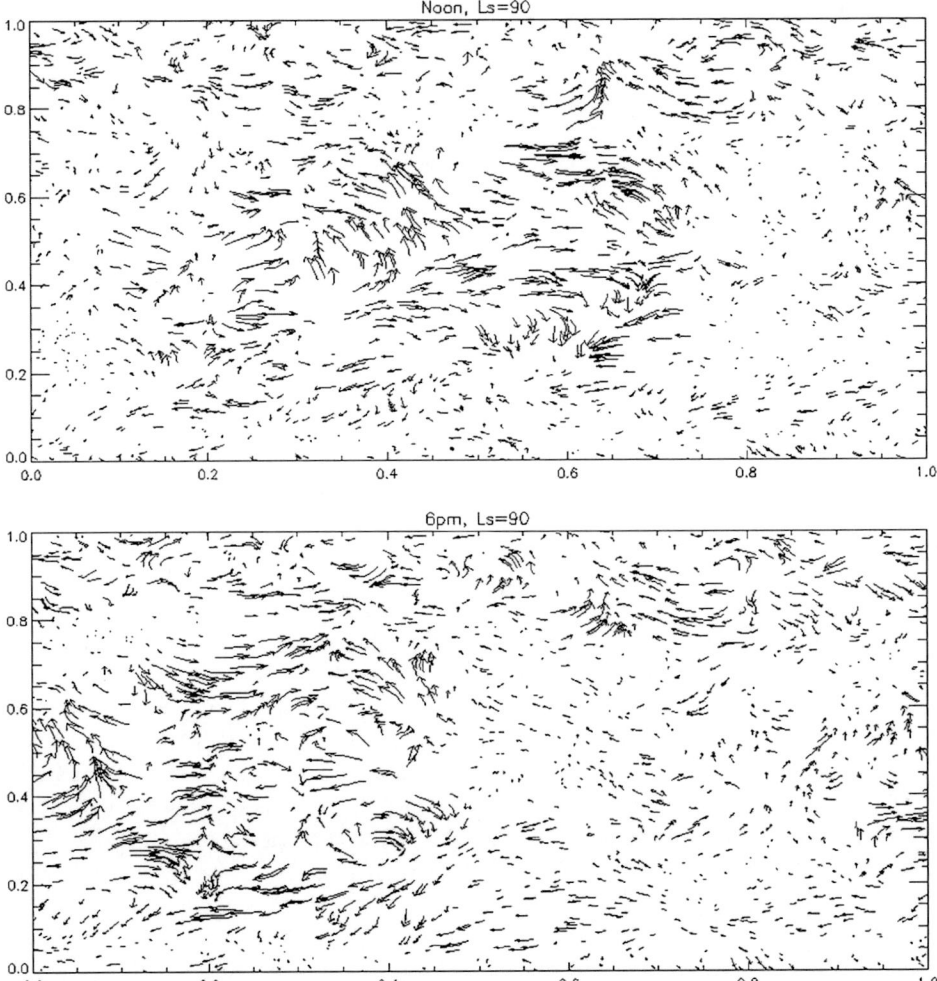

**Fig. 19.1** Surface winds on Mars generated from the LMD/Oxford Mars Climate Database, which encapsulates GCM output. Here, global noon and evening winds are shown for the northern spring equinox (Ls = 90). The arrows indicate wind direction, the arrow length indicates the speed. The fastest part of the wind pattern (corresponding to local afternoon) moves westward over the 6 h between these snapshots. Data courtesy of Laboratoire de Meteorologie Dynamique

winds and observed dunes. One global study was by Blumberg and Greeley (1996) which found (at a 4 × 5° grid resolution) a generally good agreement, except in the North American Deserts and the Arabian peninsula. They also found the GCM predictions were not effective at estimating the dune type that was formed.

In that context, the challenges encountered by Tokano (2008) and others in attempting to model the distinct dune pattern of Titan × where far fewer of the surface properties forcing a GCM are known × seem rather forbidding. Remarkably, some progress may have been made; one GCM at least appears to be able to explain the observed dune pattern as largely due to exceptional winds near the equinoxes, and the dessication of low latitudes (i.e., the formation of a wide equatorial desert belt) is a result found in several different models, suggesting it is a general property of Titan's circumstances.

At Venus, some broad predictions (Saunders et al. 1990) suggested slope winds would be the dominant factor, and some large-scale wind patterns were suggested on the basis of the known topography, but so far there has been relatively little progress on Venus GCMs as far as near-surface conditions are concerned.

One early Martian study was by Lee and Thomas (1995) who calculated the drift potential and resultant drift potential of winds generated by a Mars GCM. They noted that Martian winds provide generally unimodal transport, and thus barchan and transverse dunes should be most common (as observed) with very few star or linear dunes (they did note a couple of regions of linear dunes, in one instance inferred to be due to the funneling of wind by local topography).

Naturally, geomorphologists may be interested in wind patterns at a smaller scale. It is impractical to run a global model at a scale wherein the environment of individual dunes (say 1 km) is resolved—the model grid would just be too large to run. So a high-resolution model covering only a small part of the planet (a 'mesoscale' model) can be applied, with its boundary conditions determined by a GCM. This approach is called 'nesting'. Such models are able to capture the effects of topography, and so can be used to estimate sand transport pathways.

**Fig. 19.2** Output from a mesoscale model: the plot indicates a region about 1000 km² centered on Arsia Mons, a volcano that rises some 9 km above the surrounding plain. Here, night-time winds are shown, with a strong katabatic flow of cool air downhill. From Spiga and Lewis (2010, CC License)

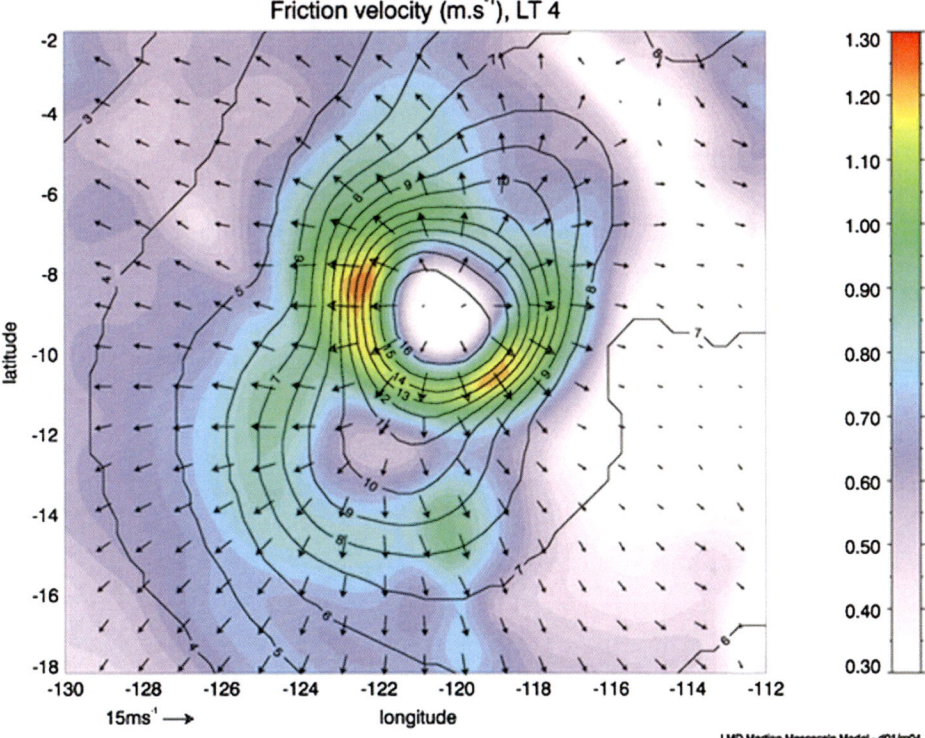

Fenton et al. (2005) used a mesoscale model to consider the airflow in Proctor crater, Mars. Kok (2010) showed that even though the hourly averaged winds are below the fluid (initiation) threshold for sand transport, the lower threshold for sustaining transport (see later) means that as long as there are occasional gusts that can cause saltation (however briefly), the probability of sand transport during winter days is in fact quite high. This explains why seasonal shifts in slip faces can be observed on Proctor dunes, even though the hourly mean speeds are below the initation threshold. Spiga and Lewis (2010) have used a mesoscale model to explore the wind variability and its effect on sediment mobilization; some example output from their model is shown in Fig. 19.2.

With a cell size that may be of the order of 1 km in such models, transient turbulent structures are not usually resolved, and subscale transport must still be parameterized. The reader will be unsurprised to learn that yet smaller-scale models also exist, so-called Large Eddy Simulations (LES) with typical grid scales of a few meters or less, which are able to explicitly model turbulent eddies such as dust devils.

Moving to an even smaller scale, with ∼0.1 m or less grids, it is possible to simulate the airflow over an individual dune. The language used to describe such models tends to be a little different, for historical reasons. Although internally it is still the Navier-Stokes equations, this scale of model was developed not for meteorology but for aeronautics, to model the drag on aeroplanes and other vehicles. The codes here are referred to as CFD (Computational Fluid Dynamics), and typically do not bother with such aspects as radiative schemes or clouds, but pay much more attention to surface shape, and the treatment of friction (Fig. 19.3). With this type of model, one can simulate such features of aeolian interest as the separation of the airflow at the crest of the dune and the recirculating eddy behind it.

A significant complication, not yet explored widely with CFD, is the effect of saltating sand. In other words, for accuracy at this small scale, the interaction of the air flow with the suspended or saltating material needs to be considered, in that the effective surface friction felt by the wind changes once sand is lofted. This coupling is an outstanding area for further work.

Finally, a rather distinct class of flow models has emerged in recent years for certain applications, called lattice-gas or Lattice-Boltzmann (LB) models, see, e.g., Frisch et al. (1986). These are somewhat similar to cellular automata (CA—see below). Like CFD, LB models consider a flowfield as a set of cells within which the flow is assumed to have uniform properties, being considered in effect as a set of particles. However, in an LB model, these properties (usually flow direction) can have one of only a small set of discrete values; for example, in a hexagonal grid, the flow direction can have only one of six values, and the speed has only one value. The interaction between cells is considered analogously to colliding particles. While this simplification sounds primitive, the aggregate behavior across large

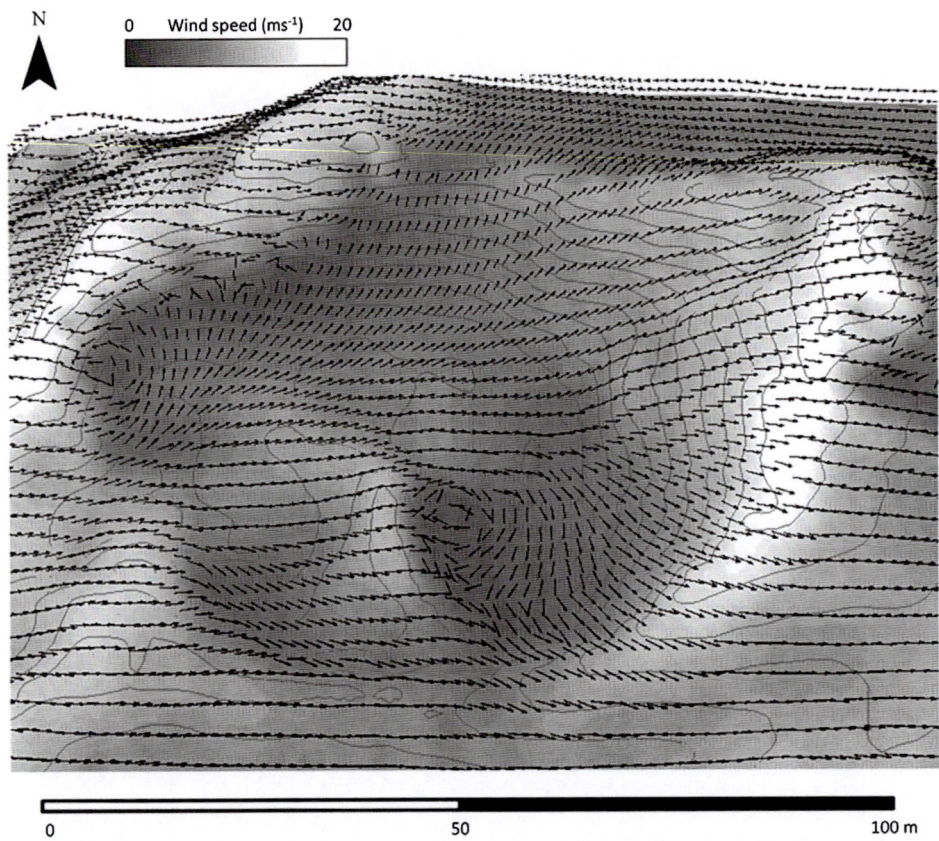

**Fig. 19.3** CFD simulation of the surface wind flow over a coastal 'blowout' dune as modeled by Smyth et al. (2012). Notice the circulating flow at the lee edges of ridges. Image courtesy Derek Jackson

**Fig. 19.4** Flow streamlines over a barchan dune, computed with a lattice gas model. The streamline compression near the crest, and the flow separation and recirculation bubble are clearly seen. Image courtesy of Clement Narteau

domains of cells can actually be simulated quite well (i.e., when there are thousands of little particles, the sum of the little vectors can have a wide range of values, as if it were a continuous variable). LB models can be parallelized effectively (i.e., configured to run efficiently on multiprocessor computer clusters) and, especially important for aeolian studies, can handle multiphase flows and complex and changing boundaries easily. LB models (one example, for snow transport, is by Masselot and Chopard 1998) are likely to see significant application to aeolian problems; recently Narteau et al. (2009) explored the coupling of a cellular automaton (see later) sand transport model to a lattice gas flow model (e.g., Fig. 19.4) and found some fascinating results, e.g., in the formation of star dunes (that arms grow upwind, but only when an odd number of wind directions is imposed—Zhang et al. 2012).

## 19.2 Modeling the Sand

At the level of individual sand grains, trajectories can be modeled as we discuss in Chap. 4, where the aerodynamic, gravitational (and, if applicable, electrostatic) forces act on a particle. Collisions between grains, the effect of particle rotation (e.g., on lift via the Robins-Magnus effect), and the variation in wind speed across the boundary layer, can all be simulated; probably one of the most important factors, especially in ripple formation, is the launch speed of a saltating particle. The recent book by Zheng (2009) discusses these factors and the application of trajectory simulations to aeolian studies; this book also discusses at some length the cellular automation and continuum dune and ripple modest that we introduce later in this chapter. A recent development is the COMSALT (Comprehensive Saltation) model (Kok and Renno 2009) which systematically treats the various factors and finds, for example, that because the kinetic energy of a particle 'splashed' from the bed is high compared with the additional energy the particle gains in the thin Martian air (unlike on Earth), that there is an order-of-magnitude difference between the wind speed needed to initiate saltation on Mars and that needed to sustain it once it has started (Kok 2010), and that this may account for the seeming paradox that predicted windspeeds on Mars are often too low to permit saltation. A crucial distinction between Mars/Earth and Venus/Titan emerges from this study, namely that splash is inefficient on the latter two worlds, where particles must be generally mobilized by direct fluid lifting (and in that sense movement is qualitatively more similar to terrestrial underwater rather than subaerial transport).

The COMSALT model finds hop distances of 0.15–0.2 m for shear velocities (friction speeds) of 0–1 m/s on Earth. On Mars, for shear velocities of 0.5–3 m/s, the hop lengths are 1–4 m. These results are more-or-less commensurate with the typical scale of ripples on the two worlds. The corresponding hop heights are 1.5–2 cm and 10–35 cm (Kok, 2010). On the other hand, for conditions of only twice the threshold on Venus and Titan, the hop length is 1 cm and 8 cm respectively, compared with $\sim$30 cm and 1 m on Earth and Mars. Thus Venus hops are invisibly short and visible ripples are unlikely to be seen (although microdunes can be formed—see Chap. 14). Ripples on Titan may be present (and conceivably could affect radar reflectivity via Bragg scattering), but would be rather smaller than on Earth.

In principle, the evolution of a dune could be calculated by applying the flowfield from a CFD model or similar as the inputs to such grain trajectory simulations. However, modeling the migration of a dune grain by explicitly-simulated excruciating grain is not an efficient way to proceed.

Various attempts to parameterize the airflow over a dune and compute the resultant sand transport with conventional equations have been made, and these efforts (see also later in this section) have tended to focus on isolated barchans, as both beautiful and somewhat simple but perhaps surprising shapes. Perhaps the first such effort, and a paper still worth reading today as it crisply outlines the problem, is that by Howard et al. (1978). A description of wind speed and direction from field measurements at different places on a Salton Sea barchan were applied to Bagnold-type sand transport formulae, and the evolution of the dune determined by book-keeping the sand. They went on to suggest what factors may control barchan form, but the first result to look for is essentially that the dune does not evolve—isolated barchans simply retain their shape and move downwind. Wipperman and Gross (1986) investigate a similar approach: one nice result (seen also in the revival of barchan models in the last decade—see later) is the evolution of a simple conical sand pile into a two-horned barchan.

For understanding the general relationship of wind direction to dune form, some quite compelling simulations of dune dynamics can be generated by applying rather simple rules over and over again to rather elementary description of the terrain (a grid of discrete heights, with perhaps only a dozen or two cells across a dune: in some cases). The seminal work in applying such Cellular Automaton (CA) models was that of Werner (1995), building on Landry and Werner (1994), and Nishimori and Ouchi (1993) have a similar approach.

The most famous example of a cellular automaton is the 'Game of Life' devised by mathematician Simon Conway. Here, there is a square grid of cells each of which can be either dead or alive (0 or 1). The status of each cell at the next timestep depends on its present status and that of the eight cells adjacent to it. If a cell has 0 or 1 or more than three live neighbours, it dies from starvation or overcrowding, but if it has two or three, it lives. A dead cell with three live neighbors will bcome live. That's it. Despite the simplicity of such rules a bewildering array of structures can exist-constant (e.g., a square of four cells), oscillating (e.g., a line of three cells), propagating (the 'glider', a set of six cells that cycles through a set of four configurations, but displaces itself by one cell over that cycle) and so on. Not only are remarkable structures possible (e.g., the glider gun, a cyclic structure which launches a glider every 30 steps) but it can be shown that the game can act as a Turing machine, i.e., in principle, any computable problem could be expressed in the game. More generally, the application of CA models to a wide range of situations has been espoused by Stephen Wolfram.

The way Werner's model works is that a grid of surface heights is set up (essentially as stacks of discrete slabs of sand), and simple rules describe the transport of slabs from one stack to another. A cell is selected at random and it is determined whether the top slab in the cell (if there are any slabs) is in the shadow of upwind slabs—i.e., it is less than 15 degrees below them (e.g., Fig. 19.5). If not, then the slab is moved in the downwind direction by some distance (the saltation hop length). Then the grid is checked to see if the angle of repose is exceeded; if so, slabs are shuffled

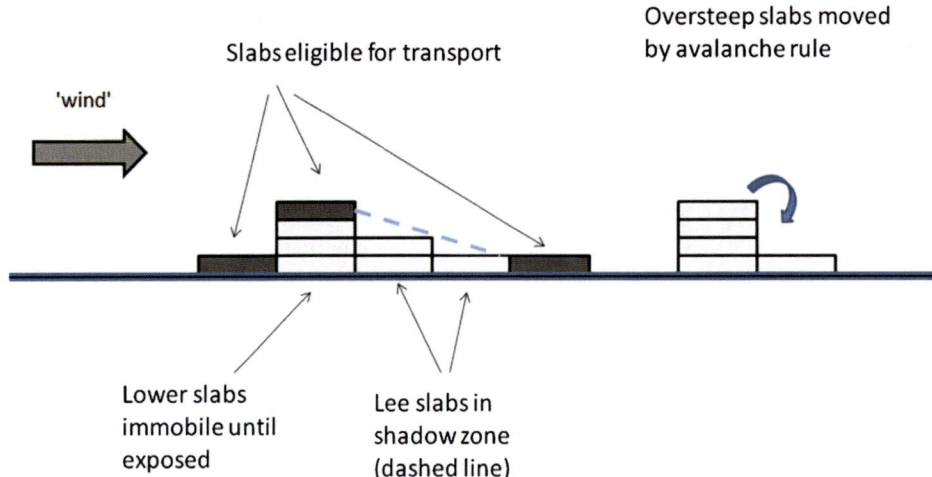

**Fig. 19.5** Schematic of the shadowing rule in a Werner-type CA model. Eligible slabs are moved at random some downwind distance (multiple slab lengths), but slabs downwind of higher stacks may be protected in their lee and so are not eligible to move. Periodically, the grid is checked to see that adjacent stacks are not too different in height; if so, the angle of response is exceeded and slabs move by avalanching to reduce the gradient

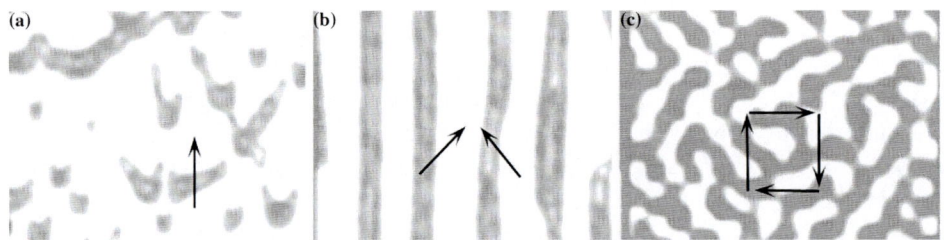

**Fig. 19.6** Results of the original classic Werner model. Unidirectional winds **a** yield crescentic forms like barchans and barchanoid ridges. Convergent winds **b** yield longitudinal forms, while winds with very diverse directions **c** lead to a network, somewhat like the akle morphology or star-type dunes. The success of such a simple model is remarkable

downhill, simulating an avalanche, to remove the instability. The grid is usually set up to be periodic, such that slabs that move off one edge are restored to the opposite edge.

This simple model not only produces features that look like dunes, but reproduce the variation of duneforms found in nature. Specifically, winds fluctuating in direction produce linear dunes, while constant winds produce barchans or barchanoid/transverse dunes, etc. The rich and familiar output of such a simple model (Fig 19.6) is a triumph.

Of course, workers since then have explored various modifications to the model, e.g., Bishop et al. (2002); rules such as differing probabilities can be applied to sand continuing to move for multiple hops on sand-free surfaces vs on cells with a non-zero number of slabs. Pelletier (2008) shows that a modification to include variable bed topography nicely reproduces the interaction of a dune field with an impact crater—a familiar situation on Mars. The most ambitious and successful approach so far is the coupling of an LB model of the airflow with the CA model of the sand (e.g., Fig. 19.7, see also Zhang et al. 2012).

One exciting possibility noted by Narteau et al. (2009) is that the last transition for a given sand slab could be bookkept—i.e., did the sand get here by saltating, or by avalanching? Thus a CA model could be explored to simulate the 3D bedding structures in a sand (or sandstone) deposit. The structures formed by various idealized analytically-specified transport conditions were simulated in cross-bedding 'atlas' by Rubin and Carter (2006): the tools are now becoming available to do the same with arbitrarily-complex forcing conditions such as wind simulations from a GCM.

Another elaboration (Fig. 19.8) is to introduce another parallel grid of vegetation (e.g., Baas and Nield 2007; Nield and Baas 2008); the sand mobility is made to depend on the vegetation, and the vegetation (a real 'Game of Life'!) depends on the sand—parabolic dunes and nebkas can be simulated this way. A number of workers have now made

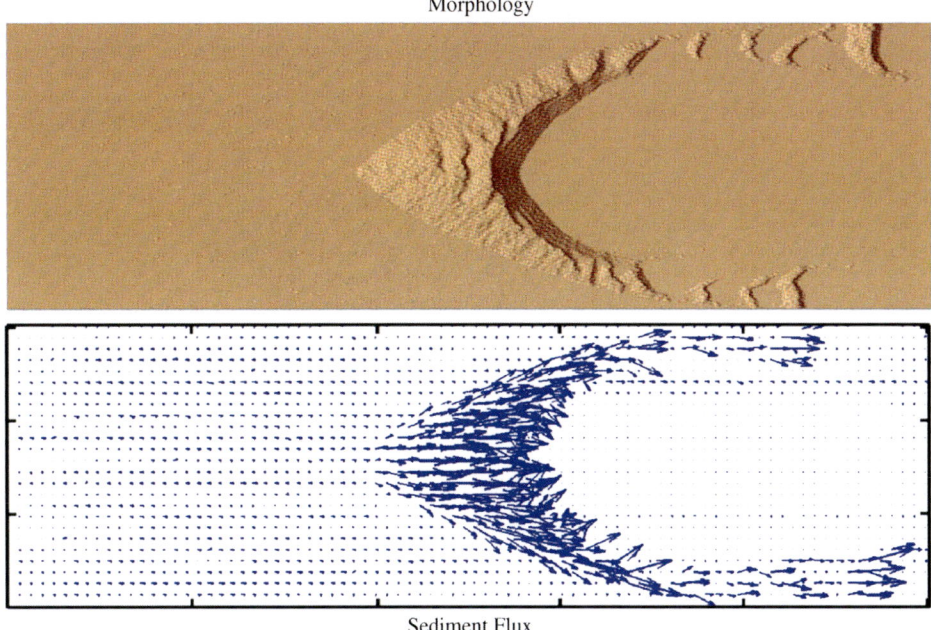

**Fig. 19.7** In a coupled model, the flow field is computed using the topography as a boundary condition, then the surface wind stress and thus the resultant sediment flux is computed. This is then used to modify the topography and the system begins the loop again. This example is courtesy of Clement Narteau

**Fig. 19.8** A cellular automaton model incorporating vegetation. These transverse and barchan dunes are moving up towards the *top right*, where vegetation modifies the transport rules (cells with vegetation have small vertical stalks, and thus appear *dark* in this representation). Image courtesy of Jo Nield

implementations of this type of CA model available on the web, e.g., Barchyn and Hugenholtz (2012).

Another approach, somewhat distinct from the cellular automaton flavor of model, is to consider the sand surface height as a continuous variable, and to model the evolution of a height profile algebraically. This type of morphodynamic model was invoked by Kroy et al. (2002); a recent review is by Duran et al. (2010). Essentially, a shear stress field is calculated, taking the surface slopes into account (i.e., the topography causes the flow to accelerate at the crest). Simplistically, this would prevent dunes from growing—sand would be least likely to accumulate at the crest. However, several factors break the symmetry of the problem. First, hydrodynamic effects cause the shear stress to be maximized upwind of the crest. This would cause deposition at the crest, leading to growth of a dune.

However, a finite distance (the 'saturation length') is needed for sand transport to build up to a steady-state value. This has the effect (see Fig. 4.17) that the sand flux reaches a peak downwind of the peak shear stress and means that a

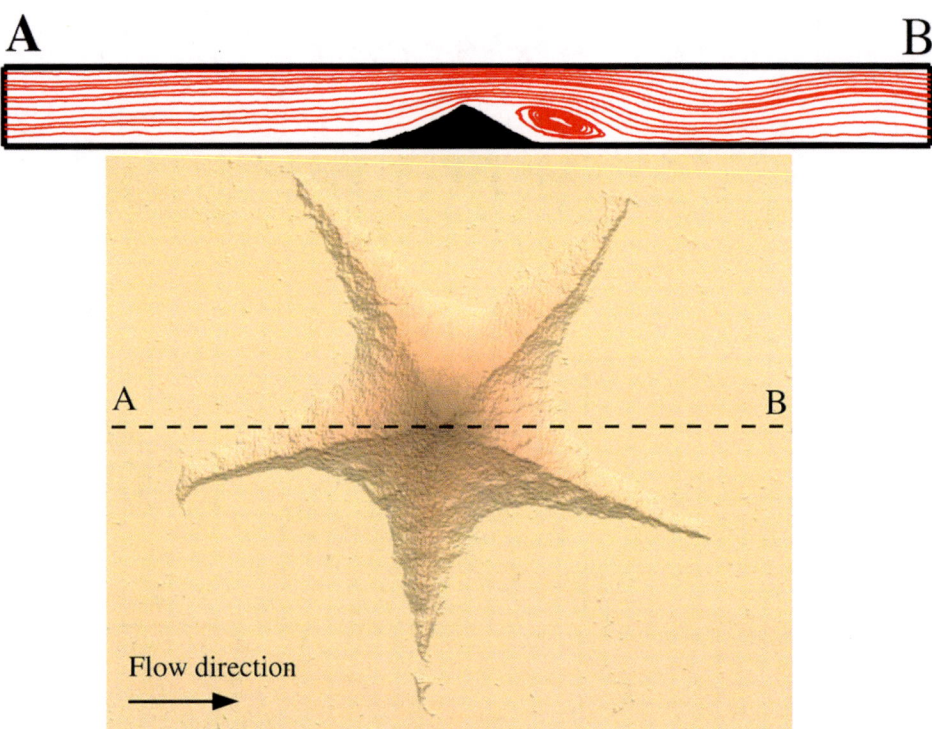

**Fig. 19.9** Flow separation in a lattice gas/cellular automaton model of a star dune. Image courtesy of Clement Narteau

bump will not grow unless its length is large compared with the saturation length. In other words, the saturation length defines a minimum scale for dunes to grow—a concept discussed in Chap. 5.

Two further effects on the downwind side of the dune are important. Sand cannot sustain slopes greater than the angle of repose. Thus a slope condition is imposed on the slip face of the model. Also (and while generally true even for rounded dunes (Fig. 5.1), it is particularly true of the sharp change of slope at the top of the slip face), if the downwind surface slope bends downwards sharply, the airflow can detach, forming a separation bubble with a recirculating flow. Flow vectors in such a bubble, estimated using CFD by Hermann et al. (2005), are shown in Fig. 5.2; the empirical size of this bubble is introduced into the continuum model. (As with CFD, lattice gas models calculate the flow and thus the bubble explicitly—e.g., Fig. 19.9.)

In the original one-dimensional incarnation of the model, an initial low sand mound can progressively grow until its height becomes such as to perturb the flow, and a slip face develops. More recently 2-dimensional variants have been explored, and models of this type have been rather successfully compared with water tank experiments (e.g., Reffet et al. 2010) and with barchan morphology on both Earth and Mars (Fig. 19.10). In particular, a range of barchan morphologies can be produced by changing the angle between bimodal winds: in unidirectional winds a conventional two-horned barchan is formed, but as bimodality is introduced the dune becomes more dome-like with the gap between the horns filling for an opening angle of ~40°.

When the opening angle is increased to 90° or more, the dune takes on a more wedge-like appearance with a single trailing tail (essentially, a linear dune with a head.) Reffet et al.'s (2010) study also shows that a single linear sand streak (whether transversely-oriented or longitudinally) breaks up into a string of barchans when bimodality (20–30°) is introduced.

These modeling approaches can be somewhat hybridized. Pelletier (2009) combines a Werner-type model, modified with Jackson and Hunt's (1975) description of the boundary layer airflow over a hill, to consider aeolian ripples, transverse dunes, and megadunes, in the same hierarchical framework. In essence, the character of the airflow and ultimately the wavelength of the bedform pattern is influenced by a roughness scale, which for ripples, he argues, is determined by the (upper 20th percentile) particle size and the length scale for the transverse dunes is in turn set by the ripple amplitude. Note, however, that very different interpretations of the determining length scales (e.g., the atmospheric boundary layer thickness as an upper limit for dune spacing and the saturation length for initial dune forms, as advocated by Andreotti and others, see Sect. 5.5).

## 19.3 Dynamics at the Dune and Ripple Level: Pattern Formation

Modeling has now reached a level where not only the formation of an individual dune morphology but the interaction between dunes is being captured. Models now can

**Fig. 19.10** Different barchan forms on Mars are compared with those in tank experiments (see Sect. 17.4) and from numerical simulations. The agreement is striking. Image courtesy of Erwan Reffet

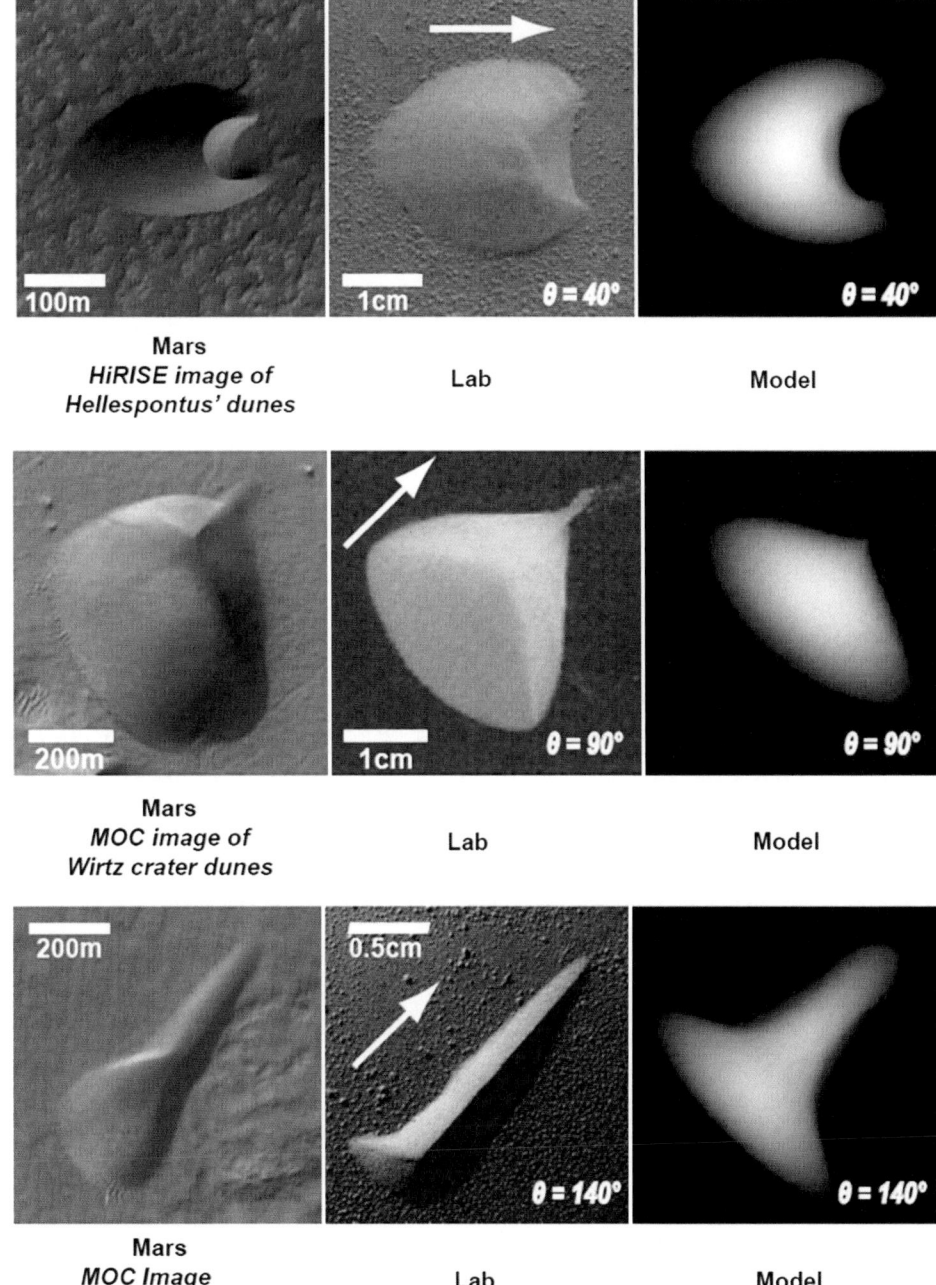

address the quantitative relationships that remote sensing data provide on the arrangement of different dunes in different settings. Some heuristic models have been applied to explain trends across a dune field (e.g., Jerolmack et al. 2012) although whether such approaches have predictive, rather than explanatory, utility remains to be seen. (This model discusses the variation of dune morphology across the White Sands dunefield in the context of a thickening boundary layer, although perhaps this may be simply another way of looking at the hierarchical roughness discussed by Pelletier (2009)—see later). The forward-modeling of dune morphology and evolution by the techniques described in the previous section has received much attention recently and has made substantial progress in reproducing much of what is observed in a wide range of settings.

One factor at work is the balance of sand fluxes on a barchan. As sand saltates across (with presumably a uniform rate across the field), then a large barchan with a larger width will intercept more sand than a small one. However, barchans 'leak' sand through their arms, and this leakage is more significant for small barchans. So, left to themselves,

**Fig. 19.11** Simulation of a barchan field. Note the trailing arms, with downstream barchans often near-centered on the sand streams from the arms of the upwind ones. Such interactions can be explored with pure CA models, continuum models, and (ideally) coupled sand/flow models. This is a CA example, courtesy of Jo Nield

a field of mixed size barchans would see the small ones shrink and the large ones grow.

However, the barchans will interact with each other. Smaller barchans propagate faster, and so will overtake, or collide with, their larger counterparts. A crucial issue is what happens during these collisions. If the barchans are of very different sizes, then the smaller one is simply absorbed into the larger one, but more interesting effects occur on more similarly-sized dunes (Fig. 19.11).

Hersen and Douady (2005) performed water tank experiments with small barchans, using glass beads of different colors to trace the fate of the sand from the two colliding dunes. The dunes could not be 'too' equal in size, as then the differential speed is too low and it takes too long to practically perform the experiment. Their elegant experiment (similar experiments in a water flume have been done by Endo et al. 2004) showed that one barchan (green) hitting another (red) about twice its size slightly off-center, merges with the larger dune but causes it to spawn another barchan downstream, not much smaller than the original 'impactor'. The sand color showed that it was not the impactor barchan propagating through the large one—the spawned dune was all red sand from the original target dune. Numerical simulations of the same 'tunneling' process were reported by Schwammle and Herrmann (2003).

These experiments can inform efforts to understand the apparent organization of barchan dunes into corridors (e.g., Hersen et al. 2004; Elbelrhiti et al. 2008), wherein across a barchan field there may be significant variation of dune size, but along the downwind direction the dunes tend to have a somewhat constant size.

Exploration of pattern formation began with ripples. Initially, ripples form at a small scale, but can progressively grow in spacing and amplitude ('pattern-coarsening'); this behavior was noted in early work by Werner and Gillespie (1993). In Pelletier's (2009) simulation, with a domain of $N = 256$ points across in each direction (each point separated by the saltation length), the pattern coarsens to a steady-state value over about 1000 $N^2$ timesteps.

This coarsening has been observed in field measurements and wind tunnel experiments: the bedform spacing (normalized by particle diameter) grows as the square root of time until the steady-state is reached. This square-root growth is also observed in the numerical model.

An interesting behavior in aeolian transport is that an initially-irregular ripple or dune pattern, with many 'defects' (i.e., free ends of dunes, or Y-junctions; see Fig. 19.12) becomes progressively more and more regular. The dynamics of individual defects were considered by Werner and Kocurek (1999) in an analytical way. That simple model showed the growth of spacing with time (initially linearly, then logarithmically, until leveling off at an asymptote) and the associated decrease in defect density. The progressive elimination of defects from the pattern is also seen in Werner-type models; e.g., in Pelletier's simulation, the defect density decreases to zero over about 10000 $N^2$ timesteps.

In this evolution scenario, all dune systems should evolve to be free of defects. Many of the oldest dune systems are relatively free of them (or at least have areas where this is the case): the Namib and Arabian deserts for one, and Titan's Belet sand sea is another example. On the other hand, when climate change (e.g., due to Croll-Milankovich cycles) leads to a change in wind regime, the dune pattern must adjust (e.g., Werner and Kocurek 1997), and in so doing (essentially a new dune/ripple system builds on top of the old one) many defects can be created, thus the defect density is seen as a kind of clock that might measure the age of a system.

General dune patterns, and dune–dune interactions, have been well-documented (e.g., Kocurek and Ewing 2005; Ewing and Kocurek 2010). However, it is only now that

**Fig. 19.12** Characteristic defect migration as elucidated by Werner and Kocurek in a transverse ripple field, flow moving from *left* to *right*. The forward branch of a Y-junction detaches, becoming a 'free end'. This shallow tip migrates faster than the rest and so catches up to the ripple ahead of it, to which it attaches forming a new 'Y' downstream of the initial one

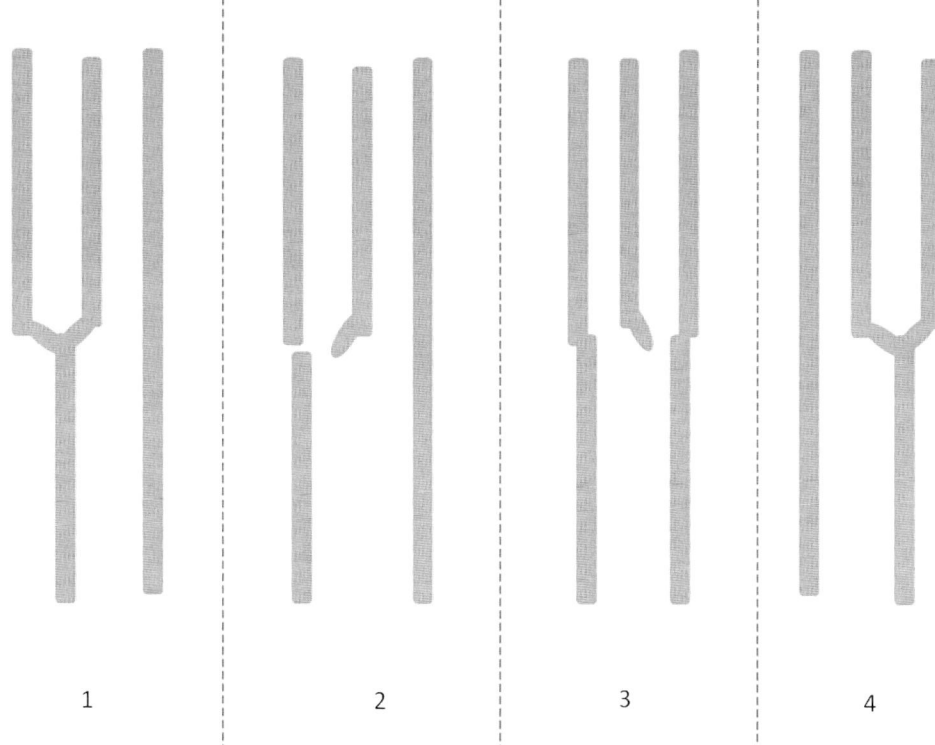

there is integration of these types of model to consider at a regional level the prevailing and seasonal variation of wind speed and direction (and possible sand supply or hydraulic effects on sand mobility), and then to apply a transport model at a smaller scale to derive what the resultant dune morphology should be, or conversely to reconstruct a wind regime history from the observed dune pattern. This is now being attempted at Mars (e.g., Ewing et al. 2010).

## 19.4 Summary

Our survey of the work on modeling dune dynamics has revealed not only a considerable body of work (and, indeed, progress) with a variety of model scales and types. Furthermore, increasingly rigorous and quantitative approaches are being brought to bear closely on the detailed data now available from laboratory, field and remote sensing data. The evolving style of investigation is revealed in the journals in which much of this work appears—from documenting and cataloging field movement in geographical journals, much of this modeling work and large amounts of field data are now appearing in the physics literature. This dramatic progress offers insightful perspectives on the underlying—often surprisingly simple—rules that lead to the formation of beautiful moving landscapes. Furthermore, the quantitative understanding of dune evolution now allows the observed landscape to be physically interpreted in terms of what wind regime and sand availability have persisted for how long, and how much of what we see today is being produced today, versus representing the degrading fossil of a past climate.

# Part V
# Why Dunes Matter

In this final section, we review some areas where the importance of dunes is demonstrated. While dunes may be seen by physicists as interesting examples of emergent systems (see Chap. 20) they can be literally a matter of life and death: Freymann and Krell (2011) describe hundreds of carcasses of dung beetles, apparently overrun by the hot dark volcanic sand of a fast-moving barchan, near Olduvai in Tanzania.

Dunes record a history of conditions, as we have noted in Part II and elsewhere. In Chap. 21 we review the implications of dunes for the study of past climates on Earth, Mars and Titan.

Deserts and dunes are significant in human history, principally as an obstacle to commerce and military operations. Indeed, some of the earliest written records suggest aeolian processes having an impact on civilization; in his Histories, Herodotus writes:

'*On the country of the Nasamonians borders that of the Psylli, who were swept away under the following circumstances. The south-wind had blown for a long time and dried up all the tanks in which their water was stored. Now the whole region within the Syrtis is utterly devoid of springs. Accordingly the Psylli took counsel among themselves, and by common consent made war upon the south wind—so at least the Libyans say, I do but repeat their words—they went forth and reached the desert; but there the south wind rose and buried them under heaps of sand: whereupon, the Psylli being destroyed, their lands passed to the Nasamonians.*'

While perhaps not literally true, the quotation demonstrates that the association of sand heaps with the wind is over two millennia old. In Chaps. 22 and 23 we discuss how moving dunes interact with human infrastructures, and how humans (and their robot proxies on Mars) have struggled to move across dunes and deserts.

Finally, dunes are a sufficiently evocative landform that they feature prominently in art and literature, and notably in our thinking about distant worlds, in science fiction. In Chap. 24, we review these fictional dune worlds and how they have been portrayed.

# Dunes as Physical Systems

The balance of complexity and regularity is something that human beings, a species adapted to exploit niches in a dynamic environment, are attuned to detect. Dunes are often rather regular landforms, but rarely exactly regular, a feature that our brains are evolved to trigger pleasure from.

One of the earliest chapters on the formation of sand dunes was by the British geographer, Vaughan Cornish, in 1895. He also studied snow drifts and waves in the sea and in rivers, clearly struck by the analogy between wind-formed waves in a sand surface and wind-driven waves on the surface of a liquid. That same analogy also inspired astronaut Story Musgrave to compose this poem upon seeing the Namib from the Space Shuttle (see Fig. 6.19):

> Now, Namibia, desert streaming into ocean,
>   waves of bright sand diving into dunes of dark water
> –visible rhythms of blue and brown,
>   sea and sand dance upon my strings.

While poetic, the analogy is not predictively useful. While some of the mechanics of airflow separation, etc. is common to the worlds of sand and sea, the ripples on those two media are defined by altogether distinct processes.

That said, the fact that ripples and dunes can form with similar patterns and arrangements over scales from ripples of a few centimeters to dunes several kilometers apart (e.g., Fig. 20.1), suggests some fundamental processes at work, and the planetary perspective—which shows, for example, very different environments on Earth and Titan producing identical landforms—only reinforces this.

As we have described in Chap. 2 to 10, and especially in Chap. 10, modern physics can now largely capture the essential

**Fig. 20.1** Ripples on the surface of dunes at the Lencois Maranhenses in Brazil. The dune surface is damp and packed hard, but bright dry sand has started to move across it, and a pattern of transverse ripples is seen at left. But towards the right, where sand is less abundant, the ripple pattern is two-dimensional, with much more crescentic morphology where the ripples are sparse. In other words, at this 10-cm scale (the case of a leatherman tool is present for scale), the pattern has many similarities with sand-starved barchan dunes at the 10–100 m scale. *Photo* R. Lorenz

processes involved, and models can reproduce—whether at the ripple scale or at the dune scale—the morphologies seen in nature. Dunes and ripples are examples of emergent systems, wherein the relentless application of simple rules can lead to complex forms, and much of the work on such systems appears in the physics rather than geography/geology literature. Interestingly, a major motif in modern physics has been the application of cellular-type models to understand such diverse phenomena (Bak 1999) as earthquakes, wildfires and stock-market crashes, wherein events can occur at a wide range of scales, usually with a power-law size distribution. While dunes themselves do not have such distributions, being rather regular, the avalanches down their slip faces do. An engaging discussion of self-organized systems which sets aeolian landforms in context is the book by Ball (1999).

# Dunes and Climate

A number of factors are required to form a dune, and thus the existence of a dune tells us that those factors are, or were once, present. If those factors are not present today, it follows that they must have occurred in the past, and so dunes can serve as windows into the past.

A principal factor in controlling whether fine particles (which in many places means soil that sustains agriculture) is rainfall, both directly (in that damp ground is sticky and sand moves less effectively) and, more importantly, by the indirect role of soil moisture on vegetation. Plants bind the soil with their roots, the organic matter from the breakdown of dead plant matter improves the sand's cohesion, and the aerodynamic shielding by the plant's branches and leaves reduces the wind stress applied to the soil. Due to all these factors, perhaps more than variations in windiness (which may yet be occurring), climate change via rainfall variation can alter the mobility of large areas of sand. The planetary perspective can help set our changing climate in context: in particular, just as Venus is the solar system's 'poster child' for greenhouse warming, Titan has a hydrological cycle that resembles an extreme version of Earth's, in the direction of which Earth's climate may be evolving. Specifically, a warmer atmosphere can hold more water vapor, and thus surface evaporation (driven by more-or-less fixed sunlight) takes longer to recharge the atmosphere with moisture after a rainstorm. But with more moisture available, the storm when it comes is more violent. Thus a warming atmosphere may see the uncomfortable double blow of heavier rainstorms with longer droughts between them. Whereas the $\sim 1$ m/yr of rainfall cycles through the $\sim 2$ cm of precipitable moisture in Earth's atmosphere, resulting in a few showers a month, Titan's thick atmosphere can hold several liquid meters' worth of methane vapor, yet the meager sunlight drives only a few centimeters per Earth year of evaporation. Thus Titan's rain is manifested as massive downpours separated by centuries. In that context, the apparently high mobility of sand on Titan is not difficult to understand.

Just as 'global warming' does not mean every single place sees a monotonic increase in temperature, there may be 'winners and losers' in the game of mobility. Some present-day deserts on Earth may become more easily cultivated, and the greening of the prairies a couple of hundred years ago, turning barchans [whose topographic signature is still present (Wolfe and Hugenholtz 2009)] into vegetated parabolic dunes is an example. But likely the majority of areas will see average soil moisture drops and consequent mobilization of sand (Fig. 21.1), which is generally bad news for people other than aeolian geomorphologists. One study in particular (Thomas et al. 2005) forecasts that pastoral and agricultural areas in the Kalahari basin which lie on presently inactive dune sands will see significantly enhanced dune activity by 2039, and that by 2099 dunefields from northern South Africa to Angola and Zambia will become highly dynamic.

The Earth likely saw less precipitation overall during the Ice Ages, when conditions were colder and the atmosphere could hold less moisture. Furthermore, because so much water was locked up in polar and continental ice sheets, the sea level was lower by around 100 m, and thus many areas such as the Persian Gulf, which are currently shallow seabeds, were in fact dry land and could act as sand sources. The overall sand dune activity on the planet was likely much more vigorous than today, forming once-active dunes which are now vegetated and stabilized (such as the Nebraska sand hills). The same periods also saw the long-distance transport of dust, resulting in the deposition of thick loess in China, the Rhine Valley, Missisipi, Alaska and elsewhere.

In all this, then, dunes give us a limited but nonetheless valuable window into the past. Models can help us decode the dune patterns and layers into prior wind regimes. The same sort of exercise has been tried at Mars (where global mapping of dunes has been done systematically, e.g., Hayward et al. 2007, 2009), although there is sufficient widespread disagreement between the models and the dunes

**Fig. 21.1** Sand trap at La Jornada Experimental Range, New Mexico, used to evaluate the agricultural impact of aeolian processes. This simple but effective vane/pivot arrangement points three flared two-part boxes upwind at different height: the triangular flare decelerates the air, which escapes through a wire mesh at top while the sand collects in the closed bin. After some sampling period (usually weeks) the boxes are opened and the time-integral soil transport is evaluated by weighing. *Photo* R. Lorenz

that many challenges remain. The wide obliquity changes in Mars recent past (100,000 year to million year timescales) set an interesting range of alternate climate possibilities, and modelers (e.g., Fenton and Richardson 2001) have explored the implications for dune formation and orientation.

Although Mars, for now, lacks the complication of vegetation, one such challenge is the extent to which sand has been immobilized by cementation ('induration'). One mechanism is to freeze material on the dunes. The giant Martian polar dune field, Olympia Undae, has dark basaltic dunes sitting on a brighter plain, but during winter the whole area is made white with carbon dioxide frost up to some tens of centimeters thick, which fixes the dunes in place. In spring, vapor released as the $CO_2$ sublimates can build up and be violently released, hosing out a spray of dark sand onto the frost. Another mechanism is to deposit dissolved minerals or salts onto sand. This happens quite naturally when the dunes have appreciable amounts of gypsum, but salt spray from the sea, or acid vapors from volcanos can also cause similar crusts to form on terrestrial sands. Both water ice and salt or acid induration had been suggested as reasons why Martian dunes had not been seen to move (although movement in at least a few places has now been documented—see Chap. 8).

On Titan so far, the character of the interaction between climate modeling and aeolian geomorphology has been a different one. Sufficiently little is known about weather on Titan that the dune pattern has actually been one of the main pieces of information with which circulation models can be refined. That has stimulated something of a resurge in interest in linear dunes and the wind regimes that form them (e.g., Reffet et al. 2010; Rubin and Hesp 2009). As confidence in these models progressively improves, and as mapping coverage has expanded to identify some 'outliers' in the dune population as possible fossils of a paleoclimate regime, Titan studies may become more like the Mars and terrestrial ones—attempts to diagnose the past, rather than dunes being diagnostic of the tools we use to simulate the wind.

Venus is probably doomed to remain in that first Titan stage, in aeolian paleoclimate terms. The couple of dune-fields known, the streaks seen on the surface, and perhaps (in future) the orientation of microdunes, serve as constraints on circulation models. But dunes are so rare overall, and so generally small (such that they may reflect only the last hours or days of wind action, rather than the tens of thousands of years) that there is little to be learned about the past from them. Its retrograde-but-equatorial rotation means that Venus likely does not see the same orbit-obliquity forced Croll-Milankovich climate change that Earth, Mars and Titan do. While Venus' climate may have convulsed in the deep past (millions to billions of years ago, as true for Titan and Earth) by volcanic forcing of its greenhouse, its few dunes will be of little help in understanding such change.

As for exoplanets, we are unlikely to see maps of their dune patterns unless they happen to have inhabitants who send or bring them to us! But we can at least speculate in a more informed manner about what scenarios are likely or possible (see Chap. 24). We might conceivably learn (from detailed study of how light from the parent star is reflected by the planet from the advanced astronomical observatories planned in coming decades) whether a world has continents and oceans, but the experience of our own watery planet, and with Titan (which the first author expected not to have dunes, because it was expected to have seas) is that planets are complex places, and seas do not rule out deserts.

# Moving on Sand

The challenges of moving people or equipment across sand can be a significant impediment to the researcher, and indeed have been important in shaping the history of commerce and warfare in desert regions. We describe in this section some of these issues—which are now no longer confined to the Earth, since humanity can now claim getting vehicles stuck on Mars among its achievements.

We hope the following may be of direct practical utility in the field. If not, since getting stuck is an inevitable part of fieldwork and often leads to extended periods of waiting, then at least the elaboration of the mechanics involved may serve as points of discussion while your colleagues dig out…

## 22.1 Walking on Dunes

We cannot tell anyone how to walk on a dune any more than a set of written instructions can tell anyone how to play tennis or execute some other motor function. However, we can offer a few observations for the reader to keep in mind once they get to a site of interest.

The challenge is that, above a certain soil loading or stress, the material fails and flows. When this happens, the foot or tire may move in an undesired way, and the person or vehicle may be obliged to perform work on the sand, thus expending energy that they would not need to on a rigid surface. That threshold stress can be rather small for a loose slip face of dry sand.

Some brief numbers are appropriate to consider. A 90 kg person with their weight on a single foot (say 200 $cm^2$) exerts a pressure of about 0.5 $N/cm^2$, or 0.05 bar. If, however, they run (perhaps doubling the instantaneous force on the leg, and the pronation of the foot means most of that force is exerted on the toes), then the pressure on the sand may increase to 10 times that, or 5 $N/cm^2$. If the failure stress of the sand (which will depend on its moisture level, how it was emplaced, etc.) lies between these values, then running will cause the toes to slip whereas slow walking may not. The phenomenon is discussed, incidentally, in the context of walking on snow in *Miss Smilla's Feeling for Snow* (Hoeg 1992).

A casual inspection will readily show what is perhaps not intuitively obvious—that the sand fails on a usually-curved surface. This surface (see Fig. 22.1) forms the basis of soil stability evaluation. Soil mechanics, the scientific study and engineering prediction of the failure of loaded soils, is of course a whole field unto itself (beyond the scope of this book, but upon which many texts exist; among free resources, the US Army Corps of Engineers has a variety of manuals that clearly describe analysis techniques). Specific measurement techniques of the strength and friction exist (we touch on them in Part III) and elaborate analyses can refine the estimation of when and how a soil will give way. But to an order of magnitude, at least, simple consideration of loading pressure and simple strength or friction estimates can give a good sense of what works and why.

Usually the stoss slopes are rather safe, with modest slopes and the sand surface somewhat packed by the saltation process. The presence of sand ripples in this regard is usually a good indicator of a somewhat stiff surface. The lee slopes are a different matter: not only are the slopes steeper, but avalanched sand is usually soft.

One quickly learns, then, to move along dunes either on the stoss slopes, or better yet, clear interdunes if they are present. It may be that an adjusted gait (walking steadily, and keeping the soles of the feet as flat as possible) will reduce the stresses applied and thus probability of soil failure, but maintaining such an unnatural gait may prove more fatiguing than accepting the occasional slip while walking normally.

Of course, if the route requires one to traverse dunes, then there is little choice but to tackle the slipface. One can choose to slowly struggle up a lee slope, making one step

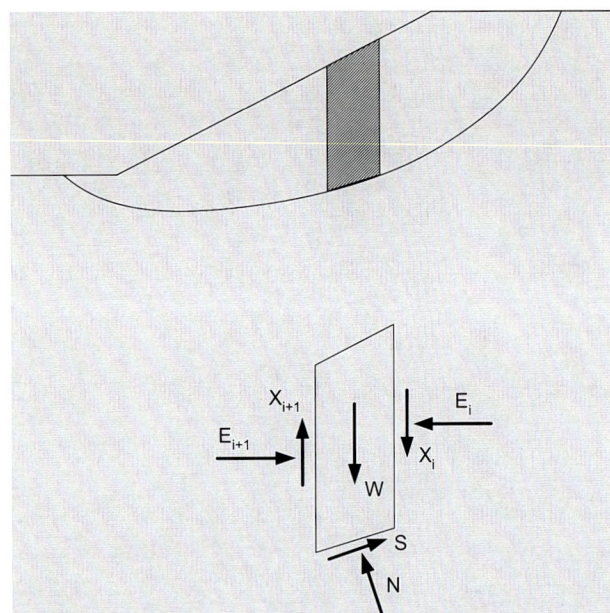

**Fig. 22.1** A schematic of how the failure criterion of a sloping soil such as a dune face can be computed by the so-called 'method of slices'. The soil is assumed to fail along a curve surface: the forces on each slice are estimated given the loading by the soil's weight and any exterior loads. When the required frictional force $S$ can no longer be provided by the soil, the material will fail. *Image* US Army Corps of Engineers

forward and ¾ step back in a frustrating but sustainable manner, or one can attempt to storm the slope, running fast enough compared with avalanche speed, that one can more efficiently ascend. The choice here is a matter of decorum, fitness, luggage and time of day; unencumbered young students on a brisk morning seeing a brief opportunity to experience dunes may enthusiastically embrace the latter approach, a senior researcher making a long traverse with a heavy GPR (see Chap. 11) on a hot afternoon may have to plod on with the former.

Of course, people are not the only entities that walk on sand. Everything from elephants (in Namibia) to beetles do so. Some species have particular adaptations to desert survival and migration, most notably the 'ship of the desert', the camel (Fig. 22.2). Its long legs keep the body elevated above the hot sand and give the stride needed for long traverses. The famous hump is filled with fatty tissue (which acts as a water store—fats contain more water per calorie than do other tissue energy stores). These are general aridlands adaptations; specific to sand, however, are the fact that the camel's wide feet splay out to minimize the ground loading, and camels have long eyelashes and can close their nostrils to cope with blowing sand and dust.

**Fig. 22.2** Camels are suited to desert travel not only because of their humps, but also because of their wide feet which reduce ground loading. First author, in Egypt

**Fig. 22.3** Where camels are unavailable, explorers made do with horses. This early US Geological Survey team struggles across a dune in Wyoming in 1907. Note that the narrow wheels on this wagon are poorly-designed for soft sand, and have sunk in appreciably. *Photo* USGS

## 22.2 Driving on Dunes

Since military encounters with Panzer divisions or hunting for sandworms are activities not to be expected for the modern researcher, one of the few areas in which they can express superiority over their peers is in driving. Some of the following observations and suggestions may be useful.

First, don't get stuck. An alert deviation to the right or left for a second or two, or a half-minute stop to survey the ground ahead may be a worthwhile investment to avoid the possibility of having to spend 20 min digging out. Beyond straightforward instinct, and avoiding obviously stupid things, the most important thing to bear in mind is that the troubles arise (sometimes unavoidably) when the bearing strength of the ground is exceeded by the stresses applied by the vehicle. Thus, depending on the terrain, one can sometimes maximize the strength of the ground (e.g., by sticking to stoss slopes). Another tip is that driving on sand in the morning can be easier because morning dew increases the sand cohesion.

Often one has little choice in where to go, and thus a low bearing strength must be accepted. Thus one must use the other half of the equation—minimize the applied stresses. Early vehicles, including those before the advent of the internal combustion engine (see Fig. 22.3) often had rather narrow wheels, which do not fare well in weak soils. Modern car tires are better, although they are generally optimized for fuel efficiency and low wear, so tend to minimize their footprint: fatter tires with lower ground pressure are better for off-road use. There is a straightforward way to lower the ground pressure on a tire, and that is to deflate it. To a first order the walls of a tire are flexible, and thus the pressure on the ground is the same as that on the inside tire wall, which is the same as the (gauge) pressure of the air. Typically for road use on a regular vehicle, this may be of the order of 20–30 psi, or 2 bar. Usually, the consideration here is to somewhat minimize the contact area (and thus maximize the pressure, since the product of the two equals the vehicle weight) and reduces the rate of wear of the tire on regular road surfaces.

**Fig. 22.4** Hybrid vehicles such as half-tracks or here, a snowmobile, are useful compromises between the low ground loading and superior traction of a long track at the rear and independently-movable steering wheels or skis at the front. Here the first author uses a snow machine in fieldwork in Barrow, Alaska. Photo by Kevin Hand, with permission

To reduce the pressure, and thereby reduce the probability of soil failure, one can let air out of the tire. How far one can do this depends on the tire—many modern tubeless tires rely on the pressure to hold the tire onto the wheel rim, so dropping it too low may cause the tire to fail. Low tire pressure will of course increase wear on the tire, and one should make sure to have a means of re-inflating the tire when returning to the road. When in the field, note also that low pressures (especially if combined with high speed) will increase the probability that large rocks or similar obstacles will damage the wheel itself (and thus also the tire).

One should, of course, consult the vehicle owner's manual or the tire documentation for guidance, and the authors and publisher of this book accept no liability for what may happen in the field. But a general indication is that one can usually go down to 10 psi (about 0.6 bar) for short periods with modest risk, but one should never go below 6 psi. To obtain ground loading comparable with a human foot (<0.2 bar), requires tracks as used on many military vehicles: specially-designed wheeled vehicles with large ('balloon') tires are an intermediate case.

Beyond the vertical reaction force exerted by the surface to stop the vehicle sinking, the ground must also supply any horizontal forces to the tires in order to effect acceleration, deceleration or turning. Therefore, moving in a straight line at a steady speed is to be preferred, and four-wheel drive vehicles will spread these side forces for acceleration over all four tires rather than two. It is difficult in sand to effect rapid speed or direction changes, and it is generally undesirable to make such changes suddenly, as transition from simple rolling may lead to a build-up of a mound of sand in front of the tires. Turns should be made with as wide a radius as possible, and it is better to coast to a halt by depressing the clutch pedal than to brake hard.

**Fig. 22.5** A stuck vehicle of Bagnold's Long Range Desert Group being prepared to unditch. Mats have been laid in front of the front wheels and a sand channel is about to be pushed under the right rear wheel. Note that the practicality of access to the underside of a vehicle leads to large numbers of people standing around while only one or two actually are doing work: this is entirely typical. Crown Copyright

On steep dunes, vehicles should be aimed along the direction of steepest slope: failure to do so means that the downhill tires will see higher load than the uphill ones and dig in further, steepening the angle. This positive feedback can lead to getting stuck or worse, a rollover.

## 22.3 Vehicles for Dune Driving

For the professional purpose of just getting to a dune site, vehicle availability, reliability and cost are considerations that may overrule performance in sand. And in many cases, long traverses on dirt roads or similar terrain may be required to get to the dunes. Thus the distinction should be recognized between vehicles used on dunes as a compromise, and vehicles designed specifically (or even only) for dunes, usually for recreation purposes. In both cases, some obvious modifications are the incorporation of convenient external access to shovels and sand channels for unsticking (see later), and carriage of spare tires, water and fuel containers.

Tracked vehicles, of course, perform well in sand—the large contact area of the tracks leads to a relatively low ground loading, and tracks can have deep tines to dig into soft ground (although this leads to heavy wear on regular road surfaces). The sand crawler in Star Wars is an interesting fictional variant. However, fuel consumption of such vehicles is very high and, because of the differential steering, their driving requires special training.

A compromise that is effective in desert environments is the half-track, wherein the rear wheel is replaced by a short track giving low ground loading and high traction, but steering uses conventional front wheels. A famous example of such a vehicle is the Sd.Kfz 250/3 half-track, nicknamed 'Greif' used as a command vehicle by Erwin Rommel in the Africa campaigns in World War II. This vehicle, festooned with 'bedframe' rails (actually, a radio antenna) was much seen in propaganda photographs (see Fig. 1.13), with Rommel using binoculars.

In fact, half-tracks made landmark advances in desert exploration. After some initial reconnaissance by regular cars and by airplane, a French expedition was dispatched

**Fig. 22.6** Military dune buggies. These are Desert Patrol Vehicles (DPVs) operated by the US Navy Seals, that permit high speed driving in desert terrain for special operations. Note the low center of gravity, the accessible spare tires, and communications. U.S. Navy photo 020413-N-5362A-013 by Photographer's Mate 1st Class Arlo Abrahamson Camp Doha, Kuwait (Feb. 13, 2002)

across the Sahara by Andre Citroen in December 1922. This first motorized crossing took 21 days, using Citroen half tracks ('autochenilles') to follow the camel tracks across the Sahara desert from Algeria to Timbuktu on the banks of the River Niger. This opened up a land connection between North Africa and the isolated Sudan—the history of this adventure, written by Ariane Audoin-Debreuil (the daughter of a key member of the expedition) was recently translated into English (Audoin-Debreuil and MacGill 2007).

The compromise design of a rear track to minimize the ground loading and provide strong traction, with separate steering at the front, is also found—for essentially the same reasons—in the snow mobile or snow machine (Fig. 22.4). These usually have only a single track for mechanical simplicity, and of course skis rather than wheels at the front. On hard icy terrain, the handling is much like a bike with rather limited steering authority, but on deep, soft sand the rear sinks down and the skis bite, and the feel becomes almost like that of a speedboat.

As for wheeled vehicles, large tires to minimize ground loading are the most obvious general modification. For military operations, the British Long Range Desert Group (LRDG, led by Bagnold in World War II) used conventional trucks (Fig. 22.5): the main innovations were in the procedures and logistics such as the amount of water carried and the use of sand channels. One addition was the use of a sun compass to permit directional fixes to be made without stopping (the ferrous metal content of the truck required one to walk some distance away to make a reliable magnetic direction determination).

Vehicles for recreational driving—thus usually more energetic than Bagnold's trucks—on dunes are often called 'Dune Buggies' or 'Sand Rails'. The term 'buggy' derives from the nickname of the Volkswagen Beetle or 'Bug', as

## 22.3 Vehicles for Dune Driving

**Fig. 22.7** Sheets of sand being thrown up by the bike's rear paddle tire—note the ridges at 45° intervals. While these provide powerful traction on loose dune sands, they would be unpleasant and unsafe for road use. Photo by Kevin Rice, Wikipedia/Creative Commons License

this was a favored vehicle to modify for dune driving, principally because of its rear-mounted engine and easily-obtainable spare parts.

Key features of a dune buggy are a low center of gravity, such that the vehicle can climb and turn on steep slipfaces without overturning. A roll cage to protect the occupants in case it does overturn is typical. Rear wheel drive is also typical. Usually a whip-like flagpole is used to display a marking pennant, to make the vehicle visible from the other side of a dune crest—in fact, such flags are often a legal requirement at many recreational sites. Dune buggies and their like have been adapted for very specific military tasks (Fig. 22.6).

In the extreme variants of dune buggies used recreationally to climb dunes, special 'paddle' tires are sometimes employed. As for half-tracks, this additional traction is usually employed only on the rear of buggies or bikes. These tires have prominent ridges to dig into the sand, and often launch sheets of sand rearwards, forming prominent 'rooster tails' of sand in the air (Fig. 22.7).

A rather accessible and very effective form of motorized transport is an All-Terrain Vehicle (ATV) or quad-bike. With four—often large—tires, these vehicles are stable and have low ground loading, making them popular for recreational driving on dunes (Fig. 22.8). The first author can attest, having driven one of these across Dumont dunes to generate a GPS shape model, that the experience of blasting up a steep dune face is exhilarating. While a more powerful quad bike may be preferred on mountain tracks and the like, in fact a less powerful but lighter bike may perform better on dunes, where ground loading is an overriding consideration.

**Fig. 22.8** A quad-bike making a steep descent on Sand Mountain, Nevada. Note the flag for safety, and the helmet. *Photo* Ralph Lorenz

## 22.4 Getting Unstuck

If one involuntarily stops moving in a wheeled vehicle, there is a chance one can recover without leaving the vehicle. One should attempt—slowly—to reverse out along the compacted tracks. If backing out fails, a number of factors may be contributing to the situation; it pays to understand them and to survey the underside of the vehicle before taking further action. Avoid sudden changes in speed or torque that may cause slip and the wheels to spin as this will just make matters worse.

First, if wheel-spinning has occurred, the tires may have dug themselves into pits. These present steeper local slopes for the wheels to climb out of, that therefore require more forward push. This barrier can most effectively be reduced by digging ramps into the forward part of the holes, although occasionally the short-cut of having helpers rock the vehicle back and forth may get you out of the pits. Indeed, whether they push or not, passengers should dismount. A 1500 kg vehicle will have its load reduced by 20 % if three passengers get out—the reduction in loading is even more significant for a small vehicle like a quadbike (Fig. 22.9). The number of bystanders in Fig. 22.5 does not necessarily indicate laziness, rather that reducing the loading on the wheels will help get unstuck.

In soft sand, reducing the slope of the wheel pits may not be enough, in that the sand may slip and the wheels spin without providing any forward force. In this instance, one needs to insert a surface that will allow the wheels to develop some friction and therefore forward push; depending on location, nearby grass or twigs may do the job or, if pushed, one can improvise with clothing or cardboard. Professionals, however, will have surfaces to hand specifically for the job: so-called 'sand channels' (Fig. 22.10) or 'tracks' are typically long steel plates a couple of tire diameters long and a bit more than a tire width wide, usually corrugated to provide stiffness and perforated to reduce weight and to improve grip. These can be inserted in front of and under the tires (to form a ramp, and to provide a nonslipping surface).

**Fig. 22.9** Dismounting (and even, as here, pulling up on the back of the bike) will reduce the wheel loading and thus offer a better chance that the sand under the wheels will hold without failing. *Photo* R. Lorenz

A much more problematic issue is if the vehicle has sunk deep enough that its lower structure is being supported by the sand—a phenomenon known as 'high-centering'. The normal force on the bottom of the vehicle means, first, that the tires carry less weight (and therefore, for a given coefficient of friction, can apply less forward push) but also that considerable friction exists on the bottom of the vehicle, meaning more tire push is needed to start moving. There is little hope for a lone vehicle to get out of this predicament unless the sand is dug out from the bottom, to put the weight back on the tires. This, of course, is somewhat arduous; since the sand is loaded, the grains are locked and it is harder to dig, and it is difficult to dig efficiently at the shallow angle necessitated by reaching in from beside the vehicle.

For both the wheel-spinning and high-centering issues (which should be recognized as separate), the ultimate recovery method is to get pulled out. A second vehicle (and for deep desert trips, multiple vehicles that are not fully loaded should always be employed in order to avoid the danger that an irreparable failure of one vehicle leaves people stranded; i.e., the people traveling in N vehicles should be able to fit in $N-1$) with wheels on more solid ground can apply a strong forward force via a towrope, which may be able to overcome high-center friction and pit slopes. Of course, digging and channels should still be employed to reduce these barriers, and care taken to prevent the second vehicle also getting stuck. Note that towropes under strain have formidable elastic energy, which can be

**Fig. 22.10** JPL scientist Essam Heggy, during a ground-penetrating radar measurement campaign in Egypt, shoving a perforated sand channel under the tire of a vehicle stuck in sand in preparation for backing out after getting stuck in soft sand moving uphill—sometimes it is easier to go back than forward. *Photo* Ralph Lorenz

lethal if the rope snaps. Also towropes that are not applied to appropriate hardpoints on the vehicle can cause damage—one author knows of at least one axle that was bent this way.

## 22.5 Getting Stuck and Unstuck on Mars or Titan

While the potential for disaster does indeed exist, getting stuck in the field is usually just an inconvenient part of the terrestrial fieldwork process, and occasionally the source of some good campfire stories. On other planets, however, there are generally fewer options: one day, when astronauts are driving on Mars, the amount of oxygen in their suits will be an important consideration in digging out.

For the present phase of unmanned exploration, there are few options indeed. Apart from possibly using a manipulator/sampling arm to push on the surface, controllers can only adjust the position and torque/speed of the wheels. The situation is analogous to a driver being imprisoned in the vehicle on Earth, they have only the steering wheel and the accelerator, and a very limited view of the wheels and the ground. There is no prospect of being dug out (short of a major wind episode) and certainly no hope of being towed.

The first rover on Mars, the tiny Sojourner rover, traversed only a few tens of meters around the Pathfinder lander, although it did encounter a couple of ripples (see Fig. 16.21). The 2003 Mars Exploration rovers Spirit and Opportunity, having operated for many years and traversed many kilometers, encountered a much wider range of terrain,

**Fig. 22.11** Rover tracks on a ripple on Meridiani Planum, Mars, identified by on-board software on Sol 603 of Opportunity's mission as having dangerous wheel slippage. The rover was recovered by backing out. (Note that due to a wheel actuator failure, the rover was driving backwards, hence the rear wheel reaches the maximum extent up the ripple.) *Image credit* NASA

including vast fields of ripples (or TARs, 'Transverse Aeolian Ridges'—see Chap. 5) in which sinkage or wheel-slippage often occurred (e.g., see Figs. 1.7 and 22.11).

On the 446th sol of operation on Mars, during which it was supposed to perform a 50 m autonomous ('blind') driving run, the Opportunity rover achieved only 2 m of forward progress, as it buried its wheels in a ripple. It then spent almost 40 days stuck on this ripple that was nicknamed, for obvious reasons, 'Purgatory' (e.g., Arvidson et al. 2011).

The wheel slippage could be documented by comparing the number of wheel revolutions measured by a counter with the forward progress measured onboard by comparing

**Fig. 22.12** NASA's Mars Exploration Rover Spirit slipped in soft ground during short backward drives on the 1886th and 1889th Martian days, or sols, of the rover's mission on Mars (April 23 and 26, 2009). Spirit used its front hazard-avoidance camera after driving on Sol 1889 to get this wide-angle view, which shows the soil disturbed by the drives. Spirit drove 1.11 m (3.6 ft) on Sol 1889 and 1.68 m (5.5 ft) on Sol 1886. The rover drags its right front wheel, which no longer rotates. For scale, the distance between the wheel tracks is about 1 m (40 in.). Image credit: NASA/JPL-Caltech

**Fig. 22.13** A screen shot from software used by the Mars Exploration Rover team for assessing movements by Spirit and Opportunity illustrates the degree to which Spirit's wheels have become embedded in soft material at the location called 'Troy'. The image simulates Spirit's position on May 8, 2009, during the 1900th Martian day, or sol, of what was originally planned as a 90-sol mission on Mars. Image credit: NASA/JPL-Caltech

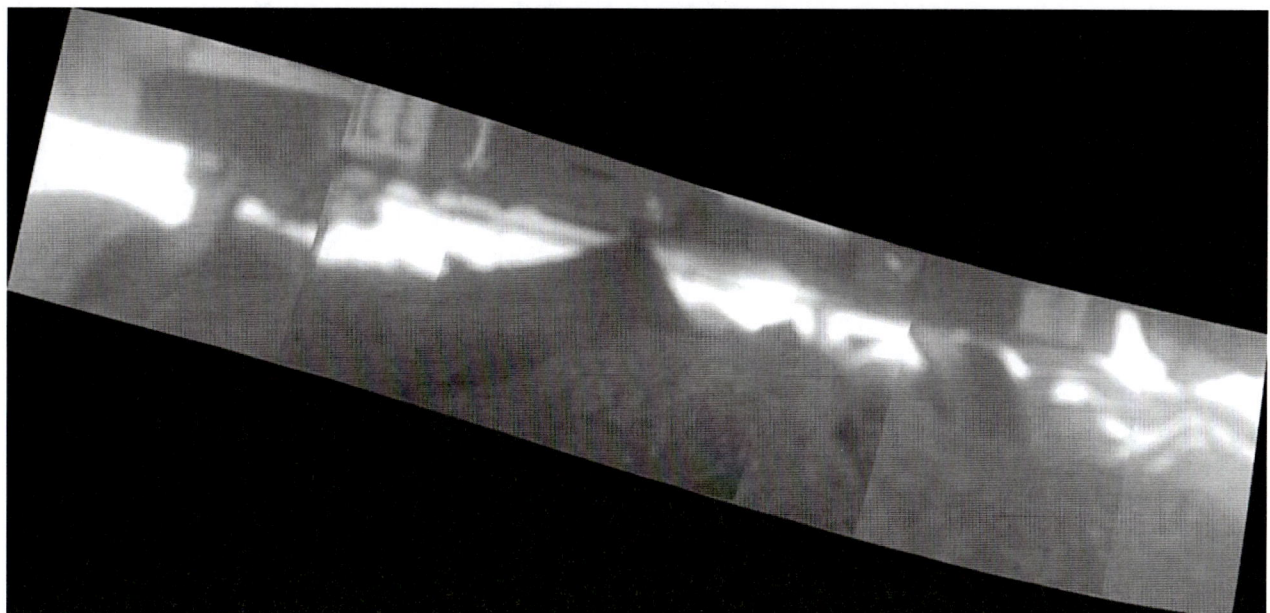

**Fig. 22.14** A mosaic of images from the articulated imager on the Spirit rover, showing the underside of the vehicle. A largely-buried wheel can be seen at the *left* edge and another towards the *right* of the image. Most problematic, however, is the sharp rock in the *middle* upon which the vehicle appears to have high-centered. *Image credit* NASA/JPL-Caltech

**Fig. 22.15** Michael Malin, *left*, a prominent Mars scientist, comments to a news reporter during tests of the 'Scarecrow' rover on Dumont Dunes in California's Mojave Desert in May 2012 to determine its performance in driving up various slopes on windward and downwind portions of dunes. This is a stripped-down mobility test rover, designed to have the same weight on Earth as the full Curiosity rover has on Mars. *Image credit* NASA/JPL-Caltech

successive camera images (i.e., visual odometry—see Biesiadecki and Maimone 2006). This allows autonomous detection of excess wheel slippage, allowing the vehicle to stop itself before it goes too far into a hazardous spot. For example, on sol 603, 5 m of blind driving made 4 m of progress before the visual odometry software detected 44 % wheel slip. The vehicle had made a shallow approach onto a ripple, resulting in all six wheels sitting on loose sediment (Fig. 22.11) but it took only one day to recover from this by turning and backing out.

The mission of the Spirit rover came to an end when it got stuck, not in a ripple, but in a patch of soft dust (Fig. 22.12). As in previous temporary stoppages, images were examined, the configuration of the rover with respect to the soil was modeled (Fig. 22.13) and extraction procedures developed and tested with an engineering model rover in the lab (Fig. 1.14).

Diagnosing the state of the rover is carried out first with telemetry on the number of wheel revolutions (which can be compared with imaging showing how much the rover has moved, and thus how much wheel slippage there has been). Second, where cameras can show the configuration of the wheels and the soil around them, images can be analyzed to evaluate the depth of sinkage. Most cameras tend to be mounted on the upper part of a rover for a good overall view, but where they are placed on articulated elements such as sampling arms, the wheels and underside can be inspected. Sometimes the news from this perspective is not good—as Fig. 22.14 shows, the wheels of the Spirit rover did not just sink into soft ground, but the vehicle got high-centered on a rock, preventing the wheels from providing as much purchase as they otherwise could.

Extensive efforts were made to recover Spirit, and there were some encouraging signs of movement in their later stages. But Martian winter drew in and, with the sun low in the sky, without being able to move to a slope oriented favorably to get power from its solar panels, Spirit's batteries became cold and depleted. Despite hopes that it might revive the following spring, Spirit did not respond to commands from Earth and thus, at least indirectly, the mission was ended by getting stuck in soft ground.

Given the unforgiving nature of getting stuck on other planets, considerable effort is devoted to understanding the limitations of the chosen vehicle's wheels, motors, mass distribution and steering performance ahead of time. The larger Curiosity rover (also known as the Mars Science Laboratory) was tested extensively on Earth to learn the limits of its mechanical systems (Fig. 22.15). Tests were done with a 'Scarecrow' mockup of the rover: although Scarecrow's wheels and motors are the same as Curiosity, Scarecrow is otherwise stripped down so that it weighs about the same on Earth as Curiosity does in the lesser gravity of Mars. At the time of writing, Curiosity had landed successfully at the Gale Crater on Mars (see Fig. 12.22) and may encounter sand dunes on its way to its target, Mt. Sharp, at the center of the crater.

With Mars, in the present epoch at least, one is relatively assured of encountering only dry sand or dust, which limits to a manageable range the types of environment in which one needs to do testing. Roving on Venus seems a rather forbidding prospect for a variety of reasons: getting stuck in what little sand there is may be the least of an engineer's problems, given the high temperatures, the corrosive environment and rugged rocky terrain. As for Titan, the chemical composition of Titan sand is of course rather different, and perhaps also its mechanical properties, especially if there has been a recent methane rainstorm. The prospect of dealing with wet organic mud presents vehicle designers and operators with a whole new set of interesting but difficult-to-test challenges, which has in part led to an emphasis in Titan exploration of considering aerial platforms such as balloons or aircraft (see Fig. 13.14).

# Dunes and People

Dunes are beautiful structures that deserve study in their own right. But lest the reader be confronted by demands to justify their study, we offer a few examples here of how dunes directly affect human livelihoods. Of course dunes have influenced human history: whether in the age of Alexander the Great, or the Gulf War or the many conflicts in between, conflict in the desert places special demands on men, animals and machines. We touch upon these in Sect. 1.3; here we discuss the ongoing war of people against sand. The fact that dunes move on timescales perceptible to humans is a remarkable one, and is much of the fascination that motivates this book. For civil engineers trying to keep roads clear, it can be a considerable inconvenience.

Settlers in the northwestern USA found some challenges with dunes that move, such as those along the Columbia river gorge (see Figs. 23.1 and. 23.2). These turned out to be poetically ephemeral, no longer existing, presumably because the sand source has been choked. Yet even in the 1960s, Oregon dunes threatened agricultural lands, and efforts to mitigate the problem inspired Frank Herbert (see Chap. 24) to author the 'Dune' novels.

Occasionally, dune movement can have a positive side: Andreotti et al. (2002) quoting a Ph.D. thesis by Oulehri, note that the difference in migration rate between small and large barchans dunes was known to Saharan people in ancient times. Barchans were used to protect goods from pillage: a small or a large dune would be chosen in which to bury a bundle of goods, depending on when it was desired to recover it. Further south in Africa, it is said that local Namib peoples will lay dead bodies in the path of a dune in order to bury them—and it will be sometime before it reappears. It is perhaps for this reason that the area is known as the Skeleton Coast.

Many oasis towns in the Sahara have long confronted sand encroachment, a prominent example being InSalah in Algeria, which has a large dune more or less in the center of town. But the Saharan sands most famously threaten Nouakchott, the capital and largest city (population today $\sim 900,000$, compared with only 75,000 in 1978) of Mauritania. The city's name is from the Berber 'Place of the Winds', which hints at its predicament. The combination of north-westerly trade winds and the harmattan, a seasonal easterly wind, causes snake-like linear dunes to march south-west towards the coast (Fig. 23.3).

Nouakchott has nowhere to go—it is pinned between the advancing dunes and the sea. In some cases, however, the advance can be accommodated as just part of the cost of living somewhere. For example, the desert oasis of Liwa in the United Arab Emirates is a major center of date cultivation. It is striking in satellite imaging as defining an arc within which large (2 km wide, 100 m tall) megabarchans are found. These megabarchans are slowly advancing southeast, at perhaps 0.1 m/year; this is apparent in that the northern boundaries of the date plantations are sharply defined by the slipfaces of the dunes (Fig. 23.4) and in some places trees can be seen in the process of being buried (Fig. 23.5). A tree may yield fruit for several decades, so the dunes advance only about one tree-length (and thus, in a typical plantation, one tree-spacing) over the useful life of the tree. So you plant the tree anyway and you get what you get.

French scientist Jean Meunier has pioneered a range of methods ('BOFIX'—of course, many of these methods have been long known, although less systematically, to desert people worldwide) to break up dunes using fences that divert the wind, spawning eddies that slice through the dune in a matter of months; other (permeable) fences can be used upstream to interrupt the sand transport to dunes threatening individual buildings (Zandonella 2003). The Food and Agriculture Organization (FAO) of the United Nations has undertaken major planting efforts, growing seedlings that might stabilize the dunes in a sustainable

**Fig. 23.1** A photograph by G.K. Gilbert in 1899. This is the last house in Biggs, Oregon, torn down shortly after this photo was taken. The town was abandoned because it was being overrun by dunes. Note the abundant ripples and the almost completely buried railway track. A similar abandoned house is still visible today as a tourist attraction at Kolmanskop near Luderitz in Namibia. *Photo credit* USGS

way. Note that in some locations, even though the near-surface is dry, once a tree can become established, its roots might reach deep enough to find moisture. Surrounding seedlings and their roots with a plastic barrier may reduce evaporation enough for artificial irrigation to be carried out for a year or so while the tree grows, and then it can keep itself going (e.g., Rognon 1999). How well these efforts succeed remains to be seen: fixing a dune by these methods, or the less sustainable approach adopted sometimes in the Middle East of spraying oil on it, does not change the fact that more sand is still coming from upwind and it will have to go somewhere.

A well-tried approach predating Meunier's approaches is to 'plant' a grid of grasses, straw or permeable fabric (Fig. 23.6). With a grid spacing of about 1 m and a fence height of about 1/10th of that, sand movement becomes substantially suppressed. This approach was introduced into China (e.g., Zheng 2009) by Russian advisers—notably to the Baotou-Lanzhou railway across the Tengger Desert.

Some of the challenges of engineering roads and other infrastructure in arid lands are discussed in Dauncey et al. (2012). In richer countries, it is of course possible to keep specific sites like roads clear of sand simply by shoveling it or bulldozing it, especially if the sand transport rate is modest (e.g., Fig. 23.7). Alghamdi and Al-Kahtani (2005) discuss these measures, quoting costs of $2/m$^3$ for manual removal and $0.5/m$^3$ for bulk (mechanized) removal. Given that small barchan dunes 2 m high wide can move at ∼50 m/year, this means a 200 m sector of road crossed by a barchan corridor may cost $10,000 a year to keep clear. This is clearly not a preferred solution, but in some settings is the most straightforward.

**Fig. 23.2** Fences installed to guide drifting sand away from the railroad in the Columbia river gorge, near Grant, Oregon. Similar fences are sometimes found elsewhere to prevent blowing snow from blocking roads. USGS *photo* by G.K. Gilbert

One approach (which can be seen applied in satellite images) (Fig. 23.8) is to have two roads, separated by a little over a dune length. Thus when one road is blocked by a dune (this usually occurs with barchans, which are the fastest-moving dunes and usually have some spacing between them), the other is not. As the months and the dune go by, the traffic is simply diverted to the other road, and then back again when the next dune hits.

This approach is of little use for irrigation canals, which can get silted up and blocked by sand flow. Still other

**Fig. 23.3** Landsat image of Nouakchott, in 2001, visible as the dark areas aligned around with somewhat straight roads. Linear dunes from the western Sahara are migrating south-westwards into the city. The Atlantic ocean is to the *left*. *Image* NASA

**Fig. 23.4** Image from the International Space Station of the villages and plantations in the Liwa Oasis, United Arab Emirates. The megabarchans here (see also Fig. 7.16) are about 2 km across, and if you look closely, you'll see the southern slip faces of the dunes generally define a sharp northern edge to the plantations, while the southern side of the plantations is often irregularly-placed in the sabkha between the dunes. *Image* NASA/JSC/EOL

**Fig. 23.5** Megabarchan slip face overrunning trees and a wall at a plantation in Liwa. Where the growth rate of trees is known, such situations place a constraint on migration rate. *Photo* R. Lorenz

**Fig. 23.6** Planting a grid of grasses can impede sand transport: here a dune near Agelat in Libya is being stabilized this way. *Photo* USGS

**Fig. 23.7** Small barchanoid dunes intrude onto the road in the Liwa oasis, United Arab Emirates. This occurs preferentially on the top of megabarchan dunes, where the wind is accelerated and there are no date palms to trap the sand. *Photo* courtesy Jani Radebaugh

**Fig. 23.8** Digital image from the International Space Station in 2011 of the oasis town of El Kharga in Egypt. The town and the dark patches of agricultural land are bracketed by two corridors of fast-moving barchans dunes coming from the north. At upper right is the airport runway (which is notably aligned in the same direction as the dune migration, so that planes can take off and land into the wind): the barchans cross the road, requiring clearing efforts. For nearby barchans viewed from the air, see Fig. 6.5. *Photo* NASA/JSC/EOL

**Fig. 23.9** The 'floating fence' at the US–Mexico border near Yuma, in the Algodones dunes. The fence is designed to maintain its height even as the sands shift beneath it. *Photo* by US Customs and Border Protection

challenges are presented to agencies charged with security. Sand blowing against a wall will simply pile up and eventually make it easy to climb the wall, thus defeating the wall's purpose. At the US–Mexico border in the Algodones dune field (Fig. 23.9), a special 5 m-high fence designed to 'float' on the sand and deform (or be moved) as the contours of the ground change underneath it has been built to prevent illegal border crossings. This 10 km fence—nicknamed the 'Sand Dragon'—was built at a cost of $40 M.

Much as was once done with the Great Wall, sand near the fence's edge is groomed periodically to generate a smooth surface on which footprints or other traces can be more easily detected. As noted by Welland, however, perhaps the biggest future battleground of man and sand will be China, where desertification is increasing the sand mobility dramatically. The scale of the problem, and the scale of measures to solve it, will be characteristically immense.

# Fictional Dune Worlds

Since human civilization requires water, and dunes imply dry conditions, there is a generally an anticorrelation of human population with dune population, and thus dunefields are an exotic, but recognizable, realm for most of us. It is therefore no surprise that dune-filled deserts figure in science fiction; indeed Arthur C. Clarke's first novel was entitled *The Sands of Mars*.

An interesting, if slightly forced, connection is that the most-read and most-translated book in French (selling over 200 million copies to date) is *Le Petit Prince* (*The Little Prince*) which is in part set on an asteroid, and thus might qualify as science fiction. This novella was written by aristocratic author Antoine de Saint-Exupéry, who worked as a mail pilot in North Africa and thus was very familiar with the Sahara. Aspects of the book are inspired by his experiences; notably, the book's narrator—a pilot—talks of being stranded in the desert next to his crashed plane. This clearly draws on Saint-Exupéry's remarkable survival in 1935 of crashing his plane in the Sahara in the middle of the night after 19 h in the air during an attempt to break the Paris-to-Saigon speed record. He and his co-pilot nearly died of dehydration wandering among the dunes in Egypt (actually only 50 km or so from the Quattaniya dunes shown in Fig. 24.1, see also Figs. 6.12 and 16.21) for 4 days before being saved by a chance encounter with a Bedouin. The writer/narrator's admiration of the desert is clear in the following excerpt, which points to a recurring theme in desert studies and fiction, of water beneath the sands:

> …without saying anything more, I looked across the ridges of sand that were stretched out before us in the moonlight. "The desert is beautiful," the little prince added. And that was true. I have always loved the desert. One sits down on a desert sand dune, sees nothing, hears nothing. Yet through the silence something throbs, and gleams… "What makes the desert beautiful," said the little prince, "is that somewhere it hides a well…

In the following text we discuss two fictional planets of note, Arrakis, the setting of the *Dune* series of novels, and Tatooine, a prominent world in the 'Star Wars' movies. The first is of historical interest, and in prompting thought about the global sand budget of a dry planet; the second is of geographical interest, in that dune migration has been observed at the film set used to make this fictional world seem real.

## 24.1 Arrakis

It is remarkable to us scientists, although perhaps not noted by most of its readers, that Frank Herbert dedicates his award-winning 1966 book *Dune* (considered the world's best-selling science fiction novel) as follows: "To the people whose labors go beyond ideas into the realm of 'real materials'—to the dry-land ecologists, wherever they may be, in whatever time they work, this effort at prediction is dedicated in humility and admiration." Herbert was inspired by US Department of Agriculture efforts to stabilize dunes in Oregon with grass.

As noted by Lorenz (2007), the novel is set on a world, Arrakis, whose coverage by sand exceeds that of any world we know. As the book describes:

> *Observe closely, Piter, and you, too, Feyd-Rautha, my darling: from sixty degrees north to seventy degrees south—these exquisite ripples. Their coloring: does it not remind you of sweet caramels? And nowhere do you see blue of lakes or rivers or seas. And these lovely polar caps—so small. Could anyone mistake this place? Arrakis! Truly unique"* (*Dune*, p. 14).

The geology—with salt flats—hints that Arrakis was once wetter, and a plot element is the prospect of terraforming it back to more benign conditions. Herbert describes the desert people 'Fremen' having clearly researched Bagnold's work and others: the Fremen wear 'stillsuits' to retain exhaled water vapor, presumably inspired by the external condenser that Bagnold introduced into the radiators of his desert trucks. The Fremen reprimand one character for their poor skills at traversing the dunes (see Chap. 22): *"We watched you come across the sand last night. You keep your force on the slip-face of the dunes. Bad"* (p. 211).

**Fig. 24.1** The Nile delta appears very prominently in this Hasselblad hand-held image taken from Gemini IV in 1965. (This was one of the first spaceborne pictures of dunes (see also Fig. 18.4): note the edges of the frame in this pre-digital-camera image.) The dark areas at *lower left* are the Birket Qarun lake and agricultural areas around Faiyum. The Mediterranean is just visible at *top left*. Bright linear dunes appear at *mid-lower left*, the two easternmost ones being the site of radar and other field studies (see Figs. 6.21 and 16.12). Saint-Exupéry crashed about a dune-length away from these. *Photo* NASA

The extent of Arrakis' dunes, covering close to 90 % of the planet if we take the quote above at face value, merits some consideration. This is far more even than Titan, the most dune-covered world we know of (a region on Titan, observed to become damp with methane rainfall, was named Arrakis Planitia in 2010). Sand formation must be somewhat self-limiting: once most of the bedrock is covered up, there is not much of it to be eroded. Some mechanism for segregating sand sources and sinks must be applied for sand production to continue. While there are doubtless exoplanets which may have had seas in the past but are dry now and may well have extensive dunefields, it seems improbable that any are so nearly completely covered in deep sand.

The thick sand deposits on Arrakis serve as the habitat of the book's iconic giant sandworms. The bioenergetics of locomotion through sand on such a large scale do not seem to have been physically evaluated—on this point we must concede artistic license. The Fremen track sandworms through surface disturbances, or 'wormsign'; were they to explore the developments in ground-penetrating radar we discuss in Chap. 16 of this book, they might find that technology superior for this application.

One other point about Arrakis deserves planetological discussion. It lacks a global magnetic field. The Fremen use local magnetic anomalies sensed with a 'paracompass' for short-range navigation—this may have been inspired by Bagnold's use of a suncompass for direction-finding on the move, since using a magnetic compass would require his Long-Range Desert Group patrols to stop and walk away from the perturbing fields of their steel trucks.

The exterior shots of the 1984 film adaptation of the story were shot in the Samalayuca desert, near Ciudad Juarez in northern Mexico. Actor Patrick Stewart alledgedly complained that the stillsuit was the most uncomfortable costume he had ever worn. Later in the production process, some additional outdoor fight-scene shots were needed: shots were made on sand against a 'blue screen' background outside the Churubusco film studio in Mexico City so that artificial painted backgrounds could be applied in post-production. In a remote-sensing connection, the red and infrared content of sunlight in these pre-digital-filming days was problematic, and specially-formulated blue paint had to be brought in from the US to make this screen spectrally pure.

## 24.2 Tatooine

The 'Star Wars' movie released in 1977 (and indeed the series) was a landmark in film-making, with many innovations in plot and style as well as in technical special effects. Movie locations include deserts as well as snow-covered and forested worlds. The movies justifiably enjoy cult status, and fan websites exist devoted to enabling fans to visit the sites at which scenes were filmed.

While most of the filming of the original 'Star Wars' movie (i.e., Episode 4) was done in Tunisia as we discuss later, it was realized after the Africa filming campaign in 1977 that some additional desert shots were required. These (where R2D2 and C3PO are wandering among dunes before encountering sand people) were made at the Stovepipe Dunes in Death Valley National Park in California, a few hours' drive from Hollywood (Fig. 24.2). In addition, an opening desert scene in the sequel 'Return of the Jedi' features the heroes' escape from crime boss Jabba the Hutt's sand-yacht. This scene was also filmed at a location inexpensively convenient for Hollywood, namely the Imperial Dunes at Yuma, Arizona (aka Algodones, Fig. 24.3).

The iconic desert scenes of planet Tatooine from the first movie, 'Episode IV, A New Hope', were shot in various locations in Tunisia, and Tatooine takes its name from the remote southern town whose Arabic name is usually transliterated as Tataouine. Many of the exotic-looking buildings are merely the local Berber style of adobe construction, such as the fortified granaries like Ksar Ouled Soltane. Curiously, on the first day of filming, Tunisia saw

## 24.2 Tatooine

**Fig. 24.2** The dunes at Stovepipe Wells in Death Valley National Park are generally star-shaped. Some are rather rounded, while others have well-developed slip faces. Note the ripples in the foreground. Looking east, the mountaintop skyline in the *left half of the image* can be matched with a screenshot from 'Star Wars' (*inset*). We might observe that the small footprint of R2D2's wheels is a rather improbable arrangement for locomotion on sand—see Chap. 22. The dunes themselves have, of course, changed between this 2008 photo and the 1977 movie. *Photo* R. Lorenz

**Fig. 24.3** The Algodones Dunes between Yuma and El Centro at the California/Arizona and US/Mexico border, seen from a commercial flight to Hollywood's Burbank airport. Part of the Algodones are administered as the Imperial Sand Dunes National Recreation Area, and are also sometimes referred to as the Glamis dunes: before the expanse of agriculture in this area (at *upper right* is a canal) this dunefield was merely an extension of the Gran Desierto de Altar to the south. Note the superposed duneforms: the emergence of this pattern is discussed by Derickson et al. (2008). A faint reflection in the window is apparent at *mid-right*, a common challenge in airliner photos-of-opportunity. A space station view of these dunes is given in Fig. 7.17 and a radar image in Fig. 18.18. *Photo* R. Lorenz

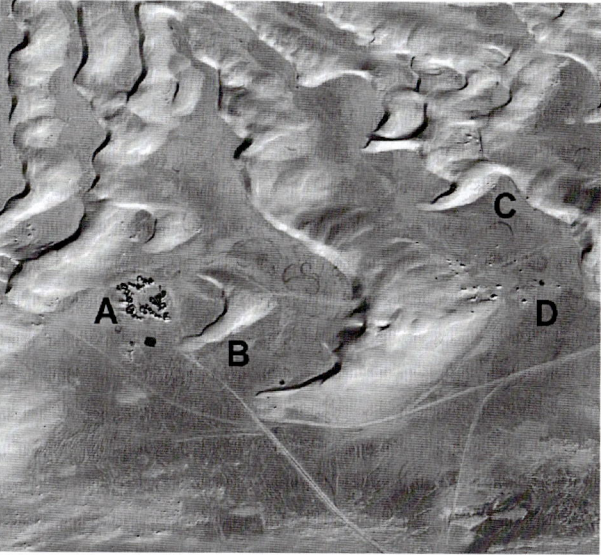

**Fig. 24.4** Commercial satellite image acquired in 2004 of the Chott El Gharsa area in Tunisia, showing (*A*) the Mos Espa film set covering about a hectare and (*B*) the menacing barchan, still some ∼100 m away. Vehicle tracks (including circular segments) are evident on the interdune pan, as well as over the dunes themselves. Also indicated are the small Repro Haddada set (*C*) seen in Fig. 24.5 and several yardangs (*D*). Quickbird satellite image courtesy of Nabil Gasmi

its first major rainstorm in 50 years. Filming near Tozeur of scenes in both Episode IV and Episode I 22 years later was disrupted by a sandstorm.

Most Episode IV scenes are on strikingly flat plains, but large tracked vehicles (the Jawa sandcrawler) merit discussion. These used giant treads, and were apparently inspired by NASA's mobile launch platform used for the

**Fig. 24.5** Remains of the Repro Haddada (Slave Quarters) set from Star Wars Episode 1. The building has been overrun by a barchan since the film was shot in the late 1997, and the roof collapsed. Another barchan is visible at *left* (note the many vehicle tracks), and in the background at right are several small yardangs. *Photo* R. Lorenz

Saturn V and later the Space Shuttle. No actual full-sized sandcrawlers were made, but working lower sections were used for close-ups. Allegedly, the Libyan government expressed concern about the large tracked vehicles near its border.

The set used to portray the Tatooine town of Mos Espa in Episode I in 1997 is of interest. It is a fairly large set of about 20 buildings (see Fig. 24.4) over a hectare in a small barchan field between the ephemeral lakes (easily visible from space) Chott el Djerid and Chott el Gharsa, and is a prominent tourist attraction for the nearby oasis town of Tozeur. Although not of concern to the film-makers, barchans of course move. A smaller adjacent set has been overrun by a barchan, essentially demolishing it (Fig. 24.5).

Another larger barchan lies to the east of the main Mos Espa set, which is the main tourist attraction, drawing in around 100,000 visitors a year. In the satellite image in Fig. 24.4, it is ~100 m away: however, its progressive approach can be observed in satellite images in Google Earth (Fig. 9.4) and the dune has essentially reached the set, as evidenced by the field picture in Fig. 1.18. A detailed study (Lorenz et al., 2013) also found a 2002 image, from which a distance of 140 m could be estimated. On average, then, this ~9 m high barchan has been moving at about 15 m per year (a typical migration rate for a dune of this size, see Chap. 9), although its rate of movement seems to have decreased in recent years, perhaps as a result of heavier-than-average rainfall (indeed, the Chott El Djerid was flooded in 2009).

The movie connection makes this site an appealing one for classroom exercises in remote sensing. It is also of broader social interest in that if the barchan overruns the Mos Espa set, its attractiveness to tourits may be substantially degraded with a consequent economic impact.

# Part VI
# Conclusions

# Conclusions 25

The geographical study of dunes and sand seas began in earnest over a century ago, but it was the work of one man, Ralph Bagnold, 70 years ago, that drew together a grand synthesis of dune morphology with the physics of the transport of sand, and brought quantitative measurements to bear on how sand and dunes move.

The breadth and depth of dune studies today, even just for the Earth, is now far beyond the compass of even the most talented individual scientist, and the references to this book (by no means comprehensive) attest to the vast number of workers in the field, most of whom must specialize in just a few of the subject areas we have attempted to survey. The space age has brought dune studies beyond being a mere Earth science, and has challenged us with familiar dune forms in exotic environments. Stimulated in part by these findings, much important progress has been made, especially in the last decade.

It is the nature of science to probe ever deeper and much work of course remains, but our impression on compiling this book is that the major factors influencing dune scale, morphology and evolution are now substantially understood, and that given a sediment and a wind description, the resultant dune landscape can be predicted. This may perhaps offers a new challenge to producers of science fiction—only a subset of all possible atmosphere–landscape combinations is geologically or meteorologically plausible!

Dunes can be found, it seems, on any world with an atmosphere thick enough to move granular material. On Venus, with the thickest atmosphere of all, large dunes are scarce because that world is starved of sand.

Mars' young dunes and ripples are generally larger than young dunes on Earth, because the thin atmosphere is associated with longer dynamic length scales (saltation trajectories and saturation length). Yet most of Mars' dunes are barchans and transverse dunes—in motion, albeit slowly. Except where dunes are confined in crater basins (as many of them are) the sand has yet to find its final resting place. In principle, Mars dunes could grow to be truly massive, since the planetary boundary layer can be tens of kilometers thick, but the sand has not accumulated to the extent needed to build dunes of that scale.

Titan presents a fascinating counterpoint and shows that the juxtaposition of seas of liquid and seas of sand is not a combination unique to the Earth. Here, young dunes are so small as to be invisible (to exploration so far) but the circulation has allowed sand to accumulate in vast equatorial sand seas, with strikingly large and regular linear dunes; the regularity of the dune pattern attests to the maturity of the sand seas. By coincidence, these dunes are the same height as large dunes on Earth, consistent with Titan's atmospheric structure. While Mars seems very much in flux, Titan's aeolian evolution seems to have reached a conclusion, perhaps with only minor astronomically-forced vacillations about an end-state.

Dunes represent a history of previous conditions, the small and uppermost features reflecting the recent past, but often with a palimpsest of climate long ago in the larger pattern underneath. This superposition and evolution (the convolution of what are often intermittent or even rare dynamic conditions) can now be explored in laboratory and computer experiments. With these and other tools, we have begun to learn the language of the dunes, and can decode their whispers of the past.

We close, then, with an image (Fig. 25.1) that mirrors the first image in this book. While neither dunes nor ripples are visible, the point is made. Planetary exploration brings a

**Fig. 25.1** The famous Earthrise image from Apollo 8 that brought an iconic perspective on our home planet. The lunar surface in the foreground is sadly bereft of bedforms. NASA image AS 08-13-2329

new perspective on our own planet, prompting a more fundamental understanding of the processes that shape it, and perhaps a growing appreciation of its diverse and beautiful landscape. In the words of T.S. Eliot:

> We shall not cease from exploration
> And the end of all our exploring
> Will be to arrive where we started
> And know the place for the first time

# References

Aber, J., Marzolff, I. and Ries, J. 2010. *Small-Format Aerial Photography; Principles, Techniques and Geoscience Applications.* Elsevier.

Avduevsky V.S., Vishnevetskii, S.L., Golov, I.A., Karpeiskii, Iu. Ia., Lavrov, A.D., Likhushin, V. Ia., Marov, M. Ia., Melnikov, D.A., Pomogin, N.I. and Pronina, N.N. 1977. Measurement of wind velocity on the surface of Venus during the operation of stations Venera 9 and Venera 10. *Cosmic Research* 14, 622–625.

Aharonson, O., Hayes, A.G., Lunine, J.I., Lorenz, R.D., Allison, M.D. and Elachi, C. 2009. An asymmetric distribution of lakes on Titan as a possible consequence of orbital forcing. *Nature Geoscience* 2, 851–854.

Ahlbrandt, T.S. 1979. Textural parameters of eolian deposits. In McKee, E.D. et al. 1979. *A Study of Global Sand Seas.* U.S. Geological Survey Professional Paper 1052.

Alghamdi, A.A.A. and Al-Kahtani, N.S. 2005. Sand control measures and sand drift fences. *ASCE Journal of Performance of Constructed Facilities* 19, 295–299.

Allen, J.R.L. 1982. *Sedimentary Structures: Their Character and Physical Basis*, Vol. I. Elsevier, (pp. 75–77, derivation of Stokes Law; pp. 512–14, aeolian bedforms).

Allison, M. 1992. A preliminary assessment of the Titan planetary boundary layer. In B. Kaldeich (Ed.), *Proceedings of the Symposium on Titan*, pp.113–118. Toulouse, France, 9–12 September 1991, ESA SP-338, European Space Agency, Noordwijk, The Netherlands.

Almeida, M.P., Parteli, E.J.R., Andrade, J.S., Hermann, H.J., 2008. Giant saltation on Mars, Proc. Nat. Acad. Sci. 105(17), 6222-6226, doi:10.1073/pnas.0800202105.

Anderson, F.S., Greeley, R., Xu, P., Lo, E., Blumberg, D., Haberle, R. and Murphy, J.R. 1999. Assessing the Martian surface distribution of aeolian sand using a Mars general circulation model. *Journal of Geophysical Research* 104, 18,991–19,002.

Anderson, R.S. 1987. A theoretical model for aeolian impact ripples. *Sedimentology* 34, 943–956.

Anderson, R.S. and Haff, P.K. 1988. Simulation of eolian saltation. *Science* 241, 820–823.

Andreotti, B. 2004. The song of dunes as a wave-particle mode locking. *Physical Review Letters* 93 (23), 238001/1–238001/4.

Andreotti, B. 2012. Sonic sands. *Reports on Progress in Physics* 75, 026602.

Andreotti, B., Claudin, P. and Douady, S. 2002. Selection of dune shapes and velocities. Part 1: Dynamics of sand, wind and barchans. *European Physical Journal B* 28, 321–339.

Andreotti, B., Bonneau, L. and Clement, E. 2008. Comment on Solving the mystery of booming sand dunes by Nathalie M. Vriend et al. *Geophys. Res. Lett.* 35, L08306.

Andreotti, B., Fourriere, A., Ould-Kaddour, F., Murray, B. and Claudin, P. 2009. Giant aeolian dune size determined by the average depth of the atmospheric boundary layer. *Nature* 457, 1120–1123.

Andreotti, B., Fourriere, A., Ould-Kaddour, F., Murray, B. and Claudin, P. 2009. Size of giant aeolian dunes limited by the average depth of the atmospheric boundarylayer. *Nature* 457, 1120–1123.

Andreotti, B., Claudin, P. and Puliquen, O. 2010. Measurements of the aeolian sand transport saturation length. *Geomorphology* 123, 343–348.

Arens, S.M. 1996. Rates of aeolian transport on a beach in a temperate humid climate. *Geomorphology* 17, 3–18.

Arvidson, R.E. et al. 2011. Opportunity Mars Rover mission: Overview and selected results from Purgatory ripple to traverses to Endeavour crater. *Journal of Geophysical Research* 116, E00F15, doi:10.1029/2010JE003746.

Audoin-Debreuil, A. and I. MacGill 2007. Crossing the Sands: The Sahara Desert Track to Timbuktu by Citroen Half Track. Dalton-Watson.

Avduevskii, V.S., Vishnevetskii, S.L., Golov, I.A., Karpeiskii, Iu. Ia., Lavrov, A.D., Likhushin, V. Ia., Marov, M. Ia., Melnikov, D.A., Pomogin, N.I. and Pronina, N.N. 1997. Measurement of wind velocity on the surface of Venus during the operation of dtations Venera 9 and Venera 10. *Cosmic Research* 14, 622–625.

Baas, A.C.W. 2002. Chaos, fractals and self-organization in coastal geomorphology: Simulating dune landscapes in vegetated environments. *Geomorphology* 48(1–3), 309–328.

Baas, A.C.W. 2004. Evaluation of saltation flux impact responders (Safires) for measuring instantaneous aeolian sand transport intensity. *Geomorphology* 59, 99–118.

Baas, A.C.W. 2008. Challenges in aeolian geomorphology—Investigating aeolian streamers. *Geomorphology* 93, 3–16.

Baas, A.C. and Nield, J.M. 2007. Modeling vegetated dune landscapes. *Geophysical Research Letters* 34, L06405.

Baas, A.C.W. and Sherman, D.J. 2005. Formation and behavior of Aeolian streamers. *Journal of Geophysical Research* 110, F03011, doi:10.1029/2004JF000270.

Bagnold, R.A. 1941. *Physics of Blown Sand and Desert Dunes.* Methuen. Reprinted by Dover Books.

Bagnold, R. 1966. The shearing and dilatation of dry sand and the "singing" mechanism. *Proceedings of the Royal Society of London, Series A—Mathematical and Pphysical Sciences* 295 (1442), 219–232.

Bak, P. 1999. How nature works: The science of self-organized criticality. ?

Ball, P. 1999. *The Self-Made Tapestry: Pattern Formation in Nature.* Oxford University Press.

Balme, M., Berman, D.C., Bourke, M.C. and Zimbelman, J.R. 2008. Transverse aeolian ridges (TARs) on Mars. *Geomorphology* 101, 703–720.

Bandfield, J.L. 2002. Global mineral distributions on Mars. *Journal of Geophysical Research* 107, 9-1 to 9–20, doi:10.1029/2001JE001510.

Bano, M. and Girard, J.-F. 2001. Radar reflections and water content estimation of aeolian sand dune. *Geophysical Research Letters* 28, 3207–3210.

Barnes, J.W. and 12 co-authors. 2008. Spectroscopy, morphometry, and photoclinometry of Titan's dunefields from Cassini/VIMS. *Icarus* 195, 400–414.

Barchyn, T. and Hugenholtz, C. 2012. A new tool for modeling dune field evolution based on an accessible, GUI version of the Werner dune model. *Geomorphology* 138, 415–419.

Barnes, J.W., Brown, R.H., Soderblom, L., et al. 2008. Spectroscopy, morphometry, and photoclinometry of Titan's dunefields from Cassini/VIMS. *Icarus* 195, 400–414.

Barnes, J. et al., 2012. AVIATR – Aerial Vehicle for In-Situ and Airborne Titan Reconnaissance. *Experimental Astronomy* 33, 55–127.

Besler, H. 2008. *The Great Sand Sea in Egypt: Formation, Dynamics and Environmental Change – A Sediment-Analytical Approach*. Elsevier, Developments in Sedimentology 59.

Bibring, J.-P. and 11 colleagues, 2005. Mars surface diversity as revealed by the OMEGA/Mars Express observations. *Science* 307, 5715, 1576–1581, doi:10.1126/science.1108806.

Biesiadecki, J. and Maimone, M. 2006. The Mars Exploration Rover surface mobility flight software: Driving ambition. Paper 06-0060, IEEE Aerospace Conference, Big Sky MT, March 2006.

Bird, M.K., Allison, M., Asmar, S.W., Atkinson, D.H., Avruch, I.M., Dutta-Roy, R., Dzierma, Y., Edenhofer, P., Folkner, W.M., Gurvits, L.I., Johnston, D.V., Plettemeier, D., Pogrebenko, S.V., Preston, R.A., Tyler, G.L. 2005. Winds on Titan: First results from the Huygens Doppler Wind Experiment. *Nature* 438, 800–802.

Bishop, S.R., Momiji, H., Carretero-González, R. and Warren, A. 2002. Modelling desert dune fields based on discrete dynamics. *Discrete Dynamics in Nature and Society* 7, 7–17.

Blom, R. and Elachi, C. 1981. Spaceborne and airborne imaging radar observations of sand dunes. *Journal of Geophys. Res.* 86, 3061–3073.

Blom, R. and Elachi, C. 1984. Spaceborne and airborne imaging radar observations of sand dunes. *Journal of Geophysical Research* 86, 3061–3073.

Blom, R. and Elachi, C. 1987. Multifrequency and multipolarization radar scatterometry of sand dunes and comparison with spaceborne and airborne radar images. *Journal of Geophysical Research* 92, 7877–7889.

Blott, S.J. and Pye, K. 2001. GRADISTAT: A grain size distribution and statistics package for the analysis of unconsolidated sediments. *Earth Surface Process. Landforms* 26, 1237–1248.

Blumberg, D.G. and Greeley, R. 1996. A comparison of general circulation model predictions to sand drift and dune orientations. *Journal of Climate* 9, 3248–3259.

Blumberg, D.G. 1998. Remote sensing of desert dune forms by polarimetric synthetic aperture radar (SAR). *Remote Sens. Environ.* 65, 204–216.

Blumberg, D.G. 2005. Analysis of large aeolian (wind-blown) bedforms using the Shuttle Radar Topography Mission (SRTM) digital elevation data. *Remote Sens. Environ.* 100, 179–189.

Bodart, C., Gassani, J., Salmon, M. and Ozer, A. 2005. Contribution of SAR interferometry (from ERS1/2) in the study of aeolian transport processes: The cases of Niger, Mauritania and Morocco. In: Proceedings of Fringe 2005 Workshop, Advances in SAR Interferometry from ENVISAT and ERS missions, Frascati November/December 2005, published on CD-ROM by the European Space Agency as ESA SP-610 in 2006.

Bougher, S., Hunten, D.M. and Phillips, R.J. 1997. *Venus II: Geology, Geophysics, Atmosphere and Solar Wind Environment*. University of Arizona.

Bourke, M. and Viles, H. 2007. *A Photographic Atlas of Rock Breakdown Features in Geomorphic Environments*. Planetary Science Institute, Tucson.

Bourke, M.C., Edgett, K.S. and Cantor, B.A. 2008. Recent aeolian dune change on Mars. *Geomorphology* 94, 247–255.

Bourke, M.C, Ewing, R.C., Finnegan, D. and McGowan, H.A. 2009. Sand dune movement in Victoria Valley, Antarctica. *Geomorphology* 109, 148–160.

Breed, C.S., Grolier, M.J. and McCauley, J.F. 1979. Morphology and distribution of common 'sand' dunes on Mars: Comparison with the Earth. *J. Geophys. Res.* 84, 8183–8204.

Bridges, N.T., Greeley, R., Haldemann, A.F.C, Herkenhoff, K.E., Kraft, M., Parker, T.J. and Ward, A.W. 1999. Ventifacts at the Pathfinder landing site. *J. Geophys. Res.* 104, 8595–8615.

Bridges, N.T., Bourke, M.C., Geissler, P.E., Banks, M.E., Colon, C., Diniega, S., Golombek, M.P., Hansen, C.J., Mattson, S., McEwen, A.S., Mellon, M.T., Stantzos, N. and Thomson, B.J. 2012a. Planet-wide sand motion on Mars. *Geology* 40, 31–34, doi:10.1130/G32373.1.

Bridges, N., Ayoub, F., Avouac, J.-P., Leprince, S., Lucas, A. and Mattson, S. 2012b. Earth-like sand fluxes on Mars. *Nature* 485, 339–342, doi:10.1028/nature11022.

Bristow, C.S., Bailey, S.D. and Lancaster, N. 2000. The sedimentary structure of linear sand dunes. *Nature* 406, 56–59.

Bristow, C.S. and Lancaster, N. 2004. Movement of a small slipfaceless dome dune in Namibia. *Geomorphology* 59, 1–4, 189–196.

Bristow, C.S., Lancaster, N. and Duller, G.A.T. 2005. Combining ground penetrating radar surveys and optical dating to determine dune migration in Namibia. *Journal of the Geological Society* 162, Part 2, 315–322.

Brock, F.V. and Richardson, S.J. 2001. *Meteorological Measurement Systems*. Oxford University Press.

Brooke, B. 2001. The distribution of carbonate eolianite. *Earth Science Reviews* 55, 135–154.

Brown, R.H., Waite, J.H. Jr. and Lebreton, J.-P. (Eds) 2010. *Titan from Cassini-Huygens*. Springer.

Brownell, P. 1977. Compressional and surface waves in sand—used by desert scorpions to locate prey. *Science* 197 (4302), 479–482.

Brückner, J., Dreibus, G., Gellert, R., Squyres, S.W., Wänke, H., Yen, A. and Zipfel, J. 2008. Mars Exploration Rovers: Chemical composition by the APXS. In: J Bell (Ed.) *The Martian Surface: Composition, Mineralogy, and Physical Properties*, pp. 58–101. Cambridge University Press.

Buick, J., Chavez-Sagarnaga, J., Zhong, Z., Ooi, J., Pankaj, Campbell, D. and Greated, C. 2005. Investigation of silo honking: Slip-stick excitation and wall vibration. *Journal of Engineering Mechanics* 131(3), 299–307.

Bullard, J. 1987. A Note on the use of the "Fryberger Method" or evaluating potential sand transport by wind. *Journal of Sedimentary Research* 67, 499–501.

Bullock, M.A. et al. 2009. Venus flagship mission: Report of the Science and Technology Definition Team. NASA Jet Propulsion Laboratory, http://vfm.jpl.nasa.gov/.

Butterfield, G.R. 1999. Near-bed mass flux profiles in aeolian sand transport: High-resolution measurements in a wind tunnel. *Earth Surface Processes and Landforms* 24, 393–412.

Calvin, W.M. and 6 colleagues. 2009. Compact reconnaissance imaging spectrometer for Mars observations of northern Martian latitudes in summer. *J. Geophys. Res.* 114, E00D11, doi:10.1029/2009JE003348.

Chalmers, M. 2006. The troubled song of the sand dunes. *Physics World*, 25–28 November.

Charnay, B. and Lebonnois, S. 2012. Two boundary layers in Titan's lower troposphere inferred from a climate model. *Nature Geoscience*, 5, 106–109.

Chojnacki, M., Burr, D.M., Moersch, J.E. and Michaels, T.I. 2011. Orbital observations of contemporary dune activity in Endeavour crater, Meridiani Planum, Mars. *Journal of Geophysical Research* 116, E00F19.

Christopherson, R.W. 2006. *Geosystems: An Introduction to Physical Geography*. Pearson Prentice-Hall.

Clancey, W.J. 2012. *Working On Mars: Voyages of Scientific Discovery with the Mars Exploration Rovers*. MIT Press.

Clark, R. et al. 2010. Detection and mapping of hydrocarbon deposits on Titan. *J. Geophys. Res.* 115, E10005.

Claudin, P. and Andreotti, B. 2006. A scaling law for aeolian dunes on Mars, Venus, Earth, and for subaqueous ripples. *Earth and Planetary Science Letters* 252, 30–44.

Collins, G.S. and Melosh, H.J. 2003. Acoustic fluidization and the extraordinary mobility of sturzstroms. *Journal of Geophyiscal Research* 108, 2473, doi:10.1029/2003JB002465.

Cooke, R.U. and Warren, A. 1973. *Geomorphology in Deserts*. University of California Press.

Cornish, V. 1897. On the formation of sand dunes. *Geographical Magazine* 9, 278–302.

Coustenis, A. and Taylor, F. 1999. *Titan: The Earth-Like Moon*. World Scientific (revised in 2008, given the new title *Titan: Exploring an Earthlike World*).

Criswell, D. and Lindsay, J. 1974. Thermal moonquakes and booming sand dunes. *Abstracts of the Lunar and Planetary Science Conference* 5, 151.

Criswell, D., Lindsay, J. and Reasoner, D. 1975. Seismic and acoustic emissions of a booming dune. *Journal of Geophysical Research* 80(35), 4963–4974.

Cutts, J.A. and Smith, R.S.U. 1973. Eolian deposits and dunes on Mars. *Journal of Geophysical Research* 78, 4139–4154.

Cutts, J.A., Blasius, K., Briggs, G.A., Carr, M.H., Greeley, R. and Masursky, H. 1976. North polar region of Mars: Imaging results from Viking 2. *Science* 194, 1329–1337.

Dagois-Bohy, S., Ngo, S., Courrech du Pont, S. and Douady, S. 2010. Laboratory singing sand avalanches. *Ultrasonics* 50, 127–132.

Dagois-Bohy, S., Courrech du Pont, S. and Douady, S. 2012. Singing-sand avalanches without dunes. *Geophysical Research Letters* 39, L20310.

Dauncey, P.C., Bates, A.D., Poole, A.B. and Engineering Group Working Party. 2012. Engineering design and construction. In: M.J. Walker et al. (Eds) *Hot Deserts: Engineering, Geology and Geomorphology*, pp. 347–392. Engineering Group Working Party Report, Geological Society, London.

Derickson, D., Kocurek, G., Ewing, R.C. and Bristow, C. 2008. Origin of a complex and spatially diverse dune-field pattern, Algodones, southeastern California. *Geomorphology* 99, 186–204.

de Silva, S.L., Zimbelman J.R., Bridges, N., Scheidt, S. and Viramonte, J.G. 2011. The coarsest gravel ripples on Earth? Preliminary observations and interpretations. *Lunar and Planetary Science XLII*, Abstract 2421, Lunar and Planetary Institute, Houston, Texas.

de Silva, S.L., Burr, D.M., Ortiz, A., Spagnuolo, M., Zimbelman, J.R. and Bridges, N.T. 2012. Dark aeolian megaripples from the Puna of Argentina: Sedimentology and implications for dark dunes on Mars. In: Lunar Planet. Sci. XLIII, Abstract 2038, Lunar and Planetary Institute, Houston.

Dhabboor, M. 2008. Sand dune tracking from satellite laser altimetry, 37th COSPAR Meeting, Montreal Canada.

Dietrich, R.V. 1977. Impact abrasion of harder by softer materials. *J. Geology* 85, 242–246.

Dobrovolskis, A.R. 1993. Atmospheric tides on Venus IV. Topographic winds and sediment transport. *Icarus* 103, 276–289.

Dong, Z., Wei, Z., Qian, G., Zhang, Z., Luo, W. and Hu, G. 2010. 'Raked' linear dunes in the Kumtagh Desert, China. *Geomorphology* 123, 122–128.

Douady, S., Manning, A., Hersen, P., Elbelrhiti, H., Protière, S., Daerr, A. and Kabbachi, B. 2006. The song of the dunes as a self-synchronized instrument. *Physical Review Letters* 97(1), 018002/1–018002/4.

Duran, O., Parteli, E.J. and Herrmann, H.J. 2010. A continuous model for sand dunes: Review, new developments and application to barchan dunes and barchan dune fields. *Earth Surface Processes and Landforms* 35, 1591–1600.

Elachi, C. 1988. *Spaceborne Radar Remote Sensing: Applications and Techniques*. IEEE.

Elachi, C., Wall, S., Janssen, M., Stofan, E., Lopes, R., Kirk, R., Lorenz, R., Lunine, J., Paganelli, F., Soderblom, L., Wood, C., Wye, L., Zebker, H., Anderson, Y., Ostro, S., Allison, M., Boehmer, R., Callahan, P., Encrenaz, P., Flameni, E., Francescetti, G., Gim, Y., Hamilton, G., Hensley, S., Johnson, W., Kelleher, K., Muhleman, D., Picardi, G., Posa, F., Roth, L., Seu, R., Shaffer, S., Stiles, B., Vetrella, S., West, R. 2006. Titan radar mapper observations from Cassini's T3 fly-by. *Nature* 441, 709–713.

Elbelrhiti, H., Claudin, P. and Andreotti, B. 2005. Field evidence for surface-wave-induced instability of sand dunes. *Nature* 437, 720–723.

Elbelrhiti, H., Andreotti, B. and Claudin, P. 2008. Barchan dune corridors: Field characterization and investigation of control parameters. *Journal of Geophysical Research* 113, F02S15, doi:10.1029/2007JF000767.

Endo, N., Taniguchi, K. and Katsuki, A. 2004. Observations of the whole process of interaction between barchans by flume experiments. *Geophysical Research Letters* 31, L12503.

Esposito, L.W., Stofan, E. R. and Cravens. T.E. (Eds) 2007. *Exploring Venus as a Terrestrial Planet*. AGU Geophysical Monograph 176, American Geophysical Union.

Etyemezian, V., Nikolich, G. et al. 2007. The Portable In-Situ Wind Erosion Laboratory (PI-SWERL): A new method to measure PM(10) potential for windblown dust properties and emissions. *Atmospheric Environment* 41(18), 3789–3796.

Ewing, H., 2007. *The Lost World of James Smithson: Science, Revolution, and the Birth of the Smithsonian*. (Smithson's proficiency at the blowpipe, pp. 65–66, 233–234). Bloombsury USA.

Ewing, R.C. and Kocurek, G. 2010. Aeolian dune interactions and dune-field pattern formation: White Sands, New Mexico. *Sedimentology* 57, 1199–1219.

Ewing, R.C., Peyret, A.P.B., Kocurek, G. and Bourke, M. 2010. Dune field pattern formation and recent transporting winds in the Olympia Undae Dune Field, north polar region of Mars. *Journal of Geophusical Research* 115, E08005.

Ewing, R.C., Hayes, A.G. and Lucas, A. 2012. Reorientation timescales of Titan's equatorial dunes. 44th Lunar and Planetary Science Conference, March 18–22, 2013, The Woodlands, Texas. LPI Contribution No. 1719, p. 1187.

Fahnestock, M.A., Scambos, T.A., Shuman, C.A., Arthern, R.J., Winebrenner, D.P. and Kwok, R. 2000. Snow megadune fields on the East Antarctic Plateau: Extreme atmosphere-ice interaction. *Geophysical Research Letters* 27, 3719–3722.

Farr, T. and Kobrick, M. 2000, Shuttle radar topography mission produces a wealth of data. *EOS Transactions* 81, 583–585.

Feldman, W.C., Bourke, M.C., Elphic, R.C., Maurice, S., Bandfield, J., Prettyman, T.H., Diez, B. and Lawrence, D.J. (2008). Hydrogen content of sand dunes within Olympia Undae. *Icarus* 196, 422–432.

Fenton, L.K. 2005. Potential sand sources for the dune fields in Noachis Terra, Mars. *Journal of Geophysical Research* 110, E11004, doi:10.1029/2005JE002436.

Fenton, L.K. 2006a. Dune migration and slip face advancement in the Rabe Crater dune field, Mars. *Geophysical Research Letters* 33, L20201, doi:10.1029/2006GL027133.

Fenton, L.K. and Hayward, R.K. 2010. Southern high latitude dune fields on Mars: Morphology, aeolian inactivity, and climate change. *Geomorphology* 121, 98–121.

Fenton, L.K and Mellon, M.T. 2006. Thermal properties of sand from Thermal Emission Spectrometer (TES) and Thermal Emission Imaging System (THEMIS): Spatial variations within the Proctor Crater dune field on Mars. *Journal of Geophysical Research* 111(E6), CiteID E06014.

Fenton. L. and Michaels, T. 2010. Characterizing the sensitivity of daytime turbulent activity on Mars with the MRAMS LES: Early results. *Mars* 5, 159–171.

Fenton, L.K. and Richardson, M.I. 2001. Martian surface winds: Insensitivity to orbital changes and implications for aeolian processes. *Journal of Geophysical Research* 108(E12), 5129, doi: 10.1029/2002JE002015.

Fenton, L.K., Bandfield, J.L. and Ward, A.W. 2003. Aeolian processes in Proctor Crater on Mars: Sedimentary history as analyzed from multiple data sets. *Journal of Geophysical Research* 108(E12), 5129, doi:10.1029/2002JE002015.

Fenton, L.K., Toigo, A. and Richardson, M.I. 2005. Aeolian processes in Proctor Crater on Mars: Mesoscale modeling of dune-forming winds. *Journal of Geophysical Research* 110, E06005, doi:10.1029/2004JE002309.

Fenton, L.K., Ewing, R.C., Bridges, N.T. and Lorenz, R. 2013. Extraterrestrial aeolian landscapes. In: John F. Shroder (Ed.) *Treatise on Geomorphology*, Volume 11, pp. 287–312. Academic Press.

Finkel, H.J. 1959. The barchans of southern Peru. *Journal of Geology* 67(6), 614–647.

Fishbaugh, K. and Head, J.W. 2005. Origin and characteristics of the Mars north polar basal unit and implications for polar geologic history. *Icarus* 174, 444–474.

Fishbaugh, K.E., Poulet, F., Chevrier, V., Langevin, Y. and Bibring, J.-P. 2007. On the origin of gypsum in the Mars north polar region. *Journal of. Geophysical Research* 112, E07002, doi:10.1029/2006JE002862.

Florensky, C.P., Ronca, L.B., Basilevsky, A.T., Burba, G.A., Nikolaeva, O.V., Pronin, A.A., Trakhtman, A.M., Volkov, V.P. and Zatetsky, V.V. 1977. The surface of Venus as revealed by Soviet Venera 9 and 10. *Geol. Soc. Am. Bull.* 88, 1537–1545.

Foley, C.N., Economou, T.E., Clayton, R.N., Brückner, J., Dreibus, G., Rieder, R. and Wänke, H. (2008) Martian surface chemistry: APXS results from the Pathfinder landing site. In: J. Bell (Ed.) *The Martian Surface: Composition, Mineralogy, and Physical Properties*, pp. 35–57. Cambridge University Press.

Folk, R.L. 1966, A review of grain-size parameters. *Sedimentology* 6(2), 73–93.

Folkner, W.M., Asmar, S.W., Border, J.S., Franklin, G.W., Finley, S.G., Gorelik, J., Johnston, D.V., Kerzhanovich, V.V., Lowe, S.T., Preston, R.A., Bird, M.K., Dutta-Roy, R., Allison, M., Atkinson, D.H., Edenhofer, P., Plettemeier, D. and Tyler, G.L. 2006. Winds on Titan from ground-based tracking of the Huygens probe. *Journal of Geophysical Research* 111, E07S02, doi:10.1029/2005JE002649.

Ford, J.P. et al. 1993. Guide to Magellan Image Interpretation. JPL Document 93-24, November 1993. (A version was also published as JPL document 89–41 in 1989).

Freymann, B.P. and Krell, F.T. 2011. Dung beetles (Coleoptera: Sarabaeidae) trapped by a moving sand dune near Olduvai Groge, Tanzania. *The Copeopterists Bulletin* 65, 422–424.

Frisch, U., Hasslacher, B. and Pomeau, Y. 1986. Lattice-gas automata for the Navier-Stokes Equation. *Physical Review Letters* 56, 1505–1508.

Fryberger, S.G. 1979. Dune forms and wind regime. In E.D. McKee (Ed.), *A Study of Global Sand Seas*. U.S. Government Printing Office, Washington, pp. 137–160.

Gardin, E., Bourke, M.C., Allemand, P. and Quantin, C. 2011. High albedo dune features suggest past dune migration and possible geochemical cementation of aeolian sediments on Mars. *Icarus* 212, 590–596.

Gardin, E., Allemmand, P., Quantin, C., Silvestro, S. and Delacourt, C. 2012. Dune fields on Mars: Recorders of a climate change? *Planetary and Space Science* 60, 314–321.

Garvin, J.B. 1990. The global budget of impact-derived sediments on Venus. *Earth, Moon, and Planets* 50–51, 175–190.

Gay, S.P. Jr. 1999. Observations regarding the movement of barchans sand dunes in the Nazca to Tanaca area of southern Peru. *Geomorphology* 27, 279–293.

Geissler, P.E., Johnson, J.R., Sullivan, R., Herkenhoff, K., Mittlefehldt, D., Fergason, R., Ming, D., Morris, R., Squyres, S., Soderblom, L. and Golombek, M. 2008. *Journal of Geophysical Research* 113, CiteID E12S31, 10.1029/2008JE003102.

Geissler, P., Sullivan, R., Golombek, M., Johnson, J., Herkenhoff, K., Bridges, N., Vaughan, A., Maki, J., Parker, T. and Bell, J. 2010. Gone with the wind: Eolian erasure of the Mars Rover tracks. *Journal of Geophysical Research* 115, E00F11, doi:10.1029/2010JE003674.

Gellert, R., Reider, R., Brückner, J., Clark, B.C., Dreibus, G., Klingelhöfer, G., Lugmair, G., Ming, D.W., Wänke, H., Yen, A., Zipfel, J. and Squyres, S.W. 2006. Alpha particle x-ray spectrometer (APXS): Results from Gusev crater and calibration report. *Journal of Geophysical Research* 111, E02S05, doi:10.1029/2005JE002555.

Gillette, D.A., Fryrear, D.W., Gill, T.E., Levy, T., Cahill, T.A. and Gearhart, E.A. 1997. Relation of vertical flux of particles smaller than 10 um to total aeolian horizontal mass flux at Owens Lake. *Journal of Geophysical Research* 102, 26009–26015.

Golitsyn, G.S. 1978. Estimates of the turbulent state of the atmosphere near the surface of Venus from the data of Venera 9 and Venera 10. *Cosmic Research* 16, 125–127.

Golombek, M., Robinson, K., McEwen, A., Bridges, N., Ivanov, B., Tornabene, L. and Sullivan, R. 2010. Constraints on ripple migration at Meridiani Planum from Opportunity and HiRISE observations of fresh craters. *Journal of Geophysical Research* 115, E00F08.

Gordon M and McKenna-Neuman C M, 2011 A study of particle splash on developing ripple forms for two bed materials Geomorphology 129, 79–91.

Goudie, A.S. 1999. The history of desert dune studies over the last 100 Years. In: A.S. Goudie, I. Livingstone and S. Stokes (Eds.) *Aeolian Environments, Sediments and Landforms*, pp. 1–13. John Wiley & Sons.

Greeley, R. (Ed.). 1978. *Aeolian Features of Southern California: A Comparative Planetary Geology Guidebook*. NASA.

Greeley, R. and Iversen, J.D. 1987. *Wind as a Geological Process on Earth, Mars, Venus, and Titan*. Cambridge University Press.

Greeley, R., Iversen, J.D., Pollack, J.B., Udovich, N. and White, B.R. 1974a. Wind tunnel studies of Martian aeolian processes. *Proceedings of the Royal Society of London*, Series A, 331–350.

Greeley, R., Iversen, J.D., Pollack, J.B., Udovich, N. and White, B. 1974b. Wind tunnel simulations of light and dark streaks on Mars. *Science* 183, 847–849.

Greeley, R., Leach, R., White, B.R., Iversen, J.D. and Pollack, J.B. 1980. Threshold windspeeds for sand on Mars: Wind tunnel simulations. *Geophysical Research Letters* 7, 121–124.

Greeley, R., Leach, R.N., Williams, S.H., Pollack, J.B. and Krinsley, D.H. 1982. Rate of wind abrasion on Mars. *Journal of Geophysical Research* 87, 10,009–10,024.

Greeley, R., Iversen, J., Leach, R., Marshall, J., White, B. and Williams, S. 1984a. Windblown sand on Venus: Preliminary results of laboratory simulations. *Icarus* 57, 112–124.

Greeley, R., Marshall, J.R. and Leach, R.N. 1984b. Microdunes and other aeolian beforms on Venus: Wind tunnel simulations. *Icarus* 60, 152–160.

Greeley, R., Marshall, J. R., Clemens, D., Dobrovolskis, A.R. and Pollack, J.B. 1991a. Venus: Concentrations of radar-reflective minerals by wind. *Icarus* 90, 123–128.

Greeley, R. Dobrovolskis, A., Gaddis, L., Iversen, J., Lancaster, N., Leach, R., Rasmussen, K., Saunders, S., VanZyl, J., Wall, S., White, B. and Zebker, H. 1991b. Radar-Aeolian Roughness Project. NASA CR 4378.

Greeley, R., Arvidson, R.E., Elachi, C., Geringer, M.A., Plaut, J.J., Saunders, R.S., Schubert, G., Stofan, E.R., Thouvenot, E.J.P., Wall, S. D. and Weitz, S.D. 1992. Aeolian features on Venus: Preliminary Magellan results. *Journal of Geophysical Research* 97, 13,319–13,345.

Greeley, R., Bender, K. Thomas, P.E., Schubert, G., Limonadi, D. and Weitz, C.M. 1995.Wind-related features and processes on Venus: Summary of Magellan results. *Icarus* 115, 399–420.

Greeley, R., Bender, K., Saunders, R.S., Schubert, G. and Weitz, C.M. 1997a. Aeolian processes and features on Venus. In: S.W. Bougher, D.M .Hunten and R.J. Phillips (Eds) *Venus II*, pp. 547–590. University of Arizona Press.

Greeley, R., Blumberg, D., McHone, J.F., Dobrovolskis, A., Iverson, J.D., Lancaster, N., Rasmussen, K., Wall, S.D. and White, B.R. 1997b. Applications of spaceborne radar laboratory data to the study of aeolian processes. *Journal of Geophysical Research* 102, 10,971–10,983.

Greeley, R., Kraft, M., Kuzmin, R.O. and Bridges, N.T. 2000. Mars Pathfinder landing site: Evidence for a change in wind regime from lander and orbiter data. *Journal of Geophysical Research* 105(E1), 1829–1840.

Greenberg, G. 2008. *A Grain of Sand: Nature's Secret Wonder*. Voyageur Press.

Grier, J. and Lunine, J.I. 1993 Speculation into possible aeolian and fluvial dune deposits on Titan. American Astronomical Society, 25th DPS Meeting, #27.04; *Bulletin of the American Astronomical Society* 25, 1105–1106.

Grundy, W., Buie, M. and Spencer, J.R. 2002. Spectroscopy of Pluto and Triton at 3–4 microns: Possible evidence for wide distribution of nonvolatile solids. *The Astronomical Journal* 124, 2273–2278.

Guinness, E.A., Leff, C. and Arvidson, R.E. 1982. Two Mars years of surface changes seen at the Viking Lander sites. *Journal of Geophysical Research* 87, 10,051–10,058.

Hack, J.T. 1941. Dunes of the Western Navajo country. *Geographical Review* 31, 240–263.

Hansen, C.J., McEwen, A.S., Ingersoll, A.P. and Terrile, R.J. 1990. Surface and airborne evidence for plumes and winds on Triton. *Science* 250(4979), 421–424.

Hansen, C.J., Bourke, M., Bridges, N.T., Byrne, S., Colon, C., Diniega, S., Dundas, C., Herkenhoff, K., McEwen, A., Mellon, M., Portyanina, G. and Thomas, N. 2011. Seasonal erosion and restoration of Mars' northern polar dunes. *Science* 331, 575–578.

Hastenrath, S.L. 1987. The barchan dunes of Peru revisited. *Zeitschrift fur Geomorphologie* 31, 167–178.

Hayes, A.G., Grotzinger, J.P., Edgar, L.A., Squyres, S.W., Watters, W.A. and Sohl-Dickstein, J. 2011. Reconstruction of eolian bed forms and paleocurrents from cross-bedded strata at Victoria Crater, Meridiani Planum, Mars. *Journal of Geophysical Research* 116, CiteID E00F21, Doi:10.1029/2010JE003688.

Hayes, A. G., Ewing, R.C., Lucas, A., McCormick, C., Troy, S. and Ballard, C. 2012. Determining Timescales of the Dune Forming Winds on Titan, Third International Planetary Dunes Workshop, Flagstaff, AZ, June 2012. Abstract #7057, Lunar and Planetary Institute, Houston TX.

Haynes, C.V. 1989. Bagnold's barchan: A 57-yr record of dune movement in the Eastern Sahara and implications for dune origin and paleoclimate since Neolithic times. *Quarternary Research* 32, 153–167.

Hayward, R.K., Mullins, K.F., Fenton, L.K., et al. 2007. Mars Global Digital Dune Database and initial science results. *Journal of Geophysical Research* 112, E11007, doi:10.1029/2007JE002943.

Hayward, R.K. Titus, T.N., Michaels, T., Fenton, L., Colaprete, A. and Christensen, P.R. 2009. Aeolian dunes as ground truth for atmospheric modeling on Mars. *Journal of Geophysical Research* 114(E11), CiteID E11012.

Hayward, R.K, Fenton, L.K., Tanaka, K.L., Titus, T.N., Colaprete, A. and Christensen, P.R. 2010. Mars Global Digital Dune Database: MC1. U.S. Geological Survey Open-File Report, 2010–1170. (http://pubs.usgs.gov/of/2010/1170/).

Hayward, R.K., Fenton, L.K., Titus, T.N., Colaprete, A. and Christensen, P.R. 2012. Mars global digital dune database: MC-30: U.S. Geological Survey Open-File Report 2012–1259, (http://pubs.usgs.gov/of/2012/1259/).

Helm, P.J. and Breed, C.S. 1999. Instrumented field studies of sediment transport by wind. US Geological Survey Professional Paper 1598-B, 31–51.

Herbert, F. 1965. *Dune*. Chilton Books.

Hermann, H., Andrade, J., Schatz, V., Sauermann, G. and Parteli, E. 2005. Calculation of the separation streamlines of barchans and transverse dunes. *Physica A* 357(1), 44–49.

Hersen, P. and Douady, S. 2005. Collision of barchan dunes as a mechanism of size regulation. *Geophysical Research Letters* 32, L21403, doi:10.1029/2005GL024179.

Hersen, P., Andersen, K.H., Elbelrhiti, H., Andreotti, B., Claudin, P. and Douady, S. 2004. Corridors of barchan dunes: Stability and size selection. *Physical Review E* 69, 011304.

Hess, S.L., Henry, R.M., Leovy, C.B., Ryan, J.A. and Tillman, J.E. 1977. Meteorological results from surface of Mars: Viking 1 and 2. *J. Geophys. Res.*, 82, 4559–4574.

Hoeg, P. 1992. *Miss Smilla's Feeling for Snow*. Rosinante. Published in English in 1996; made into a movie, 'Smilla's Sense of Snow', in 1997.

Hollands, C.B., Nanson, G.C., Jones, B.G., Bristow, C.S., Price, D.M. and Pietsch, T.J. 2006. Aeolian–fluvial interaction: Evidence for Late Quaternary channel change and wind-rift linear dune formation in the northwestern Simpson Desert, Australia. *Quaternary Science Reviews* 25, 142–162.

Horgan, B.H.N. and Bell, James F., III. 2012. Seasonally active slipface avalanches in the north polar sand sea of Mars: Evidence for a wind-related origin. *Geophysical Research Letters* 39(9), CiteID L09201.

Horgan, B.H., Bell, J.F., Noe Dobrea, E.Z., Cloutis, E.A., Bailey, D.T., Craig, M.A., Roach, L.H. and Mustard, J.F. 2009. Distribution of hydrated minerals in the north polar region of Mars. *Journal of Geophysical Research* 114(E1), CiteID E01005.

Holstein-Rathlou, C., Gunnlaugsson, H.P., Merrison, J.P., Bean, K.M., Cantor, B.A., Davis, J.A., Davy, R., Drake, N.B., Ellehoj, M.D., Goetz, W., Hviid, S.F., Lange, C.F., Larsen, S.E., Lemmon, M.T., Madsen, M.B., Malin, M., Moores, J.E., Nørnberg, P., Smith, P., xTamppari, L.K. and Taylor, P.A. 2010. Winds at the Phoenix landing site. *Journal of Geophysical Research* 115, E00E18, doi:10.1029/2009JE003411.

Howard, A.D., Morton, J.B., Gad-El-Hak, M. and Pierce, D.B. 1978. Sand transport model of barchan dune equilibrium. *Sedimentology* 25, 307–338.

Hoyle, R.B. and Woods, A.W. 1997. Analytical model of propagating sand ripples. *Physical Review E*, 56, 6861–6868.

Hugenholtz, C.H. and Barchyn, T.E. 2010. Spatial analysis of sand dunes with a new global topographic dataset: New approaches and opportunities. *Earth System Processes and Landforms* 35, 986–992.

Hunt, M.L. and Vriend, N.M. 2010. Booming sand dunes. *Annual Review of Earth and Planetary Sciences* 38, 281–301.

Hunten, D., Colin, L. and Donahue, T.M. 1983. *Venus*. University of Arizona Press.

Jackson, D.W.T. 1996. A new, instantaneous aeolian sand trap design for field use. *Sedimentology* 43, 791–796.

Jackson, D.W.T. and McKloskey, J. 1997. Preliminary results from a field investigation of aeolian sand transport using high resolution wind and transport measurements. *Geophysical Research Letters* 24, 163–166.

Jackson, D.W.T., Beyers, J.H.M., Lynch, K., Cooper, J.A.G., Baas, A.C.W. and Delgado-Fernandez, I. 2011. Investigation of three-dimensional wind flow behaviour over coastal dune morphology under offshore winds using computational fluid dynamics (CFD) and ultrasonic anemometry. *Earth Surface Processing Landforms* 36, 1113–1124.

Jackson, P.S. and Hunt, J.C.R. 1975. Turbulent wind flow over a low hill. *Quart. J. R. Met. Soc.* 101, 929–955.

Jaumann, R., Soderblom, L.A., Kirk, R., Sotin, C., Turtle, E., Lopes, R., Wood, C., Tomasko, M., Stofan, E., Lorenz, R. and Keller, H.U. 2010. Geology and surface processes on Titan. In: R. Brown et al. (Eds) *Titan after Cassini*, 75–140, Springer, New York.

Jerolmack, D.J., Mohrig, D., Grotzinger, J.P., Fike, D. and Watters, W.A. 2006. Spatial grain sorting in eolian ripples and estimation of wind conditions on planetary surfaces: Application to Meridiani Planum. Mars. *J. Geophys. Res.* 111, E12S02.

Jerolmack, D., Ewing, R., Falcini, F., Martin, R., Masteller, C., Phillips, C., Reitz, M. and Buynevich, I. 2012. Internal boundary layer model for the evolution of desert dune fields. *Nature Geoscience* 5, 206–209.

Kocurek, G. and Ewing, R. 2005. Aeolian dune field self-organization – implications for the formation of simple versus complex dune-field patterns. *Geomorphology* 72, 94–105.

Kocurek, G., Carr, M., Ewing, R., Havholm, K.G., Nagar, Y.C. and Singhvi, A.K. 2007. White Sands dune field, New Mexico: Age, dune dynamics and recent accumulations. *Sedimentary Geology* 197(3–4), 313–331.

Kok, J.F. 2010. Difference in wind speeds required for initiation versus continuation of sand transport on Mars: Implications for dunes and dust storms. *Phys. Rev. Lett.* 104, 074502, doi: 10.1103/PhysRevLett.104.074502.

Kok, J.F. and Renno, N.O. 2009. A comprehensive numerical model of steady state saltation (COMSALT). *Journal of Geophysical Research* 114, D17204, 10.1029/2009JD011702.

Kok, J., Parteli, E.R., Michaels, T. I. and Karam, D.B. 2012. The physics of wind-blown sand and dust. *Reports on Progress in Physics* 75, 106901, doi:10.1088/0034-4885/75/10/106901.

Kreslavsky, M.A. and Vdovichenko, R.V. 1998. Microdunes on Venus: Are they ubiquitous? 29th Annual Lunar and Planetary Science Conference, March 16–20, 1998, Houston, TX, Abstract no. 1166.

Krinsley, D.H. and Doorkamp, J.C. 1973. *Atlas of Quartz and Surface Textures*. Cambridge University Press.

Krinsley, D.H. and Smalley, I.J. 1972. Sand. *American Scientist* 60, 286–291.

Kroy, K., Sauermann, G. and Herrmann, H.J. 2002. Minimal model for sand dunes. *Physical Review Letters* 88, 054301.

Ksanfomality, L., Goroschkova, N. and Khondryev, V. 1983. Wind velocity near the surface of Venus from acoustic measurements. *Cosmic Research* 21, 161–167.

Lambert, B. and Long, D. 2006. A Large-Scale Ku-Band Backscatter Model of the East-Antarctic Megadune Fields, IEEE International Conference on Geoscience and Remote Sensing Symposium, 2006, pp. 3832–3834, IGARSS doi:10.1109/IGARSS.2006.982.

Lancaster, N. 1989. Star dunes. *Progress in Physical Geography* 13, 67–91.

Lancaster, N. 1995. *Geomorphology of Desert Dunes*. Routledge.

Laity, J.E. and Bridges, N.T. 2009. Ventifacts on Earth and Mars: Analytical, field, and laboratory studies supporting sand abrasion and windward feature development. *Geomorphology* 105, 202–217.

Landau, S.I. and Bogus, R.J. (Eds). 1975. *Doubleday Dictionary*. Doubleday and Co.

Landry, W. and Werner, B. 1994. Computer simulations of self-organized wind ripple patterns. *Physica D* 77, 238–260.

Lee, P. and Thomas, P.C. 1995. Longitudinal dunes on Mars: Relation to current wind regimes. *Journal of Geophysical Research* 100, 5381–5395.

Le Gall, A., Janssen, M.A., Wye, L.C., Hayes, A.G., Radebaugh, J., Savage, C., Zebker, H., Lorenz, R.D., Lunine, J.I., Kirk, R.L., Lopes, R.M., Wall, S.D., Callahan, P., Stofan, E.R., Farr, T. 2011. Cassini SAR, radiometry, scatterometry and altimetry observations of Titan's dune fields. *Icarus* 213, 608–624.

Le Gall, A., Hayes, A.G., Ewing, R., Janssen, M.A., Radebaugh, J., Savage, C., Encrenaz, P. and the Cassini Radar Team. 2012. Latitudinal and altitudinal controls of Titan's dune field morphometry. *Icarus* 217, 231–242.

Lettau, K. and Lettau, H. 1969. Bulk transport of sand by the barchans of the Pampa de La Joya in Southern Peru. *Zeitschrift fur Geomorphologie* 13, 182–195.

Levin, N., Tsoar, H., Herrmann, H.J., Maia, L.P. and Sales, V.C. 2009. Modelling the formation of residual dune ridges behind barchan dunes in North-East Brazil. *Sedimentology* 56, 1623–1641.

Li, J., Okin, G.S., Herrick, J., Belnap, J., Munson, S. and Miller, M. 2010. A simple method to estimate threshold friction velocity of wind erosion in the field. *Geophysical Research Letters* 37, L10402, doi:10.1029/2010GL043245.

Lindsay, J., Criswell, D., Criswell, T. and Criswell, B. (1976). Sound-producing dune and beach sands. *Geological Society of America Bulletin* 87(3), 463–473.

Long, J.T. and Sharp, R.P. 1964. Barchan dune movement in Imperial Valley, California. *Geological Society of America Bulletin* 75, 149–156.

Lorenz, R.D. 1996. Martian surface windspeeds, described by the Weibull distribution. *Journal of Spacecraft and Rockets* 33, 754–756.

Lorenz, R.D. 2006. Thermal interaction of the Huygens Probe with the Titan environment: Surface windspeed constraint. *Icarus* 182, 559–566.

Lorenz, R. 2007. The dunes of dune. In: K. Grazier (Ed.) *The Science of Dune*. BenBella Publishing.

Lorenz, R.D. 2009. A review of Titan mission studies. *Journal of the British Interplanetary Society* 62, 162–174.

Lorenz, R. 2010. Winds of change on Titan. *Science* 329, 519–520.

Lorenz, R.D. 2011. Observations of aeolian ripple migration on an Egyptian seif dune using an inexpensive digital timelapse camera. *Aeolian Research* 3, 229–234.

Lorenz, R.D. 2012. Edmond Halley's aeronautical calculations on the feasibility of manned flight in 1691. *Journal of Aeronautical History*, Paper 2012/02.

Lorenz, R.D. 2014. Physics of saltation and sand transport on Titan: A brief review. *Icarus* 230, 162–167.

Lorenz, R.D. and Lunine, J.I. 1996. Erosion on Titan: Past and present. *Icarus* 122, 79–91.

Lorenz, R.D. and Mitton, J. 2002. *Lifting Titan's Veil*. Cambridge University Press.

Lorenz, R. D. and Mitton, J. 2008. *Titan Unveiled.* Princeton University Press.

Lorenz, R.D. and Radebaugh, J. 2009. Global pattern of Titan's dunes: Radar survey from the Cassini Prime Mission. *Geophysical Research Letters* 36, L03202, doi:10.1029/2008GL036850.

Lorenz, R.D. and Valdez, A. 2011. Variable wind ripple migration at Great Sand Dunes National Park, observed by timelapse imagery. *Geomorphology* 133, 1–10.

Lorenz, R.D., Lunine, J.I., Grier, J.A. and Fisher, M.A. 1995. Prediction of aeolian features on planets: Application to Titan paleoclimatology. *Journal of Geophysical Research (Planets)* 88, 26, 377–26, 386.

Lorenz, R.D., Kraal, E., Eddlemon, E., Cheney, J. and Greeley, R. 2005. Sea-surface wave growth under extraterrestrial atmospheres – Preliminary wind tunnel experiments with application to Mars and Titan. *Icarus* 175, 556–560.

Lorenz, R.D. and 39 colleagues. 2006. The sand seas of Titan: Cassini radar observations of longitudinal dunes. *Science* 312(5774), 724–727, doi:10.1126/science.1123257.

Lorenz, R.D., Mitchell, K.L., Kirk, R.L., Hayes, A.G., Zebker, H.A., Paillou, P., Radebaugh, J., Lunine, J.I, Janssen, M.A., Wall, S.D., Lopes, R.M., Stiles, B., Ostro, S., Mitri, G., Stofan. E. R. and the Cassini RADAR Team. 2008. Titan's inventory of organic surface materials. *Geophysical Research Letters* 35, L02206, doi:10.1029/2007GL032118.

Lorenz, R.D., Claudin, P., Radebaugh, J., Tokano, T. and Andreotti, B. 2010. A 3km boundary layer on Titan indicated by dune spacing and Huygens data. *Icarus* 205, 719–721.

Lorenz, R.D., Newman, C.E., Tokano, T., Mitchell, J., Charnay, B., Lebonnois, S. and Achterberg, R. 2012. Formulation of an engineering wind specification for Titan Late Summer Polar Exploration. *Planetary and Space Science* 70, 73–83.

Lorenz, R., Gasmi, N., Barnes, J.W., Radebaugh, J. and Ori, G.G. 2013. Dunes on planet Tatooine: Observation of barchan migration at the Star Wars film set in Tunisia. *Geomorphology* 201, 264–271.

Lorenz, R.D., Bridges, N.T., Rosenthal, A. and Donkor, E. 2014. Elevation dependence of bedform wavelength on Tharsis Montes, Mars: Atmospheric density as a controlling parameter. *Icarus* 230, 77–80.

Lowman, P.D. 1964. A review of photography of the Earth from rockets and satellites. NASA TN D-1868.

Lowman, P.D., McDivitt, J. A. and White, E.H., II. 1967. Terrain photography on the Gemini IV Mission: Preliminary report. NASA TN D-3982.

Luna, M.C.M. de M., Parteli, E.J.R. and Herrmann, H.J. 2012. Model for a dune field with an exposed water table. *Geomorphology* 159–160, 169–177.

Lunine, J.I., Elachi, C., Wall, S.D., Allison, M.D., Anderson, Y., Boehmer, R., Callahan, P., Encrenaz, P., Flamini, E., Franceschetti, G., Gim, Y., Hamilton, G., Hensley, S., Janssen, M.A., et al. 2008. Cassini RADAR's third and fourth looks at Titan. *Icarus* 195, 414–433.

Makse, H.A. 2000. Grain segregation mechanism in aeolian sand ripples. *Eur. Phys. J. E 1* 127135.

Malin, M.C. and Edgett, K. 2001 New Mars Global Surveyor Mars Orbiter Camera: Interplanetary cruise through primary mission. *Journal of Geophysical Research* 106,(E10), 23, 429–23, 570.

Mangold, N. and 11 colleagues. 2007. Mineralogy of the Nili Fossae region with OMEGA/Mars Express data: 2. Aqueous alteration of the crust. *Journal of Geophysical Research* 112, E08S04, doi:10.1029/2006JE002835.

Marov, M. Ya. and Grinspoon, D. 1998. *The Planet Venus.* Yale University Press.

Marshall, J.R. and Greeley, R. 1992. An experimental study of aeolian structures on Venus. *Journal of Geophysical Research* 97, 1007–1016.

Marshall, J.R., Fogleman, G., Greeley, R., Hixon, R. and Tucker, D. (1991) Adhesion and abrasion of surface materials in the Venusian aeolian environment. *Journal of Geophysical Research* 96, 1931–1947.

Masselot, A. and Chopard, B. 1998. A Lattice–Boltzmann model for particle transport and deposition. *Europhysics Letters* 42, 259.

McKee, E.D. (Ed.) 1979a. *A Study of Global Sand Seas.* U.S. Geol. Survey Prof. Paper 1052, U.S. Gov. Printing Office, Washington, D.C.

McKee, E.D. 1979b. Introduction to a study of global sand seas. In: E.D. McKee (Ed.) *A Study of Global Sand Seas,* pp. 1–17. U.S. Geol. Survey Prof. Paper 1052, U.S. Gov. Printing Office, Washington, D.C.

McKenna-Neuman, C. and Nickling, W.G. 1989. A theoretical and wind-tunnel investigation of the effect of capillary water on the entrainment of sediment by wind. *Canadian Journal of Soil Science* 69, 79–96.

Mellon, M.T., Fergason, R.L. and Putzig, N.E. 2008. The thermal inertia of the surface of Mars. In: J.F. Bell (Ed.) *The Martian Surface: Composition, Mineralogy, and Physical Properties,* pp. 399–427. Cambridge University Press.

Melo, H., Parteli, E., Andrade, J. and Herrmann, H. 2012. Linear stability analysis of transverse dunes. *Physica A* 291, 4606–4614.

Melosh, H.J. and Gaffney, E. S. 1983. Acoustic fluidization and the scale dependence of impact crater morphology. *Journal of Geophyiscal Research* 88, Suppl., A830-A834.

Merrison, J.P., Gunnlaugsson, H.P., Kinch, K., Jacobsen, T.L., Jensen, A.E., Nørnberg, P. and Wahlgreen, H. 2006. An integrated laser anemometer and dust accumulator for studying wind-induced dust transport on Mars. *Planetary and Space Science* 54, 1065–1072.

Merrison, J.P., Gunnlaugsson, H.P., Knak Jensen, S. and Nørnberg, P. 2010. Mineral alteration induced by sand transport: A source for the reddish color of Martian dust. *Icarus* 205, 716–718.

Merrison, J.P., Gunnlaugsson, H.P., Hogg, M.R., Jensen, M., Lykke, J.M., Madsen, M.B., Nielsen, M.B., Nørnberg, P., Ottosen, T.A., Pedersen, R.T., Pedersen, S. and Sørensen, A.V. 2012. Factors affecting the electrification of wind-driven dust studied with laboratory simulations. *Planetary and Space Science* 60, 328–335.

Milana, J.-P. 2009. Largest wind ripples on Earth? *Geology* 37, 343–346, doi:10.1130/G25382A.1.

Ming, D.W., Gellert, R., Morris, R.V., Arvidson, R.E., Brückner, J., Clark, B.C., Cohen, B.A., d'Uston, C., Economou, T., Fleischer, I., Klingelhöfer, G., McCoy, T.J., Mittlefeldt, D.W., Schmidt, M.E., Schröder, C., Squyres, S.W., Treguier, E., Yen, A.S., Zipfel, J. 2008. Geochemical properties of rocks and soils in Gusev crater, Mars: Results of the alpha particle x-ray spectrometer from Cumberland Ridge to Home Plate. *Journal of Geophysical Research* 113, E12S39, doi:10.1029/2008JE003195.

Mitchell, J. 2008. The drying of Titan's dunes: Titan's methane hydrology and its impact on atmospheric circulation. *JGR-Planets,* ?

Momiji, H., Carretero-Gonzalez, R., Bishop, S.R. and Warren, A. 2000. Simulation of the effect of wind speedup in the formation of transverse dune fields. *Earth Surf. Proc. Landforms* 25, 905–918.

Moore, H.J., Hutton, R.E., Clow, G.D. and Spitzer, C.R. 1987. *Physical Properties of the Surface Materials at the Viking Landing Sites on Mars.* U.S. Geological Survey Professional Paper 1389.

Morris, RV., Klingelhöfer, G., Schröder, C., Fleischer, I., Ming, D.W., Yen, A.S., Gellert, R., Arvidson, R.E., Rodionov, D.S., Crumpler, L.S., Clark, B.C., Cohen, B.A., McCoy, T.J., Mittlefeldt, D.W., Schmidt, M.E., de Souza, P.A. and Squyres, S.W. 2008. Iron mineralogy and aqueous alteration from Husband Hill through Home Plate at Gusev crater, Mars: Results from the Mössbauer instrument on the Spirit Mars Exploration Rover. *Journal of Geophysical Research* 113, E12S42, doi:10.1029/2008JE003201.

Morrison, A. and Chown, M.C. 1964. Photography of the Western Sahara Desert from the Mercury MA-4 Spacecraft. NASA CR-126. Washington, DC.

Muhs, D. 2004. Mineralogical maturity in dunefields of North America, Africa and Australia. *Geomorphology* 59, 247–269.

Murchie, S. and 49 colleagues. 2007. Compact Reconnaissance Imaging Spectrometer for Mars (CRISM) on Mars Reconnaissance Orbiter (MRO). *Journal of Geophysical Research* 112, E05S03, doi:10.1029/2006JE002682.

Murphy, J.D. 1973. The geology of Eagle Cove at Bruneau, Idaho. M.A. thesis, SUNY-Buffalo, pp. 77.

Nalpanis, P., Hunt, J.C.R. and Barrett, C.F. 1993. Saltating particles over flat beds. *Journal of Fluid Mechanics* 251, 661–685.

Narteau, C., Zhang, D., Rozier, O. and Claudin, P. 2009. Setting the length and time scales of a cellular automaton dune model from the analysis of supcrimposed bed forms. *Journal of Geophysical Research* 114, F03006.

Neish, C.D., Lorenz, R.D.. Kirk, R.L. and Wye, L.C. 2010. Radarclinometry of the sand seas of Namibia and Saturn's Moon Titan. *Icarus* 208, 385–394.

Neumann, G. 2003. Polar dunes resolved by the Mars Orbiter Laser Altimeter Gridded Topography and Pulse Widths. Sixth International Conference on Mars, July 20–25, 2003, Pasadena, California, abstract no.3262.

Nield, J.M. and Baas, A.C.W. 2008. Investigating parabolic and nebkha dune formation using a cellular automaton modelling approach. *Earth Surface Processes and Landforms* 33, 724–740.

Nield, J.M. and Baas, A.C.W. 2008. The influence of different environmental and climatic conditions on vegetated aeolian dune landscape development and response. *Global and Planetary Change* 64, 76–92.

Neish, C.D., Lorenz, R.D., Kirk, R.L. andWye, L.C. 2010. Radarclinometry of the sand seas of Africa's Namibia and Saturn's moon Titan. *Icarus* 208, 385–394.

Nishimori, H. and Ouchi, N. 1993. Formation of ripple patterns and dunes by wind-blown sand. *Physical Review Letters* 71, 197–200.

Öpik, E.J. 1961. The aeolosphere and atmosphere of Venus. *Journal of Geophysical Research* 66, 2807–2819.

Patitsas, A. 2008. Singing sands, musical grains and booming sand dunes. *Journal of Physical and Natural Sciences* 2(1), 1–25.

Pelletier, J.D. 2008. *Quantitative Modeling of Earth Surface Processes*. Cambridge University Press.

Pelletier J.D. 2009. Controls on the height and spacing of eolian ripples and transverse dunes: A numerical modeling investigation. *Geomorphology* 105, 322–333.

Porco, C.C., Baker, E., Barbara, J., Beurle, K., Brahic, A., Burns, J.A., Charnoz, S., Cooper, N., Dawson, D.D., DelGenio, A.D., Denk, T., Dones, L., Dyudina, U., Evans, M.W., Fussner, S., Giese, B., Grazier, K., Helfenstein, P., Ingersoll, A.P., Jacobson, R.A., Johnson, T.V., McEwen, A., Murray, C.D., Neukum, G., Owen, W.M., Perry, J., Roatsch, T., Spitale, J., Squyres, S., Thomas, P., Tiscareno, M., Turtle, E.P., Vasavada, A.R., Veverka, J., Wagner, R., West, R. 2005. Imaging of Titan from the Cassini spacecraft. *Nature* 434, 159–168.

Potts, L.V., Akyilmaz, O., Braun, A. and Shum, C.K. 2008. Multi-resolution dune morphology using Shuttle Radar Topography Mission (SRTM) and dune mobility from fuzzy inference systems using SRTM and altimetric data. *International Journal of Remote Sensing* 29, 2879–2901.

Poynting, J. and Thompson, J. 1909. *Textbook of Physics: Sound*. Charles Griffin and Co.

Press, F. and Siever, R. 1974. *Earth*. W.H. Freeman and Co.

Putzig, N., Mellon, M.T., Herkenhoff, K.E. and Phillips, R.J. 2008. Thermophysical analysis of the North Polar Erg on Mars. Planetary Dunes Workshop, 2008, Alamogordo, NM, abstract 7028.

Pye, K. and Tsoar, H. 1990. *Aeolian Sand and Sand Dunes*. Springer. (This book was reprinted, but not revised, in 2009.).

Qong, M. 2000. Sand dune attributes estimated from SAR images. *Remote Sensing of the Environment* 74, 217–228.

Quiam, G., Dong, Z., Lup, W. and Lu, J. 2011. Mean airflow patterns upwind of topographic obstacles and their implications for the formation of echo dunes: A wind tunnel simulation of the effects of windward slope. *Journal of Geophysical Research* 116, F04026, doi:10.1029/2011JF002020.

Radebaugh, J. 2013. Dunes on Saturn's moon Titan as revealed by the Cassini Mission. *Aeolian Research* 11, 23–41.

Radebaugh, J. and 14 colleagues. 2008. Dunes on Titan observed by Cassini radar. *Icarus* 194, 690–703, doi:10.1016/j.icarus.2007.10.015.

Radebaugh, J., R. Lorenz, T. Farr, P. Paillou, C. Savage, C. Spencer, 2010. Linear Dunes on Titan and Earth: Initial Remote Sensing Comparisons, Geomorphology, 121, 122–132.

Rannou, P., Montmessin, F., Hourdin, F. and Lebonnois, S. 2006. The latitudinal distribution of clouds on Titan. *Science*, 311(5758), 201–205.

Read, P.L. and Lewis, S.R. 2004. *The Martian Climate Revisited: Atmosphere and Environment of a Desert Planet*. Springer.

Redsteer, M.H., Bogle, R.C. and Vogel, J.M., 2011. Monitoring and analysis of sand dune movement and growth on the Navajo nation, southwestern United States. USGS fact sheet 2011-3085 (pubs.usgs.gov/fs/2011/3085/).

Reffet, E., S. Courrech du Pont, S.R., Hersen, P. and Douady, S. 2010. Formation and stability of transverse and longitudinal sand dunes, *Geology* 38, 491–494.

Reiss, D., van Gasselt, S., Neukum, G. and Jaumann, R. 2004. Absolute dune ages and implications for the time of formation of gullies in Nirgal Vallis, Mars. *Journal of Geophysical Research* 109(E6), doi:10.1029/2004JE002251.

Rognon, P. 1999. Eolian sand dynamics and protection against sand drift and moving dunes. In: Proceedings of the International Symposium, 'New Technologies to Combat Desertification', Tehran, Iran 12–15 October 1998, United Nations University, Tokyo, Japan 1999.

Rubin, D.M. and Carter, C.L. 2006. *Cross-Bedding, Bedforms, and Paleocurrents*, 2nd ed. Society for Sedimentary Geology.

Rubin, David M. and Carter, Carissa L. (2006) *Bedforms and Cross-Bedding in Animation*. Society for Sedimentary Geology (SEPM), Atlas Series 2, DVD #56002.

Rubin, D.M. and Hesp, P.A. 2009. Multiple origins of linear dunes on Earth and Titan. *Nature Geoscience* 2, 653–658.

Rubin, D.M. and Hunter, R.E. 1987. Bedform alignment in directionally varying flow. *Science* 237, 276–278, doi:10.1126/science.237.4812.276.

Rubin, D.M. and Ikeda, H.1990. Flume experiments on the alignment of transverse, oblique, and longitudinal dunes in directionally varying flows. *Sedimentology* 37, 673–684, doi:10.1111/j.1365-3091.1990.tb00628.

Sabins, F.F. 1978. Thermal infrared imagery. In: F.F. Sabins (Ed.) *Remote Sensing: Principles and Interpretation*, pp. 119–175. W.H. Freeman and Co.

Sagan, C. and Chyba, C. 1990. Titan's streaks as windblown dust. *Nature* 346, 546–548.

Sagan, C. et al. 1972. Variable features on Mars: Preliminary Mariner 9 television results. *Icarus* 17, 346–372.

Sauermann, G., Kroy, K. and Herrmann, H.J. 2001. A continuum saltation model for sand dunes. *Phys. Rev. E* 64, 031305.

Saunders, R.S., Dobrovolskis, A.R., Greeley, R. and Wall, S.D. 1990. Large-scale patterns of eolian sediment transport on Venus: Predictions for Magellan. *Geophysical Research Letters* 17, 1365–1368.

Savage, C.J., Radebaugh, J., Christiansen, E.H. and Lorenz, R.D. 2013. Implications of dune pattern analysis for Titan's surface history. *Icarus*, in revision.

Schatz, V. and Hermann, H.J. 2005. Flow separation in the lee of transverse dunes. In: R. Garcia-Rojo, H.J. Herrmann and S. McNamara (Eds) *Powders and Grains*, pp. 955–958. Balkema, cond-mat/0501554.

Schmidt, D.S., Schmidt, R.A. and Dent, J.D. 1998. Electrostatic force on saltating sand. *Journal of Geophysical Research* 103, 8997–9001.

Schneider, G. 2008. *The Roadside Geology of Namibia*, 2nd Ed. Borntraeger.

Schneider, T., Graves, S.D.B., Schaller, E.L., Brown, M.E. 2012. Polar methane accumulation and rainstorms on Titan from simulations of the methane cycle. *Nature* 481, 58–61.

Schofield, J.T., Barnes, J.R., Crisp, D., Haberle, R.M., Larsen, S., Magalhaes, J.A., Murphy, J.R., Seiff, A. and Wilson, G. 1997. The Mars Pathfinder atmospheric structure investigation meteorology (ASI/MET) experiment. *Science* 278, 1752–1758, doi:10.1126/science.278.5344.1752.

Schwammle, V. and Hermann, H.J. 2003. Solitary wave behaviour of sand dunes. *Nature* 426, 619–620.

Seppälä, M and Lindé, K. 1978. Wind tunnel studies of ripple formation. *Geografiska Annaler. Series A, Physical Geography* 60, 29–42.

Sharp, R.P. 1963. Wind ripples. *Journal of Geology* 71, 617–636.

Sharp, R.P. 1964. Wind-driven in Coachella Valley, California. *Geological Society of America Bulletin* 74, 785–804.

Sharp, R.P. 1980. Wind-driven sand in Coachella Valley, California: Further data. *Geological Society of America Bulletin* 91, 724–730.

Siever, R. 1988. *Sand*. Scientific American Library, W. H. Freeman and Co.

Silvestro, S., Fenton, L., Vaz, D.A., Bridges, N.T. and Ori, G.G. 2010a. Ripple migration and dune activity on Mars: Evidence for dynamic wind processes. *Geophysical Research Letters* 37, L20203.

Silvestro, S., Di Achille, G. and Ori, G.G. 2010b. Dune morphology, sand transport pathways and possible source areas in east Thaumasia Region (Mars), *Geomorphology* 121, 84–97.

Silvestro, S., Vaz, D.A., Fenton, L.K. and Geissler, P.E. 2011. Active eeolian processes on Mars: A regional study in Arabia and Meridiani Terrae. *Geophysical Research Letters* 38, L20201.

Silvestro, S., Vaz, D.A., Ewing, R.C., Rossi, A.P., Fenton, L.K., Michaels, T.I., Flahaut, J. and Geissler, P.E. 2013. Pervasive aeolian activity along rover Curiosity's traverse in Gale crater, Mars. *Geology* 41(4), 483–486, doi:10.1130/G34162.1.

Slattery, M.C. 1990. Barchan migration on the Kuiseb river delta, Namibia. *South African Geographical Journal* 72, 5–10.

Smalley, I.J. and Krinsley, D.H. 1979. Eolian sedimentation on Earth and Mars: Some comparisons. *Icarus* 40, 276–288.

Smith, M.R. and Bandfield, J.L. 2012. Geology of quartz and hydrated silica-bearing deposits near Antoniadi crater, Mars. *Journal of Geophysical Research* 117, E06007, doi:10.1029/2011JE004038.

Smyth, AG., Jackson, D.W.T. and Cooper, J.A.G. 2012. High resolution measured and modelled three-dimensional airflow over a coastal bowl blowout. *Geomorphology* 177–178, 62–73.

Spaan,W.P. and Van den Abele, G.D. 1991. Wind borne particle measurements with acoustic sensors. *Soil Technology* 4, 51–63.

Spiga, A. and Lewis, S. 2010. Martian mesoscale and microscale wind variability of relevance for dust lifting, *Mars* 5, 146–158; doi:10.1555/mars.2010.0006.

Steidtmann, J.R. 1973. Ice and snow in eolian sand dunes of Southwestern Wyoming. *Science* 179, 796–798.

Stephen, H. and Long, D.G. 2005. Microwave backscatter modeling of erg surfaces in the Sahara Desert. *IEEE Transactions on Geoscience and Remote Sensing* 43, 238–247.

Sterk, G., Lopez, M.V. and Arrue, J.L. 1999. Saltation transport on a silt loam soil in northeast Spain. *Land Degradation & Development* 10, 545–554.

Stockton, P.H. and Gillette, D.A. 1990. Field measurement of the sheltering effect of vegetation on erodible land surfaces. *Land Degradation & Rehabilitation* 2, 77–85.

Stout, J.E. and Zobeck, T.M. 1997. Intermittent saltation. *Sedimentology* 44, 959–970.

Strangeways, I. 2003. *Measuring the Natural Environment*, 2nd Ed. Cambridge University Press.

Sullivan, R., Greeley, R., Kraft, M., Wilson,G., Golombek, M., Herkenhoff, K., Murphy, J. and Smith, P. 2000, Results of the Imager for Mars Pathfinder windsock experiment. *Journal of Geophysical Research* 105, 24,547–24,562, doi:10.1029/1999JE001234.

Sullivan, R., Bandfield, D., Bell, J.F., et al. 2005. Aeolian processes at the Mars Exploration Rover Meridiani Planum landing site. *Nature* 436, 58–61.

Sullivan, R. and 10 colleagues. 2008. Wind-driven particle mobility on Mars: Insights from Mars Exploration Rover observations at "El Dorado" and surroundings at Gusev crater. *Journal of Geophysical. Research* 113, E06S07, doi:10.1029/2008JE002541.

Sullivan, R., Anderson, R., Biesiadecki, J., Bond, T. and Stewart, H. 2011. Cohesions, friction angles, and other physical properties of Martian regolith from Mars Exploration Rover wheel trenches and wheel scuffs. *Journal of Geophysical Research* 116, E02006, doi:10.1029/2010JE003625

Sullivan, R., Zimbelman, J.R. and Greeley, R. 2012. Coarse-grained ripples on Earth and Mars: Field studies and wind tunnel experiments. In: Lunar Planet. Sci. XLIII, Abstract 2161, Lunar and Planetary Institute.

Sweeney, M., Etyemezian, V. et al. 2008. Comparison of PI-SWERL with dust emission measurements from a straight-line field wind tunnel. *Journal of Geophysical Research-Earth Surface* 113(F1).

Synkiewicz, A., Ewing, R.C., Moore, C.H., Glamoclija, M., Bustos, D. and Pratt, L.M. 2010. Origin of terrestrial gypsum dunes: Implications for Martian gypsum-rich dunes of Olympia Undae. *Geomorh.* 121, 69–83, doi:20.2026/j.geomorph.2009.02.017.

Tanaka, K.L. and Fortezzo, C.M. 2012. Geologic map of the North Polar region of Mars. U.S. Geol. Survey Sci. Invest. Map 3177, scale 1:2,000,000.

Taniguchi, K., Endo, N. and Sekiguchi, H. 2012. The effect of periodic changes in wind direction on the deformation and morphologyof isolated sand dunes based on flume experiments and field data from the Western Sahara. *Geomorphology* 179, 286–299.

Thomas, D.S., Knight, M. and Wiggs, G.F. 2005. Remobilization of southern African desert dune systems by twenty-first century global warming. *Nature* 435, 1218–1221.

Tirsch, D. 2008. Dark dunes on Mars – Analyses on origin, morphology, and mineralogical composition of the dark material in Martian craters. Ph.D. dissertation, Deutsches Zentrum fur Luft – und Raumfahrt, Berlin – Adlershof, p. 146.

Tirsch, D., Craddock, R., Platz, A., Maturilli, J., Helbert, and Jaumann, R. 2012. Spectral and petrologic analysis of basaltic sands in Ka'u Desert (Hawaii)—Implications for the dark dunes on Mars. *Earth Surface Processes and Landforms* 37, 434–448.

Tokano, T. 2008. Dune-forming winds on Titan and the influence of topography. *Icarus* 194, 243–262.

Tokano, T. 2010. Relevance of fast westerlies at equinox for the eastward elongation of Titan's dunes. *Aeolian Research* 2, 113–127.

Tokano, T. and Neubauer, F. 2002. Tidal winds on Titan caused by Saturn. *Icarus* 158, 499–515.

Toulmin, P., Baird, A.K., Clark, B.C., Keil, K., Rose, H.J., Christian, R.P., Evans, P.H. and Kelliher, W.C. 1977. Geochemical and mineralogical interpretation of the Viking inorganic chemical results. *Journal of Geophysical Research* 82(28), 4625–4634.

Tsoar, H. 1974. Desert dune morphology and dynamics, El Arish (northern Sinai). *Zeitschrift für Geomorphologie Supplementband* 20, 41–61.

Tsoar, H. 1983, Wind tunnel modeling of echo and climbing dunes. In: M.E. Brookfield and T.S. Ahlbrandt (Eds) *Eolian Sediments and Processes*, pp. 247–259. Elsevier, doi:10.1016/S0070-4571(08)70798-2.

Tsoar, H., Greeley, R. and Peterfreund, A. 1979. Mars: The North Polar sand sea and related wind patterns. *Journal of Geophysical Research* 84, 8167–8180.

Tucker, M.E. 2003. *Sedimentary Rocks in the Field*. John Wiley & Sons.

Turcotte, D.L. and Schubert, G. 1982. *Geodynamics*. John Wiley & Sons.

Twidale, C.R. 1980. The Simpson Desert, central Australia. *South African Geographic Journal* 65, 3–17.

Udden, J.A. 1898, Mechanical composition of wind deposits. *Augustana Lib. Publ.* 1, 69.

Udden, J.A. 1914, Mechanical composition of clastic sediments. *Bull. Geol. Soc. Am.* 25, 655–744.

Ulaby, F.T., Moore, R.K. and Fung, A.K. 1982. *Microwave Remote Sensing—Active and Passive—Volume I: Microwave Remote Sensing Fundamentals and Radiometry, Volume II: Radar Remote Sensing and Surface Scattering and Emission Theory, Volume III From Theory to Applications*. Longman.

Unger, J. and Haff, P.K. 1987. Steady state saltation in air. *Sedimentology* 34, 289–299.

U.S. Geological Survey, 1982. Zapata Ranch quadrangle, Colorado, 7.5 minute series (topographic) map, scale 1:24,000.

Vermeesch, P. and Drake, N. 2008. Remotely sensed dune celerity and sand flux measurements of the world's fastest barchans (Bodele, Chad). *Geophysical Research Letters* 35, L24404.

Vriend, N., Hunt, M., Clayton, R., Brennen, C., Brantley, K. and Ruiz-Angulo, A. 2007. Solving the mystery of booming sand dunes. *Geophysical Research Letters* 34(16), L16306.

Vriend, N., Hunt, M., Clayton, R., Brennen, C., Brantley, K. and Ruiz-Angulo, A. 2008. Reply to comment on solving the mystery of booming sand dunes. *Geophysical Research Letters* 35, L08307.

Wald, C. 2009. In dune map, Titan's winds seem to blow backward. *Science* 323, 1418.

Ward, A. 1979. Yardangs on Mars—Evidence of wind rrosion. *Journal of Geophysical Research* 84, 8147–8166.

Ward, A., Doyle, K.B., Helm, P., Weisman, M. and Witbeck, N. 1985. Global map of eolian features on Mars. *Journal of Geophysical Research* 90, 2038–2056.

Walker, I.J. and Nickling, W.G. 2002 Dynamics of secondary airflow and sediment transport over and in the lee of transverse dunes. *Progress in Physical Geography* 26, 47–75.

Wang, X., Dong, Z., Zhang, J. and Zhao, A. 2002. Relations between morphology, air flow, sand flux and particle size on transverse dunes, Taklimaka Sand Sea, China. *Earth Surface Processes and Landforms* 27, 515–526.

Warner, T.T. 2004. *Desert Meteorology*. Cambridge University Press.

Wasson, R.J. and Hyde, R. 1983. Factors determining desert dune type. *Nature* 304, 337–339.

Weitz, C.M., Plaut, J.I., Greeley, R. and Saunders, R.S.1994. Dunes and microdunes on Venus: Why were so few found in the Magellan Data? *Icarus* 112, 282–285.

Welland, M. 2009. *Sand: The Never-ending Story*. University of California Press. (Published in the UK as *Sand: A Journey Through Science and the Imagination*).

Wentworth, C.K. 1922, A scale of grade and class terms for clastic sediments. *Journal of Geology* 30(5), 377–392.

Werner, B.T. 1995. Aeolian dunes: Computer simulations and attractor interpretation. *Geology* 23, 1107–1110.

Werner, B.T. and Gillespie, D.T. 1993. Fundamentally discrete stochastic model for wind ripple dynamics. *Physical Review Letters* 71, 3230–3233.

Werner, B.T. and Kocurek, G. 1999. Bedform spacing from defect dynamics. *Geology* 27, 727–730.

White, B.R. 1979. Soil transport by winds on Mars. *Journal of Geophysical Research* 84, 4643–4651.

White, B.R. 1981. Venusian saltation. *Icarus* 46, 226–232.

White, B.R. and Schulz, J.C. 1977. Magnus effect in saltation. *Journal of Fluid Mechanics* 81, 497–512.

Williams, S.H. and Greeley, R. 1994. Windblown sand on Venus: The effect of high atmospheric density. *Geophysical Research Letters* 21, 2825–2828.

Wilson, I. 1972a. Sand waves. *New Scientist* 53, 634–637.

Wilson, I.G. 1972b. Aeolian bedforms: Their development and origins. *Sedimentology* 19, 173–210.

Wilson, I.G. 1973. Ergs. *Sedimentary Geology* 10, 77–106.

Wilson, S.A. and Zimbelman, J.R. 2004. The latitude-dependent nature and physical characteristics of transverse aeolian ridges on Mars. *Journal of Geophysical Research* 109, E10003, doi:10.1029/2004JE002247.

Wippermann, F.K. and Gross, G. 1986. The wind-induced shaping and migration of an isolated dune: A numerical experiment. *Boundary-Layer Meteorology* 36, 319–334.

Wolfe, S.A. and Hugenholtz, C.H. 2009. Barchan dunes stabilized under recent climate warming on the northern Great Plains. *Geology* 37, 1039–1042.

Wolfram, S. 2002. *A New Kind of Science*. Wolfram Media, Inc.

Xianwan, L., Sen, L. and Jianyou, S. 1999. Wind tunnel simulation experiment of mountain dunes. *Journal of Arid Environments* 42, 49–59.

Yang, X., Scuderi, L., Liu, T., Paillou, P., Li, H., Dong, J., Zhu, B., Jiang, W., Jochems, A, and Weissman, G. 2011. Formation of the highest sand dunes on Earth. *Geomorphology* 135, 108–116.

Yizhaq, H. 2005. A mathematical model for aeolian megaripples on Mars. *Physica A* 357, 57–63.

Yizhaq, H., Balmforth, N.J. and Provenzale, A. 2004. Blown by the wind: Nonlinear dynamics of aeolian sand ripples. *Physica D* 195, 207–228.

Zandonella, C. 2003. Shifting sands. *New Scientist* 178(2401).

Zhang, D., Narteau, C., Rozier, O. and Courrech du Pont, S. 2012. Morphology and dynamics of star dunes from numerical modeling. *Nature Geoscience* 5, 463–467.

Zheng, X. 2009. *Mechanics of Wind-blown Sand Movements*. Springer.

Zimbelman, J.R. 2010. Transverse aeolian ridges on Mars: First results from HiRISE images. *Geomorphology* 121, 22–29, doi:10.1016/j.geomorph.2009.05.012.

Zimbelman, J.R. and Williams, S.H. 2002. Chemical indicators of separate sources for eolian sands in the eastern Mojave Desert, California, and western Arizona. *Geological Society of America Bulletin* 114(4), 490–496.

Zimbelman, J.R. and Williams, S.H. 2006. Aeolian ripples on Earth and Mars: Scale diversity and implications for modes of particulate transport. In: Lunar Planet. Sci. XXXVII, Abstract 2047, Lunar and Planetary Institute.

Zimbelman, J.R. and Williams, S.H. 2007. Eolian dunes and deposits in the western United States as analogs to wind-related features on Mars. In: M. Chapman (Ed.)*The Geology of Mars*, pp. 232–264. Cambridge University Press.

Zimbelman, J.R., Irwin, R.P., Williams, S.H., Bunch, F., Valdez, A. and Stevens, S. 2009. The rate of granule ripple movement on Earth and Mars. *Icarus* 203, 71–76, doi:10.1016/j.icarus.2009.03.033.

Zimbelman, J.R., Williams, S.H. and Johnston, A.K. 2012. Cross-sectional profiles of sand ripples, megaripples, and dunes: A method for discriminating between formational mechanisms. *Earth Surf. Proc. Landforms* 37, 1120–1125, doi:10.1002/esp.3243.

# Index

**A**
Aarhus, Denmark (wind tunnel), 210, 211
Abrasion, 196, 213
Adhesion, 45
Advanced land observation satellite (Diachi), 129
Aeolian scour, 145. *See also* Blowout dune
Aeolian streamers, 37
Aerial photography, 217
Agelat, Libya, 279
Airgun, 183
Akatsuki, 175
Akle pattern, 81, 92
Albedo, 223
Algaonice, venus, 170
Algeria, 13, 91, 264, 273
Algodones, 23, 101, 102, 189, 282, 285
All-terrain vehicle, 265
Al-uzza undae, venus, 170
Amazon, 40, 122
Amboy crater, 94
Amphibole, 23
Amundsen-scott south pole station, 11
Anemometers, 191, 192
Angle of repose, 55, 238, 245
Antarctica, 11, 18, 123, 124, 131, 132, 192, 230
Aonia terra, mars, 104
APXS, 201
Arabia, 121, 124, 127
Arabian desert, *see* Rub' Al Khali
Archimedes, 19, 41
Arctic, 122, 124
Argentina, megaripples, 65, 66
Argon, 121, 138
Argyre Planitia, Mars, 228
Arica Hills, CA, 106
Arizona State University, 206
Arounga impact crater, Chad, 9
Arrakis, 283, 284
ASTER, 89, 100, 104, 220, 228
Atacama, 123, 124, 127
Atmospheres, 24, 25, 27–29, 33, 42, 44, 47
    earth, 121
    mars, 135, 138, 140
    venus, 169, 170
Atmospheric pressure, 27
Atomic Force Microscope (AFM), 203
Audoin-debreuil, Ariane, 264
Avalanching, 115, 187, 215

**B**
Badain Jaran, 125, 217
Badain Jarin desert, China, 126, 191, 218
Bagnold, Ralph, 12, 19, 115, 191, 195, 206, 289
Barchan, 77–84, 124, 129, 141, 154, 162
    forms, 245, 249
    interaction, 203
    model
    scars, 72
Barchanoid ridge, 80, 81, 93, 99, 124, 141, 147
Barking sands, Maui, HI, 115
Basalt, 23, 25
Bearing strength, 261
Belet sand sea, Titan, 88, 161, 164, 174
Besler, Helga, 23
Biotite, 23
Blake, William, 17
Blowout dune, 97, 244
BOFIX, 273
Bonneville Salt Flats, Mars, 199
Booming dunes, 116–118
Boundary layer, 43, 172, 194
Bragg scattering, 172, 205, 238, 245
Brown noise, 35
Bruneau Dunes, ID, 34, 131
Brunton Compass, *see* Transit
Bulldozer, 64, 184

**C**
Calcite, 23
Camel, 260, 261
Cape St. Mary (Victoria Crater, Mars), 70, 150
Cape St. Vincent (Victoria Crater, Mars), 70, 150
Carbon dioxide, 138, 139, 177
Cassini, 5, 32, 90, 158, 159, 161, 166, 226, 228, 231, 233, 237
Cat dune, 96
'Cat scratches', Titan, 159, 160
Cellular Automaton model, 247, 248
CFD, 243–245
Charon, 178
Chemcam, 204
Chott El Gharsa, Tunisia, 285, 286
Chott El-djerid, tunisia, 97
Chromite, 170
Ciudad juarez, 284
Clays, 22
Clementine, 122

Climate, 8, 11, 29, 121, 134, 140, 163, 175, 206, 242, 257, 258
Climbing dune, 93, 94, 96
Coachella valley, CA, 196
Cohesion, 41, 46, 157
Colorado plateau, 124
Colorado river, 104, 124
Columbia hills, 23
Compound dunes, 62, 93, 99
Complex dune, 101, 141. *See also* Compound dunes
Computational Fluid Dynamics, *see* CFD
COMSALT model, 245
Continuum model, 250
Conway, Simon, 245
Coppice dune, 95
Coral (pink) sand dunes, Utah, 21, 63
Coriolis effect, 29, 241
Cornish, Vaughan, 121, 255
Coronene, 18
COSI-Corr software, 147, 223
Cosmotheoros, 19
Creep, 39
CRISM, 145, 225, 228
Croll-Milankovich cycles, 29, 31, 134, 162
Crossbedding, 62, 137, 148–150, 155, 197, 204, 271
Curiosity (rover), 12, 62, 137, 148–150, 155, 193, 197, 200, 204, 271

**D**
Daedalia planum, Mars, 94
Danikil depression, Ethiopia, 78
Darwin, Charles, 115
Dasht-e, Kavir, Lut, 124
Death Valley (National Park), CA, 63, 285. *See also* Stovepipe dunes, Eureka dunes
Defect, 251
Density (fluid), 43, 46
Density (particle), 22, 43, 157, 195
Devil's Playground, CA, 105
Diatom, 18
Digital Elevation Model (DEM), 223, 226–229, 239
Dome dune, 77, 142
DP, *see* Drift potential
Draa, 62, 81
Drag, 43, 48, 51
　coefficient, *see* Drag
　length, 48, 51, 52, 180
Drift potential, 75, 107, 108, 242
Dumont dunes, CA, 56, 265, 271
Dunes
　buggy, 265
　novel/movie, 13, 273
　orientation, Titan, 163
　shape, *see* Barchan, Barchanoid dune, Linear dune, Star dune, Parabolic dune, Lee dune, Transverse ridge
　size, 62, 250
　stabilization, 105, 230. *See also* Vegetated dune
Dung beetles, 254
Dust, 45

**E**
Eccentricity, 31, 140, 179
Echo dune, 94
Egypt, 80, 87, 109, 185, 187, 281, 283
Eigg, Scotland, 115
El Dorado, Mars, 22, 141, 151, 152

Electrostatic force, 244
Elephant, 260
Eliot, T.S., 290
Enceladus, 177
Endeavour (crater), 150
Endurance (crater), 150
ENVISAT, 123
Eureka dunes, 63
Europa, 178
Exoplanets, 258

**F**
Falling dune, 9, 94, 96, 153
Feldspar, 23, 204, 206
Fensal-atzlan, 159
Feynman, Richard, 36
Floating fence, *see* Sand dragon
Flow separation, 55, 56
Fluid threshold, 50, 210
Flume, *see* Water flume
Fluvial, 8, 18, 39, 159, 164, 170, 175, 189
Flux instability, 52
Fortuna-meshkenet, Venus, 170, 172
Friction angle, 200. *See also* Angle of repose
Friction speed, 50, 175, 177
Friction, 267

**G**
Galapagos, 22
Gale crater, Mars, 148, 150, 152, 155, 272
Game of life, 245, 246
Ganymede, 178
Garnet, 22, 23, 58
Gemini, 220, 284
Geophone, 115
Gilbert, G.K., 79, 185, 274, 275
GIS, 223
Global circulation model (GCM), 36, 44, 140, 165, 242, 246
Gobi desert, 123, 124
Google Earth, 185, 223
GPR, 189, 191, 193, 260
GPS, 111, 131, 184–186
Grand Erg orientale (Sahara), 13
Granules, 141
Gravity, 3, 48
Great basin (USA), 123, 124
Great sand dunes national park and preserve, CO, 60, 72, 104, 112, 186, 190
Great victoria, Australia, 127
Greeley, Ronald, 242
Green sand, 22
Greenland, 123
Ground penetrating radar, 8
Guelph, Canada, 206
Gusev, 12, 187
Gypsum, 17, 23, 64, 195, 224, 225, 228, 258

**H**
Haboob, 40
Hadley circulation, 29, 165, 174
Half-Track, 11, 262. *See also* Snowmobile
Halley, 29
Halley, Edmond, 29

Hasselblad, 284
Hawaii, 27
Heat shield, 210
Hellas, 153
Hellespontus crater, Mars, 4
Hematite, 224
Herbert, Frank, 273, 283
Herodotus, 254
High-centering, 267, 271
HiRISE (camera), 14, 79–81, 83–85, 92, 104, 111, 113, 136, 138
Hodograph, 32
Hooked barchan, 79, 90, 214
Horse, 261
Hot wire, *see* Anemometer
HRSC, 228, 229
Hubble Space Telescope, 136, 158
Huygens, 158–162, 217
Huygens, Christiaan, 19
Hydrocarbon lakes, 29, 158, 166
Hydrological cycle, 158

**I**
Ice (water), 189, 258
Ice age, 257
Ilmenite, 58
Impact threshold speed, 49, 243
Imperial Sand Dunes, CA, *see* Algodones
Induration, 258
Insalah, Algeria, 273
Interferometry, 229
Intermittency, 54
International Space Station (ISS), 9, 13, 91, 101, 104, 222, 223, 277, 281
Intertropical convergence zone (ITCZ), 165
Io, 120, 177, 178–180
Iraq, 40
ISS, *see* International Space Station

**J**
Jake matijevic (rock), 204
Jupiter, 178

**K**
Kaiser (crater), 145, 146
Kalahari desert, 82, 105, 124, 126–127, 134, 257
Kara-Kum, 123, 124
Katabatic winds, 153
Kelso dunes, CA, 105
Kharga, Egypt, 80, 109, 281
Killpecker dunes, WY, 69
Kite camera, 7, 78, 82, 87, 131, 230
Knudsen number, 42, 43
Kolmogorov, Andrei, 37
Kuiper belt, 177
Kumtagh desert, China, 101
Kurtosis, 20
Kyzyl-Kum, 124

**L**
La Jornada Experimental Range, NM, 258
Laminar flow, 43
Landsat, 105, 226, 276

Lanzhou, China, 274
Large Eddy Simulation (LES), 243
Laser Induced Breakdown Spectroscopy (LIBS), 204
Laser scanning, 188–189, 230
Lateritic soil, 22
Lattice pattern of dunes, 162, 241. *See also* Raked dunes
Lattice Gas Model, 8
Lattice-Boltzmann model, 243, 244. , *See also* Lattice Gas
'Law of the Wall', 44, 46
Le Petit Prince, 283
Lee dune, 9, 93–96., *See also* Falling dune
Lencois Maranhenses, Brazil, 7, 68, 72, 124, 129, 170, 255
LIDAR, 230
Lift, 39, 49
Linear dune, 98–101, 129, 130, 132, 133, 142, 160, 185, 188, 219, 220, 238, 248
Liwa, United Arab Emirates, 57, 108, 111, 277–280
LiwaLobes (avalanche), 4, 58
Logarithmic axes, 20, 46
Long Range Desert Group, 12, 263, 284
Lucite, 195
Luderitz, 274
Lunette, 94
Lyot (crater), 145

**M**
Mach number, 42, 43
Magellan (spacecraft), 24, 94, 144–150, 232–239
Magnetite, 18, 58, 63, 186, 200
Magnus effect, *see* Robins-Magnus effect
Malin, Michael, 196, 271
Mariner 4, 220
Mariner 9, 4, 135, 136, 138, 144, 220, 226
Mars Exploration Rovers, *see* Spirit, Opportunity
Mars Express, 137. *See also* HRSC
Mars Global Surveyor, 136–138, 146, 220. *See also* MOC
Mars Odyssey, *see* THEMIS
Mars Pathfinder lander, *see* Pathfinder
Mars Reconnaissance Orbiter (MRO), 59, 113, 137, 138, 143, 220, 228. *See also* HiRISE and CRISM
Mars rover, 183, 199, 208, 209. *See also* Sojourner, Spirit, Opportunity, Curiosity
Mars, 169, 171, 209–211, 268, 272
MARSWIT (Mars Wind Tunnel), 208–212
Medano Creek, 130
Megabarchan, 108, 111, 124, 278, 280
Megaripple, 66, 150, 186–188, 190
Menat Undae, 170
Meridiani Planum, Mars, 269
Meridiani Terra 95, 64. *See also* Opportunity
Mesoscale model, 242, 243
Meteorology, 194, 243
Methane, 178
Method of slices, 260
Meunier, Jean, 273
Microdunes, 165, 178, 241
Microphone, 115
Microscope, 25
Microtektite, 25
Migration rates, 111–113
Milankovich, *See* Croll-Milankovich
Mingsha San (Singing Sand Mountain), China, 115
MOC camera, 143, 145, 152
MODIS, 133
Moenkopi dunes, AZ, 217

Moisture, effect on friction speed, 47
Mojave desert, 96, 129, 271. *See also* Kelso dunes
MOLA, 229
Mongolia, 222
Moon, 47
Morocco, 8, 80, 115, 116
Mos Espa ('Star Wars'), 12
Motor current, 200
Muhs, Daniel, 23
Muscovite, 23
Musgrave, Story, 255

**N**
Namib desert, 31, 109, 125, 160, 165, 166, 221, 233, 239, 250, 255, 273, 274
Namibia, 255, 260, 274. *See also* Namib
Navajo, 93
Navier-Stokes equations, 241, 243
Nebkha, 97, 134
New Horizons, 178
Newton, Isaac, 41
Nili Patera, 23, 113, 147–149
Nirgal Vallis, Mars, 152, 153
Nitrogen, 121, 177, 178
Niveo-aeolian processes, 69
Noachis terra, Mars, 104
Normandy, 18
North Polar Erg (NPE), 153. *See also* Olympia Undae
Nouakchott, 276

**O**
Obliquity, 31, 33, 153, 174
Ol Doinyo Lengai, Tanzania, 18
Olduvai, Tanzania, 253
Olivine, 22, 23, 225
Olympia undae, Mars, 73, 136, 258
Oman, 127
Ooid, 18
Opportunity (rover), 3, 70, 97, 113, 275, 285
Oregon, 79, 273–275
OSL (Optically Stimulated Luminescence), 183, 206
Oxygen, 121

**P**
Paddle tire, 265
PAHs (Polycyclic Aromatic Hydrocarbons), 23
Palen, 56
Parabolic deposit, Venus, 25
Parabolic dune, 72, 95, 96, 98, 99, 106, 130, 190, 246, 257
Parga chasma, Venus, 94
Patagonia, 124
Pathfinder (lander), 35, 45, 192, 201, 217, 268
Pattern analysis, 99, 161
Penetrometer, 183
Peru, 18, 109
Petra, Jordan, 64
Phenanthrene, 18
Phoenix (lander), 36, 44, 127, 137, 140, 197–198, 202, 203
Piezoelectric, 195
PI-SWERL, 183
Pitot tube, *see* Anemometer

Planetary Boundary Layer (PBL), 28, 29, 61, 166
Planetary rotation, 29, 241
Plumes, 178–180
Pluto, 177, 178
Polo, Marco, 115
Power-Law, 256
Prandtl, Ludwig, 44, 45
Prandtl-von Karman Equation, *see* 'Law of the Wall'
Pressure
  atmosphere, 210
  tire, 262
Proctor crater, Mars, 6, 138, 143, 145, 226, 243
Prometheus, 179
Pudgy barchan, 81
Pumice, 65, 66
Puna, Argentina, 18
Purgatory dune, 150, 151
Purulla, Argentina, 66
Pyrite, 170
Pyroxene, 23

**Q**
Quad bike, 265, 266
Quartz, 22, 23, 58, 183, 210, 224
Quattaniya, Egypt, 185, 193, 260, 268, 283
Quill (radar), 285

**R**
R2D2 (robot), 285
Rabe (crater), 197, 219
Radar, 229–236, 238, 245, 268, 284. *See also* SAR, GPR
Radarclinometry, 233, 235, 239
RADARSAT, 133
Raked dunes, 101
Rayleigh (crater), 35, 61
Rayleigh Distribution, 35
RDP, *see* Resultant drift potential
Red noise, 35
Reptation, 49, 59
Resultant drift potential, 107, 108
Reversing dune, 131, 152
Reynolds number, 46
Ripples, 248, 255, 269
Riverbed, Titan, 162
Robins, Benjamin, 42
Robins-Magnus Effect, 41, 42, 49
Rocket exhaust, 197–198
Rocknest (Mars Rock formation), 149, 151
Rommel, Erwin, 10, 263
Rotation, planetary, 29, 241
Roter Kamm impact crater, Namibia, 220, 274
Roughness length, 44, 45
Rub' Al Khali ('Empty Quarter'), 31, 76, 81, 89, 98–101, 104, 108, 123, 124, 186, 219
Rubin, David, 64

**S**
Sabkha, 94, 277
SAFIRE detector, 35, 195
Sahara desert, 264 *See also* Algeria, Morocco, Tunisia
Sahel, 123

Saint-Exupery, Antoine, 283, 284
Saltation, 50 et seq, 39, 245, 250
    cloud, 49
    flux, 36, 195 *See also* Sand transport
    length, 47, 61, 250
    trajectory, 48, 49, 206, 245
Salton Sea barchan, 245
Samalayuca desert, Mexico, 284
Sand channels, 263, 264, 266
Sand Dragon border fence, 101
Sand Mountain, NV, 115, 188, 266
Sand
    number of grains, 19
    removal, 171
    rose, 107, 108
    sheet, 190
    streak, 9, 93, 105, 150
    supply, *see* sand transport
    tracks, *see* sand channels
    transport, 134, 141, 145, 274
    trap, 258, 280
    volume, 164
Sandworm (fiction), 284
Sangre de Cristo mountains, 130
SAR (Synthetic Aperture Radar), 159, 230, 239
Saturation length, 50–53, 55, 289
Saturn, 232, 286
Scale Height, 27
Scanning Electron Microscope (SEM), 203
Scarecrow rover, 271
Scatterometer (radar), 231, 238
Scoop, 199
Seasat, 230, 231
Sebkha, *see* Sabkha
Sechura, 124
Sediments, Mars, 140, 154
Seif dune, 87 *See also* Linear dune
Seismometer
Selective Availability, 185
Senkyo, Titan, 159
SENSIT, 195
Separation bubble, 94, 248
Shangri-La, Titan, 159
Shape, particle, 19, 21, 55
Sharp, Robert, 195
Sharurah, 31
Shear
    stress, 45, 183, 247
    vane, 183
Sheet, *see* Sand sheet
Shields curve, 42
Sieve, 183
Silo honking, 117
Simpson desert, Australia, 82, 127
Singing sand, 115, 215
SIR-C/X-SAR, 230, 231
Slip face, 77, 81–84, 283
Slope failure, 64, 263
Slippage, *see* Wheel slippage
Slope winds, 33, 174, 242
Smoking dune, 56
Snow, 17, 18, 97, 189
    mobile, 185
Sojourner (rover), 6, 136, 202, 204, 268

Sound barrier, 42
South Pole Station, 11
Space Shuttle, 230 *See also* SIR-C
Spirit Rover, 12, 200, 272
Splash, 59, 65
SPOT, 148, 175, 183, 221
SRTM, 229, 238–239
Star dune, 8, 13, 103, 143, 244
Star Wars, 12, 283–284
Stewart, Patrick, 284
Stimpy (rock, Mars), 97
Stokes, George, 43
Stokes' Law, 42, 43
Stovepipe Dunes (Stovepipe Wells, Death Valley, CA), 32, 284
Streak, *see* Sand streak
Streamers, *see* Aeolian streamers
Sudan, 109
Sulphur dioxide frost, 179
Superposed dunes, 99–102
Superstition Dunes, CA, 95
Suspension, 59, 62

**T**
Taklamakan desert (Tarim Basin), China, 115, 125
Tatooine, 12, 109, 283–286
Teardrop barchan, 80, 144
TEGA instrument, 202
Tengger desert, China, 274
Terkezi Oasis, Libya, 105
Terminal velocity, 42
Thar, 124
Thamasia Terra, Mars, 104
The Sand Recknoner, 19
THEMIS (Mars Odyssey), 144
Thermal imaging, 187
Thermal inertia, 141, 166, 226
Thermoluminescence, 205
Threshold
    friction speed, 49, 172, 177 *See also* Impact threshold speed
    parameter, 46
Timbuktu, 264
Timelapse, 109, 187
Tire, 265, 267
Titan, 157–170
Topography, influence on winds, 31–33, 152, 228
Troy, 270
Tozeur, 12
Traction (creep), 39
Trade wind, 241
Transit (instrument), 184, 194
Transit (navigation satellite), 184
Transverse Aeolian Ridges (TARs), 59, 61, 62, 64, 83, 85, 131, 142, 143, 152, 153, 269
Transverse ridge (also Transverse dune), 81, 84, 99, 101, 108, 141, 125, 242, 249
Trellis pattern, 100 *See also* Lattice pattern
Trent University, Canada, 206–208
Tribochemistry, 25
Triboelectricity, 47, 215
Trikanter, 196
Triton, 47, 157, 177, 178
Troy dust deposit, Mars, 10
Tularosa Basin, 130

Tunisia, 12, 223, 284
Turbulence, 36–37
Turbulent
　flow, 37, 42
　intensity, 36
Twin Peaks, 6

## U

Ubari Sand Sea, Libya (*also* Erg Awbari, 125, 218
Udden, Jon, 19
Udden-Wentworth scale, 19, 20
Ultrasound, 194
United Arab Emirates, 37, 57, 58, 76, 82, 90, 99, 111, 273, 277, 280
　*See also* Rub' Al Khali
Unsticking (vehicle), 263–267

## V

V-2, 218
Vegetated dunes, 75, 82, 230, 257 *See also* Parabolic dunes
Vegetation, 105, 106, 124–127, 131, 134
Venera, 169, 171, 175
　landers, 192
Ventifact, 214
Venus Express, 175
Venus, 169–176
　microdunes, 53, 58
Venus Wind Tunnel, 212
Victoria Crater, Mars, 64, 104, 150
Victoria Valley, Antarctica, 112, 113
Viking, 136, 138, 144, 158, 169, 193, 197
Viking Lander, 22, 32, 35, 56
VIMS (Cassini), 158, 162, 167
Viscosity, 27, 28, 42–44
Volcanism, 175
Von Karman constant, 44
Voyager, 157, 158, 177–179

## W

Wadell, 22
Wadi Rum, Jordan, 96
Walnut shell, 210

Water
　flume, 80, 213, 250
　vapor, 140, 169, 257
Water Tank Experiment, 6, 213
Wavelength, ripple, 59, 61, 62, 65
Weibull Distribution, 34, 35
Wentworth, Charles, 19
Werner model, *see* Cellular automaton
Wheel slippage, 7, 269, 271, 272
White Sands, NM, 67, 69, 72, 127, 130, 131, 184, 230
White, Ed, 220
Wind
　rose, 36, 139
　statistics, 35
　tunnel, 13, 36, 42, 44, 46–51, 54, 55, 59
Wind tunnel, Venus, 172, 174
Windsock experiment, 45
Windspeed, Titan, 158
Windy Point, CA, 196
Winged barchan, 77, 84, 143
Wolfram, Stephen, 245
Wyoming, 69, 261

## X

Xanadu, 165
X-ray diffraction, 204, 205
X-ray fluorescence, 183, 202, 203

## Y

Yardang, 160, 170, 196, 285, 286

## Z

Zion National Park, Utah, 69